Lecture Notes in Computer Science   6139

Commenced Publication in 1973
Founding and Former Series Editors:
Gerhard Goos, Juris Hartmanis, and Jan van Leeuwen

Haim Kaplan (Ed.)

Algorithm Theory – SWAT 2010

12th Scandinavian Symposium and Workshops
on Algorithm Theory
Bergen, Norway, June 21-23, 2010
Proceedings

 Springer

Volume Editor

Haim Kaplan
School of Computer Science
Tel Aviv University
Tel Aviv, Israel
E-mail: haimk@post.tau.ac.il

Library of Congress Control Number: Applied for

CR Subject Classification (1998): F.2, E.1, I.3.5, G.2, F.1, F.2.2

LNCS Sublibrary: SL 1 – Theoretical Computer Science and General Issues

ISSN   0302-9743
ISBN-10   3-642-13730-X Springer Berlin Heidelberg New York
ISBN-13   978-3-642-13730-3 Springer Berlin Heidelberg New York

springer.com

© Springer-Verlag Berlin Heidelberg 2010
Printed in Germany

Typesetting: Camera-ready by author, data conversion by Scientific Publishing Services, Chennai, India
Printed on acid-free paper   06/3180

Preface

This volume contains the papers presented at SWAT 2010, the 12th Scandinavian Symposium on Algorithm Theory. Since 1988 SWAT has been held biennially in the Nordic countries; it has a loose association with WADS (Workshop on Algorithms and Data Structures) that is held on odd-numbered years in North America. This 12th SWAT was held during June 21–23, at the University of Bergen in Norway.

The conference focuses on algorithms and data structures. The call for papers invited contributions in all areas of algorithms and data structures, including approximation algorithms, computational biology, computational geometry, distributed algorithms, external-memory algorithms, graph algorithms, online algorithms, optimization algorithms, parallel algorithms, randomized algorithms, string algorithms and algorithmic game theory. A total of 78 papers were submitted, out of which the Program Committee selected 36 for presentation at the symposium. In addition, invited lectures were given by Sanjeev Arora from Princeton University, Prabhakar Raghavan from Yahoo! Research Labs, and Dana Randall from Georgia Institute of Technology.

We would like to thank all the people who contributed to making SWAT 2010 a success. In particular, we thank the Program Committee and all of our many colleagues who helped the committee evaluate the submissions. We also thank the Norwegian Research Council and the University of Bergen for their support.

April 2010   Haim Kaplan

Conference Organization

Program Chair

Haim Kaplan   Tel Aviv University, Israel

Program Committee

Ittai Abraham	Microsoft Research, Silicon Valley, USA
Pankaj Agarwal	Duke University, USA
Hans Bodlaender	University of Utrecht, The Netherlands
Gerth Brodal	Aarhus University, Denmark
Guy Even	Tel Aviv University, Israel
Fabrizio Grandoni	University of Rome, Italy
Roberto Grossi	University of Pisa, Italy
Juha Karkkainen	University of Helsinki, Finland
Matya Katz	Ben-Gurion University, Israel
Valerie King	University of Victoria, Canada
Christos Levcopoulos	Lund University, Sweden
Stefano Leonardi	University of Rome, Italy
Moshe Lewenstein	Bar-Ilan University, Israel
Ulrich Meyer	Goethe University Frankfurt, Germany
Rob van Stee	Max Planck Institut fur Informatik, Germany
Martin Strauss	University of Michigan, USA
Maxim Sviridenko	IBM T.J. Watson Research Center, USA
Chaitanya Swamy	University of Waterloo, Canada
Robert E. Tarjan	Princeton University, USA
Yngve Villanger	University of Bergen, Norway
Neal Young	University of California, Riverside, USA

Local Organizing Committee

The organizing committee comprised the following members of the Department of Informatics at the University of Bergen:
Jean Blair
Binh-Minh Bui-Xuan
Frederic Dorn
Fedor Fomin
Pinar Heggernes
Daniel Lokshtanov
Fredrik Manne (Co-chair)
Jesper Nederlof

Jan Arne Telle (Co-chair)
Martin Vatshelle
Yngve Villanger

Steering Committee

Lars Arge   University of Aarhus, Denmark
Magnus M. Halldorsson   University of Iceland
Rolf Karlsson   Lund University, Sweden
Andrzej Lingas   Lund University, Sweden
Jan Arne Telle   University of Bergen, Norway
Esko Ukkonen   University of Helsinki, Finland

External Reviewers

Deepak Ajwani	Greg Aloupis	Aris Anagnostopoulos
Elliot Anshelevich	Jan Arne Telle	Van Bang Le
Gill Barequet	Manu Basavaraju	MohammadHossein Bateni
Giovanni Battaglia	Luca Becchetti	Andreas Beckmann
Daniel Berend	Anna Bernasconi	Therese Biedl
Nicolas Bonichon	Jaroslaw Byrka	Alberto Caprara
Jean Cardinal	Paz Carmi	Diego Ceccarelli
Chandra Chekuri	Flavio Chierichetti	Sebastien Collette
Jose Correa	Marek Cygan	Ovidiu Daescu
Thomas van Dijk	Feodor Dragan	Stephane Durocher
Khaled Elbassioni	Michael Elkin	Leah Epstein
Thomas Erlebach	Esther Ezra	Rui Ferreira
Amos Fiat	Fedor Fomin	Tobias Friedrich
Stefan Funke	Shashidhar Ganjugunte	Serge Gaspers
Filippo Geraci	Anna Gilbert	Petr Golovach
Martin Golumbic	Zvi Gotthilf	Alexander Grigoriev
Joachim Gudmundsson	Magnus Halldorsson	Sariel Har Peled
Elad Hazan	Danny Hermelin	John Hershberger
Thore Husfeldt	Robert Irving	Bengt Nilsson
Vincent Jost	Bart Jansen	Gudmundsson Joachim
Satyen Kale	Michael Kaufmann	Mark Keil
Balazs Keszegh	Andrew King	Philip Klein
Tsvi Kopelowitz	Janne Korhonen	Guy Kortsarz
Nitish Korula	Adrian Kosowski	Annamaria Kovacs
Stefan Kratsch	Mathieu Liedloff	Andrzej Lingas
Yang Liu	Daniel Lokshtanov	Panu Luosto
Aleksander Madry	Veli Mäkinen	Bodo Manthey

Daniel Marx
Nicole Megow
Victor Milenkovic
Gila Morgenstern
Seffi Naor
Igor Nitto
Svetlana Olonetsky
Giuseppe Ottaviano
Geevarghese Philip
Balaji Raghavachari
Dror Rawitz
Johan van Rooij
Mohammad Salavatipour
Piotr Sankowski
Jiri Sgall
Somnath Sikdar
Shakhar Smorodinsky
Ioan Todinca
Marinus Veldhorst
Antoine Vigneron
Volker Weichert
Udi Wieder
Thomas Wolle
Afra Zomorodian
David Cohen-Steiner

Jiri Matousek
Giulia Menconi
Shuichi Miyazaki
Gabriel Moruz
Andrei Negoescu
Tim Nonner
Sebastian Ordyniak
Sang-il Oum
Jeff Phillips
Venkatesh Raman
Igor Razgon
Adi Rosen
Leena Salmela
Saket Saurabh
Hadas Shachnai
Jouni Sirén
Karol Suchan
Jarkko Toivonen
Angelina Vidali
Yoshiko Wakabayashi
Carola Wenk
Gerhard Woeginger
Sergey Yekhanin
Uri Zwick

Moti Medina
Julian Mestre
Thomas Molhave
Viswanath Nagarajan
Ofer Neiman
Yahav Nussbaum
Alessio Orlandi
Pekka Parviainen
Marcin Pilipczuk
R. Ravi
Liam Roditty
Kunihiko Sadakane
Laura Sanità
Danny Segev
Shimon Shahar
Michiel Smid
Mohammad Hajiaghayi
Geza Toth
Giovanni Viglietta
Yusu Wang
Renato Werneck
Paul Wollan
Hamid Zarrabi-Zadeh

Table of Contents

Optimal Exploration of Terrains with Obstacles

Jurek Czyzowicz[1,*], David Ilcinkas[2,**],
Arnaud Labourel[2,**,***], and Andrzej Pelc[1,†]

[1] Département d'informatique, Université du Québec en Outaouais,
Gatineau, Québec J8X 3X7, Canada
`jurek@uqo.ca, pelc@uqo.ca`
[2] LaBRI, CNRS & Université de Bordeaux, 33405 Talence, France
`david.ilcinkas@labri.fr, labourel.arnaud@gmail.com`

Abstract. A mobile robot represented by a point moving in the plane
has to explore an unknown flat terrain with impassable obstacles. Both
the terrain and the obstacles are modeled as arbitrary polygons. We
consider two scenarios: the *unlimited vision*, when the robot situated
at a point p of the terrain explores (sees) all points q of the terrain
for which the segment pq belongs to the terrain, and the *limited vision*,
when we require additionally that the distance between p and q be at
most 1. All points of the terrain (except obstacles) have to be explored
and the performance of an exploration algorithm, called its complexity,
is measured by the length of the trajectory of the robot.

For unlimited vision we show an exploration algorithm with complexity $O(P + D\sqrt{k})$, where P is the total perimeter of the terrain (including
perimeters of obstacles), D is the diameter of the convex hull of the terrain, and k is the number of obstacles. We do not assume knowledge
of these parameters. We also prove a matching lower bound showing
that the above complexity is optimal, even if the terrain is known to
the robot. For limited vision we show exploration algorithms with complexity $O(P + A + \sqrt{Ak})$, where A is the area of the terrain (excluding
obstacles). Our algorithms work either for arbitrary terrains, if one of
the parameters A or k is known, or for c-fat terrains, where c is any constant (unknown to the robot) and no additional knowledge is assumed.
(A terrain $\mathcal{T}$ with obstacles is c-fat if $R/r \leq c$, where R is the radius
of the smallest disc containing $\mathcal{T}$ and r is the radius of the largest disc
contained in $\mathcal{T}$.) We also prove a matching lower bound $\Omega(P + A + \sqrt{Ak})$
on the complexity of exploration for limited vision, even if the terrain is
known to the robot.

Keywords: Mobile robot, exploration, polygon, obstacle.

* Partially supported by NSERC discovery grant.
** Partially supported by the ANR project ALADDIN, the INRIA project CEPAGE
and by a France-Israel cooperation grant (Multi-Computing project).
*** This work was done during this author's stay at the Université du Québec en
Outaouais as a postdoctoral fellow.
† Partially supported by NSERC discovery grant and by the Research Chair in Dis-
tributed Computing at the Université du Québec en Outaouais.

H. Kaplan (Ed.): SWAT 2010, LNCS 6139, pp. 1–12, 2010.
© Springer-Verlag Berlin Heidelberg 2010

1   Introduction

The background and the problem. Exploring unknown terrains by mobile robots has important applications when the environment is dangerous or of difficult access for humans. Such is the situation when operating in nuclear plants or cleaning toxic wastes, as well as in the case of underwater or extra-terrestrial operations. In many cases a robot must inspect an unknown terrain and come back to its starting point. Due to energy and cost saving requirements, the length of the robot's trajectory should be minimized.

We model the exploration problem as follows. The terrain is represented by an arbitrary polygon $\mathcal{P}_0$ with pairwise disjoint polygonal obstacles $\mathcal{P}_1, ..., \mathcal{P}_k$, included in $\mathcal{P}_0$, i.e., the terrain is $\mathcal{T} = \mathcal{P}_0 \setminus (\mathcal{P}_1 \cup \cdots \cup \mathcal{P}_k)$. We assume that borders of all polygons $\mathcal{P}_i$ belong to the terrain. The robot is modeled as a point moving along a polygonal line inside the terrain. It should be noted that the restriction to polygons is only to simplify the description, and all our results hold in the more general case where polygons are replaced by bounded subsets of the plane homeotopic with a disc (i.e., connected and without holes) and regular enough to have well-defined area and boundary length. Every point of the trajectory of the robot is called *visited*. We consider two scenarios: the *unlimited vision*, when the robot visiting a point p of the terrain $\mathcal{T}$ *explores* (sees) all points q for which the segment pq is entirely contained in $\mathcal{T}$, and the *limited vision*, when we require additionally that the distance between p and q be at most 1. In both cases the task is to explore all points of the terrain $\mathcal{T}$. The cost of an exploration algorithm is the length of the trajectory of the robot, which should be as small as possible. The *complexity* of an algorithm is the order of magnitude of its cost. We assume that the robot does not know the terrain before starting the exploration, but it has unbounded memory and can record the portion of the terrain seen so far and the already visited portion of its trajectory.

Our results. For unlimited vision we show an exploration algorithm with complexity $O(P + D\sqrt{k})$, where P is the total perimeter of the terrain (including perimeters of obstacles), D is the diameter of the convex hull of the terrain, and k is the number of obstacles. We do not assume knowledge of these parameters. We also prove a matching lower bound for exploration of some terrains (even if the terrain is known to the robot), showing that the above complexity is worst-case optimal.

For limited vision we show exploration algorithms with complexity $O(P + A + \sqrt{Ak})$, where A is the area of the terrain[1]. Our algorithms work either for arbitrary terrains, if one of the parameters A or k is known, or for *c-fat* terrains, where c is any constant larger than 1 (unknown to the robot) and no additional knowledge is assumed. (A terrain $\mathcal{T}$ is *c-fat* if $R/r \leq c$, where R is the radius of

---

[1] Since parameters D, P, A are positive reals that may be arbitrarily small, it is important to stress that complexity $O(P + A + \sqrt{Ak})$ means that the trajectory of the robot is at most $c(P + A + \sqrt{Ak})$, for some constant c and *sufficiently large* values of P and A. Similarly for $O(P + D\sqrt{k})$. This permits to include, e.g., additive constants in the complexity, in spite of arbitrarily small parameter values.

the smallest disc containing $\mathcal{T}$ and r is the radius of the largest disc contained in $\mathcal{T}$.) We also prove a matching lower bound $\Omega(P + A + \sqrt{Ak})$ on the complexity of exploration, even if the terrain is known to the robot.

The main open problem resulting from our research is whether exploration with asymptotically optimal cost $O(P + A + \sqrt{Ak})$ can be performed in arbitrary terrains without *any* a priori knowledge. Another interesting open problem is whether such worst-case performance can be obtained by an $O(k)$-competitive algorithm. (Our algorithms are a priori not competitive).

Related work. Exploration of unknown environments by mobile robots was extensively studied both for the unlimited and for the limited vision. Most of the research in this domain concerns the competitive framework, where the trajectory of the robot not knowing the environment is compared to that of the optimal exploration algorithm having full knowledge.

One of the most important works for unlimited vision is [8]. The authors gave a 2-competitive algorithm for rectilinear polygon exploration without obstacles. The case of non-rectilinear polygons (without obstacles) was also studied in [7,15,12] and competitive algorithms were given.

For polygonal environments with an arbitrary number of polygonal obstacles, it was shown in [8] that no competitive strategy exists, even if all obstacles are parallelograms. Later, this result was improved in [1] by giving a lower bound in $\Omega(\sqrt{k})$ for the competitive ratio of any on-line algorithm exploring a polygon with k obstacles. This bound remains true even for rectangular obstacles. On the other hand, there exists an algorithm with competitive ratio in $O(k)$ [7,15]. Moreover, for particular shapes of obstacles (convex and with bounded aspect ratio) the optimal competitive ratio $\Theta(\sqrt{k})$ has been proven in [15].

Exploration of polygons by a robot with limited vision has been studied, e.g., in [9,10,11,13,14,16]. In [9] the authors described an on-line algorithm with competitive ratio $1+3(\Pi S/A)$, where Π is a quantity depending on the perimeter of the polygon, S is the area seen by the robot, and A is the area of the polygon. The exploration in [9,10] fails on a certain type of polygons, such as those with narrow corridors. In [11], the authors consider exploration in discrete steps. The robot can only explore the environment when it is motionless, and the cost of the exploration algorithm is measured by the number of stops during the exploration. In [13,14], the complexity of exploration is measured by the trajectory length, but only terrains composed of identical squares are considered. In [16] the author studied off-line exploration of the boundary of a terrain with limited vision.

An experimental approach was used in [2] to show the performance of a greedy heuristic for exploration in which the robot always moves to the frontier between explored and unexplored area. Practical exploration of the environment by an actual robot was studied, e.g., in [6,19]. In [19], a technique is described to deal with obstacles that are not in the plane of the sensor. In [6] landmarks are used during exploration to construct the skeleton of the environment.

Navigation is a closely related task which consists in finding a path between two given points in a terrain with unknown obstacles. Navigation in a $n \times n$ square containing rectangular obstacles aligned with sides of the square was considered in [3,4,5,18]. It was shown in [3] that the navigation from a corner to the center of a room can be performed with a competitive ratio $O(\log n)$, only using tactile information (i.e., the robot modeled as a point sees an obstacle only when it touches it). No deterministic algorithm can achieve better competitive ratio, even with unlimited vision [3]. For navigation between any pair of points, there is a deterministic algorithm achieving a competitive ratio of $O(\sqrt{n})$ [5]. No deterministic algorithm can achieve a better competitive ratio [18]. However, there is a randomized approach performing navigation with a competitive ratio of $O(n^{\frac{4}{9}} \log n)$ [4]. Navigation with little information was considered in [20]. In this model, the robot cannot perform localization nor measure any distances or angles. Nevertheless, the robot is able to learn the critical information contained in the classical shortest-path roadmap and perform locally optimal navigation.

2   Unlimited Vision

Let S be a smallest square in which the terrain $\mathcal{T}$ is included. Our algorithm constructs a *quadtree decomposition* of S. A quadtree is a rooted tree with each non-terminal node having four children. Each node of the quadtree corresponds to a square. The children of any non-terminal node v correspond to four identical squares obtained by partitioning the square of v using its horizontal and vertical symmetry axes. This implies that the squares of the terminal nodes form a partition of the root[2]. More precisely,

1. $\{S\}$ is a quadtree decomposition of S
2. If $\{S_1, S_2, \ldots, S_j\}$ is a quadtree decomposition of S, then
   $\{S_1, S_2, \ldots, S_{i-1}, S_{i_1}, S_{i_2}, S_{i_3}, S_{i_4}, S_{i+1}, \ldots, S_j\}$, where $S_{i_1}, S_{i_2}, S_{i_3}, S_{i_4}$ form a partition of S_i using its vertical and horizontal symmetry axes, is a quadtree decomposition of S

The trajectory of the robot exploring $\mathcal{T}$ will be composed of parts which will follow the boundaries of $\mathcal{P}_i$, for $0 \leq i \leq k$, and of straight-line segments, called *approaching segments*, joining the boundaries of $\mathcal{P}_i$, $0 \leq i \leq k$. Obviously, the end points of an approaching segment must be visible from each other. The quadtree decomposition will be dynamically constructed in a top-down manner during the exploration of $\mathcal{T}$. At each moment of the exploration we consider the set $\mathcal{Q}_S$ of all squares of the current quadtree and the set $\mathcal{Q}_\mathcal{T}$ of squares being the terminal nodes of the current quadtree. We will also construct dynamically a bijection $f : \{\mathcal{P}_0, \mathcal{P}_1, \ldots, \mathcal{P}_k\} \longrightarrow \mathcal{Q}_S \setminus \mathcal{Q}_\mathcal{T}$.

When a robot moves along the boundary of some polygon $\mathcal{P}_i$, it may be in one of two possible modes: the *recognition mode* - when it goes around the entire boundary of a polygon without any deviation, or in the *exploration mode* - when, while moving around the boundary, it tries to detect (and approach) new obstacles. When the decision to approach a new obstacle is made at some point

---

[2] In order to have an exact partition we assume that each square of the quadtree partition contains its East and South edges but not its West and North edges.

r of the boundary of $\mathcal{P}_i$ the robot moves along an approaching segment to reach the obstacle, processes it by a recursive call, and (usually much later), returning from the recursive call, it moves again along this segment in the opposite direction in order to return to point r and to continue the exploration of $\mathcal{P}_i$. However, some newly detected obstacles may not be immediately approached. We say that, when the robot is in position r, an obstacle $\mathcal{P}_j$ is *approachable*, if there exists a point $q \in \mathcal{P}_j$, belonging to a square $S_t \in \mathcal{Q}_\mathcal{T}$ of diameter $D(S_t)$ such that $|rq| \leq 2D(S_t)$. It is important to state that if exactly one obstacle becomes approachable at moment t, then it is approached immediately and if more than one obstacle become approachable at a moment t, then one of them (chosen arbitrarily) is approached immediately and the others are approached later, possibly from different points of the trajectory. Each time a new obstacle is visited by the robot (i.e., all the points of its boundary are visited in the recognition mode) the terminal square of the current quadtree containing the first visited point of the new obstacle is partitioned. This square is then associated to this obstacle by function f. The trajectory of the robot is composed of three types of sections: *recognition sections*, *exploration sections* and *approaching sections*. The boundary of each polygon will be traversed twice: first time contiguously during a recognition section and second time through exploration sections, which may be interrupted several times in order to approach and visit newly detected obstacles. We say that an obstacle is *completely explored*, if each point on the boundary of this obstacle has been traversed by an exploration section. We will prove that the sum of the lengths of the approaching sections is $O(D\sqrt{k})$.

Algorithm. ExpTrav (polygon R, starting point r^* on the boundary of R)
1 Make a recognition traversal of the boundary of R
2 Partition square $S_t \in \mathcal{Q}_\mathcal{T}$ containing r^* into four identical squares
3 $f(R) := S_t$
4 **repeat**
5   Traverse the boundary of R until, for the current position r, there exists
       a visible point q of a new obstacle Q belonging to square $S_t \in \mathcal{Q}_\mathcal{T}$,
       such that $|rq| \leq 2D(S_t)$
6   Traverse the segment rq
7   ExpTrav(Q, q)
8   Traverse the segment qr
9 **until** R is completely explored

Before the initial call of ExpTrav, the robot reaches a position r_0 at the boundary of the polygon $\mathcal{P}_0$. This is done as follows. At its initial position v, the robot chooses an arbitrary half-line α which it follows as far as possible. When it hits the boundary of a polygon $\mathcal{P}$, it traverses the entire boundary of $\mathcal{P}$. Then, it computes the point u which is the farthest point from v in $\mathcal{P} \cap \alpha$. It goes around $\mathcal{P}$ until reaching u again and progresses on α, if possible. If this is impossible, the robot recognizes that it went around the boundary of $\mathcal{P}_0$ and it is positioned on this boundary. It initialises the quadtree decomposition to a smallest square S containing $\mathcal{P}_0$. This square is of size $O(D(\mathcal{P}_0))$. The length of the above walk is less than $3P$.

Lemma 1. *Algorithm* ExpTrav *visits all boundary points of all obstacles of the terrain* $\mathcal{T}$.

Lemma 2. *Function* f *is a bijection from* $\{\mathcal{P}_0, \mathcal{P}_1, \ldots, \mathcal{P}_k\}$ *to* $\mathcal{Q}_\mathcal{S} \setminus \mathcal{Q}_\mathcal{T}$, *where* $\mathcal{Q}_\mathcal{S}$ *and* $\mathcal{Q}_\mathcal{T}$ *correspond to the final quadtree decomposition produced by Algorithm* ExpTrav.

Lemma 3. *For any quadtree* T, *rooted at a square of diameter* D *and having* x *non-terminal nodes, the sum* $\sigma(T)$ *of diameters of these nodes is at most* $2D\sqrt{x}$.

Theorem 1. *Algorithm* ExpTrav *explores the terrain* $\mathcal{T}$ *of perimeter* P *and convex hull diameter* D *with* k *obstacles in time* $O(P + D\sqrt{k})$.

Proof. Take an arbitrary point p inside $\mathcal{T}$ and a ray outgoing from p in an arbitrary direction. This ray reaches the boundary of $\mathcal{T}$ at some point q. Since, by Lemma 1 point q was visited by the robot, p was visible from q during the robot's traversal, and hence p was explored.

To prove the complexity of the algorithm, observe that the robot traverses twice the boundary of each polygon of $\mathcal{T}$, once during its recognition in step 1 and the second time during the iterations of step 5. Hence the sum of lengths of the recognition and exploration sections is $2P$. The only other portions of the trajectory are produced in steps 6 and 8, when the obstacles are approached and returned from. According to the condition from step 5, an approaching segment is traversed in step 6 only if its length is shorter than twice the diameter of the associated square. If $k = 0$ then the sum of lengths of all approaching segments is 0, due to the fact that exploration starts at the external boundary of the terrain. In this case the length of the trajectory is at most $5P$ (since the length of at most $3P$ was traversed before the initial call). Hence we may assume that $k > 0$. By Lemma 2 each obstacle is associated with a different non-terminal node of the quadtree and the number x of non-terminal nodes of the quadtree equals $k + 1$. Hence the sum of lengths of all approaching segments is at most $2\sigma(T)$. By Lemma 3 we have $\sigma(T) \leq 2D\sqrt{x} = 2D\sqrt{k+1}$, hence the sum of lengths of approaching segments is at most $2\sigma(T) \leq 4D\sqrt{k+1} \leq 4D\sqrt{2k} \leq 6D\sqrt{k}$. Each segment is traversed twice, so the total length of this part of the trajectory is at most $12D\sqrt{k}$. It follows that the total length of the trajectory is at most $5P + 12D\sqrt{k}$. □

Theorem 2. *Any algorithm for a robot with unlimited visibility, exploring polygonal terrains with* k *obstacles, having total perimeter* P *and the convex hull diameter* D, *produces trajectories in* $\Omega(P + D\sqrt{k})$ *in some terrains, even if the terrain is known to the robot.*

Proof. In order to prove the lower bound, we show two families of terrains: one for which $P \in \Theta(D)$ (P cannot be smaller), D and k are unbounded and still the exploration cost is $\Omega(D\sqrt{k})$, and the other in which P is unbounded, D is arbitrarily small, $k = 0$ and still the exploration cost is $\Omega(P)$. Consider the terrain from Figure 1(a) where k identical tiny obstacles are distributed evenly at the $\sqrt{k} \times \sqrt{k}$ grid positions inside a square of diameter D. The distance between

Fig. 1. Lower bound for unlimited visiblity

obstacles is at least $\frac{D\sqrt{2}}{2(\sqrt{k}+1)} - \epsilon$ where $\epsilon > 0$ may be as small as necessary by choosing obstacles sufficiently small. The obstacles are such that to explore the small area inside the convex hull of the obstacle the robot must enter this convex hull. Since each such area must be explored, the trajectory of the robot must be of size at least $(k-1)\left(\frac{D\sqrt{2}}{2(\sqrt{k}+1)} - \epsilon\right)$, which is clearly in $\Omega(D\sqrt{k})$. Note that the perimeter P is in $\Theta(D)$.

The terrain from Fig. 1(b) is a polygon of arbitrarily small diameter (without obstacles), whose exploration requires a trajectory of size $\Omega(P)$, where P is unbounded (as the number of "corridors" can be unbounded). Indeed, each "corridor" must be traversed almost completely to explore points at its end. The two families of polygons from Fig. 1 lead to the $\Omega(P + D\sqrt{k})$ lower bound. □

3   Limited Vision

In this section we assume that the vision of the robot has range 1. The following algorithm is at the root of all our positive results on exploration with limited vision. The idea of the algorithm is to partition the environment into small parts called *cells* (of diameter at most 1) and to visit them using a depth-first traversal. The local exploration of cells can be performed using Algorithm ExpTrav, since the vision inside each cell is not limited by the range 1 of the vision of the robot. The main novelty of our exploration algorithm is that the robot completely explores *any* terrain. This should be contrasted with previous algorithms with limited visibility, e.g. [9,10,13,14] in which only a particular class of terrains with obstacles is explored, e.g., terrains without narrow corridors or terrains composed of complete identical squares. This can be done at cost $O(A)$. Our lower bound shows that exploration complexity of arbitrary terrains depends on the perimeter and the number of obstacles as well. The complete exploration of arbitrary terrains achieved by our algorithm significantly complicates both the exploration process and its analysis.

Our algorithms LET$_A$ and LET$_k$, and the tourist algorithm described in [15] share a similar approach to exploration, i.e., using several square decompositions of the terrain with different side lengths to figure out the characteristics of the terrain and achieve efficient exploration. However, our algorithms differ from the tourist algorithm in two important ways : (1) the exploration of the inside of each square is done with an optimal algorithm (See section 2) instead of a greedy one and (2) the limited visibility forces an upper bound on the side of the square (significantly complicating $\mathtt{LET_k}$). More importantly, due to the numerous differences between our model and the one of [15], the analyses of the complexities of the algorithms are unrelated.

Algorithm. $\mathtt{LimExpTrav}$ (LET, for short)
INPUT: A point s inside the terrain $\mathcal{T}$ and a positive real $F \leq \sqrt{2}/2$.
OUTPUT: An exploration trajectory of $\mathcal{T}$, starting and ending at s.
Tile the area with squares of side F, such that s is on the boundary of a square. The connected regions obtained as intersections of $\mathcal{T}$ with each tile are called *cells*. For each tile S, maintain a quadtree decomposition Q_S initially set to $\{S\}$. Then, arbitrarily choose one of the cells containing s to be the starting cell C and call $\mathtt{ExpCell}(C, s)$.

Procedure $\mathtt{ExpCell}$(current cell C, starting point $r^* \in C$)
1 Record C as visited
2 $\mathtt{ExpTrav}(C, r^*)$ using the quadtree decomposition Q_S; S is the tile s.t. $C \subseteq S$
3 **repeat**
4   Traverse the boundary of C until the current position r belongs to
         an unvisited cell U
5     $\mathtt{ExpCell}(U, r)$
      (if r is in several unvisited cells, choose arbitrarily one to be processed)
6 **until** the boundary of C is completely traversed

It is worth to note that, at the beginning of the exploration of the first cell belonging to a tile S, the quadtree of this tile is set to a single node. However, at the beginning of explorations of subsequent cells belonging to S, the quadtree of S may be different. So the top-down construction of this quadtree may be spread over the exploration of many cells which will be visited at different points in time.

Consider a tile T and a cell $C \subseteq T$. Let A_C be the area of C and B_C the length of its boundary. Let P_C be the length of the part of the boundary of C included in the boundary of the terrain $\mathcal{T}$, and let R_C be the length of the remaining part of the boundary of C, i.e., $R_C = B_C - P_C$.

Lemma 4. *There is a positive constant c, such that $R_C \leq c(A_C/F + P_C)$, for any cell C.*

The following is the key lemma for all upper bounds proved in this section. Let $\mathcal{S} = \{T_1, T_2, \ldots, T_n\}$ be the set of tiles with non-empty intersection with $\mathcal{T}$ and $\mathcal{C} = \{C_1, C_2, \cdots, C_m\}$ be the set of cells that are intersections of tiles from $\mathcal{S}$ with $\mathcal{T}$. For each $T \in \mathcal{S}$, let k_T be the number of obstacles of $\mathcal{T}$ entirely contained in T.

Lemma 5. *For any $F \leq \sqrt{2}/2$, Algorithm LET explores the terrain $\mathcal{T}$ of area A and perimeter P, using a trajectory of length $O(P + A/F + F \sum_{i=1}^{n} \sqrt{k_{T_i}})$.*

Proof. First, we show that Algorithm *LET* explores the terrain $\mathcal{T}$. Consider the graph G whose vertex set is $\mathcal{C}$ and edges are the pairs $\{C, C'\}$ such that C and C' have a common point at their boundaries. The graph G is connected, since $\mathcal{T}$ is connected. Note that for any cell C and point r on the boundary of C, `ExpTrav` on C and r and thus `ExpCell` on C and r starts and ends on r. Therefore, Algorithm *LET* performs a depth first traversal of graph G, since during the execution of `ExpCell`$(C, \dots)$, procedure `ExpCell`$(U, \cdots)$ is called for each unvisited cell U adjacent to C. Hence, `ExpCell`$(C, \dots)$ is called for each cell $C \in \mathcal{C}$, since G is connected. During the execution of `ExpCell`(C, r), C is completely explored by `ExpTrav`(C, r) by the same argument as in the proof of Lemma 1, since the convex hull diameter of C is less than one.

It remains to show that the length of the *LET* trajectory is $O(P + A/F + F \sum_{i=1}^{n} \sqrt{k_{T_i}})$. For each $j = 1, \dots, m$, the part of the *LET* trajectory inside the cell C_j is produced by the execution of `ExpCell`$(C_j, \dots)$. In step 2 of `ExpCell`$(C_j, \dots)$, the robot executes `ExpTrav` with $D = \sqrt{2}F$ and $P = P_{C_j} + R_{C_j}$. The sum of lengths of recognition and exploration sections of the trajectory in C_j is at most $2(P_{C_j} + R_{C_j})$. The sum of lengths of approaching sections of the trajectory in T_i is at most $6\sqrt{2}F\sqrt{k_{T_i}}$ and each approaching section is traversed twice (cf. proof of Theorem 1). In step 3 of `ExpCell`$(C_j, \dots)$, the robot only makes the tour of the cell C_j, hence the distance traveled by the robot is at most $P_{C_j} + R_{C_j}$. It follows that:

$$|LET| \leq 3 \sum_{j=1}^{m} (P_{C_j} + R_{C_j}) + 12\sqrt{2}F \sum_{i=1}^{n} \sqrt{k_{T_i}}$$

$$\leq 3 \sum_{j=1}^{m} ((1+c)P_{C_j} + cA_{C_j}/F) + 12\sqrt{2}F \sum_{i=1}^{n} \sqrt{k_{T_i}} \quad \text{by Lemma 4}$$

$$\leq 3(c+1)P + 3cA/F + 12\sqrt{2}F \sum_{i=1}^{n} \sqrt{k_{T_i}}. \qquad \square$$

In view of Lemma 5, exploration of a particular class of terrains can be done at a cost which will be later proved optimal.

Theorem 3. *Let $c > 1$ be any constant. Exploration of a c-fat terrain of area A, perimeter P and with k obstacles can be performed using a trajectory of length $O(P + A + \sqrt{Ak})$ (without any a priori knowledge).*

Proof. The robot executes Algorithm *LET* with $F = \sqrt{2}/2$. By Lemma 5, the total cost is $O(P + A + \sum_{i=1}^{n} \sqrt{k_{T_i}})$. Recall that n is the number of tiles that have non-empty intersection with the terrain. We have $\sum_{i=1}^{n} \sqrt{k_{T_i}} \leq \sum_{i=1}^{n} \sqrt{\frac{k}{n}} = \sqrt{nk}$. Hence, it remains to show that $n = O(A)$ to prove that the cost is $O(P + A + \sqrt{Ak})$. By definition of a c-fat terrain, there is a disk D_1 of radius r included in the terrain and a disk D_2 of radius R that contains the terrain, such that $\frac{R}{r} \leq c$. There are $\Theta(r^2)$ tiles entirely included in D_1 and hence in the terrain. So, we have $A = \Omega(r^2)$. $\Theta(R^2)$ tiles are sufficient to cover D_2 and hence the terrain. So $n = O(R^2)$. Hence, we obtain $n = O(A)$ in view of $R \leq cr$. $\qquad \square$

Consider any terrain $\mathcal{T}$ of area A, perimeter P and with k obstacles. We now turn attention to the exploration problem if some knowledge about the terrain is available a priori. Notice that if A and k are known before the exploration, Lemma 5 implies that Algorithm LET executed for $F = \min\{\sqrt{A/k}, \sqrt{2}/2\}$ explores *any* terrain at cost $O(A + P + \sqrt{Ak})$. (Indeed, if $F = \sqrt{A/k}$ then $A/F = \sqrt{Ak}$ and $kF = \sqrt{Ak}$, while $F = \sqrt{2}/2$ implies $A/F = \Theta(A)$ and $kF = O(A)$.) This cost will be later proved optimal. It turns out that a much more subtle use of Algorithm LET can guarantee the same complexity assuming only knowledge of A *or* k. We present two different algorithms depending on which value, A or k, is known to the robot. Both algorithms rely on the same idea. The robot executes Algorithm LET with some initial value of F until either the terrain is completely explored, or a certain stopping condition, depending on the algorithm, is satisfied. This execution constitutes the first stage of the two algorithms. If exploration was interrupted because of the stopping condition, then the robot proceeds to a new stage by executing Algorithm LET with a new value of F. Values of F decrease in the first algorithm and increase in the second one. The exploration terminates at the stage when the terrain becomes completely explored, while the stopping condition is never satisfied. In each stage the robot is oblivious of the previous stages, except for the computation of the new value of F that depends on the previous stage. This means that in each stage exploration is done "from scratch", without recording what was explored in previous stages. In order to test the stopping condition in a given stage, the robot maintains the following three values: the sum A^* of areas of explored cells, updated after the execution of ExpTrav in each cell; the length P^* of the boundary traversed by the robot, continuously updated when the robot moves along a boundary for the first time (i.e., in the recognition mode); and the number k^* of obstacles approached by the robot, updated when an obstacle is approached. The values of A^*, P^* and k^* at the end of the i-th stage are denoted by A_i, P_i and k_i, respectively. Let F_i be the value of F used by Algorithm LET in the i-th stage. Now, we are ready to describe the stopping conditions and the values F_i in both algorithms.

Algorithm $\mathrm{LET_A}$, for A known before exploration
The value of F used in Algorithm LET for the first stage is $F_1 = \sqrt{2}/2$. The value of F for subsequent stages is given by $F_{i+1} = \frac{A}{k_i F_i}$. The stopping condition is $\{k^*F_i \geq 2A/F_i$ and $k^*F_i \geq P^* + 1\}$.

Algorithm $\mathrm{LET_k}$, for k known before exploration
The value of F used in Algorithm LET for the first stage is $F_1 = \frac{1}{k+\sqrt{2}}$. The value of F for subsequent stages is given by $F_{i+1} = \min\left\{\frac{A_i}{kF_i}, \frac{\sqrt{2}}{2}\right\}$. The stopping condition is $\{A^*/F_i \geq 2kF_i$ and $A^*/F_i \geq P^* + 1$ and $F_i < \sqrt{2}/2\}$.

Consider a moment t during the execution of Algorithm LET. Let C_t be the set of cells recorded as visited by Algorithm LET at moment t, and let $\mathcal{O}_t$ be the set of obstacles approached by the robot until time t. For each $C \in C_t$, let B_C be the length of the intersection of the exterior boundary of cell C with the boundary of the terrain. For each $O \in \mathcal{O}_t$, let $|O|$ be the perimeter of obstacle O and let $k_t = |\mathcal{O}_t|$. The following proposition is proved similarly as Lemma 5.

Proposition 1. *There is a positive constant d such that the length of the trajectory of the robot until any time t, during the execution of Algorithm LET, is at most $d(\sum_{C \in C_t}(B_C + A_C/F) + (k_t + 1) \cdot F + \sum_{O \in \mathcal{O}_t} |O|)$.*

The following lemma establishes the complexity of exploration if either the area of the terrain or the number of obstacles is known a priori.

Lemma 6. *Algorithm LET$_A$ (resp. LET$_k$) explores a terrain T of area A, perimeter P and with k obstacles, using a trajectory of length $O(P + A + \sqrt{Ak})$, if A (resp. k) is known before exploration.*

The following theorem shows that the lengths of trajectories in Lemma 6 and in Theorem 3 are asymptotically optimal.

Theorem 4. *Any algorithm for a robot with limited visibility, exploring polygonal terrains of area A, perimeter P and with k obstacles, produces trajectories of length $\Omega(P + A + \sqrt{Ak})$ in some terrains, even if the terrain is known to the robot.*

Lemma 6 and Theorem 4 imply

Theorem 5. *Consider terrains of area A, perimeter P and with k obstacles. If either A or k is known before the exploration, then the exploration of any such terrain can be performed using a trajectory of length $\Theta(P + A + \sqrt{Ak})$, which is asymptotically optimal.*

Notice that in order to explore a terrain at cost $O(P + A + \sqrt{Ak})$, it is enough to know the parameter A or k up to a multiplicative constant, rather than the exact value. This can be proved by a carefull modification of the proof of Lemma 6. For the sake of clarity, we stated and proved the weaker version of Lemma 6, with knowledge of the exact value.

Suppose now that no a priori knowledge of any parameters of the terrain is available. We iterate Algorithm LET$_A$ or LET$_k$ for A (resp. k) equal $1, 2, 4, 8, \dots$ interrupting the iteration and doubling the parameter as soon as the explored area (resp. the number of obstacles seen) exceeds the current parameter value. The algorithm stops when the entire terrain is explored (which happens at the first probe exceeding the actual unknown value of A, resp. k). We get an exploration algorithm using a trajectory of length $O((P + A + \sqrt{Ak}) \log A)$, resp. $O((P + A + \sqrt{Ak}) \log k)$. By interleaving the two procedures we get the minimum of the two costs. Thus we have the following corollary.

Corollary 1. *Consider terrains of area A, perimeter P and with k obstacles. Exploration of any such terrain can be performed without any a priori knowledge at cost differing from the worst-case optimal cost with full knowledge only by a factor $O(\min\{\log A, \log k\})$.*

References

1. Albers, S., Kursawe, K., Schuierer, S.: Exploring unknown environments with obstacles. Algorithmica 32, 123–143 (2002)
2. Bandyopadhyay, T., Liu, Z., Ang, M.H., Seah, W.K.G.: Visibility-based exploration in unknown environment containing structured obstacles. Advanced Robotics, 484–491 (2005)
3. Bar-Eli, E., Berman, P., Fiat, A., Yan, R.: On-line navigation in a room. Journal of Algorithms 17, 319–341 (1994)
4. Berman, P., Blum, A., Fiat, A., Karloff, H., Rosen, A., Saks, M.: Randomized robot navigation algorithms. In: Proc. 7th ACM-SIAM Symp. on Discrete Algorithms, pp. 74–84 (1996)
5. Blum, A., Raghavan, P., Schieber, B.: Navigating in unfamiliar geometric terrain. SIAM Journal on Computing 26, 110–137 (1997)
6. Cuperlier, N., Quoy, M., Giovanangelli, C.: Navigation and planning in an unknown environment using vision and a cognitive map. In: Proc. Workshop: Reasoning with Uncertainty in Robotics, pp. 48–53 (2005)
7. Deng, X., Kameda, T., Papadimitriou, C.H.: How to learn an unknown environment. In: Proc. 32nd Symp. on Foundations of Comp. Sci. (FOCS 1991), pp. 298–303 (1991)
8. Deng, X., Kameda, T., Papadimitriou, C.H.: How to learn an unknown environment I: the rectilinear case. Journal of the ACM 45, 215–245 (1998)
9. Gabriely, Y., Rimon, E.: Spanning-tree based coverage of continuous areas by a mobile robot. In: Proc. Int. Conf. of Robotics and Automaton (ICRA 2001), pp. 1927–1933 (2001)
10. Gabriely, Y., Rimon, E.: Competitive on-line coverage of grid environments by a mobile robot. Computational Geometry: Theory and Applications 24(3), 197–224 (2003)
11. Ghosh, S.K., Burdick, J.W., Bhattacharya, A., Sarkar, S.: Online algorithms with discrete visibility - exploring unknown polygonal environments. Robotics & Automation Magazine 15, 67–76 (2008)
12. Hoffmann, F., Icking, C., Klein, R., Kriegel, K.: The polygon exploration problem. SIAM J. Comput. 31, 577–600 (2001)
13. Icking, C., Kamphans, T., Klein, R., Langetepe, E.: Exploring an unknown cellular environment. In: Abstracts of the 16th European Workshop on Computational Geometry, pp. 140–143 (2000)
14. Kolenderska, A., Kosowski, A., Małafiejski, M., Żyliński, P.: An Improved Strategy for Exploring a Grid Polygon. In: SIROCCO, pp. 222–236 (2009)
15. Kalyanasundaram, B., Pruhs, K.: A Competitive Analysis of Algorithms for Searching Unknown Scenes. Comput. Geom. 3, 139–155 (1993)
16. Ntafos, S.: Watchman routes under limited visibility. Comput. Geom. Theory Appl. 1, 149–170 (1992)
17. Osserman, R.: The isoperimetric inequality. Bull. Amer. Math. Soc. 84, 1182–1238 (1978)
18. Papadimitriou, C.H., Yannakakis, M.: Shortest paths without a map. Theor. Comput. Sci. 84, 127–150 (1991)
19. Sim, R., Little, J.J.: Autonomous vision-based exploration and mapping using hybrid maps and Rao-Blackwellised particle filters. Intelligent Robots and Systems, 2082–2089 (2006)
20. Tovar, B., Murrieta-Cid, R., Lavalle, S.M.: Distance-optimal navigation in an unknown environment without sensing distances. IEEE Transactions on Robotics 23, 506–518 (2007)

Reconstructing a Simple Polygon from Its Angles

Yann Disser, Matúš Mihalák, and Peter Widmayer

Institute of Theoretical Computer Science, ETH Zürich
{ydisser,mmihalak,widmayer}@inf.ethz.ch

Abstract. We study the problem of reconstructing a simple polygon from angles measured at the vertices of the polygon. We assume that at each vertex, a sensing device returns the sequence of angles between each pair of vertices that are visible. We prove that the sequence of angle measurements at all vertices of a simple polygon in cyclic order uniquely determines the polygon up to similarity. Furthermore, we propose an algorithm that reconstructs the polygon from this information in polynomial time.

1 Introduction

The reconstruction of geometric objects from measurement data has attracted considerable attention over the last decade [7,11,13]. In particular, many variants of the problem of reconstructing a polygon with certain properties have been studied. For different sets of data this polygon reconstruction problem has been shown to be NP-hard [4,8,10]. Recently, data from rather novel sensing devices like range-finding scanners has been considered, and most of the reconstruction problems that such devices naturally induce have been shown to be NP-hard as well, while a few others are polynomial time solvable [1]. We study the reconstruction problem induced by sensors that measure a sequence of angles. Specifically, we assume that at each vertex v of a simple polygon, the sequence of vertices *visible* from v is perceived in counterclockwise (ccw) order as seen around v, starting at the ccw neighbor vertex of v on the polygon boundary. As usual, we call two polygon vertices (mutually) *visible*, if the straight line segment connecting them lies entirely in the polygon. In addition to seeing visible vertices the angle sensor measures angles between adjacent rays from v to the vertices it sees, and it returns the ccw sequence of these measured angles (cf. Figure 1). Note that such an angle measurement indirectly also yields the angles between any pair of rays to visible vertices (not only adjacent pairs). Our polygon reconstruction problem takes as input a ccw sequence of angle measurements, one measurement at each vertex of a simple polygon, and asks for a simple polygon that fits the measured angles; we call this problem the *polygon reconstruction problem from angles* (cf. Figure 2).

Our contribution. We propose an algorithm that solves the polygon reconstruction problem from angles in polynomial time, and we show that the solution is unique (up to similarity). More precisely, we focus on the visibility graph, i.e.,

H. Kaplan (Ed.): SWAT 2010, LNCS 6139, pp. 13–24, 2010.

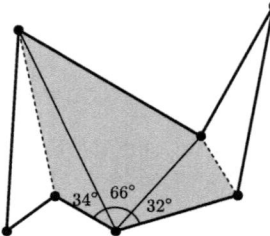

Fig. 1. Illustration of an angle measurement: the sensor returns the vector $(32°, 66°, 34°)$

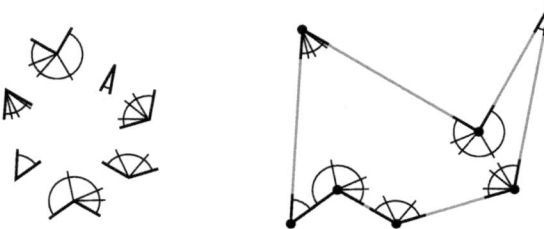

Fig. 2. Given a sequence of angle measurements in ccw order along the boundary (left); the goal is to find a polygon that fits these angles (right)

the graph with a node for every vertex of the polygon and an edge between two nodes if the corresponding vertices see each other. It is sufficient to reconstruct the visibility graph of a polygon, as, together with the angle data, it is then easy to infer the shape of the polygon up to similarity.[1] We show that only the visibility graph of the original polygon $\mathcal{P}$ is compatible with the information contained in the angle data measured in $\mathcal{P}$. Our algorithm finds this unique visibility graph in polynomial time and thus reconstructs the original polygon up to similarity in polynomial time. Note that if only the set of angle measurements is given, i.e. the order of the vertices along the boundary is not known, it is impossible to uniquely reconstruct the visibility graph of a polygon in general.[2] While we assume that the measured angles come from a simple polygon, our algorithm as a side effect is also capable of detecting false inputs, i.e., measurements that do not fit any simple polygon.

---

[1] The shape of the polygon can be obtained in linear time from the visibility graph and angle data. We can achieve this by first computing a triangulation and then fixing the length of one edge. All other lengths in the triangulation can then be computed in linear time.

[2] To see this, consider a square and "attach" to every corner of it a triangle. Make the shapes of the triangles all different and such that the vertices of a triangle that are not at the corner of the square only see the corner vertices of the square they are attached to (plus the vertices of the triangle of course). Now any permutation of the triangles results in the same set of angle measurements but in different polygons.

The key difficulty of the reconstruction of the visibility graph lies in the fact that vertices in our setting have no recognizable labels, i.e., an angle measurement at a vertex returns angles between distant vertices but does not identify these distant vertices globally. Instead, our algorithm needs to identify these vertices in a consistent way across the whole input. In this sense, our problem has a similar flavor as the turnpike reconstruction problem (also known as one dimensional partial digest problem), whose complexity is still open [12].

Related Work. For our purposes, the combinatorial nature of a polygon is encoded in its visibility graph. Solving the visibility graph reconstruction problem for certain data may be a step towards understanding visibility graphs in general. Their characterization has been an open problem for many years that has attracted considerable attention [5,6].

A question closely related to the offline reconstruction of the visibility graph of a polygon appears in the area of robotic exploration, namely what sensory and motion capabilities enable simple robots inside a polygon to reconstruct the visibility graph [2,14]. The idea to reconstruct it from angle data was first mentioned in this context [2], but was also discussed earlier [9]. In all these models a simple robot is assumed to sense visible vertices in ccw order (but does not sense the global identity of visible vertices). In [2], the problem of reconstructing the visibility graph of a polygon was solved for simple robots that can measure angles and additionally are equipped with a compass. In the case of robots that can only distinguish between angles smaller and larger than π, it was shown in the same study that adding the capability of retracing their movements empowers the robots to reconstruct the visibility graph (even if they do not know n, the number of vertices of the unknown polygon). In both cases a polynomial-time algorithm was given. Recently, it was shown that the ability to retrace their movements alone already enables simple robots to reconstruct the visibility graph (when at least an upper bound on the number of vertices of the polygon is given), even though only an exponential algorithm was given [3]. Our result implies that measuring angles alone is also already sufficient. On the other hand, it is known that the inner angles (the angles along the boundary) of the polygon do not contain sufficient information to uniquely reconstruct the visibility graph, even when combined with certain combinatorial information [2].

The general problem of reconstructing polygons from measurement data has mainly been studied in two variants. The first variant asks to find some polygon $\mathcal{P}^\star$ that is consistent with the data measured in the original polygon $\mathcal{P}$. For example, it was studied how a polygon $\mathcal{P}^\star$ compatible with the information obtained from "stabbing" $\mathcal{P}$ or compatible with the set of intersection points of $\mathcal{P}$ with some lines can be constructed [7,11]. The problem we consider falls in the second variant in which generally the problem is to reconstruct $\mathcal{P}$ itself uniquely from data measured in $\mathcal{P}$, i.e., we have to show that only $\mathcal{P}$ is compatible with the data. A previous study in this area shows that the inner angles of $\mathcal{P}$ together with the cross-ratios of its triangulation uniquely determine $\mathcal{P}$ [13].

Outline. We introduce the visibility graph reconstruction problem in detail in Section 2. In Section 3 we show that there is a unique solution to the problem, and give an algorithm that finds the unique solution in polynomial time.

2   The Visibility Graph Reconstruction Problem

Let $\mathcal{P}$ be a simple polygon with visibility graph $G_{\text{vis}} = (V, E_{\text{vis}})$, where V denotes the set of vertices of $\mathcal{P}$ and $n = |V|$. We fix a vertex $v_0 \in V$ and denote the other vertices of $\mathcal{P}$ by $v_1, v_2, \ldots, v_{n-1}$ in ccw order along the boundary starting at v_0's ccw neighbor. For ease of presentation only, we assume polygons to be in general position, i.e. no three vertices are allowed to lie on a line. All definitions and results can be adapted to be valid even without this assumption (note that our definition of *visible* vertices implies that the line segment connecting two mutually visible vertices can have more than two points on the boundary of the polygon in this case). The degree of a vertex $v_i \in V$ in G_{vis} is denoted by $d(v_i)$ and the sequence $\text{vis}(v_i) = \left(v_i, u_1, u_2, \ldots, u_{d(v_i)}\right)$ is defined to enumerate the vertices visible to v_i ordered in ccw order along the boundary starting with v_i itself. We write $\text{vis}_0(v_i)$ to denote v_i itself and $\text{vis}_k(v_i)$, $1 \le k \le d(v_i)$ to denote u_k. For two distinct vertices $v_i, v_j \in V$, $\text{chain}(v_i, v_j)$ denotes the sequence $(v_i, v_{i+1}, \ldots, v_j)$ of the vertices between v_i and v_j along the boundary in ccw order. Similarly, $\text{chain}_v(v_i, v_j)$ denotes the subsequence of $\text{chain}(v_i, v_j)$ that contains only the vertices that are visible to v. Note that here and in the following all indices are understood modulo n.

We define the *visibility segments* of v to be the segments $\overline{vu_1}, \overline{vu_2} \ldots, \overline{vu_{d(v)}}$ in this order. Similarly, we define the *visibility angles* of v to be the ordered sequence of angles between successive visibility segments, such that the i-th visibility angle is the angle between $\overline{vu_i}$ and $\overline{vu_{i+1}}$, for all $1 \le i \le d(v) - 1$.

Let $v, v_i, v_j \in V$. We write $\measuredangle_v(v_i, v_j)$ to denote the angle between the lines $\overline{vv_i}$ and $\overline{vv_j}$ (in that order) even if v, v_i, v_j do not mutually see each other. Let $1 \le l < r \le d(v)$. We write $\measuredangle_v(l, r)$ to denote $\measuredangle_v(\text{vis}_l(v), \text{vis}_r(v))$. We will need the notion of the approximation $\measuredangle_v^{\uparrow}(v_i, v_j)$ of the angle $\measuredangle_v(v_i, v_j)$, which is defined as follows (cf. Figure 3): Let $v_{i'}$ be the last vertex in $\text{chain}_v(v_{i+1}, v_i)$ and $v_{j'}$ be the first vertex in $\text{chain}_v(v_j, v_{j-1})$. We then define $\measuredangle_v^{\uparrow}(v_i, v_j) = \measuredangle_v(v_{i'}, v_{j'})$. Observe that if $\{v, v_i\}, \{v, v_j\} \in E_{\text{vis}}$, we have $\measuredangle_v^{\uparrow}(v_i, v_j) = \measuredangle_v(v_i, v_j)$. Also note that knowing the visibility angles of a vertex v is equivalent to knowing $\measuredangle_v(l_v, r_v)$ for all $1 \le l_v < r_v \le d(v)$.

In terms of the above definitions, the goal of the visibility graph reconstruction problem is to find E_{vis} when we are given n, V, $d(v)$ for all $v \in V$, and $\measuredangle_v(l_v, u_v)$ for all $v \in V$ and all $1 \le l_v < u_v \le d(v)$, as well as the (ccw) order in which the vertices appear along the boundary.

3   Triangle Witness Algorithm

The key question when trying to reconstruct the visibility graph of a polygon is how to identify a vertex u visible to some known vertex v. Knowing all angles at

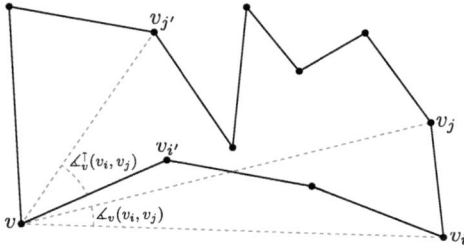

Fig. 3. Illustration of the approximation $\angle_v^\uparrow(v_i, v_j) = \angle_v(v_{i'}, v_{j'})$ of the angle $\angle_v(v_i, v_j)$

every vertex may seem to be far too much information and the reconstruction problem may thus seem easily solvable by some greedy algorithm. Before we actually present the triangle witness algorithm that solves the reconstruction problem, we show that some natural greedy algorithms do not work in general.

Greedy Approach. It is a natural idea to first orient all angles w.r.t. a single, global orientation (e.g. the line $\overline{v_{n-1}v_0}$) by summing angles around the polygon boundary. Then, if a vertex v sees some other vertex u under a certain global angle α, u must see v under the inverse angle $\alpha + \pi$, as the line $\overline{uv}$ has a single orientation. A simple greedy approach to identify the vertex u in the view from v could be to walk from v along the boundary and find the first vertex that sees some other vertex under the global angle $\alpha + \pi$. The example in Fig. 4 however shows that this approach does not work in general.

A similar but somewhat stronger approach is to allow global angles to go beyond $[0, 2\pi)$ while summing up around the polygon boundary, cf. Figure 4 (there, for instance, vertex v_1 sees the second visible vertex in ccw order under the angle $\alpha - \pi$ which is less than 0). This would prevent the pairing of vertex v_0 with vertex v_1 in that example. Nevertheless, there are still examples where this strategy fails and in fact it is not possible to greedily match angles:[3] inspect Figure 5 for an example of two polygons for which no matter how a greedy algorithm chooses to pair vertices, it has to fail for one of the two.

Triangle Witness Algorithm. We now give an algorithm for the reconstruction of the visibility graph from the visibility angles of all vertices. Note that from now on we map all angles to the range $[0, 2\pi)$. Our algorithm considers all vertices at once and gradually identifies edges connecting vertices that lie further and further apart along the boundary. Intuitively, once we know all vertices in $\{v_{i+1}, v_{i+2}, \ldots, v_{k-1}\}$ that are visible to v_i, there is only one candidate vertex which might be v_k, namely the next unidentified vertex in $\mathrm{vis}(v_i)$. Our algorithm thus only needs to decide whether v_i sees v_k. The key ingredient here is the use of a *triangle witness* vertex that indicates whether two other vertices see each other. Because any polygon can be triangulated, we know that for every two vertices $\{v_i, v_j\} \in E_{\mathrm{vis}}$ with $v_j \neq v_{i+1}$, there is a "witness" vertex $v_l \in \mathrm{chain}(v_{i+1}, v_{j-1})$

---

[3] We do not aim, however, to give complete proof or to fully characterize all failing greedy algorithms based on the idea of angle matching.

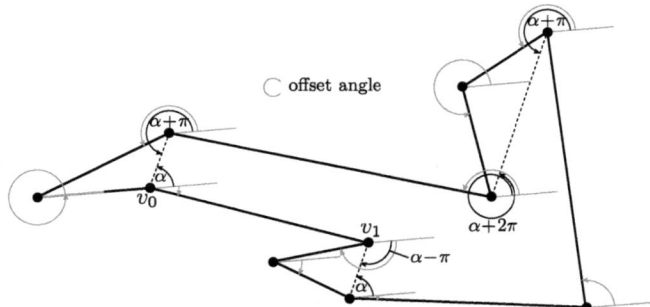

Fig. 4. Illustration of the idea behind the greedy pairing algorithm for a single angle α and starting vertex v_0. If we map angles to the range $[0, 2\pi)$, we allow v_0 and v_1 to be paired which is obviously impossible.

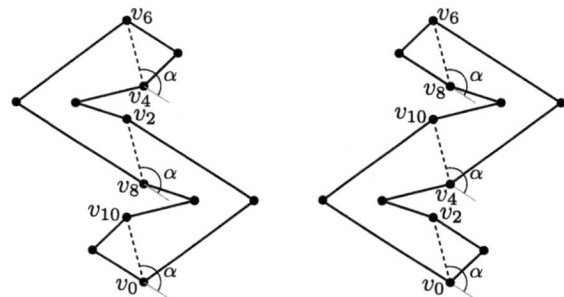

Fig. 5. An example in which only one visibility graph can correctly be reconstructed by any greedy pairing algorithm

that they both see, such that v_i, v_l, v_j form a triangle (of angle sum π). We now extend this notion to the case where $\{v_i, v_j\} \notin E_{\text{vis}}$.

Definition 1. *Let $v_i, v_j \in V$ be two different vertices and $v_j \neq v_{i+1}$. Let further $v_l \in \text{chain}(v_{i+1}, v_{j-1})$ with $\{v_i, v_l\}, \{v_j, v_l\} \in E_{\text{vis}}$. Then we say v_l is a triangle witness of (v_i, v_j), if it fulfills the generalized angle-sum condition*

$$\angle_{v_i}^{\uparrow}(v_l, v_j) + \angle_{v_j}^{\uparrow}(v_i, v_l) + \angle_{v_l}(v_j, v_i) = \pi.$$

In the following we motivate the definition of a triangle witness (cf. Figure 6). As before, we know that if two vertices $v_i, v_j \in V$, $v_j \neq v_{i+1}$ see each other, there must be a vertex $v_l \in \text{chain}(v_{i+1}, v_{j-1})$ which sees both of them. For any choice of v_l, the condition $\angle_{v_i}(v_l, v_j) + \angle_{v_j}(v_i, v_l) + \angle_{v_l}(v_j, v_i) = \pi$ is trivially fulfilled. In the case that v_i does not see v_j, the only difference from v_i's perspective is that for any choice of v_l, $\angle_{v_i}(v_l, v_j)$ does not appear in its visibility angles. We want to modify the condition to capture this difference. The idea is to replace v_j in $\angle_{v_i}(v_l, v_j)$ by an expression that happens to be v_j, if and only if v_i sees v_j. We choose "the first vertex in $\text{chain}_{v_i}(v_j, v_{j-1})$", which is v_j, exactly if v_i sees v_j. If,

similarly, we also replace v_i in $\angle_{v_j}(v_i, v_l)$ by "the last vertex in chain$_{v_j}(v_{i+1}, v_i)$",
we obtain the generalized angle-sum condition of Definition 1. We will later see
(stated as Lemma 4) that there is a triangle witness for a pair (v_i, v_j), if and
only if $\{v_i, v_j\} \in E_{\text{vis}}$.

We can now describe the triangle witness algorithm. It iterates through increasing number of steps k along the boundary, focusing at step k on all edges
of the form $\{v_i, v_{i+k}\}$. Throughout it maintains two maps F, B that store for
every vertex all the edges identified so far that go at most k steps forward or
backward along the boundary, respectively. We write $F[v_i][v_j] = s$, if $\{v_i, v_j\}$ is
the s-th edge incident to v_i in ccw order, and $B[v_i][v_j] = s$, if $\{v_i, v_j\}$ is the s-th
edge incident to v_i in ccw order. Note that $B[v_i]$ is filled in cw order during the
algorithm, i.e. its first entry will be $B[v_i][v_{i-1}] = d(v_i)$. Whenever convenient,
we use $F[v_i]$ and $B[v_i]$ like a set, e.g. we write $v_l \in F[v_i]$ to denote that there is
an entry v_l in $F[v_i]$ and write $|F[v_i]|$ to denote the number of entries in $F[v_i]$.
Observe also that $|F[v_i]| + 1$ is the index of the first vertex (in ccw order) in
$\text{vis}(v_i)$ that is not yet identified. It is clear that once we completed the maps for
k up to $\lceil \frac{n}{2} \rceil$, we essentially have computed E_{vis}.

The initialization of the maps for $k = 1$ is simple as every vertex sees its
neighbors on the boundary. In later iterations for every vertex v_i there is always
exactly one candidate vertex for v_{i+k}, namely the $(|F[v_i]| + 1)$-th vertex visible
to v_i. We decide whether v_i and v_{i+k} see each other by going over all vertices
between v_i and v_{i+k} in ccw order along the boundary and checking whether there
is a triangle witness $v_l \in \text{chain}(v_{i+1}, v_{i+k-1})$. If and only if this is the case, we
update E_{vis}, F, B with the edge $\{v_i, v_{i+k}\}$. For a listing of the triangle witness
algorithm see Algorithm 1.

In the following we prove the correctness of the triangle witness algorithm.
For this we mainly have to show that having a triangle witness is necessary and
sufficient for a pair of vertices to see each other. To show this, we will need the
notion of *blockers* and *shortest paths* in polygons.

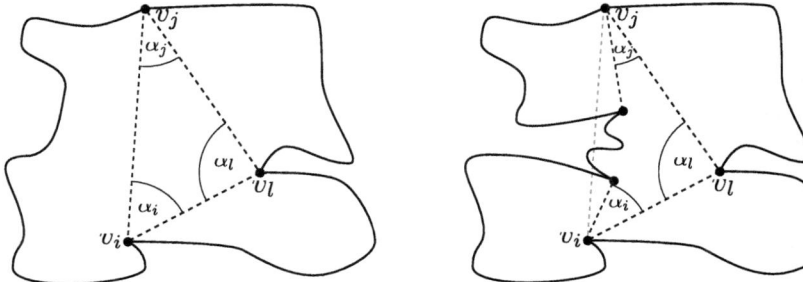

Fig. 6. Illustration of the generalized angle sum condition of Definition 1. On the left
$\{v_i, v_j\} \in E_{\text{vis}}$ and the angles α_i, α_j and α_l of the condition sum up to π. On the right,
$\{v_i, v_j\} \notin E_{\text{vis}}$ and the sum of the angles is strictly less than π.

Algorithm 1. Triangle witness algorithm
_____
input: n, $d(\cdot)$, $\angle.(\cdot, \cdot)$
output: E_{vis}

  1. $F \leftarrow$ [array of n empty maps], $B \leftarrow$ [array of n empty maps], $E_{\mathrm{vis}} \leftarrow \emptyset$
  2. **for** $i \leftarrow 0, \ldots, n-1$
  3. $E_{\mathrm{vis}} \leftarrow E_{\mathrm{vis}} \cup \{v_i, v_{i+1}\}$
  4. $F[v_i][v_{i+1}] \leftarrow 1$
  5. $B[v_{i+1}][v_i] \leftarrow d(v_i)$
  6. **for** $k \leftarrow 2, \ldots, \lceil \frac{n}{2} \rceil$
  7.   **for** $i \leftarrow 0, \ldots, n-1$
  8. $j \leftarrow i + k$
  9.   **for** $l \leftarrow i+1, \ldots j-1$
10.   **if** $v_l \in F[v_i] \wedge v_l \in B[v_j]$
11. $\alpha_i \leftarrow \angle_{v_i}(F[v_i][v_l], |F[v_i]| + 1)$ $(= \angle_{v_i}^{\uparrow}(v_l, v_j)$, cf. proof of Th. 1$)$
12. $\alpha_j \leftarrow \angle_{v_j}(d(v_j) - |B[v_j]|, B[v_j][v_l])$ $(= \angle_{v_j}^{\uparrow}(v_i, v_l)$, cf. proof of Th. 1$)$
13. $\alpha_l \leftarrow \angle_{v_l}(F[v_l][v_j], B[v_l][v_i])$ $(= \angle_{v_l}(v_j, v_i)$, cf. proof of Th. 1$)$
14.   **if** $\alpha_i + \alpha_j + \alpha_l = \pi$
15. $E_{\mathrm{vis}} \leftarrow E_{\mathrm{vis}} \cup \{v_i, v_j\}$
16. $F[v_i][v_j] = |F[v_i]| + 1$
17. $B[v_j][v_i] = d(j) - |B[v_j]|$
18.   abort innermost loop
_____

Definition 2. *Let* $v_i, v_j \in V$. *We say* $v_b \in \mathrm{chain}(v_{i+1}, v_{j-1})$ *is a* blocker *of* (v_i, v_j), *if for all* $u \in \mathrm{chain}(v_i, v_{b-1})$, $v \in \mathrm{chain}(v_{b+1}, v_j)$ *we have* $\{u, v\} \notin E_{\mathrm{vis}}$ *(cf. Figure 7 (left)).*

Note that if v_b is a blocker of (v_i, v_j), v_b also is a blocker of (u, v) for all $u \in \mathrm{chain}(v_i, v_{b-1})$, $v \in \mathrm{chain}(v_{b+1}, v_j)$.

    A *path* between two vertices $a, b \in V$ of a polygon $\mathcal{P}$ is defined to be a curve that lies entirely in $\mathcal{P}$ and has a and b as its endpoints. A *shortest path* between a and b is a path of minimum Euclidean length among all the paths between the two vertices.

Lemma 1 (Lemmas 3.2.3 and 3.2.5. in [5]). *Let* $v_i, v_j \in V$. *The shortest path between* v_i *and* v_j *is unique and is a chain of straight line segments that connect at vertices.*

We can therefore write $(a, u_0, u_1, \ldots, b)$ to denote a shortest path, where we refer to the u_i's as the path's *interior vertices*. The following statements motivate the term 'blocker'.

Lemma 2. *Let* $v_i, v_j \in V$ *with* $\{v_i, v_j\} \notin E_{\mathrm{vis}}$. *Every interior vertex of the shortest path from* v_i *to* v_j *is a blocker of either* (v_i, v_j) *or* (v_j, v_i).

Proof. Consult Figure 7 (right) along with the proof. For the sake of contradiction assume that $v_b \in V$ is an interior vertex of the shortest path p_{ij} from v_i to v_j that is not a blocker of either (v_i, v_j) or (v_j, v_i). W.l.o.g. assume

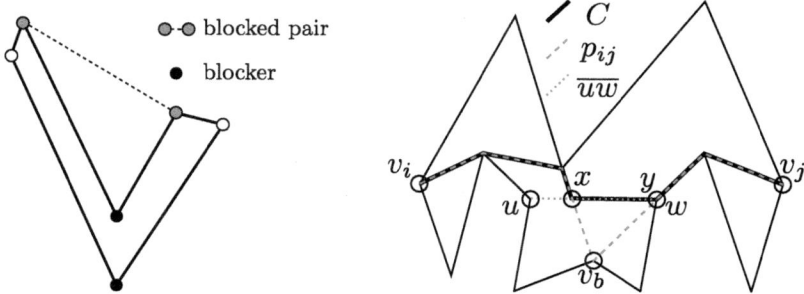

Fig. 7. Left: a pair of vertices can have blockers on both sides. Right: illustration of the objects in the proof of Lemma 2.

$v_b \in$ chain(v_{i+1}, v_{j-1}). As v_b is not a blocker of (v_i, v_j), there are two vertices $u \in$ chain(v_{i+1}, v_{b-1}), $w \in$ chain(v_{b+1}, v_{j-1}) with $\{u, w\} \in E_{\text{vis}}$. Thus, the segment $\overline{uw}$ is entirely in the polygon and separates it in two parts, one part containing v_b and the other containing both v_i and v_j. As p_{ij} visits v_b, it must cross $\overline{uw}$ at least twice. Let x, y be the first and last intersection points of $\overline{uw}$ with p_{ij}. Consider the curve C that follows p_{ij} until x, then follows $\overline{uw}$ until y and finally follows p_{ij} until v_j. Because of the triangle inequality, C is strictly shorter than p_{ij} which is a contradiction with the assumption that p_{ij} is a shortest path. $\square$

Corollary 1. *Let $v_i, v_j \in V$. If $\{v_i, v_j\} \notin E_{\text{vis}}$, there is either a blocker of (v_i, v_j) or of (v_j, v_i).*

We now relate the definition of a blocker to the geometry of the polygon.

Lemma 3. *Let $v_i, v_j \in V$ with $i = j + 2$, $\{v_i, v_j\} \notin E_{\text{vis}}$. If $w := v_{j+1} = v_{i-1}$ is convex (inner angle $< \pi$), then $v_{i'} = \arg\min_{v_b \in \text{chain}_{v_i}(v_{i+1}, v_{j-1})} \angle_{v_i}(v_b, w)$ and $v_{j'} = \arg\min_{v_b \in \text{chain}_{v_j}(v_{i+1}, v_{j-1})} \angle_{v_j}(w, v_b)$ are blockers of (v_i, v_j) that lie left of the oriented line $\overline{v_i v_j}$.*

Proof. As w is convex, the shortest path p_{ij} from v_i to v_j only contains vertices of chain(v_i, v_j). As p_{ij} only makes right turns (i.e. any three consecutive vertices on p_{ij} form a ccw triangle), all interior vertices of p_{ij} lie left of the oriented line $\overline{v_i v_j}$. Furthermore $v_{i'}$ and $v_{j'}$ are the first and the last interior vertices of p_{ij} respectively. By Lemma 2 we thus know that both $v_{i'}$ and $v_{j'}$ are blockers of (v_i, v_j). From before we also know that they both lie left of the oriented line $\overline{v_i v_j}$. $\square$

We now get to the central lemma that essentially states that the existence of a triangle witness is necessary and sufficient for a pair of vertices to see each other.

Lemma 4. *Let $v_i, v_j \in V$ with $|\text{chain}(v_i, v_j)| > 2$. There is a triangle witness v_l for (v_i, v_j), if and only if $\{v_i, v_j\} \in E_{\text{vis}}$.*

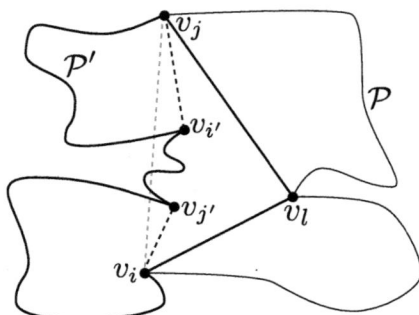

Fig. 8. Sketch of the definitions in the proof of Lemma 4

Proof. If $\{v_i, v_j\} \in E_{\mathrm{vis}}$, because any polygon can be triangulated, there must be a vertex $v_l \in \mathrm{chain}(v_{i+1}, v_{j-1})$ for which both edges $\{v_i, v_l\}$ and $\{v_l, v_j\}$ are in E_{vis}. For this vertex we have $\angle_{v_i}^{\uparrow}(v_l, v_j) + \angle_{v_j}^{\uparrow}(v_i, v_l) + \angle_{v_l}(v_j, v_i) = \angle_{v_i}(v_l, v_j) + \angle_{v_j}(v_i, v_l) + \angle_{v_l}(v_j, v_i) = \pi$ as all three relevant edges are in E_{vis} and the sum over the angles of any triangle is π.

For the converse implication assume there is a triangle witness v_l of (v_i, v_j). For the sake of contradiction, assume $\{v_i, v_j\} \notin E_{\mathrm{vis}}$.

Consider the polygon $\mathcal{P}'$ induced by the vertices $v_i, v_l, v_j, \mathrm{chain}(v_{j+1}, v_{i-1})$, cf. Figure 8. As $\{v_i, v_l\}, \{v_l, v_j\} \in E_{\mathrm{vis}}$, $\mathcal{P}'$ is simple and well defined. In $\mathcal{P}'$, v_l is a convex vertex, as it fulfills the generalized angle-sum condition of Definition 1 and thus $\angle_{v_l}(v_j, v_i) < \pi$, because all angles are positive. We can therefore apply Lemma 3 (on v_j, v_i) w.r.t. $\mathcal{P}'$ and conclude that both $v_{j'}$ and $v_{i'}$ block (v_j, v_i), where $v_{j'} = \arg\min_{v_b \in \mathrm{chain}_{v_i}(v_{j+1}, v_{i-1})} \angle_{v_i}(v_l, v_b)$ and $v_{i'} = \arg\min_{v_b \in \mathrm{chain}_{v_j}(v_{j+1}, v_{i-1})} \angle_{v_j}(v_b, v_l)$. This is then also true in our original polygon $\mathcal{P}$ and thus $v_{i'} \in \mathrm{chain}(v_j, v_{j'})$ as otherwise $v_{j'}$ would block $(v_j, v_{i'})$ and $v_{i'}$ would block $(v_{j'}, v_i)$ contradicting the definition of $v_{j'}$ and $v_{i'}$, respectively. Observe that $v_{i'}$ is the last vertex in $\mathrm{chain}(v_{i+1}, v_i)$ visible to v_j and $v_{j'}$ is the first vertex in $\mathrm{chain}(v_j, v_{j-1})$ visible to v_i.

By applying Lemma 3 to $\mathcal{P}'$, we know that both $v_{j'}$ and $v_{i'}$ are left of the oriented line $\overline{v_j v_i}$. This means $\angle_{v_i}^{\uparrow}(v_l, v_j) = \angle_{v_i}(v_l, v_{j'}) < \angle_{v_i}(v_l, v_j)$ and $\angle_{v_j}^{\uparrow}(v_i, v_l) = \angle_{v_j}(v_{i'}, v_l) < \angle_{v_j}(v_i, v_l)$ and thus $\angle_{v_i}^{\uparrow}(v_l, v_j) + \angle_{v_j}^{\uparrow}(v_i, v_l) + \angle_{v_l}(v_j, v_i) < \angle_{v_i}(v_l, v_j) + \angle_{v_j}(v_i, v_l) + \angle_{v_l}(v_j, v_i) = \pi$, which is a contradiction with our assumption that v_l is a triangle witness of (v_i, v_j). $\square$

Theorem 1. *The triangle witness algorithm is correct, computes a unique solution, and can be implemented with a running time of $O(n^3 \log n)$.*

Proof. As the edges in E_{vis} are the same as the edges stored in F and the same as the edges stored in B throughout the algorithm, it is sufficient to show that after step k of the iteration both F and B contain exactly the edges between vertices that are at most k steps apart along the boundary. As no two vertices can be further apart than $\lceil \frac{n}{2} \rceil$ steps along the boundary, this immediately implies that

E_{vis} eventually contains exactly the edges of the visibility graph. More precisely, we inductively show that after step k of the iteration, $F[v_i]$ contains the vertices of $\text{chain}_{v_i}(v_{i+1}, v_{i+k})$ and $B[v_i]$ contains the vertices of $\text{chain}_{v_i}(v_{i-k}, v_{i-1})$ for all $v_i \in V$. For sake sake of simplicity we write $F[v_i] = \text{chain}_{v_i}(v_{i+1}, v_{i+k})$ and $B[v_i] = \text{chain}_{v_i}(v_{i-k}, v_{i-1})$.

The discussion for $k = 1$ is trivial as every vertex has an edge to both its neighbors. The algorithm initializes F and B to consist of these edges. It remains to show for all $0 \leq i < n$ that assuming $F[v_i] = \text{chain}_{v_i}(v_{i+1}, v_{i+k-1})$ and $B[v_i] = \text{chain}_{v_i}(v_{i-k+1}, v_{i-1})$ after step $k-1$, we have $F[v_i] = \text{chain}_{v_i}(v_{i+1}, v_{i+k})$ and $B[v_i] = \text{chain}_{v_i}(v_{i-k}, v_{i-1})$ after step k.

The algorithm adds an edge between two vertices v_i and v_{i+k}, if and only if there is a vertex $v_l \in \text{chain}(v_{i+1}, v_{i+k-1})$ with $v_l \in F[v_i] \wedge v_l \in B[v_{i+k}]$ for which $\alpha_i + \alpha_j + \alpha_l = \pi$, where $\alpha_i, \alpha_j, \alpha_l$ are defined as in Algorithm 1. As v_i and v_l are less than k steps apart on the boundary, the induction assumption implies that $F[v_i] = \text{chain}_{v_i}(v_{i+1}, v_{i+k-1})$ and $B[v_{i+k}] = \text{chain}_{v_{i+k}}(v_{i+1}, v_{i+k-1})$. Thus, $v_l \in F[v_i] \wedge v_l \in B[v_{i+k}]$ is equivalent to $\{v_i, v_l\}, \{v_{i+k}, v_l\} \in E_{\text{vis}}$ and by Lemma 4 it suffices to show that $\alpha_i = \measuredangle_{v_i}^{\uparrow}(v_l, v_{i+k}), \alpha_j = \measuredangle_{v_{i+k}}^{\uparrow}(v_i, v_l), \alpha_l = \measuredangle_{v_l}(v_{i+k}, v_i)$ for all $v_l \in F[v_i] \cap B[v_{i+k}]$. Again by induction $F[v_i] = \text{chain}_{v_i}(v_{i+1}, v_{i+k-1})$ and thus $\text{vis}_{F[v_i][v_l]}(v_i) = v_l$ and $\text{vis}_{|F[v_i]|+1}(v_i) = \arg\min_{v_b \in \text{chain}_{v_i}(v_{i+k}, v_{i-1})} \measuredangle_{v_i}(v_{i+1}, v_b)$ and we thus get $\alpha_i = \measuredangle_{v_i}(F[v_i][v_l], |F[v_i]|+1) = \measuredangle_{v_i}^{\uparrow}(v_l, v_{i+k})$. Similarly as v_l and v_{i+k} are less than k steps apart on the boundary, we get $\alpha_j = \measuredangle_{v_{i+k}}^{\uparrow}(v_i, v_l)$. Further, with the induction assumption we also have $\text{vis}_{F[v_l][v_{i+k}]}(v_l) = v_{i+k}$ and $\text{vis}_{B[v_l][v_i]}(v_l) = v_i$ and thus $\alpha_l = \measuredangle_{v_l}(F[v_l][v_j], B[v_l][v_i]) = \measuredangle_{v_l}(v_{i+k}, v_i)$.

The uniqueness of the algorithm's solution follows immediately from the fact that the existence of a triangle witness is necessary and sufficient for two vertices to see each other.

For every vertex v_i and every $k = 1, 2, \ldots, \lceil \frac{n}{2} \rceil$, the algorithm has to iterate over all candidates $v_l \in \text{chain}(v_{i+1}, v_{i+k-1})$ of a triangle witness of (v_i, v_{i+k}). In total at most $O(n^3)$ such combinations have to be examined. In order to decide whether a particular v_l is a triangle witness of (v_i, v_{i+k}), first the algorithm has to decide whether v_l is visible to both v_i and v_{i+k}. If we use a self-balancing tree data structure for $F[v_i]$ and $B[v_{i+k}]$ for all choices of i and k, this decision requires $O(\log n)$ time. Summing the corresponding angles and comparing the result to π takes constant time. Hence the total running time is $O(n^3 \log n)$.   □

Note that as the triangle witness algorithm computes a unique solution, it provides an immediate way of identifying inconsistent input, i.e. angle data that does not belong to any polygon. If upon termination of the algorithm $|F[v_i] \cup B[v_i]| \neq d(v_i)$ for some vertex v_i, the input must be inconsistent. Otherwise, we can compute a triangulation of the visibility graph and infer the shape of it in the plane. Then the input was consistent if and only if the computed shape is valid (i.e., if this gives a simple polygon that is consistent with the input sequence of angle measurements).

References

1. Biedl, T., Durocher, S., Snoeyink, J.: Reconstructing polygons from scanner data. In: Dong, Y., Du, D.-Z., Ibarra, O. (eds.) ISAAC 2009. LNCS, vol. 5878, pp. 862–871. Springer, Heidelberg (2009)
2. Bilò, D., Disser, Y., Mihalák, M., Suri, S., Vicari, E., Widmayer, P.: Reconstructing visibility graphs with simple robots. In: Proceedings of the 16th International Colloquium on Structural Information and Communication Complexity, pp. 87–99 (2009)
3. Chalopin, J., Das, S., Disser, Y., Mihalák, M., Widmayer, P.: How simple robots benefit from looking back. In: Proceedings of the 7th International Conference on Algorithms and Complexity (to appear)
4. Formann, M., Woeginger, G.: On the reconstruction of simple polygons. Bulletin of the EATCS 40, 225–230 (1990)
5. Ghosh, S.K.: Visibility Algorithms in the Plane. Cambridge University Press, Cambridge (2007)
6. Ghosh, S.K., Goswami, P.P.: Unsolved problems in visibility graph theory. In: Proceedings of the India-Taiwan Conference on Discrete Mathematics, pp. 44–54 (2009)
7. Jackson, L., Wismath, K.: Orthogonal polygon reconstruction from stabbing information. Computational Geometry 23(1), 69–83 (2002)
8. Jansen, K., Woeginger, G.: The complexity of detecting crossingfree configurations in the plane. BIT Numerical Mathematics 33(4), 580–595 (1993)
9. Kameda, T., Yamashita, M.: On the reconstruction of polygons with (simple) robots. Personal communication (2009)
10. Rappaport, D.: On the complexity of computing orthogonal polygons from a set of points. Technical Report SOCS-86.9, McGill University, Montréal, Canada (1986)
11. Sidlesky, A., Barequet, G., Gotsman, C.: Polygon reconstruction from line crosssections. In: Proceedings of the 18th Annual Canadian Conference on Computational Geometry, pp. 81–84 (2006)
12. Skiena, S., Smith, W., Lemke, P.: Reconstructing sets from interpoint distances. In: Proceedings of the Sixth Annual Symposium on Computational Geometry, pp. 332–339 (1990)
13. Snoeyink, J.: Cross-ratios and angles determine a polygon. In: Proceedings of the 14th Annual Symposium on Computational Geometry, pp. 49–57 (1998)
14. Suri, S., Vicari, E., Widmayer, P.: Simple robots with minimal sensing: From local visibility to global geometry. International Journal of Robotics Research 27(9), 1055–1067 (2008)

Semidefinite Programming and Approximation Algorithms: A Survey

Sanjeev Arora

Computer Science Dept.
& Center for Computational Intractability,
Princeton University
arora@cs.princeton.edu

Computing approximately optimal solutions is an attractive way to cope with NP-hard optimization problems. In the past decade or so, *semidefinite programming* or SDP (a form of convex optimization that generalizes linear programming) has emerged as a powerful tool for designing such algorithms, and the last few years have seen a profusion of results (worst-case algorithms, average case algorithms, impossibility results, etc).

This talk will be a survey of this area and these recent results. We will survey three generations of SDP-based algorithms. Each generation involves new modes of analysis, which have also led to new results in mathematics. At the end we will touch upon work that draws upon the so-called *Matrix Multiplicative weight method*, which greatly improves the running time of SDP-based algorithms, making them potentially quite practical. These algorithms have also found surprising application in Quantum computing, especially the recent result QIP=PSPACE.

The survey will be essentially self-contained.

H. Kaplan (Ed.): SWAT 2010, LNCS 6139, p. 25, 2010.

Strictly-Regular Number System and Data Structures*

Amr Elmasry[1], Claus Jensen[2], and Jyrki Katajainen[3]

[1] Max-Planck Institut für Informatik, Saarbrücken, Germany
[2] The Royal Library, Copenhagen, Denmark
[3] Department of Computer Science, University of Copenhagen, Denmark

Abstract. We introduce a new number system that we call the strictly-regular system, which efficiently supports the operations: digit-increment, digit-decrement, cut, concatenate, and add. Compared to other number systems, the strictly-regular system has distinguishable properties. It is superior to the regular system for its efficient support to decrements, and superior to the extended-regular system for being more compact by using three symbols instead of four. To demonstrate the applicability of the new number system, we modify Brodal's meldable priority queues making deletion require at most $2 \lg n + O(1)$ element comparisons (improving the bound from $7 \lg n + O(1)$) while maintaining the efficiency and the asymptotic time bounds for all operations.

1 Introduction

Number systems are powerful tools of the trade when designing worst-case-efficient data structures. As far as we know, their usage was first discussed in the seminar notes by Clancy and Knuth [1]. Early examples of data structures relying on number systems include finger search trees [2] and binomial queues [3]. For a survey, see [4, Chapter 9]. The problem with the normal binary number representation is that a single increment or decrement may change all the digits in the original representation. In the corresponding data structure, this may give rise to many changes that would result in weak worst-case performance.

The characteristics of a positional number system $\mathcal{N}$ are determined by the constraints imposed on the digits and the weights corresponding to them. Let $rep(d, \mathcal{N}) = \langle d_0, d_1, \ldots, d_{r-1} \rangle$ be the sequence of digits representing a positive integer d in $\mathcal{N}$. (An empty sequence can be used to represent zero.) By convention, d_0 is the least-significant digit and $d_{r-1} \neq 0$ is the most-significant digit. The value of d in $\mathcal{N}$ is $val(d, \mathcal{N}) = \sum_{i=0}^{r-1} d_i \cdot w_i$, where w_i is the weight corresponding to d_i. As a shorthand, we write $rep(d)$ for $rep(d, \mathcal{N})$ and $val(d)$ for $val(d, \mathcal{N})$. In a redundant number system, it is possible to have $val(d) = val(d')$ while $rep(d) \neq rep(d')$. In a b-ary number system, $w_i = b^i$.

* The work of the authors was partially supported by the Danish Natural Science Research Council under contract 09-060411 (project "Generic programming—algorithms and tools"). A. Elmasry was supported by the Alexander von Humboldt Foundation and the VELUX Foundation.

H. Kaplan (Ed.): SWAT 2010, LNCS 6139, pp. 26–37, 2010.

A sequence of digits is said to be *valid* in $\mathcal{N}$ if all the constraints imposed by $\mathcal{N}$ are satisfied. Let d and d' be two numbers where $rep(d) = \langle d_0, d_1, \ldots, d_{r-1} \rangle$ and $rep(d') = \langle d'_0, d'_1, \ldots, d'_{r'-1} \rangle$ are valid. The following operations are defined.

increment(d, i): Assert that $i \in \{0, 1, \ldots, r\}$. Perform $\texttt{++}d_i$ resulting in d', i.e. $val(d') = val(d) + w_i$. Make d' valid without changing its value.

decrement(d, i): Assert that $i \in \{0, 1, \ldots, r-1\}$. Perform $\texttt{--}d_i$ resulting in d', i.e. $val(d') = val(d) - w_i$. Make d' valid without changing its value.

cut(d, i): Cut $rep(d)$ into two valid sequences having the same value as the numbers corresponding to $\langle d_0, d_1, \ldots, d_{i-1} \rangle$ and $\langle d_i, d_{i+1}, \ldots, d_{r-1} \rangle$.

concatenate(d, d'): Concatenate $rep(d)$ and $rep(d')$ into one valid sequence that has the same value as $\langle d_0, d_1, \ldots, d_{r-1}, d'_0, d'_1, \ldots, d'_{r'-1} \rangle$.

add(d, d'): Construct a valid sequence d'' such that $val(d'') = val(d) + val(d')$.

One should think that a corresponding data structure contains d_i components of rank i, where the meaning of rank is application specific. A component of rank i has size $s_i \leq w_i$. If $s_i = w_i$, we see the component as perfect. In general, the size of a structure corresponding to a sequence of digits need not be unique.

The regular system [1], called the segmented system in [4], comprises the digits $\{0, 1, 2\}$ with the constraint that every 2 is preceded by a 0 possibly having any number of 1's in between. Using the syntax for regular expressions (see, for example, [5, Section 3.3]), every regular sequence is of the form $(0 \mid 1 \mid 01^*2)^*$. The regular system allows for the increment of any digit with $O(1)$ digit changes [1,6], a fact that can be used to modify binomial queues to accomplish *insert* at $O(1)$ worst-case cost. Brodal [7] used a zeroless variant of the regular system, comprising the digits $\{1, 2, 3\}$, to ensure that the sizes of his trees are exponential with respect to their ranks. For further examples of structures that use the regular system, see [8,9]. To be able to perform decrements with $O(1)$ digit changes, an extension was proposed in [1,6]. Such an extended-regular system comprises the digits $\{0, 1, 2, 3\}$ with the constraint that every 3 is preceded by a 0 or 1 possibly having any number of 2's in between, and that every 0 is preceded by a 2 or 3 possibly having any number of 1's in between. For examples of structures that use the extended-regular system, see [6,10,11].

In this paper, we introduce a number system that we call the strictly-regular system. It uses the digits $\{0, 1, 2\}$ and allows for both increments and decrements with $O(1)$ digit changes. The strictly-regular system contains less redundancy and is more compact, achieving better constant factors while supporting a larger repertoire of operations. We expect the new system to be useful in several other contexts in addition to the applications we mention here.

Utilizing the strictly-regular system, we introduce the strictly-regular trees. Such trees provide efficient support for adding a new subtree to the root, detaching an existing one, cutting and concatenating lists of children. We show that the number of children of any node in a strictly-regular tree is bounded by $\lg n$, where n is the number of descendants of such node.

A *priority queue* is a fundamental data structure which stores a dynamic collection of elements and efficiently supports the operations *find-min*, *insert*, and *delete*. A *meldable* priority queue also supports the operation *meld* efficiently. As

Table 1. Known results on the worst-case comparison complexity of priority-queue operations when *decrease* is not considered and *find-min* has $O(1)$ cost. Here n and m denote the sizes of priority queues.

Source	insert	delete	meld
[12]	$O(1)$	$\lg n + O(1)$	–
[11]	$O(1)$	$\lg n + O(\lg \lg n)$	$O(\lg(\min\{n, m\}))$
[7] (see Section 3.1)	$O(1)$	$7 \lg n + O(1)$	$O(1)$
[13]	$O(1)$	$3 \lg n + O(1)$	$O(1)$
this paper	$O(1)$	$2 \lg n + O(1)$	$O(1)$

a principal application of our number system, we implement an efficient meldable priority queue. Our best upper bound is $2 \lg n + O(1)$ element comparisons per *delete*, which is achieved by modifying the priority queue described in [7]. Table 1 summarizes the related known results.

The paper is organized as follows. We introduce the number system in Section 2, study the application to meldable priority queues in Section 3, and discuss the applicability of the number system to other data structures in Section 4.

2 The Number System

Similar to the redundant binary system, in our system any digit d_i must be 0, 1, or 2. We call 0 and 2 *extreme digits*. We say that the representation is *strictly regular* if the sequence from the least-significant to the most-significant digit is of the form $\left(1^+ \mid 01^*2\right)^* \left(\varepsilon \mid 01^+\right)$. In other words, such a sequence is a combination of zero or more interleaved 1^+ and 01^*2 blocks, which may be followed by at most one 01^+ block. We use $w_i = 2^i$, implying that the weighted value of a 2 at position i is equivalent to that of a 1 at position $i + 1$.

2.1 Properties

An important property that distinguishes our number system from other systems is what we call the *compactness* property, which is defined in the next lemma.

Lemma 1. *For any strictly-regular sequence, $\sum_{i=0}^{r-1} d_i$ is either $r - 1$ or r.*

Proof. The sum of the digits in a 01^*2 block or a 1^* block equals the number of digits in the block, and the sum of the digits in the possibly trailing 01^+ block is one less than the number of digits in that block.   □

Note that the sum of digits $\sum_{i=0}^{r-1} d_i$ for a positive integer in the regular system is between 1 and r; in the zeroless system, where $d_i \in \{1, 2, \ldots h\}$, the sum of digits is between r and $h \cdot r$; and in the zeroless regular representation, where $d_i \in \{1, 2, 3\}$ [7], the sum of digits is between r and $2r$.

An important property, essential for designing data structures with exponential size in terms of their rank, is what we call the *exponentiality* property. Assume $s_i \geq \theta^i / c$ and $s_0 = 1$, for fixed real constants $\theta > 1$ and $c > 0$. A number system has such property if for each valid sequence $\sum_{i=0}^{r-1} d_i \cdot s_i \geq \theta^r / c - 1$ holds.

Lemma 2. *For the strictly-regular system, the exponentiality property holds by setting $\theta = c = \Phi$, where Φ is the golden ratio.*

Proof. Consider a sequence of digits in a strictly-regular representation, and think about $d_i = 2$ as two 1's at position i. It is straightforward to verify that there exists a distinct 1 whose position is at least i, for every i from 0 to $r - 2$. In other words, we have $\sum_{i=0}^{r-1} d_i \cdot s_i \geq \sum_{i=0}^{r-2} s_i$. Substituting with $s_i \geq \Phi^{i-1}$ and $s_0 = 1$, we obtain $\sum_{i=0}^{r-1} d_i \cdot s_i \geq 1 + \sum_{i=0}^{r-3} \Phi^i \geq \Phi^{r-1} - 1$. $\square$

The exponentiality property holds for any zeroless system by setting $\theta = 2$ and $c = 1$. The property also holds for any θ when $d_{r-1} \geq \theta$; this idea was used in [8], by imposing $d_{r-1} \geq 2$, to ensure that the size of a tree of rank r is at least 2^r. On the other hand, the property does not hold for the regular system.

2.2   Operations

It is convenient to use the following subroutines that change two digits but not the value of the underlying number.

fix-carry(d, i): Assert that $d_i \geq 2$. Perform $d_i \leftarrow d_i - 2$ and $d_{i+1} \leftarrow d_{i+1} + 1$.
fix-borrow(d, i): Assert that $d_i \leq 1$. Perform $d_{i+1} \leftarrow d_{i+1} - 1$ and $d_i \leftarrow d_i + 2$.

Temporarily, a digit can become a 3 due to $++d_i$ or *fix-borrow*, but we always eliminate such a violation before completing the operations. We demonstrate in Algorithm *increment* (*decrement*) how to implement the operation in question with at most one *fix-carry* (*fix-borrow*), which implies Theorem 1. The correctness of the algorithms follows from the case analysis of Table 2.

Theorem 1. *Given a strictly-regular representation of d, increment(d, i) and decrement(d, i) incur at most three digit changes.*

Algorithm *increment*(d, i)

1: $++d_i$
2: Let d_b be the first extreme digit before d_i, $d_b \in \{0, 2, \text{undefined}\}$
3: Let d_a be the first extreme digit after d_i, $d_a \in \{0, 2, \text{undefined}\}$
4: **if** $d_i = 3$ **or** ($d_i = 2$ **and** $d_b \neq 0$)
5:    *fix-carry*(d, i)
6: **else if** $d_a = 2$
7:    *fix-carry*(d, a)

Algorithm *decrement*(d, i)

1: Let d_b be the first extreme digit before d_i, $d_b \in \{0, 2, undefined\}$
2: Let d_a be the first extreme digit after d_i, $d_a \in \{0, 2, undefined\}$
3: **if** $d_i = 0$ **or** ($d_i = 1$ **and** $d_b = 0$ **and** $i \neq r - 1$)
4:     *fix-borrow*(d, i)
5: **else if** $d_a = 0$
6:     *fix-borrow*(d, a)
7: $--d_i$

By maintaining pointers to all extreme digits in a circular doubly-linked list, the extreme digits are readily available when increments and decrements are carried out at either end of a sequence.

Corollary 1. *Let $\langle d_0, d_1, \ldots, d_{r-1} \rangle$ be a strictly-regular representation of d. If such sequence is implemented as two circular doubly-linked lists, one storing all the digits and another all extreme digits, any of the operations increment$(d, 0)$, increment$(d, r - 1)$, increment(d, r), decrement$(d, 0)$, and decrement$(d, r - 1)$ can be executed at $O(1)$ worst-case cost.*

Theorem 2. *Let $\langle d_0, d_1, \ldots, d_{r-1} \rangle$ and $\langle d'_0, d'_1, \ldots, d'_{r'-1} \rangle$ be strictly-regular representations of d and d'. The operations cut(d, i) and concatenate(d, d') can be executed with $O(1)$ digit changes. Assuming without loss of generality that $r \leq r'$, add(d, d') can be executed at $O(r)$ worst-case cost including at most r carries.*

Proof. Consider the two sequences resulting from a cut. The first sequence is strictly regular and requires no changes. The second sequence may have a preceding 1^*2 block followed by a strictly-regular subsequence. In such case, we perform a *fix-carry* on the 2 ending such block to reestablish strict regularity. A catenation requires a fix only if $rep(d)$ ends with a 01^+ block and $rep(d')$ is not equal to 1^+. In such case, we perform a *fix-borrow* on the first 0 of $rep(d')$. An addition is implemented by adding the digits of one sequence to the other starting from the least-significant digit, simultaneously updating the pointers to the extreme digits in the other sequence, while maintaining strict regularity. Since each increment propagates at most one *fix-carry*, the bounds follow.   □

2.3   Strictly-Regular Trees

We recursively define a *strictly-regular tree* such that every subtree is as well a strictly-regular tree. For every node x in such a tree

– the rank, in brief $rank(x)$, is equal to the number of the children of x;
– the *cardinality sequence*, in which entry i records the number of children of rank i, is strictly regular.

The next lemma directly follows from the definitions and Lemma 1.

Lemma 3. *Let $\langle d_0, d_1, \ldots d_{r-1} \rangle$ be the cardinality sequence of a node x in a strictly-regular tree. If the last block of this sequence is a 01^+ block, then $rank(x) = r - 1$; otherwise, $rank(x) = r$.*

Table 2. d_i is displayed in bold. d_a is the first extreme digit after d_i, k is a positive integer, α denotes any combination of 1^+ and 01^*2 blocks, and ω any combination of 1^+ and 01^*2 blocks followed by at most one 01^+ block.

Initial configuration	Action	Final configuration
$\alpha 01^*\mathbf{2}$	$d_i \leftarrow 3;\ \textit{fix-carry}(d,i)$	$\alpha 01^*11$
$\alpha 01^*\mathbf{2}1^k\omega$	$d_i \leftarrow 3;\ \textit{fix-carry}(d,i)$	$\alpha 01^*121^{k-1}\omega$
$\alpha 01^*\mathbf{2}01^*2\omega$	$d_i \leftarrow 3;\ \textit{fix-carry}(d,i)$	$\alpha 01^*111^*2\omega$
$\alpha 01^*\mathbf{2}01^k$	$d_i \leftarrow 3;\ \textit{fix-carry}(d,i)$	$\alpha 01^*111^k$
$\alpha \mathbf{1}$	$d_i \leftarrow 2;\ \textit{fix-carry}(d,i)$	$\alpha 01$
$\alpha \mathbf{1}1^k\omega$	$d_i \leftarrow 2;\ \textit{fix-carry}(d,i)$	$\alpha 021^{k-1}\omega$
$\alpha \mathbf{1}01^*2\omega$	$d_i \leftarrow 2;\ \textit{fix-carry}(d,i)$	$\alpha 011^*2\omega$
$\alpha \mathbf{1}01^k$	$d_i \leftarrow 2;\ \textit{fix-carry}(d,i)$	$\alpha 011^k$
$\alpha 01^*1\mathbf{1}^*2$	$d_i \leftarrow 2;\ \textit{fix-carry}(d,a)$	$\alpha 01^*21^*01$
$\alpha 01^*1\mathbf{1}^*21^k\omega$	$d_i \leftarrow 2;\ \textit{fix-carry}(d,a)$	$\alpha 01^*21^*021^{k-1}\omega$
$\alpha 01^*1\mathbf{1}^*201^*2\omega$	$d_i \leftarrow 2;\ \textit{fix-carry}(d,a)$	$\alpha 01^*21^*011^*2\omega$
$\alpha 01^*1\mathbf{1}^*201^k$	$d_i \leftarrow 2;\ \textit{fix-carry}(d,a)$	$\alpha 01^*21^*011^k$
$\alpha 0\mathbf{1}^*2$	$d_i \leftarrow 1;\ \textit{fix-carry}(d,a)$	$\alpha 11^*01$
$\alpha 0\mathbf{1}^*21^k\omega$	$d_i \leftarrow 1;\ \textit{fix-carry}(d,a)$	$\alpha 11^*021^{k-1}\omega$
$\alpha 0\mathbf{1}^*201^*2\omega$	$d_i \leftarrow 1;\ \textit{fix-carry}(d,a)$	$\alpha 11^*011^*2\omega$
$\alpha 0\mathbf{1}^*201^k$	$d_i \leftarrow 1;\ \textit{fix-carry}(d,a)$	$\alpha 11^*011^k$
$\alpha 01^*1\mathbf{1}^*$	$d_i \leftarrow 2$	$\alpha 01^*21^*$
$\omega \mathbf{0}$	$d_i \leftarrow 1$	$\omega 1$
$\alpha \mathbf{0}1^k$	$d_i \leftarrow 1$	$\alpha 11^k$

(a) Case analysis for $increment(d,i)$.

Initial configuration	Action	Final configuration
$\alpha \mathbf{0}2\omega$	$\textit{fix-borrow}(d,i);\ d_i \leftarrow 1$	$\alpha 11\omega$
$\alpha \mathbf{0}1^k2\omega$	$\textit{fix-borrow}(d,i);\ d_i \leftarrow 1$	$\alpha 101^{k-1}2\omega$
$\alpha \mathbf{0}1^k$	$\textit{fix-borrow}(d,i);\ d_i \leftarrow 1$	$\alpha 101^{k-1}$
$\alpha 0\mathbf{1}^*12\omega$	$\textit{fix-borrow}(d,i);\ d_i \leftarrow 2$	$\alpha 01^*21\omega$
$\alpha 0\mathbf{1}^*11^k2\omega$	$\textit{fix-borrow}(d,i);\ d_i \leftarrow 2$	$\alpha 01^*201^{k-1}2\omega$
$\alpha 0\mathbf{1}^*11^k$	$\textit{fix-borrow}(d,i);\ d_i \leftarrow 2$	$\alpha 01^*201^{k-1}$
$\alpha 1\mathbf{1}^*02\omega$	$\textit{fix-borrow}(d,a);\ d_i \leftarrow 0$	$\alpha 01^*21\omega$
$\alpha 1\mathbf{1}^*01^k2\omega$	$\textit{fix-borrow}(d,a);\ d_i \leftarrow 0$	$\alpha 01^*201^{k-1}2\omega$
$\alpha 1\mathbf{1}^*01^k$	$\textit{fix-borrow}(d,a);\ d_i \leftarrow 0$	$\alpha 01^*201^{k-1}$
$\alpha 01^*2\mathbf{1}^*02\omega$	$\textit{fix-borrow}(d,a);\ d_i \leftarrow 1$	$\alpha 01^*11^*21\omega$
$\alpha 01^*2\mathbf{1}^*01^k2\omega$	$\textit{fix-borrow}(d,a);\ d_i \leftarrow 1$	$\alpha 01^*11^*201^{k-1}2\omega$
$\alpha 01^*2\mathbf{1}^*01^k$	$\textit{fix-borrow}(d,a)\ ;\ d_i \leftarrow 1$	$\alpha 01^*11^*201^{k-1}$
$\alpha 1\mathbf{1}^*$	$d_i \leftarrow 0$	$\alpha 01^*$
$\alpha 01^*\mathbf{1}$	$d_i \leftarrow 0$	$\alpha 01^*$
$\alpha 01^*2\mathbf{1}^*$	$d_i \leftarrow 1$	$\alpha 01^*11^*$

(b) Case analysis for $decrement(d,i)$.

The next lemma illustrates the exponentiality property for such trees.

Lemma 4. *A strictly-regular tree of rank r has at least 2^r nodes.*

Proof. The proof is by induction. The claim is clearly true for nodes of rank 0. Assume the hypothesis is true for all the subtrees of a node x with rank r. Let y be the child of x with the largest rank. From Lemma 3, if the last block of the cardinality sequence of x is a 01^+ block, then $rank(x) = rank(y)$. Using induction, the number of nodes of y's subtree is at least 2^r, and the lemma follows. Otherwise, the cardinality sequence of x only contains 01^*2 and 1^+ blocks. We conclude that there exists a distinct subtree of x whose rank is at least i, for every i from 0 to $r - 1$. Again using induction, the size of the tree rooted at x must be at least $1 + \sum_{i=0}^{r-1} 2^i = 2^r$.   □

The operations that we would like to efficiently support include: adding a subtree whose root has rank at most r to the children of x; detaching a subtree from the children of x; splitting the sequence of the children of x, those having the highest ranks and the others; and concatenating a strictly-regular subsequence of trees, whose smallest rank equals r, to the children of x.

   In accordance, we need to support implementations corresponding to the subroutines *fix-carry* and *fix-borrow*. For these, we use *link* and *unlink*.

link(T_1, T_2): Assert that the roots of T_1 and T_2 have the same rank. Make one root the child of the other, and increase the rank of the surviving root by 1.

unlink(T): Detach a child with the largest rank from the root of tree T. If T has rank r, the resulting two trees will have ranks either $r - 1, r - 1$ or $r - 1, r$.

Subroutine *fix-carry*(d, i), which converts two consecutive digits $d_i = 2$ and $d_{i+1} = q$ to $0, q + 1$ is realizable by subroutine *link*. Subroutine *fix-borrow*(d, i), which converts two consecutive digits $d_i = 0$ and $d_{i+1} = q$ to $2, q - 1$ is realizable by subroutine *unlink* that results in two trees of equal rank. However, unlinking a tree of rank r may result in one tree of rank $r - 1$ and another of rank r. In such case, a *fix-borrow* corresponds to converting the two digits $0, q$ to $1, q$. For this scenario, as for Table 2(b), it is also easy to show that all the cases following a decrement lead to a strictly-regular sequence. We leave the details for the reader to verify.

3   Application: Meldable Priority Queues

Our motivation is to investigate the worst-case bound for the number of element comparisons performed by *delete* under the assumption that *find-min*, *insert*, and *meld* have $O(1)$ worst-case cost. From the comparison-based lower bound for sorting, we know that if *find-min* and *insert* only involve $O(1)$ element comparisons, *delete* has to perform at least $\lg n - O(1)$ element comparisons, where n is the number of elements stored prior to the operation.

3.1   Brodal's Meldable Priority Queues

Our development is based on the priority queue presented in [7]. In this section, we describe this data structure. We also analyse the constant factor in the bound on the number of element comparisons performed by *delete*, since the original analysis was only asymptotic.

The construction in [7] is based on two key ideas. First, *insert* is supported at $O(1)$ worst-case cost. Second, *meld* is reduced to *insert* by allowing a priority queue to store other priority queues inside it. To make this possible, the whole data structure is a tree having two types of nodes: □-nodes (read: *square* or *type-I nodes*) and ⊙-nodes (read: *circle* or *type-II nodes*). Each node stores a locator to an element, which is a representative of the descendants of the node; the representative has the smallest element among those of its descendants.

Each node has a non-negative integer *rank*. A node of rank 0 has no ⊙-children. For an integer $r > 0$, the ⊙-children of a node of rank r have ranks from 0 to $r - 1$. Each node can have at most one □-child and that child can be of arbitrary rank. The number of ⊙-children is restricted to be at least one and at most three per rank. More precisely, the *regularity constraint* posed is that the cardinality sequence is of the form $\left(1 \mid 2 \mid 12^*3\right)^*$. This regular number system allows for increasing the least significant digit at $O(1)$ worst-case cost. In addition, because of the zeroless property, the size of a subtree of rank r is at least 2^r and the number of children of its root is at most $2r$. The rank of the root is required to be zero. So, if the tree holds more than one element, the other elements are held in the subtree rooted at the □-child of the root.

To represent such multi-way tree, the standard child-sibling representation can be used. Each node stores its rank as an integer, its type as a Boolean, a pointer to its parent, a pointer to its sibling, and a pointer to its ⊙-child having the highest rank. The children of a node are kept in a circular singly-linked list containing the ⊙-children in rank order and the □-child after the ⊙-child of the highest rank; the □-child is further connected to the ⊙-child of rank 0. Additionally, each node stores a pointer to a linked list, which holds pointers to the first ⊙-node in every group of three consecutive nodes of the same rank corresponding to a 3 in the cardinality sequence.

A basic subroutine used in the manipulation of these trees is *link*. For node u, let *element*(u) denote the element associated with u. Let u and v be two nodes of the same rank such that *element*$(u) \leq$ *element*(v). Now, *link* makes v a ⊙-child of u. This increases the rank of u by one. Note that *link* has $O(1)$ worst-case cost and performs one element comparison.

The minimum element is readily found by accessing the root of the tree, so *find-min* is easily accomplished at $O(1)$ worst-case cost.

When inserting a new element, a node is created. The new element and those associated with the root and its □-child are compared; the two smallest among the three are associated with the root and its □-child, and the largest is associated with the created node. Hereafter, the new node is added as a ⊙-child of rank 0 to the □-child of the root. Since the cardinality sequence of that node was regular before the insertion, only $O(1)$ structural changes are necessary to restore the regularity constraint. That is, *insert* has $O(1)$ worst-case cost.

To meld two trees, the elements associated with the root and its □-child are taken from both trees and these four elements are sorted. The largest element is associated with a ⊡-child of the root of one tree. Let T be that tree, and let S be the other tree. The two smallest elements are then associated with the root of S and its ⊡-child. Accordingly, the other two elements are associated with the root of T and its ⊡-child. Subsequently, T is added as a rank-0 ⊙-child to the ⊡-child of the root of S. So, also *meld* has $O(1)$ worst-case cost.

When deleting an element, the corresponding node is located and made the current node. If the current node is the root, the element associated with the ⊡-child of the root is swapped with that associated with the root, and the ⊡-child of the root is made the current node. On the other hand, if the current node is a ⊙-node, the elements associated with the current node and its parent are swapped until a ⊡-node is reached. Therefore, both cases reduce to a situation where a ⊡-node is to be removed.

Assume that we are removing a ⊡-node z. The actual removal involves finding a node that holds the smallest element among the elements associated with the children of z (call this node x), and finding a node that has the highest rank among the children of x and z (call this node y). To reestablish the regularity constraint, z is removed, x is promoted into its place, y is detached from its children, and all the children previously under x and y, plus y itself, are moved under x. This is done by performing repeated linkings until the number of nodes of the same rank is one or two. The rank of x is updated accordingly.

In the whole deletion process $O(\lg n)$ nodes are handled and $O(1)$ work is done per node, so the total cost of *delete* is $O(\lg n)$. To analyse the number of element comparisons performed, we point out that a node with rank r can have up to $2r$ ⊙-children (not $3r$ as stated in [7]). Hence, finding the smallest element associated with a node requires up to $2 \lg n + O(1)$ element comparisons, and reducing the number of children from $6 \lg n + O(1)$ to $\lg n + O(1)$ involves $5 \lg n + O(1)$ element comparisons (each *link* requires one). To see that this bound is possible, consider the addition of four numbers 1, 1232^k, 2222^k, and 1232^k (where the least significant digits are listed first), which gives $1211^{k+1}2$.

Our discussion so far can be summarized as follows.

Theorem 3. *Brodal's meldable priority queue, as described in [7], supports find-min, insert, and meld at $O(1)$ worst-case cost, and delete at $O(\lg n)$ worst-case cost including at most $7 \lg n + O(1)$ element comparisons.*

3.2   Our Improvement

Consider a simple mixed scheme, in which the number system used for the children of ⊙-nodes is perfect, following the pattern 1^*, and that used for the children of ⊡-nodes is regular. This implies that the ⊙-nodes form binomial trees [3]. After this modification, the bounds for *insert* and *meld* remain the same if we rely on the delayed melding strategy. However, since each node has at most $\lg n + O(1)$ children, the bound for *delete* would be better than that reported in Theorem 3. Such an implementation of *delete* has three bottlenecks: finding the minimum, executing a delayed *meld*, and adding the ⊙-children of a ⊡-node to another node. In this mixed system, each of these three procedures requires at most $\lg n + O(1)$ element comparisons. Accordingly, *delete* involves at most $3 \lg n + O(1)$ element comparisons. Still, the question is how to do better!

The major change we make is to use the strictly-regular system instead of the zeroless regular system. We carry out *find-min*, *insert*, and *meld* similar to [7]. We use subroutine *merge* to combine two trees. Let y and y' be the roots of these trees, and let r and r' be their respective ranks where $r \leq r'$. We show how to *merge* the two trees at $O(r)$ worst-case cost using $O(1)$ element comparisons. For this, we have to locate the nodes representing the extreme digits closest to r in the cardinality sequence of y'. Consequently, by Theorems 1 and 2, a cut or an increment at that rank is done at $O(1)$ worst-case cost. If $element(y') \leq element(y)$, add y as a $\odot$-child of y', update the rank of y' and stop. Otherwise, cut the $\odot$-children of y' at r. Let the two resulting sublists be C and D, C containing the nodes of lower rank. Then, concatenate the lists representing the sequence of the $\odot$-children of y and the sequence D. We regard y' together with the $\odot$-children in C and y''s earlier $\boxdot$-child as one tree whose root y' is a $\odot$-node. Finally, place this tree under y and update the rank of y.

Now we show how to improve *delete*. If the node to be deleted is the root, we swap the elements associated with the root and its $\boxdot$-child, and let that $\boxdot$-node be the node z to be deleted. If the node to be deleted is a $\odot$-node, we repeatedly swap the elements associated with this node and its parent until the current node is a $\boxdot$-node (Case 1) or the rank of the current node is the same as that of its parent (Case 2). When the process stops, the current node z is to be deleted.

Case 1: z is a $\boxdot$-node. Let x denote the node that contains the smallest element among the children of z (if any). We remove z, lift x into its place, and make x into a $\boxdot$-node. Next, we move all the other $\odot$-children of z under x by performing an addition operation, and update the rank of x. Since z and x may each have had a $\boxdot$-child, there may be two $\boxdot$-children around. In such case, *merge* such two subtrees and make the root of the resulting tree the $\boxdot$-child of x.

Case 2: z is a $\odot$-node. Let p be the parent of z. We remove z and move its $\odot$-children to p by performing an addition operation. As $rank(p) = rank(z)$ before the addition, $rank(p) = rank(z)$ or $rank(z) + 1$ after the addition. If $rank(p) = rank(z) + 1$, to ensure that $rank(p)$ remains the same as before the operation, we detach the child of p that has the highest rank and *merge* the subtree rooted at it with the subtrees rooted at the $\boxdot$-children of p and z (there could be up to two such subtrees), and make the root of the resulting tree the $\boxdot$-child of p.

Let r be the maximum rank of a node in the tree under consideration. Climbing up the tree to locate a node z has $O(r)$ cost, since after every step the new current node has a larger rank. In Case 1, a $\boxdot$-node is deleted at $O(r)$ cost involving at most r element comparisons when finding its smallest child. In Cases 1 and 2, the addition of the $\odot$-children of two nodes has $O(r)$ cost and requires at most r element comparisons. Additionally, applying the *merge* operation on two trees (Case 1) or three trees (Case 2) has $O(r)$ cost and requires $O(1)$

element comparisons. Thus, the total cost is $O(r)$ and at most $2r + O(1)$ element comparisons are performed. Using Lemma 4, $r \leq \lg n$, and the claim follows.

In summary, our data structure improves the original data structure in two ways. First, by Lemma 4, the new system reduces the maximum number of children a node can have from $2 \lg n$ to $\lg n$. Second, the new system breaks the bottleneck resulting from delayed melding, since two subtrees can be merged with $O(1)$ element comparisons. The above discussion implies the following theorem.

Theorem 4. *Let n denote the number of elements stored in the data structure prior to a deletion. There exists a priority queue that supports find-min, insert, and meld at $O(1)$ worst-case cost, and delete at $O(\lg n)$ worst-case cost including at most $2 \lg n + O(1)$ element comparisons.*

4   Other Applications

Historically, it is interesting to note that in early papers a number system supporting increments and decrements of an arbitrary digit was constructed by putting two regular systems back to back, i.e. $d_i \in \{0, 1, 2, 3, 4, 5\}$. It is relatively easy to prove the correctness of this system. This approach was used in [14] for constructing catenable deques, in [9] for constructing catenable finger search trees, and in [8] for constructing meldable priority queues. (In [8], $d_i \in \{2, 3, 4, 5, 6, 7\}$ is imposed, since an extra constraint that $d_i \geq 2$ was required to facilitate the violation reductions and to guarantee the exponentiality property.) Later on, it was realized that the extended-regular system, $d_i \in \{0, 1, 2, 3\}$, could be utilized for the same purpose (see, for example, [6]). The strictly-regular system may be employed in applications where these more extensive number systems have been used earlier. This replacement, when possible, would have two important consequences:

1. The underlying data structures become simpler.
2. The operations supported may become a constant factor faster.

While surveying papers that presented potential applications to the new number system, we found that, even though our number system may be applied, there were situations where other approaches would be more favourable. For example, the relaxed heap described in [11] relies on the zeroless extended-regular system to support increments and decrements. Naturally, the strictly-regular system could be used instead, and this would reduce the number of trees that have to be maintained. However, the approach of using a two-tier structure as described in [11] makes the reduction in the number of trees insignificant since the amount of work done is proportional to the logarithm of the number of trees. Also, a fat heap [6] uses the extended-regular binary system for keeping track of the potential violation nodes and the extended-regular ternary system for keeping track of the trees in the structure. However, we discovered that a priority queue with the same functionality and efficiency can be implemented with simpler tools without using number systems at all. The reader is warned: number systems are powerful tools but they should not be applied haphazardly.

Up till now we have ignored the cost of accessing the extreme digits in the vicinity of a given digit. When dealing with the regular or the extended-regular systems this can be done at $O(1)$ cost by using the guides described in [8]. In contrary, for our number system, accessing the extreme digits in the vicinity of any digit does not seem to be doable at $O(1)$ cost. However, the special case of accessing the first and last extreme digits is soluble at $O(1)$ cost.

In some applications, like fat heaps [6] and the priority queues described in [8], the underlying number system is ternary. We have not found a satisfactory solution to extend the strictly-regular system to handle ternary numbers efficiently; it is an open question whether such an extension exists.

References

1. Clancy, M., Knuth, D.: A programming and problem-solving seminar. Technical Report STAN-CS-77-606, Dept. of Computer Science, Stanford University (1977)
2. Guibas, L.J., McCreight, E.M., Plass, M.F., Roberts, J.R.: A new representation for linear lists. In: Proceedings of the 9th Annual ACM Symposium on Theory of Computing, pp. 49–60. ACM Press, New York (1977)
3. Vuillemin, J.: A data structure for manipulating priority queues. Communications of the ACM 21(4), 309–315 (1978)
4. Okasaki, C.: Purely Functional Data Structures. Cambridge University Press, Cambridge (1998)
5. Aho, A.V., Lam, M.S., Sethi, R., Ullman, J.D.: Compilers: Principles, Techniques, & Tools, 2nd edn. Pearson Education, Inc., London (2007)
6. Kaplan, H., Shafrir, N., Tarjan, R.E.: Meldable heaps and Boolean union-find. In: Proceedings of the 34th Annual ACM Symposium on Theory of Computing, pp. 573–582. ACM Press, New York (2002)
7. Brodal, G.S.: Fast meldable priority queues. In: Sack, J.-R., Akl, S.G., Dehne, F., Santoro, N. (eds.) WADS 1995. LNCS, vol. 955, pp. 282–290. Springer, Heidelberg (1995)
8. Brodal, G.S.: Worst-case efficient priority queues. In: Proceedings of the 7th Annual ACM-SIAM Symposium on Discrete Algorithms, pp. 52–58. ACM/SIAM (1996)
9. Kaplan, H., Tarjan, R.E.: Purely functional representations of catenable sorted lists. In: Proceedings of the 28th Annual ACM Symposium on Theory of Computing, pp. 202–211. ACM, New York (1996)
10. Elmasry, A.: A priority queue with the working-set property. International Journal of Foundations of Computer Science 17(6), 1455–1465 (2006)
11. Elmasry, A., Jensen, C., Katajainen, J.: Two-tier relaxed heaps. Acta Informatica 45(3), 193–210 (2008)
12. Elmasry, A., Jensen, C., Katajainen, J.: Multipartite priority queues. ACM Transactions on Algorithms, Article 14 5(1) (2008)
13. Jensen, C.: A note on meldable heaps relying on data-structural bootstrapping. CPH STL Report 2009-2, Department of Computer Science, University of Copenhagen (2009), http://cphstl.dk
14. Kaplan, H., Tarjan, R.E.: Persistent lists with catenation via recursive slow-down. In: Proceedings of the 27th Annual ACM Symposium on Theory of Computing, pp. 93–102. ACM, New York (1995)

An $O(\log\log n)$-Competitive Binary Search Tree with Optimal Worst-Case Access Times

Prosenjit Bose[1,*], Karim Douïeb[1,*], Vida Dujmović[1,*], and Rolf Fagerberg[2]

[1] School of Computer Science, Carleton University
{jit,karim,vida}@cg.scs.carleton.ca
[2] Department of Mathematics and Computer Science,
University of Southern Denmark
rolf@imada.sdu.dk

Abstract. We present the *zipper tree*, an $O(\log\log n)$-competitive on-line binary search tree that performs each access in $O(\log n)$ worst-case time. This shows that for binary search trees, optimal worst-case access time and near-optimal amortized access time can be guaranteed simultaneously.

1 Introduction

A *dictionary* is a basic data structure for storing and retrieving information. The *binary search tree* (BST) is a well-known and widely used dictionary implementation which combines efficiency with flexibility and adaptability to a large number of purposes. It constitutes one of the fundamental data structures of computer science.

In the past decades, many BST schemes have been developed which perform element accesses (and indeed many other operations) in $O(\log n)$ time, where n is the number of elements in the tree. This is the optimal single-operation worst-case access time in a comparison based model. Turning to *sequences* of accesses, it is easy to realize that for specific access sequences, there may be BST algorithms which serve m accesses in less than $\Theta(m \log n)$ time. A common way to evaluate how well the performance of a given BST algorithm adapts to individual sequences, is *competitive analysis*: For an access sequence X, define $\mathrm{OPT}(X)$ to be the minimum time needed by any BST algorithm to serve it. A given BST algorithm A is then said to be $f(n)$-*competitive* if it performs X in $O(f(n)\,\mathrm{OPT}(X))$ time for all X. To make this precise, a more formal definition of a BST model and of the sequences X considered is needed—standard in the area is to use the binary search tree model (BST model) defined by Wilber [12], in which the only existing non-trivial lower bounds on $\mathrm{OPT}(X)$ have been proven [3,12].

In 1985, Sleator and Tarjan [10] developed a BST called *splay trees*, which they conjectured to be $O(1)$-competitive. Much of the research on the efficiency of BSTs on individual input sequences has grown out of this conjecture. However,

---

[*] The authors are partially supported by NSERC and MRI.

H. Kaplan (Ed.): SWAT 2010, LNCS 6139, pp. 38–49, 2010.

despite decades of research, the conjecture is still open. More generally, it is unknown if there exist asymptotically optimal BST data structures. In fact, for many years the best known competitive ratio for any BST structure was $O(\log n)$, which is achieved by a plain static balanced tree.

This situation was recently improved by Demaine *et al.*, who in a seminal paper [3] developed an $O(\log \log n)$-competitive BST structure, called the *tango tree*. This was the first improvement in competitive ratio for BSTs over the trivial $O(\log n)$ upper bound. Being $O(\log \log n)$-competitive, tango trees are always at most a factor $O(\log \log n)$ worse than $\mathrm{OPT}(X)$. On the other hand, they may actually pay this multiplicative overhead at each access, implying that they have $\Theta(\log \log n \log n)$ worst-case access time, and use $\Theta(m \log \log n \log n)$ time on some access sequences of length m. In comparison, any balanced BST (even static) has $O(\log n)$ worst-case access time and spends $O(m \log n)$ on every access sequence.

The problem we consider in this paper is whether it is possible to combine the best of these bounds—that is, whether an $O(\log \log n)$-competitive BST algorithms that performs each access in optimal $O(\log n)$ worst-case time exists. We answer it affirmatively by presenting a data structure achieving these complexities. It is based on the overall framework of tango trees—however, where tango trees use red-black trees [6] for storing what is called preferred paths, we develop a specialized BST representation of the preferred paths, tuned to the purpose. This representation is the main technical contribution, and its description takes up the bulk of the paper.

In the journal version of their seminal paper on tango trees, Demaine *et al.* suggested that such a structure exists. Specifically, in the further work section, the authors gave a short sketch of a possible solution. Their suggested approach, however, relies on the existence of a BST supporting dynamic finger, split and merge in $O(\log r)$ worst-case time where r is the rank difference between the accessed element and the previously accessed element. Such a BST could indeed be used for the auxiliary tree representation of preferred paths. However, the existence of such a structure (in the BST-model) is an open problem. Consequently, since the publication of their work, the authors have revised their stance and consider the problem solved in this paper to be an open problem [7]. Recently, Woo [13] made some progress concerning the existence of a BST having the dynamic finger property in worst-case. He developed a BST algorithm satisfying, based on empirical evidence, the dynamic finger property in worst-case. Unfortunately this BST algorithm does not allow insertion/deletion or split/merge operations, thus it cannot be used to maintain the preferred paths in a tango tree.

After the publication of the tango tree paper, two other $O(\log \log n)$-competitive BSTs have been introduced by Derryberry *et al.* [4,11] and Georgakopoulos [5]. The multi-splay trees [4] are based on tango trees, but instead of using red-black trees as auxiliary trees, they use splay trees [10]. As a consequence, multi-splay trees can be shown [4,11] to satisfy additional properties, including the scanning and working-set bounds of splay trees, while maintaining $O(\log \log n)$-competitiveness. Georgakopoulos uses the interleave lower bound of

Demaine *et al.* to develop a variation of splay trees called *chain-splay* trees that achieves $O(\log \log n)$-competitiveness while not maintaining any balance condition explicitly. However, neither of these two structures achieves a worst-case single access time of $O(\log n)$. A data structure achieving the same running time as tango trees alongside $O(\log n)$ worst-case single access time was developed by Kujala and Elomaa [8], but this data structure does not adhere to the BST model (in which the lower bounds on OPT(X) are proved).

The rest of this paper is organized as follows: In Section 2, we formally define the model of BSTs and the access sequences considered. We state the lower bound on OPT(X) developed in [3,12] for analyzing the competitive ratio of BSTs. We also describe the central ideas of tango trees. In Section 3, we introduce a preliminary data structure called *hybrid trees*, which does not fit the BST model proper, but which is helpful in giving the main ideas of our new BST structure. Finally in Section 4, we develop this structure further to fit the BST model. This final structure, called *zipper trees*, is a BST achieving the optimal worst-case access time while maintaining the $O(\log \log n)$-competitiveness property.

2 Preliminaries

2.1 BST Model

In this paper we use the binary search tree model (BST model) defined by Wilber [12], which is standard in the area. Each node stores a key from a totally ordered universe, and the keys obey in-order: at any node, all of the keys in its left subtree are less than the key stored in the node, and all of the keys in its right subtree are greater (we assume no duplicate keys appear). Each node has three pointers, pointing to its left child, right child, and parent. Each node may keep a constant[1] amount of additional information, but no further pointers may be used.

To perform an access, we are given a pointer initialized to the root. An access consists of moving this pointer from a node to one of its adjacent nodes (through the parent pointer or one of the children pointers) until it reaches the desired element. Along the way, we are allowed to update the fields and pointers in any nodes that the pointer touches. The access cost is the number of nodes touched by the pointer. As is standard in the area, we only consider sequences consisting of element accesses on a fixed set S of n elements. In particular, neither unsuccessful searches nor updates take place.

2.2 Interleave Lower Bound

The interleave bound is a lower bound on the time taken by any binary search tree in the BST model to perform an access sequence $X = \{x_1, x_2, \ldots, x_m\}$. The interleave bound was developed by Demaine *et al.* [3] and was derived from a previous bound of Wilber [12].

---

[1] According to standard conventions, $O(\log_2 n)$ bits are considered as constant.

Let P be a static binary search tree of minimum height, built on the set of keys S. We call P the *reference* tree. For each node y in P, we consider the accesses of X that are to keys in nodes in the subtree of P rooted at y (including y). Each access of this subsequence is then labelled "left" or "right", depending on whether the accessed node is in the left subtree of y (including y), or in its right subtree, respectively. The *amount of interleaving through y* is the number of alternations between left and right labels in this subsequence. The interleave bound $\mathrm{IB}(X)$ is the sum of these interleaving amounts over all nodes y in P. The exact statement of the lower bound from [3] is as follows:

Theorem 1 (From [3]). *For any access sequence X, $\mathrm{IB}(X)/2 - n$ is a lower bound on $\mathrm{OPT}(X)$.*

2.3   Tango Trees

We outline the main ideas of tango trees [3]. As in the previous subsection, denote by the reference tree P a static binary search tree of height $O(\log n)$ built on a set of keys S. At a given point in time, the *preferred child* of an internal node y in P is defined as its left or right child depending on whether the last access to a node in the subtree rooted at y (including y) was in the left subtree of y (including y) or in its right subtree respectively. We call a maximal chain of preferred children a *preferred path*. The set of preferred paths partitions P into disjoint parts of size $O(\log n)$. Remember that P is a static tree, only the preferred paths may evolve over time (namely, after each access).

The ingenious idea of tango trees is to represent the nodes on a preferred path as a balanced tree of height $O(\log \log n)$, called an *auxiliary* tree. The tango tree can be seen as a collection of auxiliary trees linked together. The leaves of an auxiliary tree representing a preferred path p link to the roots of auxiliary trees representing the paths immediately below p in P, with the links uniquely determined by the inorder ordering. The auxiliary tree containing the root of P constitutes the top-part of the tango tree. In order to distinguish auxiliary trees within the tango tree, the root of each auxiliary tree is marked (using one bit).

Note that the reference tree P is not an explicit part of the structure, it just helps to explain and understand the concept of tango trees. When an access is performed, the preferred paths of P may change. This change is actually a combination of several cut and concatenation operations involving subpaths. Auxiliary trees in tango tree are implemented as red-black trees [6], and [3] show how to implement these cut and concatenation operations using standard split and join operations on red-black tree. Here are the main two operations used to maintain tango trees:

CUT-TANGO(A, d) – divide the red-black tree A into two red-black trees, one storing all nodes in the preferred path having depth at most d in P, and another storing all nodes having depth greater than d.

CONCATENATE-TANGO(A, B) – merge two red-black trees that store two disjoint paths for which in P the bottom of one path (stored in A) is the parent of the top of the other path (stored in B). I.e., in the tango tree, the root of B is attached to a leaf of A.

These operations can be implemented by splits and joins in $O(\log k)$ time for trees of size k, if adding extra information in nodes: Besides the key value and the depth in P, each node also stores the minimum and maximum depth over the nodes in its subtree within its auxiliary tree. This additional data is easily maintained in red-black trees. As the trees store paths in P, we have $k = O(\log n)$. Hence, if an access passes i different preferred paths in P, the necessary change in the tango tree will be $O(i)$ cut and concatenation operations, which is performed in $O(i \log \log n)$ time. Over any access sequence X the total number of cut and concatenation operations performed in P corresponds to the interleave bound $\mathrm{IB}(X)$, thus tango trees serve X in $O(\log \log n \, \mathrm{IB}(X))$ time, which by Thm. 1 makes them $O(\log \log n)$-competitive.

3 Hybrid Trees

In this section, we introduce a data structure called *hybrid trees*, which has the right running time, but which does not fit the BST model proper. However, it is helpful intermediate step which contains the main ideas of our final BST structure.

3.1 Path Representation

For all preferred paths in P, we keep the top $\Theta(\log \log n)$ nodes exactly as they appear on the path. We call this the *top path*. The remaining nodes (if any) of the path we store as a red-black tree, called the *bottom tree*, which we attach below the top path. Since a preferred path has size $O(\log n)$, this bottom tree has height $O(\log \log n)$. More precisely, we will maintain the invariant that a top path has length in $[\log \log n, 3 \log \log n]$, unless no bottom tree appears, in which case the constraint is $[0, 3 \log \log n]$. (This latter case, where no bottom tree appears, will induce simple and obvious variants of the algorithms in the remainder of the paper, variants which we for clarity of exposition will not mention further.)

A *hybrid tree* consists of all the preferred paths of P, represented as above, linked together to form one large tree, analogous to tango trees. The required worst-case search complexity of hybrid trees is captured by the following lemma.

Lemma 1. *A hybrid tree T satisfies the following property:* $d_T(x) = O(d_P(x)) \quad \forall x \in S$, *where $d_T(x)$ and $d_P(x)$ is defined as the depth of the node x in the tree T and in the reference tree P, respectively. In particular, T has $O(\log n)$ height.*

Proof. Consider a preferred path p in P and its representation tree h. The distance in h, in terms of number of edges to follow, from the root of h to one of its nodes or leaves x is no more than a constant times the distance in p between x and the root of p. Indeed, if x is part of the top path, then the distance to the root of the path by construction is the same in h and p. Otherwise, this distance increases by at most a constant factor, since h has a height of $O(\log \log n)$ and the distance in p is already $\Omega(\log \log n)$. To reach a node x in the reference tree P, we have to traverse (parts of) several preferred paths. As the above argument holds for all paths, the first statement $d_T(x) = O(d_P(x))$ of the lemma follows by summing over the paths traversed. Finally, since the depth of any node in P is $O(\log n)$, this implies that T has height $O(\log n)$. □

3.2   Maintaining Hybrid Trees under Accesses

The path p traversed in P to reach a desired node may pass through several preferred paths. During this access the preferred paths in P must change such that p becomes the new preferred path containing the root. This entails *cut* and *concatenate* operations on the preferred paths passed by p: When p leaves a preferred path, the path must be cut at the depth in P of the point of leave, and the top part cut out must be concatenated with the next preferred path to be traversed.

We note that one may as well perform only the cutting while traversing p, producing a sequence of cut out parts hanging below each other, which can then be concatenated in one go at the end of the access, producing the new preferred path starting at the root. We will use this version below.

In this subsection, we will show how to maintain the hybrid tree representation of the preferred paths after an access. Our main goal is to to give methods for performing the operations *cut* and *concatenate* on our path representations in the following complexities: When the search passes only the top path of a path representation (Case 1 cut), the cut operation takes $O(k)$ time, where k is the number of nodes traversed in the top path. When the search passes the entire top path and ends up in the bottom tree (Case 2 cut), the cut operation takes $O(\log \log n)$ time. The concatenation operation, which glues together all the cut out path representation parts at the end of the access, is bounded by the time used by the search and the cut operations performed during the access.

Assuming these running times, it follows, by the invariant that all top paths (with bottom trees below them) have length $\Theta(\log \log n)$, that the time of an access involving i cut operations in P is bounded both by the number of nodes on the search path p, and by $i \log \log n$. By Lemma 1, this is $O(\min\{\log n, i \log \log n\})$ time. Hence, we by Thm. 1 will have achieved optimal worst-case access time while maintaining $O(\log \log n)$-competitiveness.

Cut: Case 1: We only traverse the top path of a path representation. Let k be the number of nodes traversed in this top path and let x be the last traversed node. The cut operation marks the node succeeding x on the top path as the new root of the path representation, and unmarks the other child of x (to make the cut out part join the pieces that after the final concatenation will constitute the new preferred path induced by the search).

The cut operation now has removed k nodes from the top path of the path representation. This implies that we possibly have to update the representation, since the $\Theta(\log \log n)$ bound on the size of its top path has to be maintained. Specifically, if the size of the top path drops below $2 \log \log n$, we will move some nodes from the bottom tree to the top path. The nodes should be those from the bottom tree having smallest depth (in P), i.e., the next nodes on the preferred path in P. It is for small k (smaller than $\log\log n$) not clear how to extract the next k nodes from the bottom tree in $O(k)$ time. Instead, we use an *extraction process*, described below, which extracts the next $\log\log n$ nodes from the bottom tree in $O(\log\log n)$ steps and run this process incrementally: Whenever i nodes are cut from the top path, the extraction process is advanced by $\Theta(i)$ steps, and then the process is stopped until the next cut at this path occurs. Thus, the work of the extraction process is spread over several Case 1 cuts (if not stopped before by a Case 2 cut, see below). The speed of the process is chosen such that the extraction of $\log\log n$ nodes is completed before that number of nodes have been cut away from the top path, hence it will raise the size of the top path to at least $2\log\log n$ again. In general, we maintain the additional invariant that the top path has size at least $2\log\log n$, unless an extraction process is ongoing.

Case 2: We traverse the entire top path of path representation A, and enter the bottom tree. Let x be the last traversed node in A and let y be the marked child of x that is the root of the next path representation on the search path. First, we finish any pending extraction process in A, so that its bottom tree becomes a valid red-black tree. Then we rebuild the top path into a red-black tree in linear time (see details under the description of concatenate below), and we join it with the bottom tree using CONCATENATE-TANGO. Then we perform CUT-TANGO(A', d) where A' is the combined red-black tree, and $d = d_P(y) - 1$. After this operation, all nodes of depth greater than d are removed from the path representation A to form a new red-black tree B attached to A (the root of B is marked in the process). To make the tree B a valid path representation, we perform an extraction process twice, which extracts $2\log\log n$ nodes from it to form a top path. Finally we unmark y. This takes $O(\log\log n)$ time in total.

Concatenate: What is cut out during an access is a sequence of top paths (Case 1 cuts) and red-black trees (Case 2 cuts) hanging below each other. We have to concatenate this sequence into a single path representation. We first rebuild each consecutive sequence of top paths (which can be found as maximum sequences of nodes which have one marked child) into valid red-black trees, in time linear in the number of nodes in each sequence (details below). This leaves a sequence of at most $2j + 1$ valid red-black trees hanging below each other where j is the number of Case 2 cuts performed during the access. Then we iteratively perform CONCATENATE-TANGO(A,B), where A is the current highest red-black tree and B is the tree hanging below A, until there is one remaining red-black tree (this is done in $O(j\log\log n)$ time, which we have already spent on the Case 2 cuts). Finally we extract $2\log\log n$ nodes from the obtained red-black tree to construct the top path of the path representation. The time used for concatenate is bounded by the time used already during the search and cut part of the access.

To convert a path of length k into a red-black tree in $O(k)$ time, we can simply traverse the path downwards, inserting the nodes one by one into a growing (and initially empty) red-black tree. During each insertion, the remaining path will be considered a leaf of the tree. Since the next node to be inserted will become either the successor or the predecessor of the node just inserted, and since the amount of rebalancing in red-black trees is amortized $O(1)$ per update, the entire process is linear in the number of nodes inserted.

Extract: We now show how to perform the central process of our structure, namely extracting the next part of a top path from a bottom tree. Specifically, we will extract a subpath of $\log \log n$ nodes of minimum depth (in P) from the bottom tree A' of a given path representation A, using $O(\log \log n)$ time.

Let x be the deepest node on the top path of A, such that the unmarked child of x corresponds to the root of the bottom tree A'. The extraction process will separate the nodes of depth (in P) smaller than $d = d_P(x) + \log \log n$ from the bottom tree A'. Let a *zig* segment of a preferred path p be a maximal sequence of nodes such that each node in the sequence is linked to its right child in p. A *zag* segment is defined similarly such that each node on the segment is linked to its left child (see Fig. 1).

The key observation we exploit is the following: the sequence of all zig segments, ordered by their depth in the path, followed by the sequence of all reversed zag segments, ordered reversely by their depth in the path, is equal to the ordering of the nodes in key space (see Fig. 1). This implies that to extract the nodes of depth smaller than d (in P) from a bottom tree, we can cut the extreme ends (in key space) of the tree, linearize them to two lists, and then combine them by a binary merge procedure using depth in P as the ordering. This forms the core of the extract operation. We have to do this using rotations, while maintaining a tree at all times. We now give the details of how to do this, with Fig. 3 illustrating the process.

Using extra fields in each node storing the minimum and maximum depth value (in P) of nodes inside its subtree, we can find the node ℓ' of minimum key

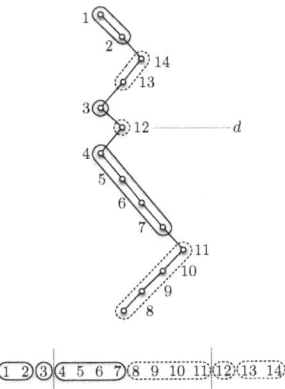

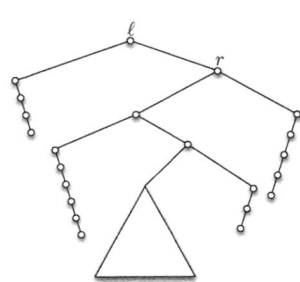

Fig. 1. A path, its decomposition into zig (solid regions) and zag (dashed regions) segments, and its layout in key order

Fig. 2. The path representation in the zipper trees of Sect. 4

value that has a depth greater than d in $O(\log \log n)$ time, by starting at the root of A' and repeatedly walking to the leftmost child whose subtree has a node of depth greater than d. Then define ℓ as the predecessor of ℓ'. Symmetrically, we can find the node r' of maximum key value that has depth greater than d and define r as the successor of r'.

First we split A' at ℓ to obtain two subtrees B and C linked to the new root ℓ where B contains a first sequence of nodes at depth smaller than d. Then we split C at r to obtain the subtrees D and E where E contains a second sequence of nodes at depth smaller than d.

In $O(\log \log n)$ time we convert the subtrees B and E into paths as shown in Fig. 3. To do so we perform a left rotation at the root of B until its right subtree is a leaf (i.e., when its right child is a marked node). Then we repeat the following: if the left child of the root has no right child the we perform a right rotation at the root of B (adding one more node to right spine, which will constitute the final path). Otherwise we perform a left rotation at the left child of the root of B, moving its right subtree into the left spine. The entire process takes time linear in the size of B, since each node is involved in a rotation at most 3 times (once a node enters the left spine, it can only leave it by being added to the right spine). A symmetric process is performed to convert the subtree E into a path.

The last operation, called a *zip*, merges (based on depths in P) the two paths B and E, in order to form the next part of the top path. We repeatedly select the root of B or E that has the smallest depth in the tree P. The selected root is brought to the bottom of the top path using $O(1)$ rotations. The zip operation stops when the subtrees B and E are both empty. Eventually, we perform a left rotation at the node ℓ if needed, i.e., if r has a smaller depth in P than ℓ. The time taken is linear in the extracted number of nodes, i.e., $O(\log \log n)$. The process consists of a series of rotations, hence can stopped and resumed without problems. Therefore, the discussion presented in this section allows us to conclude with the following theorem.

Theorem 2. *Our hybrid tree data structure is $O(\log \log n)$-competitive and performs each access in $O(\log n)$ worst-case time.*

3.3   Hybrid Trees and the BST Model

We specify in the description of the cut operation (more precisely, in Case 1) that the extraction process is executed incrementally, i.e., the work is spread over several cut operations. In order to efficiently revive an extraction process which has been stopped at some point in the past, we have to return to the position where its next rotation should take place. This location is always somewhere in the bottom tree, so traversing the top path to reach the bottom tree would be too costly for the analysis of Case 1. Instead, we store in the marked node (the first node of the top path) appropriate information on the state of the process. Additionally, we store an extra pointer pointing to the node where the next rotation in the process should take place. This allows us to revive an extraction process in constant time. Unfortunately, the structure so obtained is not in the BST model (see Section 2.1), due to the extra pointer. In the next section we show how to further develop the idea

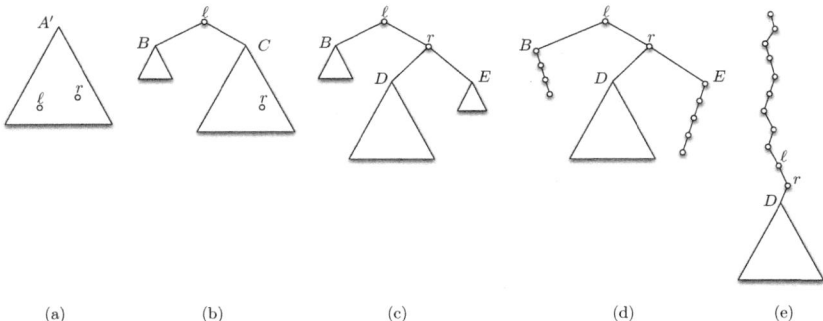

Fig. 3. (a) Tree A'. (b) Split A' at ℓ. (c) Split C at r. (d) Convert the subtrees B and E into paths. (e) Zip the paths B and E.

from this section into a data structure fitting the BST model. Still, we note that the structure of this section can be implemented in the comparison-based model on a pointer machine, with access sequences X being served in $O(\log \log n \, \mathrm{OPT}(X))$ time, and each access taking $O(\log n)$ time worst-case.

4   Zipper Trees

The data structure described in the previous section is a BST, except that each marked node has an extra pointer facilitating constant time access to the point in the path representation where an extraction process should be revived. In this section, we show how to get rid of this extra pointer and obtain a data structure with the same complexity bounds, but now fitting the BST model described in Section 2.1. To do so, we develop a more involved version of the representation of preferred paths and the operations on them. The goal of this new path representation is to ensure that all rotations of an extraction process are located within distance $O(1)$ of the root of the tree of the representation. The two main ideas involved are: 1) storing the top path as lists, hanging to the sides of the root, from which the top path can be generated incrementally by merging as it is traversed during access, and 2) using a version of the split operation that only does rotations near the root. The time complexity analysis follows that of hybrid trees, and will not be repeated.

4.1   Path Representation

For all preferred paths in P we decompose its highest part into two sequences, containing its zig and its zag segments, respectively. These are stored as two paths of nodes, of increasing and decreasing key values, respectively. As seen in Section 3.2 (cf. Fig. 1), both will be ordered by their depth in P. Let ℓ and r be the lowest (in terms of depth in P) node in the zig and zag sequence respectively. The node ℓ will be the root of the auxiliary tree (the marked node). The remainder of the zig sequence is the left subtree of ℓ, r is its right child, and the remainder of the zag sequence is the right subtree of r. We call this upper part of the tree a *zipper*. We repeat this decomposition once again for the next part of the path to obtain another zipper which is the left subtree of r. Finally the remaining of the nodes on the path are stored as a red-black tree of height $O(\log \log n)$, hanging below the lowest zipper. Fig. 2 illustrates the construction. The two zippers constitute the *top path*, and the red-black tree the *bottom tree*. Note that the root of the bottom tree is reachable in $O(1)$ time from the root of the path representation. We will maintain the invariant that individually, the two zippers contain at most $\log \log n$ nodes each, while (if the bottom tree is non-empty) they combined contain at least $(\log \log n)/2$ nodes. A *zipper tree* consists of all the preferred paths of P, represented as above, linked together to form one large tree.

4.2   Maintaining Zipper Trees under Accesses

We give the differences, relative to Section 3.2, of the operations during an access.

Cut: When searching a path representation, we incrementally perform a zip operation (i.e., a merge based on depth order) on the top zipper, until it outputs either the node searched for, or a node that leads to the next path representation. If the top zipper gets exhausted, the lower zipper becomes the upper zipper, and an incremental creation of a new lower zipper by an extraction operation on the bottom tree is initiated (during which the lower zipper is defined to have size zero). Each time one more node from the top zipper is being output (during the current access, or during a later access passing through this path representation), the extraction advances $\Theta(1)$ steps. The speed of the extraction process is chosen such that it finishes with $\log \log n$ nodes extracted before $(\log \log n)/2$ nodes have been output from the top zipper. The new nodes will make up a fresh lower zipper, thereby maintaining the invariant.

If the access through a path representation overlaps (in time) at most one extraction process (either initiated by itself or by a previous access), it is defined as a Case 1 cut. No further actions takes place, besides the proper remarkings of roots of path representations, as in Section 3.2. If a second extraction process is about to be initiated during an access, we know that $\Theta(\log \log n)$ nodes have been passed in this path representation, and we define it as a Case 2 cut. Like in Section 3.2 this now ends by converting the path representation to a red-black tree, cutting it like in tango trees, and then converting the red-black tree remaining into a valid path representation (as defined in the current section), all in $\Theta(\log \log n)$ time.

Concatenate: There is no change from Section 3.2, except that the final path representation produced is as defined in the current section.

Extract: The change from Section 3.2 is that the final zip operation is not performed (the process stops at step (d) in Fig. 3), and that we use a split operation on red-black trees where all structural changes consist of rotations a distance $O(1)$ from the root (of the bottom tree, which is itself at a distance $O(1)$ from the root of the zipper tree). In the full version of this paper [1], we describe such a split operation. Searching for the splitting point takes place incrementally as part of this operation.

5   Conclusion

The main goal in this area of research is to improve on the currently best competitive ratio of $O(\log \log n)$. Here we have been able to tighten other bounds, namely the worst-case search time, thereby broadening our understanding of competitive BSTs. One natural question is to what extent competitiveness is compatible with optimal balance maintenance. We have given a positive answer for $O(\log \log n)$-competitiveness. On the other hand, splay-trees [10] and GreedyFuture trees [2,9], the two BSTs conjectured to be dynamically optimal, do not guarantee optimal worst-case search time. Thus, even if dynamically optimal trees should be proven to exist, the present result could still be a relevant alternative with optimal worst-case performance.

References

1. Bose, P., Douïeb, K., Dujmović, V., Fagerberg, R.: An O(loglog n)-competitive binary search tree with optimal worst-case access times, arXiv 1003.0139 (2010)
2. Demaine, E., Harmon, D., Iacono, J., Kane, D.M., Patrascu, M.: The geometry of binary search trees. In: Proc. of the 10th ACM-SIAM Symp. on Disc. Alg. (SODA), pp. 496–505 (2009)
3. Demaine, E., Harmon, D., Iacono, J., Pătraşcu, M.: Dynamic optimality—almost. SICOMP 37(1), 240–251 (2007)
4. Derryberry, J., Sleator, D.D., Wang, C.C.: $O(\log \log n)$-competitive dynamic binary search trees. In: Proc. of the 7th ACM-SIAM Symp. on Disc. Alg. (SODA), pp. 374–383 (2006)
5. Georgakopoulos, G.F.: Chain-splay trees, or, how to achieve and prove loglogn-competitiveness by splaying. Inf. Process. Lett. 106(1), 37–43 (2008)
6. Guibas, L.J., Sedgewick, R.: A dichromatic framework for balanced trees. In: Proc. of the 19th Found. of Comp. Sci. (FOCS), pp. 8–21 (1978)
7. Iacono, J.: Personal communication (July 2009)
8. Kujala, J., Elomaa, T.: Poketree: A dynamically competitive data structure with good worst-case performance. In: Asano, T. (ed.) ISAAC 2006. LNCS, vol. 4288, pp. 277–288. Springer, Heidelberg (2006)
9. Munro, J.I.: On the competitiveness of linear search. In: Paterson, M. (ed.) ESA 2000. LNCS, vol. 1879, pp. 338–345. Springer, Heidelberg (2000)
10. Sleator, D.D., Tarjan, R.E.: Self-adjusting binary search trees. J. ACM 32(3), 652–686 (1985)
11. Wang, C.C.: Multi-Splay Trees. PhD thesis, Computer Science Department, School of Computer Science, Carnegie Mellon University (2006)
12. Wilber, R.: Lower bounds for accessing binary search trees with rotations. SICOMP 18(1), 56–67 (1989)
13. Woo, S.L.: Heterogeneous Decomposition of Degree-Balanced Search Trees and Its Applications. PhD thesis, Computer Science Department, School of Computer Science, Carnegie Mellon University (2009)

The Emergence of Sparse Spanners and Greedy Well-Separated Pair Decomposition

Jie Gao and Dengpan Zhou

Department of Computer Science, Stony Brook University,
Stony Brook, NY 11794, USA
{jgao,dpzhou}@cs.sunysb.edu

Abstract. A spanner graph on a set of points in $\mathbb{R}^d$ provides shortest paths between any pair of points with lengths at most a constant factor of their Euclidean distance. A spanner with a sparse set of edges is thus a good candidate for network backbones, as in transportation networks and peer-to-peer network overlays. In this paper we investigate new models and aim to interpret why good spanners 'emerge' in reality, when they are clearly built in pieces by agents with their own interests and the construction is not coordinated. Our main result is to show that the following algorithm generates a $(1 + \varepsilon)$-spanner with a linear number of edges. In our algorithm, the points build edges at an *arbitrary* order. A point p will only build an edge pq if there is no existing edge $p'q'$ with p' and q' at distances no more than $\frac{1}{4(1+1/\varepsilon)} \cdot |pq|$ from p, q respectively. Eventually when all points finish checking edges to all other points, the resulted collection of edges forms a sparse spanner as desired. As a side product, the spanner construction implies a greedy algorithm for constructing linear-size well-separated pair decompositions that may be of interest on its own.

Keywords: Spanner, Well-separated pair decomposition, Greedy algorithm.

1 Introduction

A geometric graph G defined on a set of points $P \subseteq \mathbb{R}^d$ with all edges as straight line segments of weight equal to the length is called a *Euclidean spanner*, if for any two points $p, q \in P$ the shortest path in G has length at most $s \cdot |pq|$ where $|pq|$ is the Euclidean distance. The factor s is called the *stretch factor* of G and the graph G is called an s-spanner. Spanners with a sparse set of edges provide good approximations for the pairwise Euclidean distances and are good candidates for network backbones. Thus, there has been a lot of work on the construction of sparse Euclidean spanners in both the centralized [19,33] and distributed settings [34].

In this paper we are interested in the emergence of good Euclidean spanners formed by uncoordinated agents. Many real-world networks, such as the transportation network and the Internet backbone network, are good spanners — one can typically drive from any city to any other city in the U.S. with the total travel distance at most a small constant times their straight line distance. The same thing happens with the Internet backbone graph as well. However, these large networks are not owned or built by any single authority. They are often assembled with pieces built by different governments

H. Kaplan (Ed.): SWAT 2010, LNCS 6139, pp. 50–61, 2010.

or different ISPs, at different points in time. Nevertheless altogether they provide a convenient sparse spanner. The work in this paper is motivated by this observation of the lack of coordination in reality and we would like to interpret why a good Euclidean spanner is able to 'emerge' from these agents incrementally.

Prior work that attempt to remove centralized coordination has been done, as in the network creation game [21,15,26,4,31], first introduced by Fabrikant *et al.* [21] to understand the evolution of network topologies maintained by selfish agents. A cost function is assigned to each agent, capturing the cost paid to build connections to others minus the benefit received due to the resulting network topology. The agents play a game by minimizing their individual costs and one is interested in the existence and the price of anarchy of Nash equilibria. Though being theoretically intriguing, there are two major open questions along this direction. First, the choice of cost functions is heuristic. Almost all past literatures use a unit cost for each edge and they deviate in how the benefit of 'being connected to others' is modeled. There is little understanding on what cost function best captures the reality yet small variation in the cost function may result in big changes in the network topologies at Nash equilibria. There is also not much understanding of the topologies at Nash equilibria, some of them are simplistic topologies such as trees or complete graphs, that do not show up often in the real world. It remains open whether there is a natural cost model with which the Nash equilibrium is a sparse spanner.

The game theoretic model also has limitations capturing the reality: selfish agents may face deadlines and have to decide on building an edge or not immediately; once an edge is built, it probably does not make sense to remove it (as in the case of road networks); an agent may not have the strategies of all other agents making the evaluation of the cost function difficult. In this paper, we take a different approach and ask whether there is any simple rule, with which each agent can determine on its own, and collectively build and maintain a sparse spanner topology without any necessity of coordination or negotiation. The simple rule serves as a 'certificate' of the sparse spanner property that warrants easy spanner maintenance under edge dynamics and node insertion. We believe such models and good algorithms under these models worth further exploration and this paper makes a first step along this line.

Our contribution. We consider in this paper the following model that abstracts the scenarios explained earlier. There are n points in the plane. Each point represents a separate agent and may consider to build edges from itself to other points. These decisions can happen at different points in time. When an agent p plans on an edge pq, p will only build it if whether there does not exist a 'nearby' edge $p'q'$ in the network, where $|pp'|$ and $|qq'|$ are within $\frac{1}{4(1+1/\varepsilon)} \cdot |p'q'|$ from p and q respectively. This strategy is very intuitive — if there is already a cross-country highway from Washington D.C. to San Francisco, it does not make economical sense to build a highway from New York to Los Angeles. We assume that each agent will eventually check on each possible edge from itself to all other points, but the order on who checks which edge can be *completely arbitrary*. With this strategy, the agents only make decisions with limited information and no agent has full control over how and what graph will be constructed. It is not obvious that this strategy will lead to a sparse spanner. It is not clear that the graph is even connected.

The main result in this paper is to prove that with the above strategy executed in *any* arbitrary order, the graph built at the end of the process is a sparse spanner:

- The stretch factor of the spanner is $1 + \varepsilon$.
- The number of edges is $O(n)$.
- The total edge length of the spanner is $O(|\mathrm{MST}| \cdot \log \alpha)$, where α is the *aspect ratio*, i.e., the ratio of the distance between the furthest pair and the closest pair, and $|\mathrm{MST}|$ is the total edge length of the minimum spanning tree of the point set.
- The degree of each point is $O(\log \alpha)$ in the worst case and $O(1)$ on average.

To explain how this result is proved, we first obtain as a side product the following *greedy* algorithm for computing a well-separated pair decomposition. A pair of two sets of points, (A, B), is called *s-well-separated* if the smallest distance between any two points in A, B is at least s times greater than the diameters of A and B. An *s-well-separated pair decomposition* (*s*-WSPD for short) for P is a collection of *s*-well-separated pairs $\mathcal{W} = \{(A_i, B_i)\}$ such that for any pair of points $p, q \in P$ there is a pair $(A, B) \in \mathcal{W}$ with $p \in A$ and $q \in B$. The size of an *s*-WSPD is the number of point set pairs in $\mathcal{W}$. Well-separated pair decomposition (WSPD) was first introduced by Callahan and Kosaraju [12] and they developed algorithms for computing an *s*-WSPD with linear size for points in $\mathbb{R}^d$. Since then WSPD has found many applications in computing k-nearest neighbors, n-body potential fields, geometric spanners and approximate minimum spanning trees [9,10,12,11,6,5,32,29,24,20].

So far there are two algorithms for computing optimal size WSPD, one in the original paper [12] and one in a later paper [22]. Both of them use a hierarchical organization of the points (e.g., the fair split tree in [12] and the discrete center hierarchy in [22]) and output the well-separated pairs in a recursive way. In this paper we show the following simple *greedy* algorithm also outputs an *s*-WSPD with linear size. We take an *arbitrary* pair of points p, q that is not yet covered in any existing well-separated pair, and consider the pair of subsets $(B_r(p), B_r(q))$ with $r = |pq|/(2s + 2)$ and $B_r(p)$ $(B_r(q))$ as the set of points of P within distance r from p (q). Clearly $(B_r(p), B_r(q))$ is an *s*-well-separated pair and all the pairs of points (p', q') with $p' \in B_r(p)$ and $q' \in B_r(q)$ are covered. The algorithm continues until all pairs of points are covered. We show that, no matter in which order the pairs are selected, the greedy algorithm will always output a linear number of well-separated pairs. Similarly, this greedy algorithm can be executed in an environment when coordination is not present, while the previous algorithms (in [12,22]) cannot.

WSPD is deeply connected to geometric spanners. Any WSPD will generate a spanner if one puts an edge between an arbitrary pair of points p, q from each well-separated pair $(A, B) \in \mathcal{W}$ [6,5,32,29]. The number of edges in the spanner equals the size of $\mathcal{W}$. In the other direction, the deformable spanner proposed in [22] implies a WSPD of linear size. The connection is further witnessed in this paper: our spanner emergence algorithm implies a WSPD generated in a greedy manner. Thus the well-separated pairs and spanner edges are one-to-one correspondence.

Last, this paper focuses on the Euclidean case when the points are distributed in the plane. The basic idea extends naturally to points in higher dimensions as well as metrics with constant doubling dimensions [25] (thus making the results applicable in non-Euclidean settings), as the main technique involves essentially various forms of geometric packing arguments.

Applications. The results can be applied in maintaining nice network overlay topologies for P2P file sharing applications [30]. Such P2P overlay networks are often constructed in a distributed manner without centralized control, to achieve robustness, reliability and scalability. One important issue is reducing routing delay by making the overlay topology aware of the underlying network topology [14,35,28,39,40]. But all these work are heuristics without any guarantee. A spanner graph would be a good solution for the overlay construction, yet there is no centralized authority in the P2P network that supervises the spanner construction and the peers may join or leave the network frequently. The work in this paper initiates the study of the emergence of good spanners in the setting when there is little coordination between the peers and the users only need a modest amount of incomplete information of the current overlay topology.

We show that the spanner can be constructed under a proper model such that only $O(n \log \alpha)$ messages need to be delivered. The spanner topology is implicitly stored on the nodes with each node's storage cost bounded by $O(\log \alpha)$. With such partial information stored at each node, there is a local distributed algorithm that finds a $(1+\varepsilon)$-stretch path between any two nodes.

We remark that the idea of the greedy spanner resembles, on an abstract level, the 'highway hierarchy' in transportation networks. It has been shown that to find a shortest path to a destination, one only needs to search for a 'highway entrance' within certain radius, and search only on the highways beyond that. This turns out to substantially reduce the time to find shortest paths on such graphs [8,36]. Our work provides a possible explanation of how the road system evolved to the way it is today. We propose to verify this in our future work.

Related work. In the vast amount of prior literature on geometric spanners, there are three main ideas: Θ-graphs, the greedy spanners, and the WSPD-induced spanners [33]. Please refer to the book for a nice survey [33]. We will review two spanner construction ideas that are most related to our approach. The first idea is the path-greedy spanner construction [13,16,17,18]. All pairwise edges are ordered with non-decreasing lengths and checked in that order. An edge is included in the spanner if the shortest path in the current graph is longer than s times the Euclidean distance, and is discarded otherwise. Variants of this idea generate spanners with constant degree and total weight $O(|\text{MST}|)$. This idea cannot be applied in our setting as edges constructed in practice may not be in non-decreasing order of their lengths. The second idea is to use the gap property [13] — the sources and sinks of any two edges in an edge set are separated by a distance at least proportional to the length of the shorter of the two edges and their directions are differed no more than a given angle. The gap-greedy algorithm [7] considers pairs of points, again, in order of non-decreasing distances, and includes an edge in the spanner if and only if it does not violate the gap property. The spanner generated this way has constant degree and total weight $O(|\text{MST}|)$. Compared with our algorithm, our strategy is a relaxation of the gap property in the way that the edges in our spanner may have one of their endpoints arbitrarily close (or at the same points) and we have no restriction on the direction of the edges and the ordering of the edges to be considered. The proof for the gap greedy algorithm requires heavily plane geometry tools and our proof technique only uses packing argument and can be extended to the general metric setting as long as a similar packing argument holds. To get these benefit our algorithm has slightly worse upper bounds on the spanner weight by a logarithmic factor.

Prior work on compact routing [37,23,3,27,2] usually implies a $(1 + \varepsilon)$-spanner explicitly or implicitly. Again, these spanners are constructed in a coordinated setting.

2   Uncoordinated Spanner Construction Algorithm

Given n points in $\mathbb{R}^d$, each point p will check whether an edge pq should be built. p builds pq only if there does not exist an edge $p'q'$ such that p and q are within distance $\frac{|p'q'|}{2(s+1)}$ from p', q' respectively.

This incremental construction of edges is executed by different agents in a completely uncoordinated manner. We assume that no two agents perform the above strategy at exactly the same time. Thus when any agent conducts the above process, the decision is based on the current network already constructed. The algorithm terminates when all agents finish checking the edges from themselves to all other points. We first examine the properties of the constructed graph G by these uncoordinated behaviors. We will discuss in Section 4 a proper complexity model for the uncoordinated construction in a distributed environment and also bound the computing cost of this spanner.

Before we proceed, we first realize the following invariant is maintained by the graph G. The proof follows immediately from the construction of G.

Lemma 1.   *1. For any edge pq that is not in G, there is another edge $p'q'$ in G such that $|pp'| \leq |p'q'|/(2s + 2)$, $|qq'| \leq |p'q'|/(2s + 2)$.*
   2. For any two edges pq, $p'q'$ in the constructed graph G, suppose that pq is built before $p'q'$, then one of the following is true: $|pp'| > |pq|/(2s + 2)$ or $|qq'| > |pq|/(2s + 2)$.

To show that the algorithm eventually outputs a good spanner, we first show the connection of G with the notion of *well-separated pair decomposition*.

Definition 1 (Well-separated pair). *Let $t > 0$ be a constant, and a pair of sets of points A, B are well-separated with respect to t (or t-separated), if $d(A, B) \geq t \cdot \max(diam(A), diam(B))$, where $diam(A)$ is the diameter of the point set A, and $d(A, B) = \min\limits_{p \in A, q \in B} |pq|$.*

Definition 2 (Well-separated pair decomposition). *Let $t > 0$ be a constant, and P be a point set. A well-separated pair decomposition (WSPD) with respect to t of P is a set of pairs $\mathcal{W} = \{\{A_1, B_1\}, \ldots, \{A_m, B_m\}\}$, such that*

   1. $A_i, B_i \subseteq P$, and the pair sets A_i and B_i are t-separated for every i.
   2. For any pair of points $p, q \in P$, there is at least one pair (A_i, B_i) such that $p \in A_i$ and $q \in B_i$.

Here m is called the size of the WSPD.

It is not hard to see that the uncoordinated spanner is equivalent to the following greedy algorithm that computes an s-WSPD.

1. Choose an arbitrary pair (p, q), not yet covered by existing pairs in $\mathcal{W}$.
2. Include the pair of point sets $B_r(p)$ and $B_r(q)$ in the WSPD $\mathcal{W}$, with $r = |pq|/(2 + 2s)$, where $B_r(p)$ is the collection of points within distance r from point p.
3. Label the point pair (p_i, q_i) with $p_i \in B_r(p)$ and $q_i \in B_r(q)$ as being covered.
4. Repeat the above steps until every pair of points is covered.

Clearly the above algorithm produces a WSPD, as each pair $(B_r(p), B_r(q))$ is well-separated and all pairs of points are covered. The spanner edge (p, q) is one-to-one correspondence to the well-separated pair $(B_r(p), B_r(q))$ in the above algorithm — the rule in Lemma 1 prevented two edges from the same well-separated pair in $\mathcal{W}$ to be constructed. Thus the number of edges in the spanner G is the same as the size of the greedy WSPD. It is already known that for any well-separated pair decomposition, if one edge is taken from each well-separated pair, then the edges will become a spanner on the original point set [6,5,32,29]. For our specific greedy s-WSPD, we are able to get a slightly better stretch. The proof is omitted due to space constraint.

Theorem 1. *From the greedy s-WSPD, one build a graph G that includes each pair (p, q) when it is selected by the greedy algorithm. Thus G is a spanner with stretch factor $(s + 1)/(s - 1)$.*

To make the stretch factor as $1 + \varepsilon$, we just take $s = 1 + 2/\varepsilon$ in our spanner construction. Next, we show that the greedy WSPD algorithm will output a linear number of well-separated pairs.

3 A Greedy Algorithm for Well-Separated Pair Decomposition

We show the connection of the greedy WSPD with a specific WSPD constructed by the deformable spanner [22], in the way that at most a constant number of pairs in $\mathcal{W}$ is mapped to each well-separated pair constructed by the deformable spanner. To be consistent, the greedy WSPD is denoted by $\mathcal{W}$ and the WSPD constructed by the deformable spanner is denoted by $\hat{\mathcal{W}}$.

Deformable spanner. Given a set of points P in the plane, a set of *discrete centers* with radius r is defined to be the maximal set $S \subseteq P$ that satisfies the *covering* property and the *separation* property: any point $p \in P$ is within distance r to some point $p' \in S$; and every two points in S are of distance at least r away from each other. In other words, all the points in P can be covered by balls with radius r, whose centers are exactly those points in the discrete center set S. And these balls do not cover other discrete centers.

We now define a hierarchy of discrete centers in an recursive way. S_0 is the original point set P. S_i is the discrete center set of S_{i-1} with radius 2^i. Without loss of generality we assume that the closest pair has distance 1 (as we can scale the point set and do not change the combinatorial structure of the discrete center hierarchy). Thus the number of levels of the discrete center hierarchy is $\log \alpha$, where α is the aspect ratio of the point set P, defined as the ratio of the maximum pairwise distance to the minimum pairwise distance, that is, $\alpha = \max_{u,v \in P} |uv| / \min_{u,v \in P} |uv|$. Since a point p may stay in multiple consecutive levels and correspond to multiple nodes in the discrete center hierarchy, we denote by $p^{(i)}$ the existence of p at level i. For each point $p^{(i-1)} \in S_{i-1}$ on level

i, it is within distance 2^i from at least one other point on level $i + 1$. Thus we assign to $p^{(i-1)}$ a *parent* $q^{(i)}$ in S_i such that $|p^{(i-1)}q^{(i)}| \leq 2^i$. When there are multiple points in S_i that cover $p^{(i-1)}$, we choose one as its parent arbitrarily. We denote by $P(p^{(i-1)})$ the parent of $p^{(i-1)}$ on level i. We denote by $P^{(i)}(p) = P(P^{(i-1)}(p))$ the *ancestor* of p at level i.

The deformable spanner is based on the hierarchy, with all edges between two points u and v in S_i if $|uv| \leq c \cdot 2^i$, where c is a constant equal to $4 + 16/\varepsilon$. We restate some important properties of the deformable spanner below.

Lemma 2 (Packing Lemma [22]). *In a point set $S \subseteq R^d$, if every two points are at least distance r away from each other, then there can be at most $(2R/r + 1)^d$ points in S within any ball with radius R.*

Lemma 3 (Deformable spanner properties [22]). *For a set of n points in R^d with aspect ration α,*

1. *For any point $p \in S_0$, its ancestor $P^{(i)}(p) \in S_i$ is of distance at most 2^{i+1} away from p.*
2. *Any point $p \in S_i$ has at most $(1 + 2c)^d - 1$ edges with other points of S_i.*
3. *The deformable spanner $\hat{G}$ is a $(1 + \varepsilon)$-spanner G with $O(n/\varepsilon^d)$ edges.*
4. *$\hat{G}$ has total weight $O(|MST| \cdot \lg \alpha/\varepsilon^{d+1})$, where $|MST|$ is the weight of the minimal spanning tree of the point set S.*

As shown in [22], the deformable spanner implies a well-separated pair decomposition $\hat{\mathcal{W}}$ by taking all the 'cousin pairs'. Specifically, for a node $p^{(i)}$ on level i, we denote by P_i the collection of points that are descent of $p^{(i)}$ (including $p^{(i)}$ itself), called the *decedents*. Now we take the pair (P_i, Q_i), the sets of decedents of a *cousin pair* $p^{(i)}$ and $q^{(i)}$, i.e., $p^{(i)}$ and $q^{(i)}$ are *not* neighbors in level i but their parents are neighbors in level $i + 1$. This collection of pairs constitutes a $\frac{4}{\varepsilon}$-well-separated pair decomposition. The size of $\hat{\mathcal{W}}$ is bounded by the number of cousin pairs and is $O(n/\varepsilon^d)$.

Size of greedy WSPD. The basic idea is to map the pairs in the greedy WSPD $\mathcal{W}$ to the pairs in $\hat{\mathcal{W}}$ and show that at most a constant number of pairs in $\mathcal{W}$ map to the same pair in $\hat{\mathcal{W}}$.

Theorem 2. *The greedy s-WSPD $\mathcal{W}$ has size $O(ns^d)$.*

Proof. Choose $c = 4(s + 1)$ (or, $s = c/4 - 1$) in the deformable spanner DS. The size of $\hat{\mathcal{W}}$ is $O(ns^d)$. Now we will construct a map from $\mathcal{W}$ to $\hat{\mathcal{W}}$. Each pair $\{P, Q\}$ in $\mathcal{W}$ is created by considering the points inside the balls $B_r(p), B_r(q)$ with radius $r = |pq|/(2 + 2s)$ around p, q. Now we consider the ancestors of p, q in the spanner DS respectively. There is a unique level i such that the ancestor $u_i = P^{(i)}(p)$ and $v_i = P^{(i)}(q)$ do not have a spanner edge in between but the ancestor $u_{i+1} = P^{(i+1)}(p)$ and $v_{i+1} = P^{(i+1)}(q)$ have an edge in between. The pair u_i, v_i is a cousin pair by definition and thus the decedents of them correspond to an s-well-separated pair in $\hat{\mathcal{W}}$. We say that the pair $(B_r(p), B_r(q)) \in \mathcal{W}$ maps to the descendant pair $(P_i, Q_i) \in \hat{\mathcal{W}}$.

By the discrete center hierarchy (Lemma 3), we show that,

$$|pq| \geq |u_i v_i| - |pu_i| - |qv_i| \geq |u_i v_i| - 2 \cdot 2^{i+1} \geq (c - 4) \cdot 2^i.$$

The last inequality follows from that fact that u_i, v_i do not have an edge in the spanner and $|u_i v_i| > c \cdot 2^i$. On the other hand,

$$|pq| \le |pu_{i+1}| + |u_{i+1}v_{i+1}| + |qv_{i+1}| \le 2 \cdot 2^{i+2} + c \cdot 2^{i+1} = 2(c+4) \cdot 2^i.$$

The last inequality follows from the fact that u_{i+1}, v_{i+1} have an edge in the spanner and $|u_{i+1}v_{i+1}| \le c \cdot 2^{i+1}$. Similarly, we have

$$c \cdot 2^i < |u_i v_i| \le |u_i u_{i+1}| + |u_{i+1}v_{i+1}| + |v_i v_{i+1}| \le 2 \cdot 2^{i+1} + c \cdot 2^{i+1} = 2(c+2) \cdot 2^i.$$

Therefore the distance between p and q is $c' \cdot |u_i v_i|$, where $(c-4)/(2c+4) \le c' \le (2c+8)/c$.

Now suppose two pair $(B_{r_1}(p_1), B_{r_1}(q_1))$, $(B_{r_2}(p_2), B_{r_2}(q_2))$ in $\mathcal{W}$ map to the same pair u_i and v_i by the above process. Without loss of generality suppose that p_1, q_1 are selected before p_2, q_2 in our greedy algorithm. Here is the observation:

1. $|p_1 q_1| = c'_1 \cdot |u_i v_i|$, $|p_2 q_2| = c'_2 \cdot |u_i v_i|$, $r_1 = |p_1 q_1|/(2+2s) = c'_1 \cdot |u_i v_i|/(2+2s)$, $r_2 = c'_2 \cdot |u_i v_i|/(2+2s)$, where $(c-4)/(2c+4) \le c'_1, c'_2 \le (2c+8)/c$, and r_1, r_2 are the radius of the balls for the two pairs respectively.
2. The reason that (p_2, q_2) can be selected in our greedy algorithm is that at least one of p_2 or q_2 is outside the balls $B(p_1), B(q_1)$, by Lemma 1. This says that at least one of p_2 or q_2 is of distance r_1 away from p_1, q_1.

Now we look at all the pairs (p_ℓ, q_ℓ) that are mapped to the same ancestor pair (u_i, v_i). The pairs are ordered in the same order as they are constructed, i.e., p_1, q_1 is the first pair selected in the greedy WSPD algorithm. Suppose r_{min} is the minimum among all radius r_i. $r_{min} \ge c/(2c+8) \cdot |u_i v_i|/(2+2s) = |u_i v_i|/(4s+8)$. We group these pairs in the following way. The first group H_1 contains (p_1, q_1) and all the pairs (p_ℓ, q_ℓ) that have p_ℓ within distance $r_{min}/2$ from p_1. We say that (p_1, q_1) is the representative pair in H_1 and the other pairs in H_1 are *close* to the pair (p_1, q_1). The second group H_2 contains, among all remaining pairs, the pair that was selected in the greedy algorithm the earliest, and all the pairs that are close to it. We repeat this process to group all the pairs into k groups, $H_1, H_2, \cdots, H_k$. For all the pairs in each group H_j, we have one representative pair, denoted by (p_j, q_j) and the rest of the pairs in this group are close to it.

We first bound the number of pairs belonging to each group by a constant with a pack argument. With our group criteria and the above observations, all p_ℓ in the group H_j are within radius r_{min} away from each other. This means that the q_ℓ's must be far away — the q_ℓ's must be at least distance r_{min} away from each other, by Lemma 1. On the other hand, all the q_ℓ's are descendant of the node v_i, so $|v_i q_\ell| \le 2^{i+1}$ by Theorem 3. That is, all the q_ℓ's are within a ball of radius 2^{i+1} centered at v_i. By the packing Lemma 2, the number of such q_ℓ's is at most $(2 \cdot 2^{i+1}/r_{min} + 1)^d \le (2 \cdot 2^{i+1}(4s+8)/|u_i v_i| + 1)^d \le (4(s+2)/(s+1) + 1)^d$. This is also the bound on the number of pairs inside each group.

Now we bound the number of different groups, i.e., the value k. For the representative pairs of the k groups, $(p_1, q_1), (p_2, q_2), \cdots, (p_k, q_k)$, all the p_i's must be at least distance $r_{min}/2$ away from each other. Again these p_i's are all descendant of u_i

and thus are within distance 2^{i+1} from u_i. By a similar packing argument, the number of such p_i's is bounded by $(4 \cdot 2^{i+1}/r_{min} + 1)^d \le (8(s + 2)/(s + 1) + 1)^d$. So the total number of pairs mapped to the same ancestor pair in $\mathcal{W}$ will be at most $(4(s + 2)/(s + 1) + 1)^d \cdot (8(s + 2)/(s + 1) + 1)^d = (O(1 + 1/s))^d$. Thus the total number of pairs in W is at most $O(ns^d)$. This finishes the proof.

With the connection of the greedy WSPD with the uncoordinated spanner construction in Section 2, we immediately get the following theorem (with proofs omitted).

Theorem 3. *The uncoordinated spanner with parameter s is a spanner with stretch factor $(s+1)/(s-1)$ and has $O(ns^d)$ number of edges, a maximal degree of $O(\lg \alpha \cdot s^d)$, average degree $O(s^d)$, and total weight $O(\lg \alpha \cdot |MST| \cdot s^{d+1})$.*

4   Spanner Construction and Applications

The uncoordinated spanner construction can be applied for peer-to-peer system design, to allow users to maintain a spanner in a distributed manner. For that, we will first extend our spanner results to a metric with constant doubling dimension. The doubling dimension of a metric space (X, d) is the smallest value γ such that each ball of radius R can be covered by at most 2^γ balls of radius $R/2$ [25].

Theorem 4. *For n points and a metric space defined on them with constant doubling dimension γ, the uncoordinated spanner construction outputs a spanner G with stretch factor $(s + 1)/(s - 1)$, has total weight $O(\gamma^2 \cdot \lg \alpha \cdot |MST| \cdot s^{O(\gamma)})$ and has $O(\gamma^2 \cdot n \cdot s^{O(\gamma)})$ number of edges. Also it has a maximal degree of $O(\gamma \cdot \lg \alpha \cdot s^{O(\gamma)})$ and average degree $O(\gamma \cdot s^{O(\gamma)})$.*

Distributed construction. Now we would like to discuss the model of computing for P2P overlay design as well as the construction cost of the uncoordinated spanner. We assume that there is already a mechanism maintained in the system such that any node x can obtain the distance to any node y in $O(1)$ time. For example, this can be done by a TRACEROUTE command executed by x to the node y. We also assume that there is a service answering near neighbor queries: given a node p and a distance r, return the neighbors within distance r from p. Such an oracle is often maintained in a distributed file sharing system. Various structured P2P system support such function with low cost [30]. Even in unstructured system such as BitTorrent, the Ono plugin is effective at locating nearby peers, with vanishingly small overheads [1].

   The spanner edges are recorded in a distributed fashion so that no node has the entire picture of the spanner topology. After each edge pq is constructed, the peers p, q will inform their neighboring nodes (those in $B_r(p)$ and $B_r(q)$ with $r = |pq|/(2s + 2)$) that such an edge pq exists so that they will not try to connect to one another. We assume that these messages are delivered immediately so that when any newly built edge is informed to nodes of relevance. The number of messages for this operation is bounded by $|B_r(p)| + |B_r(q)|$. The amount of storage at each node x is proportional to the number of well-separated pairs that include x. The following theorem bounds the total number of such messages during the execution of the algorithm and the amount of storage at each node.

Theorem 5. *For the uncoordinated spanner G and the corresponding greedy WSPD $\mathcal{W} = \{(P_i, Q_i)\}$ with size m, each node x is included in at most $O(s^d \lg \alpha)$ well-separated pairs in $\mathcal{W}$. Thus, $\sum_{i=1}^{m}(|P_i| + |Q_i|) = O(ns^d \cdot \lg \alpha)$.*

Distributed routing. Although the spanner topology is implicitly stored on the nodes with each node only knows some piece of it, we are actually able to do a distributed and local routing on the spanner with only information available at the nodes such that the path discovered has maximum stretch $(s + 1)/(s - 1)$. In particular, for any node p who has a message to send to node q, it is guaranteed that (p, q) is covered by a well-separated pair $(B_r(p'), B_r(q'))$ with $p \in B_r(p')$ and $q \in B_r(q')$. By the construction algorithm, the edge $p'q'$, after constructed, is informed to all nodes in $B_r(p') \cup B_r(q')$, including p. Thus p includes in the packet a partial route with $\{p \rightsquigarrow p', p'q', q' \rightsquigarrow q\}$. The notation $p \rightsquigarrow p'$ means that p will need to first find out the low-stretch path from p to the node p' (inductively), from where the edge $p'q'$ can be taken, such that with another low-stretch path to be found out from q' to q, the message can be delivered to q. This way of routing with partial routing information stored with the packet is similar to the idea of source routing [38] except that we do not include the full routing path at the source node. By the same induction as used in the proof of spanner stretch (Theorem 1), the final path is going to have stretch at most $(s + 1)/(s - 1)$.

Nearest neighbor search. We remark that with the spanner each node can easily find its nearest neighbor. Recall that each point x keeps all the pairs (p, q) that create a 'dumb-bell' pair set covering x. Then we claim, among all these p, one of them must be the nearest neighbor of x. Otherwise, suppose y is the nearest neighbor of x, and y is not one of p. But in the WSPD, (x, y) will belong to one of the pair set (P_i, Q_i), which correspond to a pair (p', q'). Then there is a contradiction, as $|xp'| < |xy|$ implies that y is not the nearest neighbor of x. Thus one's nearest neighbor is locally stored at this node already. According to Theorem 5, x will belong to at most $O(s^d \lg \alpha)$ different pair sets. So the nearest neighbor search can be finished in $O(s^d \lg \alpha)$ time by using just the local information.

References

1. http://www.aqualab.cs.northwestern.edu/projects/Ono.html
2. Abraham, I., Gavoille, C., Goldberg, A.V., Malkhi, D.: Routing in networks with low doubling dimension. In: Proc. of the 26th International Conference on Distributed Computing Systems (ICDCS) (July 2006)
3. Abraham, I., Malkhi, D.: Compact routing on euclidian metrics. In: PODC '04: Proceedings of the twenty-third annual ACM symposium on Principles of distributed computing, pp. 141–149. ACM, New York (2004)
4. Albers, S., Eilts, S., Even-Dar, E., Mansour, Y., Roditty, L.: On nash equilibria for a network creation game. In: SODA '06: Proceedings of the seventeenth annual ACM-SIAM symposium on Discrete algorithm, pp. 89–98. ACM, New York (2006)
5. Arya, S., Das, G., Mount, D.M., Salowe, J.S., Smid, M.: Euclidean spanners: short, thin, and lanky. In: Proc. 27th ACM Symposium on Theory Computing, pp. 489–498 (1995)
6. Arya, S., Mount, D.M., Smid, M.: Randomized and deterministic algorithms for geometric spanners of small diameter. In: Proc. 35th IEEE Symposium on Foundations of Computer Science, pp. 703–712 (1994)

7. Arya, S., Smid, M.: Efficient construction of a bounded-degree spanner with low weight. Algorithmica 17, 33–54 (1997)
8. Bast, H., Funke, S., Matijevic, D., Sanders, P., Schultes, D.: In transit to constant time shortest-path queries in road networks. In: Applegate, D., Brodal, G. (eds.) 9th Workshop on Algorithm Enginneering and Experiments (ALENEX'07), New Orleans, USA, pp. 46–59. SIAM, Philadelphia (2007)
9. Callahan, Kosaraju: Faster algorithms for some geometric graph problems in higher dimensions. In: Proc. 4th ACM-SIAM Symposium on Discrete Algorithms, pp. 291–300 (1993)
10. Callahan, P.B.: Optimal parallel all-nearest-neighbors using the well-separated pair decomposition. In: Proc. 34th IEEE Symposium on Foundations of Computer Science, pp. 332–340 (1993)
11. Callahan, P.B., Kosaraju, S.R.: Algorithms for dynamic closest-pair and n-body potential fields. In: Proc. 6th ACM-SIAM Symposium on Discrete Algorithms, pp. 263–272 (1995)
12. Callahan, P.B., Kosaraju, S.R.: A decomposition of multidimensional point sets with applications to k-nearest-neighbors and n-body potential fields. J. ACM 42, 67–90 (1995)
13. Chandra, B., Das, G., Narasimhan, G., Soares, J.: New sparseness results on graph spanners. Internat. J. Comput. Geom. Appl. 5, 125–144 (1995)
14. Chu, Y., Rao, S., Seshan, S., Zhang, H.: Enabling conferencing applications on the internet using an overlay muilticast architecture. SIGCOMM Comput. Commun. Rev. 31(4), 55–67 (2001)
15. Corbo, J., Parkes, D.: The price of selfish behavior in bilateral network formation. In: PODC '05: Proceedings of the twenty-fourth annual ACM symposium on Principles of distributed computing, pp. 99–107. ACM, New York (2005)
16. Das, G., Heffernan, P., Narasimhan, G.: Optimally sparse spanners in 3-dimensional Euclidean space. In: Proc. 9th Annu. ACM Sympos. Comput. Geom., pp. 53–62 (1993)
17. Das, G., Narasimhan, G.: A fast algorithm for constructing sparse Euclidean spanners. Internat. J. Comput. Geom. Appl. 7, 297–315 (1997)
18. Das, G., Narasimhan, G., Salowe, J.: A new way to weigh malnourished Euclidean graphs. In: Proc. 6th ACM-SIAM Sympos. Discrete Algorithms, pp. 215–222 (1995)
19. Eppstein, D.: Spanning trees and spanners. In: Sack, J.-R., Urrutia, J. (eds.) Handbook of Computational Geometry, pp. 425–461. Elsevier Science Publishers B.V., North-Holland (2000)
20. Erickson, J.: Dense point sets have sparse Delaunay triangulations. In: Proc. 13th ACM-SIAM Symposium on Discrete Algorithms, pp. 125–134 (2002)
21. Fabrikant, A., Luthra, A., Maneva, E., Papadimitriou, C.H., Shenker, S.: On a network creation game. In: PODC '03: Proceedings of the twenty-second annual symposium on Principles of distributed computing, pp. 347–351 (2003)
22. Gao, J., Guibas, L., Nguyen, A.: Deformable spanners and their applications. Computational Geometry: Theory and Applications 35(1-2), 2–19 (2006)
23. Gottlieb, L.-A., Roditty, L.: Improved algorithms for fully dynamic geometric spanners and geometric routing. In: SODA '08: Proceedings of the nineteenth annual ACM-SIAM symposium on Discrete algorithms, Philadelphia, PA, USA, pp. 591–600. Society for Industrial and Applied Mathematics (2008)
24. Gudmundsson, J., Levcopoulos, C., Narasimhan, G., Smid, M.: Approximate distance oracles for geometric graphs. In: Proc. 13th ACM-SIAM Symposium on Discrete Algorithms, pp. 828–837 (2002)
25. Gupta, A., Krauthgamer, R., Lee, J.R.: Bounded geometries, fractals, and low-distortion embeddings. In: FOCS '03: Proceedings of the 44th Annual IEEE Symposium on Foundations of Computer Science, pp. 534–543 (2003)

26. Jansen, T., Theile, M.: Stability in the self-organized evolution of networks. In: GECCO '07: Proceedings of the 9th annual conference on Genetic and evolutionary computation, pp. 931–938. ACM, New York (2007)
27. Konjevod, G., Richa, A.W., Xia, D.: Optimal-stretch name-independent compact routing in doubling metrics. In: PODC '06: Proceedings of the twenty-fifth annual ACM symposium on Principles of distributed computing, pp. 198–207 (2006)
28. Kwon, M., Fahmy, S.: Topology-aware overlay networks for group communication. In: NOSSDAV '02: Proceedings of the 12th international workshop on Network and operating systems support for digital audio and video, pp. 127–136. ACM, New York (2002)
29. Levcopoulos, C., Narasimhan, G., Smid, M.H.M.: Improved algorithms for constructing fault-tolerant spanners. Algorithmica 32(1), 144–156 (2002)
30. Lua, K., Crowcroft, J., Pias, M., Sharma, R., Lim, S.: A survey and comparison of peer-to-peer overlay network schemes. IEEE Communications Surveys & Tutorials, 72–93 (2005)
31. Moscibroda, T., Schmid, S., Wattenhofer, R.: On the topologies formed by selfish peers. In: PODC '06: Proceedings of the twenty-fifth annual ACM symposium on Principles of distributed computing, pp. 133–142. ACM, New York (2006)
32. Narasimhan, G., Smid, M.: Approximating the stretch factor of Euclidean graphs. SIAM J. Comput. 30, 978–989 (2000)
33. Narasimhan, G., Smid, M.: Geometric Spanner Networks. Cambridge University Press, Cambridge (2007)
34. Peleg, D.: Distributed Computing: A Locality Sensitive Approach. In: Monographs on Discrete Mathematics and Applications. SIAM, Philadelphia (2000)
35. Ratnasamy, S., Handley, M., Karp, R., Shenker, S.: Topologically-aware overlay construction and server selection. In: Proceedings of the 21th Annual Joint Conference of the IEEE Computer and Communications Societies (INFOCOM'05), vol. 3, pp. 1190–1199 (2002)
36. Sanders, P., Schultes, D.: Engineering highway hierarchies. In: Azar, Y., Erlebach, T. (eds.) ESA 2006. LNCS, vol. 4168, pp. 804–816. Springer, Heidelberg (2006)
37. Slivkins, A.: Distance estimation and object location via rings of neighbors. In: PODC '05: Proceedings of the twenty-fourth annual ACM symposium on Principles of distributed computing, pp. 41–50 (2005)
38. Tanenbaum, A.S.: Computer networks (3rd ed.). Prentice-Hall, Inc., Upper Saddle River (1996)
39. Wang, W., Jin, C., Jamin, S.: Network overlay construction under limited end-to-end reachability. In: Proceedings of the 24th Annual Joint Conference of the IEEE Computer and Communications Societies (INFOCOM'05), March 2005, vol. 3, pp. 2124–2134 (2005)
40. Zhang, X., Li, Z., Wang, Y.: A distributed topology-aware overlays construction algorithm. In: MG '08: Proceedings of the 15th ACM Mardi Gras conference, pp. 1–6. ACM, New York (2008)

A Bottom-Up Method and Fast Algorithms for
MAX INDEPENDENT SET[*]

Nicolas Bourgeois[1], Bruno Escoffier[1],
Vangelis Th. Paschos[1], and Johan M.M. van Rooij[2]

[1] LAMSADE, CNRS FRE 3234 and Université Paris-Dauphine, France
{bourgeois,escoffier,paschos}@lamsade.dauphine.fr
[2] Department of Information and Computing Sciences
Universiteit Utrecht, The Netherlands
johanvr@cs.uu.nl

Abstract. We first propose a new method, called "bottom-up method", that, informally, "propagates" improvement of the worst-case complexity for "sparse" instances to "denser" ones and we show an easy though non-trivial application of it to the MIN SET COVER problem. We then tackle MAX INDEPENDENT SET. Following the bottom-up method we propagate improvements of worst-case complexity from graphs of average degree d to graphs of average degree greater than d. Indeed, using algorithms for MAX INDEPENDENT SET in graphs of average degree 3, we tackle MAX INDEPENDENT SET in graphs of average degree 4, 5 and 6. Then, we combine the bottom-up technique with measure and conquer techniques to get improved running times for graphs of maximum degree 4, 5 and 6 but also for general graphs. The best computation bounds obtained for MAX INDEPENDENT SET are $O^*(1.1571^n)$, $O^*(1.1918^n)$ and $O^*(1.2071^n)$, for graphs of maximum (or more generally average) degree 4, 5 and 6 respectively, and $O^*(1.2127^n)$ for general graphs. These results improve upon the best known polynomial space results for these cases.

Keywords: Bottom-Up Method, Max Independent Set, Exact Algorithms.

1 Introduction

Very active research has been recently conducted around the development of exact algorithms for **NP**-hard problems with non-trivial worst-case complexity (see the seminal paper [10] for a survey on both methods used and results obtained). Among the problems studied in this field, MAX INDEPENDENT SET (and particular versions of it) is one of those that have received a very particular attention and mobilized numerous researchers.

Here, we propose in Section 2 a generic method that propagates improvements of worst-case complexity from "sparse" instances to "denser" (less sparse) ones, where the density of an instance is proper to the problem handled and refers

[*] Research supported by the French Agency for Research under the DEFIS program TODO, ANR-09-EMER-010.

H. Kaplan (Ed.): SWAT 2010, LNCS 6139, pp. 62–73, 2010.
© Springer-Verlag Berlin Heidelberg 2010

to the average value of some parameter of its instance. We call this method "bottom-up method". The basic idea here has two ingredients: (i) the choice of the recursive measure of the instance and (ii) a way to ensure that on "denser" instances, a good branching (wrt. the chosen measure) occurs.

We then illustrate our method to MIN SET COVER. Given a finite ground set $\mathcal{U}$ and a set-system $\mathcal{S}$ over $\mathcal{U}$, MIN SET COVER consists of determining a minimum-size subsystem $\mathcal{S}'$ covering $\mathcal{U}$. Here, the density of an instance is defined to be the average cardinality of the sets in the set-system $\mathcal{S}$. Application of the method to MIN SET COVER is rather direct but it produces quite interesting results. As we show in Section 2 it outperforms the results of [9] in instances with average-set sizes 6, 7, 8, ... Note that we are not aware of results better than those given here.

We next handle the MAX INDEPENDENT SET problem. Given a graph $G = (V, E)$, MAX INDEPENDENT SET consists of finding a maximum-size subset $V' \subseteq V$ such that for any $(v_i, v_j) \in V' \times V'$, $(v_i, v_j) \notin E$. For this problem, [4] proposes an algorithm with worst-case complexity bound $O^*(1.2201^n)$[1]. All the results we present here are polynomial space algorithms. We also quote the $O(1.2108^n)$ time bound in [7] using exponential space (claimed to be improved down to $O(1.1889^n)$ in the technical report [8], still using exponential space). Dealing with MAX INDEPENDENT SET in graphs of maximum degree 3, faster and faster algorithms have been devised for optimally solving this problem. Let us quote the recent $O^*(1.0892^n)$ time algorithm in [6], and the $O^*(1.0854^n)$ time algorithm by the authors of the article at hand [3]. For MAX INDEPENDENT SET density of a graph is measured by its average degree. So, the bottom-up method here extends improvements of the worst-case complexity in graphs of average degree d to graphs of average degree greater than d.

In order to informally sketch our bottom-up method in the case of MAX INDEPENDENT SET, suppose that one knows how to solve the problem on graphs with average degree d in time $O^*(\gamma_d^n)$. Solving the problem on graphs with average degree $d' > d$ is based upon two ideas: we first look for a running time expression of the form $\alpha^m \beta^n$, where α and β depend both on the input graph (namely on its average degree), *and on the value* γ_d (see Section 2). In other words, the form of the running time we seek is parameterized by what we already know on graphs with smaller average degrees. Next, according to this form, we identify particular values d_i (not necessarily integer) on the average degree that ensure that a "good" branching occurs. This allows us to determine a good running time for increasing values of the average degree. Note also that a particular interest of this method lies in the fact that any improvement on the worst-case complexity on graphs of average degree 3 immediately yields improvements for higher average degrees. A direct application of this method leads for instance for MAX INDEPENDENT SET in graphs with average degree 4 to an upper complexity bound that already slightly outperforms the best known time

---

[1] In a very recent article [5] a $O^*(1.2132^n)$ time algorithm for MAX INDEPENDENT SET is proposed. Plugging this new result allows to further improve our results. We give the corresponding bounds at the end of Section 4.

bound of $O^*(1.1713^n)$ of [1] (Section 2). This result is further improved down to $O^*(1.1571^n)$ (Section 3) with a more refined case analysis. We also provide bounds for (connected) graphs with average degree at most 5 and 6.

In section 4, we combine measure and conquer with bottom-up to show that MAX INDEPENDENT SET can be optimally solved in time $O^*(1.2127^n)$ in general graphs, thus improving the $O^*(1.2201^n)$ bound [4]. Furthermore, in graphs of maximum degree at most 5 and 6, we provide time bounds of $O^*(1.1918^n)$ and $O^*(1.2071^n)$, respectively, that improve upon the respective $O^*(1.2023^n)$ and $O^*(1.2172^n)$ time bounds of [4][2].

We give the results obtained using the $O^*(1.0854^n)$ time bound of [3] for graphs of degree 3. Note that using previously known bounds for solving MAX INDEPENDENT SET in graphs of degree 3 (worse than $O^*(1.0854^n)$), the bottom-up method would also lead to improved results (with respect to those known in the literature). We illustrate this point in Table 1.

Table 1. MAX INDEPENDENT SET results for graphs of degree 4, 5, 6 and general graphs with starting points several complexity bounds for graphs of degree 3

Degree 3	Degree 4	Degree 5	Degree 6	General graphs
1.08537	1.1571	1.1918	1.2071	1.2127
1.0892 [6]	1.1594	1.1932	1.2082	1.2135
1.0977 [2]	1.1655	1.198	1.213	1.217

Maybe more interesting than the improvements themselves is the fact that they are obtained via an original method that, once some, possibly long, case analysis has been performed on instances of small density, it is directly applicable for getting results higher density instances, even for general ones. We think that this method deserves further attention and insight, since it might be used to solve also other problems.

Throughout this paper, we will denote by $N(u)$ and $N[u]$ the neighborhood and the closed neighborhood of u, respectively ($N(u) = \{v \in V : (u,v) \in E\}$ and $N[u] = N(u) \cup \{u\}$).

2   The Bottom-Up Method

In this section we present the method that relies on the following two stepping stones: (i) the choice of the recursive complexity measure, applied in a very simple way in Section 2.1 to the MIN SET COVER to get non trivial time bounds for instances of bounded average set-cardinalities, and (ii) a way to ensure that on instances of density greater that d, a good branching (wrt. the chosen complexity measure) occurs. This second point is illustrated in Section 2.2 for MAX INDEPENDENT SET.

---

[2] The bound in graphs of degree at most d is obtained as $O^*(1.2201^{n w_d})$, where w_d is the weight associated to vertices of degree d in the measure and conquer analysis in [4]; better bounds could maybe be achieved with this method, but this is not straightforward, since reduction rules may create vertices of arbitrarily large degree.

2.1 The Importance of Recursive Complexity Measure: The Case of MIN SET COVER

Let us consider the MIN SET COVER problem with ground set $\mathcal{U} = \{x_1, \cdots, x_n\}$ and set system $\mathcal{S} = \{S_1, \cdots, S_m\}$. We show that the bottom-up method easily applies to get algorithm with improved time bounds for instance with sets of average (or maximum) size d, for any $d \geq 5$. In what follows $p = \sum_{i=1}^{m} |S_i|$.

Lemma 1. *The algorithm in [9] solves* MIN SET COVER *in time resp.* $O^*(1.55^m)$, $O^*(1.63^m)$, $O^*(1.71^m)$, $O^*(1.79^m)$ *in instances with sets of average size 5, 6, 7 and 8.*

Proof. Denoting by n_i the number of sets of size i and m_j the number of elements of frequency j, [9] gives an algorithm working in time $O^*(1.2302^{k(I)})$ where $k(I) = \sum_{i \geq 1} w_i n_i + \sum_{j \geq 1} v_j m_j$. Here w_i is the weight associated to a set of size i, and v_j is the weight associated to an element of frequency j. It is easy to see that, by convexity[3], if the average size of sets is an integer d, then $\sum_{i \geq 1} w_i n_i \leq m w_d$. Moreover, note that $v_j/j \leq v_3/3$ for all $j \geq 1$[4], hence $\sum_{j \geq 1} v_j m_j \leq v_3/3 \sum_{j \geq 1} j m_j = dm v_3/3$. We get $k(I) \leq m(w_d + v_3 d/3)$ (this bound being tight if all sets have size d and all elements have frequency 3). □

Let us consider an instance with $p > dm$. The bottom-up method assumes that we know how to solve the problem in instances with sets of average size d in time $O^*(\gamma_d^m)$. It seeks a complexity of the form $O^*(\gamma_d^m y^{p-dm})$; indeed, it is valid by hypothesis for $p = dm$, ie. on instances with sets of average size d. Let us consider a set S of size $s \geq d + 1$. We branch on S. If we do not take it, we remove one set and $s \geq d + 1$ edges; if we take it, suppose that each element in S has frequency at least 3. Then we remove 1 set and (at least) $3s \geq 3(d+1)$ edges. Hence, for the complexity to be valid we have to choose y such that $\gamma_d^m y^{p-dm} \geq \gamma_d^{m-1} y^{p-(d+1)-d(m-1)} + \gamma_d^{m-1} y^{p-3(d+1)-d(m-1)}$, or equivalently $1 \geq \gamma_d^{-1} y^{-1} + \gamma_d^{-1} y^{-(2d+3)}$.

Taking for instance $\gamma_5 = 1.55$ (Lemma 1), this is true for $y = 1.043$. Let us check the other cases:

- If there is an element j of frequency 1 in S, we have to take S and we remove one set and $s \geq d + 1$ edges. This does not create any problem as long as $\gamma_d^{m-1} y^{(p-d-1)-d(m-1)} \leq \gamma_d^m y^{p-dm}$, i.e., $y \leq \gamma_d$.
- Otherwise, if there is an element j of frequency 2 which is in S and S', then either we take S and remove 1 set and at least $2(d + 1)$ edges, or we remove S and take S', and we remove 2 sets and at least $d + 2$ edges. So we have to check that the value of y computed in the general case verifies $1 \geq \gamma_d^{-1} y^{-(d+2)} + \gamma_d^{-2} y^{-2+d}$.

Then if sets have average size $d + 1$, since $p = (d + 1)m$, the complexity is $O^*(\gamma_{d+1}^m)$ with $\gamma_{d+1} = \gamma_d \times y$. Starting from $\gamma_5 = 1.55$, the recurrences give $\gamma_6 = 1.61$, $\gamma_7 = 1.66$ (with $y = 1.031$) and $\gamma_8 = 1.70$ (with $y = 1.024$).

---

[3] w_i's are $(0,0.3755,0.7509,0.9058,0.9720,0.9982)$ for $i = 1, \cdots, 6$ and 1 for $i \geq 7$.

[4] v_j's are $(0, 0.2195, 0.6714, 0.8766, 0.9569, 0.9882)$ for $i = 1, \cdots, 6$ and 1 for $i \geq 7$.

Theorem 1. MIN SET COVER *is solvable in times* $O^*(1.61^m)$, $O^*(1.66^m)$ *and* $O^*(1.70^m)$ *in instances with sets of average size* 6, 7 *and* 8, *respectively.*

Interestingly enough, the time bound of $O^*(1.2302^{m(w_d+v_3 d/3)})$ obtained in Lemma 1 is bigger than 2^m for $d \geq 11$ while using the previous method, it can be shown that $\gamma_d < 2$ for any d (for instance $\gamma_{100} < 1.98$). Of course, the analysis conducted in [9] is not oriented towards the bounded size case, so better results might be obtained; we will say a few fords in conclusion on the links between this complexity measure and the one of measure and conquer.

2.2 Making a Good Branching: The Case of MAX INDEPENDENT SET

In order to use efficiently the previous complexity measure for MAX INDEPENDENT SET, we prove that, in a graph whose average degree is bounded from below by some constant (possibly not an integer), we are sure that there exists a rather dense local configuration we can branch on. More precisely, if the average degree is greater than d, this implies that we can find a vertex v with at least $f(d)$ edges incident to some neighbor of v, for some increasing function f. Let us give an example. In the independent set problem, there are two well known reductions rules that allow to remove without branching vertices of degree at most 2^5. In the following we will assume that these reductions rules has been performed, i.e., the graph does not contain any vertex of degree at most 2. Then, trivially, if the graph has average degree greater than $d \in \mathbb{N}$, we know there exists a vertex v of degree at least $d + 1$. If we assume that no vertex is dominated[6], then there exist at least $f(d) = 2(d+1) + \lceil (d+1)/2 \rceil$ edges incident to some neighbor of v. Indeed, there exist $d+1$ edges incident to v, $d+1$ edges between a neighbor of v and a vertex not neighbor of v (one for each neighbor of v, to avoid domination) and, since each vertex has degree at least 3, at least $\lceil (3(d+1) - 2(d+1))/2 \rceil = \lceil (d+1)/2 \rceil$ other edges. Note that such relationships may be established even if d is non integer. For instance, we will see that if $d > 24/7$, then there exists a vertex of degree 5 or two adjacent vertices having degree 4, leading to $f(d) = 11$. This property linking the average degree to the quality of the branching is given in Lemma 2.

Then, for a given d, either the average degree is greater than d, and we can make an efficient branching (i.e., a branching that induces a recurrence relation leading to a lower time-bound), or it is not and we can use an algorithm tailored for low-degree graphs. Thus, Lemma 2 fixes a set of critical degrees (d_i) and we define step-by-step (from the smallest to the highest) algorithms STABLE(d_i), that work on graphs of average degree d_i or less. With this lemma, we analyse the running time of these algorithms thanks to a measure allowing to fruitfully use the existence of the dense local configurations mentioned above. As for MIN

---

[5] A vertex of degree at most 1 should be taken in the solution. If a vertex v has two neigbors u_1 and u_2, take v if u_1 and u_2 are adjacent, otherwise remove v, u_1, u_2 from the graph and add a new vertex u_{12} whose neighborhood is $N(u_1) \cup N(u_2) \setminus \{v\}$ (this reduction is called vertex folding, see for instance [4]).

[6] u dominates v if $N[u] \subseteq N[v]$. In this case, it is never interesting to take v.

SET COVER, if we know how to solve the problem in time $O^*(\gamma_d^n)$ in graphs with average degree d, and that when the average degree is greater than d a good branching occurs, then we seek a complexity of the form $O^*(\gamma_d^n y^{2m-dn})$. This complexity measure is chosen because it is by hypothesis valid in graphs with average degree d. The recurrences given by the branching will give the best possible value for y. This bottom-up analysis (from smaller to higher average degree) is detailed in Proposition 1.

At the end of the section, we mention some results obtained by a direct application of this method for graphs of average degree 4, 5 and 6.

Lemma 2. *There exists a specific sequence* $(\epsilon_{i,j}, f_{i,j})_{i \geq 4, j \leq i-2}$ *such that, if the input graph has average degree more than* $i - 1 + \epsilon_{i,j}$, *then the following branching is possible: either remove 1 vertex and* i *edges, or* $i + 1$ *vertices and (at least)* $f_{i,j}$ *edges. For any* i, $\epsilon_{i,0} = 0$. *The following table gives the beginning of the sequence* $(\epsilon_{i,j}, f_{i,j})$:

$(\epsilon_{i,j}, f_{i,j})$	$j = 0$	$j = 1$	$j = 2$	$j = 3$	$j = 4$
$i = 4$	$(0, 10)$	$(3/7, 11)$	$(3/5, 12)$		
$i = 5$	$(0, 15)$	$(4/9, 16)$	$(4/7, 17)$	$(4/5, 18)$	
$i = 6$	$(0, 20)$	$(5/23, 21)$	$(5/11, 22)$	$(20/37, 23)$	$(5/7, 24)$

Before giving the proof of the lemma, let us give an example, with $i = 5$ and $j = 2$. This lemma states that if the average degree is greater than $4 + \epsilon_{5,2} = 4 + 4/7$, then we can branch on a vertex v and either remove this vertex and 5 edges, or 6 vertices and (at least) 17 edges.

Proof. Fix some vertex v_0 of maximum degree d, such that, for any other vertex v of degree d in the graph, $\sum_{w \in N(v)} d(w) \leq \sum_{w \in N(v_0)} d(w)$, and set $\delta = \sum_{w \in N(v_0)} d(w)$.

For $k \leq d$, let n_k be the number of vertices of degree k and m_{kd} be the number of edges (u, v) such that $d(u) = k$ and $d(v) = d$. For $k \leq d - 1$, set $\alpha_k = m_{kd}/n_d$ and $\alpha_d = 2m_{dd}/n_d$. In other words, α_k is the average number of vertices of degree k that are adjacent to a vertex of degree d. Since folding or reduction rules remove vertices of degree at most 2, we fix $\alpha_k = 0$ for $k \leq 2$. Summing up inequalities on any vertex of degree d, we get (details are omitted):

$$\sum_{k \leq d} k\alpha_k \leq \delta \tag{1}$$

$$\sum_{k \leq d} \alpha_k = d \tag{2}$$

Fix now $\epsilon = 2m/n - (d - 1) \in]0, 1[$. Then, $\epsilon = \frac{\sum_{k \leq d}(k+1-d)n_k}{\sum_{k \leq d} n_k}$. This function is decreasing with n_k, $\forall k < d$. Use some straightforward properties: $n_k \geq \frac{m_{kd}}{k}$, $\forall k < d$ and $dn_d = \sum_{k < d} m_{kd} + 2m_{dd}$. This leads to:

$$\epsilon \leq \frac{n_d - \sum_{k<d}(d - 1 - k)m_{kd}/k}{n_d + \sum_{k<d} m_{kd}/k} = \frac{1 - \sum_{k<d}(d - 1 - k)\alpha_k/k}{1 + \sum_{k<d} \alpha_k/k} \tag{3}$$

Clearly, when we discard v_0, we remove from the graph one vertex and d edges; when we add it, $d+1$ vertices are deleted. Now, let μ_2 be the minimal number of edges we delete when we add v_0 to the solution. Since there are at least $2d(v_0)$ edges between $N(v_0)$ and the remaining of the graph, and thanks to inequalities (1) and (2), we get:

$$\mu_2 \geq 2d + \left\lceil \frac{\delta - 2d}{2} \right\rceil \geq 2d + \left\lceil \frac{\sum_{k \leq d}(k-2)\alpha_k}{2} \right\rceil. \tag{4}$$

For $0 \leq j \leq i - 2$, we now consider the following programs $(P_{i,j})$: $\max(\epsilon)$ under constraints (1),(2),(3),(4) and $\mu_2 \leq f_{i,j} - 1$. In other words, we look for the maximal value $\epsilon_{i,j}$ for ϵ such that it is possible that any vertex in the graph verifies $\mu_2 \leq f_{i,j} - 1$. Equivalently, if the graph has degree higher than $i - 1 + \epsilon_{i,j}$, we remove at least $f_{i,j}$ edges (when taking some well chosen vertex). □

Let $d_{i,j} = i - 1 + \epsilon_{i,j}$. Now we use Lemma 2 to recursively define an algorithm **STABLE**$(d_{i,j})$ solving MAX INDEPENDENT SET in a graph of average degree at most $d_{i,j}$. **STABLE**$(d_{i,j})$ performs the usual preprocessing (described above at the beginning of the section) and branches on a vertex that maximizes the number of edges incident to its neighborhood. It repeats this step until the average degree is at most $d_{i,j-1}$, then it applies algorithm **STABLE**$(d_{i,j-1})$[7].

Suppose that **STABLE**$(d_{i,j-1})$ has a running time bounded by $\gamma_{i,j-1}^n$. Let $\nu_1 = 1$, $\mu_1 = i$, $\nu_2 = i + 1$ and $\mu_2 = f_{i,j-1}$.

Proposition 1. *STABLE($d_{i,j}$) runs in time $T(m,n) = O^*\left(\gamma_{i,j-1}^n y_{i,j}^{2m - d_{i,j-1}n}\right)$, where $y_{i,j}$ is the smallest solution of the inequality:*

$$1 \geq \gamma_{i,j-1}^{-\nu_1} y^{-2\mu_1 + d_{i,j-1}\nu_1} + \gamma_{i,j-1}^{-\nu_2} y^{-2\mu_2 + d_{i,j-1}\nu_2}$$

In particular, $T(m,n) = O^*(\gamma_{i,j}^n)$ where $\gamma_{i,j} = \gamma_{i,j-1} y_{i,j}^{\epsilon_{i,j} - \epsilon_{i,j-1}}$.

Proof. The running time claimed is valid for graphs of average degree $d_{i,j-1}$. If the graph has average degree greater than $d_{i,j-1}$, thanks to Lemma 2 we branch on a vertex where we remove either ν_1 vertices and μ_1 edges or ν_2 vertices and (at least) μ_2 edges. Then the running time is valid as long as y is such that $T(m,n) \geq T(m - \mu_1, n - \nu_1) + T(m - \mu_2, n - \nu_2) + p(m,n)$ (for some polynomial p). This gives the recurrence relation claimed by proposition's statement.

Since $2m \leq d_{i,j}n$ in a graph of average degree at most $d_{i,j}$, the running time $O^*(y_{i,j}^n)$ follows. Of course, we need to initialize the recurrence, for example with $\gamma_{4,0} = 1.0854$ in graphs of average degree $d_{4,0} = 3$ (i.e., with the basis of the running time in [3]).

For completeness, we need to pay attention to the fact that, once the branching has been performed, reduction rules might be applied in order to remove vertices of degree at most 2 in the remaining graph. For instance, if a separated tree on ν vertices is created, reduction rules will remove this tree hence, in all, ν vertices

---

[7] If $j = 0$ of course we use **STABLE**$(d_{i-1,i-3})$ in graphs of average degree at most $d_{i-1,i-3}$ and the same result as in Proposition 1 holds.

and $\nu - 1$ edges. Note that all reduction rules remove $\nu \geq 1$ vertices and at least $\nu - 1$ edges. We have to check that these operations do not increase $T(m, n)$, i.e., $T(m, n) \geq T(m - \nu + 1, n - \nu)$, or $y^{d_{i,j}-1-2+2/\nu} \leq \gamma_{i,j-1}$. $\square$

To conclude this section, let us note that as direct applications of Proposition 1 we obtain (details are omitted) an algorithm running in time $O^*(1.1707^n)$ for graphs of average degree 4 (slightly outperforming the bound of $O^*(1.1713^n)$ by [1]), and algorithms running in times $O^*(1.2000^n)$ and $O^*(1.2114^n)$ for graphs of average degrees 5 and 6, respectively (based upon the algorithm in time $O^*(1.1571^n)$ for graphs of average degree 4 of Theorem 1). It is worth noticing that these results are obtained by a direct application of the method proposed; they will be further improved in the rest of the paper, using more involved case analysis or techniques, but already outperform the best known bounds so far.

3   Refined Case Analysis for Graphs of Average Degree 4

In Lemma 2 we have shown the existence of local dense configurations when the graph has average degree more than 3. For instance, we have seen that if it has average degree at least $4 + 4/7$, then we can branch on a vertex v and either remove this vertex and 5 edges, or 6 vertices and (at least) 15 edges. In this section, we apply a similar method, by performing a deeper analysis, to compute the running time of an algorithm for graphs of average degree 4, in order to prove the following theorem.

Theorem 1. *It is possible to solve* MAX INDEPENDENT SET *on graphs with maximum (or even average) degree 4 with running time* $O^*(1.1571^n)$.

Proof (Sketch). Based upon what has been discussed in Section 2, we seek a complexity of the form $O^*(\gamma^n y^{2m-3n})$, where $\gamma = 1.0854$, valid for graphs of average degree 3. We assume that our graph has $m > 3n/2$ edges. In particular, there is a vertex of degree at least 4.

Assume that we perform a branching that reduces the graph by either ν_1 vertices and μ_1 edges, or by ν_2 vertices and μ_2 edges. Then, by recurrence, our complexity formula is valid for y being the largest root of the following equality: $1 = \gamma^{-\nu_1} y^{-2\mu_1+3\nu_1} + \gamma^{-\nu_2} y^{-2\mu_2+3\nu_2}$. Then, either there exists a vertex of degree at least 5, or the maximum degree is 4. In the former case, we reduce the graph either by $\nu_1 = 1$ vertex and $\mu_1 = 5$ edges, or by $\nu_2 = 6$ vertices and $\mu_2 \geq 13$ edges, leading to $y = 1.0596$.

In what follows, we consider the latter case, *i.e.*, the graph has maximum degree 4, and we denote u_1, u_2 u_3 and u_4 the four neighbors of some vertex v. We call inner edge an edge between two vertices in $N(v)$, and outer edge an edge (u_i, x) where $x \notin \{v\} \cup N(v)$. We study 4 cases, depending on the configuration of $N(v)$. Here, we consider that no trees are created while branching (the case of trees is not detailed here due to lack of space).

Case 1. All the neighbors of v have degree 4.

This case is easy. Indeed, if there are at least 13 edges incident to vertices in $N(v)$, by branching on v we get $\nu_1 = 1$, $\mu_1 = 4$, $\nu_2 = 5$ and $\mu_2 \geq 13$. This gives $y = 1.0658$.

But there is only one possibility with no domination and only 12 edges incident to vertices in $N(v)$: when u_1, u_2, u_3, u_4 is a 4-cycle. In this case, we can reduce the graph before branching. Any optimal solution cannot contain more than two vertices from the cycle. If it contains only one vertex, then replacing it by v does not change its size. Finally, there exist only three disjoint possibilities: keep u_1 and u_3, keep u_2 and u_4 or keep only v. Hence, we can replace $N(v) \cup \{v\}$ by only two adjacent vertices $u_1 u_3$ and $u_2 u_4$, such that u is adjacent to $u_1 u_3$ (resp., $u_2 u_4$) if and only if u is adjacent to u_1 or to u_3 (resp., to u_2 or to u_4).

Case 2. All the neighbors of v have at least 2 outer edges.

If one of them has degree 4, then there are at least 13 edges removed when taking v, and we get again $\nu_1 = 1$, $\mu_1 = 4$, $\nu_2 = 5$ and $\mu_2 \geq 13$.

Otherwise, once v is removed, any u_i now has degree 2. Note that when folding a vertex of degree 2, we reduce the graph by 2 vertices and 2 edges (if the vertex dominates another one, this is even better). Since any two vertices u_i cannot be adjacent to each other, we can remove 8 vertices and at least 8 edges by folding u_1, u_2, u_3, u_4. Indeed, if for instance, u_1 dominates its neighbors (its two neighbors being adjacent), we remove 3 vertices and at least 5 edges which is even better. Removing 8 vertices and at least 8 edges is very interesting since it leads to $\nu_1 = 9$, $\mu_1 = 12$, $\nu_2 = 5$, $\mu_2 = 12$, and $y = 1.0420$.

Case 3. u_1 has degree 3 and only one outer edge.

u_1 has one inner edge, say (u_1, u_2). Let y be the third neighbor of u_1. We branch on y. Suppose first that u_2 has degree 3. If we take y we remove 4 vertices and (at least) 8 edges (there is at most one inner edge in $N(y)$); if we don't take y, then we remove also v and we remove globally 2 vertices and 7 edges.

Obviously, this is not sufficient. There is an easily improvable case, when a neighbor of y has degree 4 (or when y itself has degree 4), or when the neighbors of y are not adjacent to each other. Indeed, in this case, there are at least 9 edges in $N(y)$, and we get $\nu_1 = 4$, $\mu_1 = 9$, $\nu_2 = 2$ and $\mu_2 \geq 7$, leading to $y = 1.0661$. Now, we can assume that y has degree 3, its three neighbors have also degree 3, and that the same holds for z, the neighbor of u_2. Furthermore, we assume that they are both part of a triangle.

We reason with respect to the quantity $|N(y) \cap N(z)|$. If $|N(y) \cap N(z)| = 2$, then either some neighbor of y has degree 4, or else v is a separator of size 1. If $|N(y) \cap N(z)| = 1$, then their common neighbor has degree 4. Finally, if $|N(y) \cap N(z)| = 0$, then at least a neighbor of, say, z is neither u_3 nor u_4. Hence, when discarding y, we take u_1, so remove u_2 and then add z to the solution. We get $\nu_1 = 4$, $\mu_1 = 8$, $\nu_2 = 7$ and $\mu_2 \geq 13$, leading to $y = 1.0581$.

Suppose now that u_2 has degree 4. Then, when we don't take y, since we don't take v, u_1 has degree 1. Then, we can take it and remove u_2 and its incident edges. Then, when we don't take y, we remove in all 4 vertices and 10 edges. In other words, $\nu_1 = 4$, $\mu_1 = 8$, $\nu_2 = 4$ and $\mu_2 \geq 10$. This gives $y = 1.0642$.

Case 4. u_1 has degree 4 and only one outer edge.

Since Case 1 does not occur, we can assume that there is a vertex (say u_4) of degree 3. Since Case 3 does not occur, u_4 has no inner edge. Hence, u_1 is adjacent to both u_2 and u_3. Then, there are only two possibilities. If there are no other inner edges, since Case 3 does not occur, u_2 and u_3 have two outer edges, and we have in all 13 edges. This gives once again $\nu_1 = 1$, $\mu_1 = 4$, $\nu_2 = 5$ and $\mu_2 \geq 13$. Otherwise, there is an edge between u_2 and u_3. Then, v, u_1, u_2, u_3 form a 4-clique. We branch on u_4. If we take u_4, we delete $\nu_1 = 4$ vertices and (at least) $\mu_1 = 9$ edges (v has degree 4 and is not adjacent to other neighbors of u_4). If we discard u_4 then, by domination, we take v, and delete $\nu_2 = 5$ vertices and at least $\mu_2 = 12$ edges. So, $y = 1.0451$ and Case 4 is concluded.

The remaining part of the proof (not given here) has to deal with the case when some trees are created while branching and to verify that performing a reduction rule (such as a vertex folding) does not increase the measure.   □

4   When Bottom-Up Meets Measure and Conquer: Final Improvements and an Algorithm in General Graphs

In this section we devise algorithms for graphs of maximum degree 5, 6 and general graphs. The algorithms follow the same line as the one devised in [4]. There, a branching is performed on a vertex of maximum degree, and a measure and conquer technique is used to analyse the running time: vertices of degree d receive a weight $w_d \geq 0$ which is non-decreasing with d (with $w_d = 1$ for $d \geq 7$ and $w_1 = w_2 = 0$). The running time of the algorithm is measured as a function of the total weight of the graph (initially smaller than n). In other words, running times are expressed as $T(n) = O^*(c^{\sum_{i \in V} w_i})$, where $\sum_{i \in V} w_i \leq n$. When a branching on a vertex of degree d is done, the decreasing δ_d of the total weight of the graph is measured. Weights are then optimized in such a way that δ_d leads to the same complexity (neglecting polynomial terms) for any d. Weights are subject to some constraints, such as, for example, the fact that reduction rules must not increase the total weight of the graph.

Here, we modify the algorithm above and its analysis in two ways in order to improve the running time for MAX INDEPENDENT SET. First, we incorporate the fact that we have efficient algorithms able to solve MAX INDEPENDENT SET in graphs of maximum degree 3 and 4. We will use these algorithms when the input graph has degree at most 4, modifying the set of constraints in the measure and conquer analysis (see Proposition 2) and leading to a better running time. Then we improve the analysis of measure and conquer in graphs of maximum degree 5, by taking into account that creating a vertex of high degree by vertex folding may decrease the total weight. As a final result, we combine the two previous ideas to get a general algorithm in time $O^*(1.2127^n)$ in Theorem 3.

Proposition 2. MAX INDEPENDENT SET *can be solved in time* $O^*(1.2135^n)$.

Proof. According to former sections, we know that it is possible to compute MAX INDEPENDENT SET on graphs of maximum degree Δ with running time bounded above by $O^*(\gamma_\Delta^n)$, where $\gamma_3 = 1.0854$ and $\gamma_4 = 1.1571$.

Our algorithm works as follows: while $\Delta \geq 5$, run the preprocessing (in particular, fold appropriate vertices and reduce mirrors) described in [4] and branch on any vertex of maximum degree. Once $\Delta \leq 4$, run the algorithms described in [3] (case of degree 3) and in Section 3 (case of degree 4).

We analyse the running time of this algorithm with the same measure and conquer techniques as in [4] modified as follows. First, we do not need to consider branching on vertices of degree 3 or 4, and this allows of course to choose much more efficient weights. On the other hand, we have to consider two additional constraints. Indeed, the bound on effective running time of the algorithm we use for degree 3 and 4 must be lower than the complexity we claim. Since we provided in Theorem 1 an algorithm in $O^*(\gamma_3^n (\gamma_4/\gamma_3)^{2m-3n})$, we have to verify that for any graph of maximum degree 4 or less and average degree $d = 3 + n_4/n$ (note that $n_3 + n_4 = n$) $\gamma_3^n \left(\frac{\gamma_4}{\gamma_3}\right)^{(d-3)n} \leq \gamma^{w_3 n_3 + w_4 n_4}$, or equivalently $\log \gamma_3 + (d-3)(\log \gamma_4 - \log \gamma_3) \leq \log \gamma((4-d)w_3 + (d-3)w_4)$. Notice that $4-d$ and $d-3$ are nonnegative, thus this inequality is a consequence of $w_3 \geq \frac{\log \gamma_3}{\log \gamma}$ and $w_4 \geq \frac{\log \gamma_4}{\log \gamma}$. The best values we have found are $w_3 = 0.493$, $w_4 = 0.765$, $w_5 = 0.914$, $w_6 = 0.9777$, satisfying all the constraints and leading to $\gamma = 1.2135$. $\qquad \square$

As consequences we get running times of $O^*(1.1935^n)$ and $O^*(1.2083^n)$ in graphs of maximum degree 5 and 6 (by the fact that $\sum_{v \in V} w(v) \leq w_\Delta n$). Moreover, as mentioned before, it is possible to improve them slightly.

Theorem 2. *On graphs of maximum degree 5,* MAX INDEPENDENT SET *can be solved with running time $O^*(1.1918^n)$.*

Proof. In graphs of maximum degree Δ, inequalities $w_d \leq 1$ for any $d > \Delta$ are not relevant anymore (since initially $\sum_{v \in V} w(v) \leq n$ as soon as $w_d \leq 1$ for $d \leq \Delta$). However, the values w_d must respect the constraint that folding does not increase the total weight of the graph. If we use weights $w_3^5 = 0.52161$, $w_4^5 = 0.83161$, $w_5^5 = 1$ and $w_6^5 = 1.16839$ and $w_d^5 = 1.33678$ for $d > 6$, all the constraints are satisfied and we get $\gamma_5 = 1.1918$. $\qquad \square$

Theorem 3. *It is possible to solve* MAX INDEPENDENT SET *in time $O^*(1.2127^n)$.*

Proof. As discussed above, MAX INDEPENDENT SET is solvable on graphs of maximum degree Δ in time $O^*(\gamma_\Delta^n)$, where $\gamma_3 = 1.0854$, $\gamma_4 = 1.1571$ and $\gamma_5 = 1.1918$. Our algorithm works as follows: while $\Delta \geq 6$, run the same algorithm as in [4]. Once $\Delta \leq 5$, run algorithm described in Theorem 2 based upon our improvements for $\Delta \leq 5$. The additional constraint is now $\gamma_5^{w_3 n_3 + w_4 n_4 + n_5} \leq \gamma^{w_3 n_3 + w_4 n_4 + w_5 n_5}$, which is a consequence of $w_i \geq w_i^5 \frac{\log \gamma_5}{\log \gamma}$, $\forall i \leq 5$. The weights: $w_3 = 0.47459$, $w_4 = 0.75665$, $w_5 = 0.90986$, $w_6 = 0.9757$, $w_7 = 0.9994$ and $w_d = 1$ for $d \geq 8$ satisfy all the constraints and lead to $\gamma = 1.2127$. $\qquad \square$

As a consequence of the proof of Theorem 3, we solve MAX INDEPENDENT SET in graphs of degree at most 6 in time $O^*(1.2127^{w_6 n}) = O^*(1.2071^n)$.

As mentioned in the introduction, a recent article provides an algorithm solving MAX INDEPENDENT SET in time $O^*(1.2132^n)$. Considering this new result instead of the bound of [4], we obtain the following time bounds: $O^*(1.1895^n)$, $O^*(1.2050^n)$ and $O^*(1.2114^n)$ for respectively graphs of maximum degree 5, 6, and for general graphs.

5   Conclusion

The complexity measure of the method proposed in the paper may appear, at least for the two problems considered, as an adaptation of measure and conquer for bounded degree instances. Indeed, for example, $\gamma_d^m y^{2m-dn}$ can be easily written as $2^{\sum_i w_i n_i}$ where n_i is the number of vertices of degree i. However, first, the way we seek the complexity in the bottom up method actually specifies the strong links that have to be verified between weights in order to use as efficiently as possible an algorithm for lower degree graphs (or, more generally for low density instances). The second point is to exhibit and to fruitfully use the link between density and branching. A recursive application of the method then allows to take into account situations where a good branching necessarily occurs to derive good complexity bounds.

Though the precise links between bottom-up and measure and conquer is not very clear yet, at least these two points seem not to be considered in a usual measure and conquer analysis. Furthermore, the results obtained for MAX INDEPENDENT SET by using it, the best known until now, are interesting per se.

References

1. Beigel, R.: Finding maximum independent sets in sparse and general graphs. In: Proc. SODA'99, pp. 856–857 (1999)
2. Bourgeois, N., Escoffier, B., Paschos, V.T.: An $O^*(1.0977^n)$ exact algorithm for MAX INDEPENDENT SET in sparse graphs. In: Grohe, M., Niedermeier, R. (eds.) IWPEC 2008. LNCS, vol. 5018, pp. 55–65. Springer, Heidelberg (2008)
3. Bourgeois, N., Escoffier, B., Paschos, V.T., van Rooij, J.M.M.: Fast Algorithms for Max Independent Set in Graphs of Small Average Degree. CoRR, abs/0901.1563 (2009)
4. Fomin, F.V., Grandoni, F., Kratsch, D.: A measure and conquer approach for the analysis of exact algorithms. Journal of the ACM 56(5) (2009)
5. Kneis, J., Langer, A., Rossmanith, P.: A Fine-grained Analysis of a Simple Independent Set Algorithm. In: Proc. FSTTCS 2009, pp. 287–298 (2009)
6. Razgon, I.: Faster computation of maximum independent set and parameterized vertex cover for graphs with maximum degree 3. J. Discrete Algorithms 7, 191–212 (2009)
7. Robson, J.M.: Algorithms for maximum independent sets. J. Algorithms 7(3), 425–440 (1986)
8. Robson, J.M.: Finding a maximum independent set in time $O(2^{n/4})$. Technical Report 1251-01, LaBRI, Université de Bordeaux I (2001)
9. Van Rooij, J.M.M., Bodlaender, H.L.: Design by measure and conquer a faster algorithm for dominating set. In: STACS 2008, pp. 657–668 (2008)
10. Woeginger, G.J.: Exact algorithms for NP-hard problems: a survey. In: Jünger, M., Reinelt, G., Rinaldi, G. (eds.) Combinatorial Optimization - Eureka!, You Shrink! LNCS, vol. 2570, pp. 185–207. Springer, Heidelberg (2003)

Capacitated Domination Faster Than $O(2^n)^\star$

Marek Cygan[1], Marcin Pilipczuk[1], and Jakub Onufry Wojtaszczyk[2]

[1] Dept. of Mathematics, Computer Science and Mechanics,
University of Warsaw, Poland
[2] Institute of Mathematics, Polish Academy of Sciences
{cygan,malcin,onufry}@mimuw.edu.pl

Abstract. In this paper we consider the CAPACITATED DOMINATING SET problem — a generalisation of the DOMINATING SET problem where each vertex v is additionally equipped with a number $c(v)$, which is the number of other vertices this vertex can dominate. We provide an algorithm that solves CAPACITATED DOMINATING SET exactly in $O(1.89^n)$ time and polynomial space. Despite the fact that the CAPACITATED DOMINATING SET problem is quite similar to the DOMINATING SET problem, we are not aware of any published algorithms solving this problem faster than the straightforward $O^*(2^n)^1$ solution prior to this paper. This was stated as an open problem at Dagstuhl seminar 08431 in 2008 and IWPEC 2008.

We also provide an exponential approximation scheme for CAPACITATED DOMINATING SET which is a trade-off between the time complexity and the approximation ratio of the algorithm.

1   Introduction

The field of exact exponential-time algorithms for NP-hard problems has attracted a lot of attention in recent years (see Woeginger's survey [1]). Many difficult problems can be solved much faster than by the obvious brute-force algorithm; examples include INDEPENDENT SET [2], DOMINATING SET [2,3] , CHROMATIC NUMBER [4] and BANDWIDTH [5,6]. A few powerful techniques have been developed, including Measure & Conquer [2] and inclusion/exclusion principle applications [4,7,8]. However, there is still a bunch of problems for which no faster solution than the obvious one is known. These include SUBGRAPH ISOMORPHISM and CHROMATIC INDEX which are mentioned as open problems in [9,10].

In order to define the CAPACITATED DOMINATING SET problem let us introduce some basic notions. Let $G = (V, E)$ be an undirected graph. Given $F \subset E$ we write $V(F)$ to denote the set of all endpoints of the edges in F. Given $W \subset V$ by $G[W]$ we denote the subgraph induced by W. We say a vertex $v \in V$ dominates $u \in V$ if $u = v$ or $uv \in E$, i.e. a vertex dominates itself and all its neighbours. By $N[v] = \{v\} \cup \{u : uv \in E\}$ we denote the set of vertices dominated by v. We extend this notation to any

---

* This work was partially supported by the Polish Ministry of Science grants N206 491038 and N206 491238.

^{1} By O^* we denote standard big O notation but omitting polynomial factors.

H. Kaplan (Ed.): SWAT 2010, LNCS 6139, pp. 74–80, 2010.
© Springer-Verlag Berlin Heidelberg 2010

subset $W \subset V$ by putting $N[W] = \bigcup_{v \in W} N[v]$. We say that a set W dominates a vertex u if $u \in N[W]$. The set $N[W]$ is called the *closed neighbourhood* of W.

The DOMINATING SET problem asks for the smallest set that dominates the whole V. In the CAPACITATED DOMINATING SET problem each vertex v is additionally equipped with a number $c(v)$, which is the number of other vertices this vertex can dominate. Formally, we say that a set $S \subset V$ is a capacitated dominating set if there exists $f_S : V \setminus S \to S$ such that $f_S(v)$ is a neighbour of v for each $v \in V \setminus S$ and $|f_S^{-1}(v)| \le c(v)$ for each $v \in S$. The function f_S is called a *dominating function* for the set S. The CAPACITATED DOMINATING SET problem asks for the smallest possible size of a capacitated dominating set. Note that for a given set S checking whether it is a capacitated dominating set is a polynomial–time problem which can be solved using max–flow or maximum matching techniques.

Finding an algorithm faster than $O^*(2^n)$ for DOMINATING SET was an open problem until 2004. Currently the fastest algorithm by van Rooij et al. runs in $O(1.5048^n)$ time [3]. The problem of solving CAPACITATED DOMINATING SET faster than the check-all-subsets $O^*(2^n)$ time algorithm was posted by van Rooij in 2008 at Dagstuhl seminar [10] and on IWPEC 2008 open problem list. In this paper we present an algorithm providing the positive answer to this question.

Note that at first glance breaking $O^*(2^n)$ barrier for the CAPACITATED DOMINATING SET problem seems a hard task, since even the brute-force $O^*(2^n)$ time algorithm involves matching or max–flow techniques. A standard approach to improving exponential time algorithms is through fixed parameter tractability (see eg. [11]). However from the parameterized point of view, Dom et al. [12] showed that this problem is $W[1]$-hard when parameterized by both treewidth and solution size, and Bodlaender et al. [13] showed that even the planar version parameterized by the solution size is also $W[1]$-hard.

Superpolynomial approximation was recently considered as a way of coping with hardness of approximation of different NP-hard problems. Results in this field include a subexponential approximation algorithm for BANDWIDTH on trees [14] and exponential approximation schemes for CHROMATIC NUMBER [15,16] or BANDWIDTH on arbitrary graphs [17,18].

Our results. In Section 2 we provide an algorithm which solves the CAPACITATED DOMINATING SET problem in $O(1.89^n)$. The algorithm constructs $O^*(\binom{n}{n/3}) = O(1.89^n)$ reductions of the input graph into a CONSTRAINED CAPACITATED DOMINATING SET problem instance (defined in Section 2.1), each solvable in polynomial time. Section 2.3 tackles the exponential approximation of CAPACITATED DOMINATING SET. More precisely, we provide an approximation algorithm that for a given $c \in (0, \frac{1}{3})$, in time

$$O^*\left(\binom{n}{cn}\right) = O^*\left(\left(1 / \left(c^c (1-c)^{1-c}\right)\right)^n\right),$$

computes a $(\frac{1}{4c} + c)$-approximation in the case of $c < \frac{1}{4}$ or a $(2 - 3c)$-approximation in the case of $\frac{1}{4} \le c < \frac{1}{3}$. This result should be compared to the trivial approximation scheme that works in $O^*(\binom{n}{cn})$ time too: iterate over all subsets of V that have size at most cn or at least $(1 - c)n$ and return the smallest feasible solution found. However, this algorithm has an approximation factor of $\frac{1}{c} - 1$, which is between $2\times$ and $4\times$ worse than our ratio. All algorithms in this paper require polynomial space.

Our algorithm for CAPACITATED DOMINATING SET is somewhat similar to one of the first algorithms to break $O^*(2^n)$ for the classical DOMINATING SET problem, namely the algorithm of Randerath and Schiermeyer [19]. Their algorithm also involves matching arguments and our algorithm, applied to DOMINATING SET, can be viewed as a simplification of their algorithm. However we do not know whether their algorithm could be used to solve the CAPACITATED DOMINATING SET problem.

2   CAPACITATED DOMINATING SET

2.1   CONSTRAINED CAPACITATED DOMINATING SET

In this section we introduce a constrained version of the CAPACITATED DOMINATING SET problem, namely the CONSTRAINED CAPACITATED DOMINATING SET problem, which can be solved in polynomial time.

The input of CONSTRAINED CAPACITATED DOMINATING SET is an undirected graph $G = (V, E)$, a set $U \subseteq V$ and a capacity function $c : V \to \{0, \ldots, n - 1\}$. We ask for a smallest capacitated dominating set $S \subseteq V$ containing U such that each vertex outside U dominates at most one other vertex. Formally we ask for a dominating function f_S satisfying

$$|f_S^{-1}(v)| \leq 1 \quad \text{for each} \quad v \in S \setminus U$$
$$|f_S^{-1}(v)| \leq c(v) \quad \text{for each} \quad v \in U \tag{1}$$

Let $G = (V, E)$ with $U \subseteq V$ and a capacity function $c : V \to \{0, \ldots, n - 1\}$ be a CONSTRAINED CAPACITATED DOMINATING SET instance. Consider the following graph $G' = (V', E')$:

- for any $v \in V \setminus U$ we have $v \in V'$;
- for any $v \in U$ we have $c(v)$ copies $v_1, v_2, \ldots, v_{c(v)}$ of v in V';
- for any $v \in V \setminus U$ and $u \in U$ the edge $u_i v \in E'$ for all i iff $uv \in E$;
- for $v, w \in V \setminus U$ we have $vw \in E'$ iff $vw \in E$ and $c(v) + c(w) > 0$;
- there are no edges of the form $v_i w_j$ or $v_i v_j$ for $v, w \in U$.

We show a correspondence between feasible solutions of CONSTRAINED CAPACITATED DOMINATING SET in G and matchings in G'.

Lemma 1. *Let S be a capacitated dominating set in G with a dominating function f_S satisfying condition 1. Then one may construct in polynomial time a matching $\phi(S, f_S)$ in G' satisfying $|\phi(S, f_S)| = |V| - |S|$.*

Proof. Let us define the matching $\phi(S, f_S)$ as follows:

- for each $v \notin S$ such that $f_S(v) \notin U$ add $vf_S(v)$ to $\phi(S, f_S)$;
- for each $v \notin S$ such that $u = f_S(v) \in U$ add vu_i to $\phi(S, f_S)$, where u_i is a copy of u in G' and different copies u_i are chosen for different vertices v with $f_S(v) = u$ (note that $|f_S^{-1}(u)| \leq c(u)$, so there are enough vertices u_i).

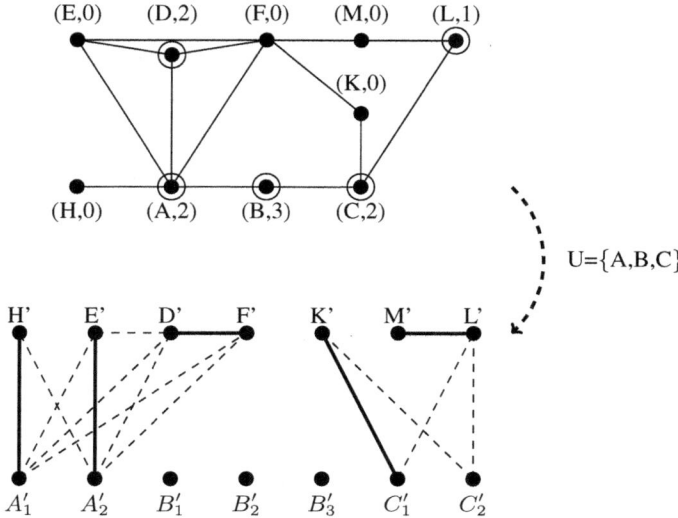

Fig. 1. From the constrained capacitated dominating set $\{A, B, C, D, L\}$ (where $U = \{A, B, C\}$) to a matching. By a pair (X, i) we denote a vertex X with its capacity $c(X) = i$.

Note that every vertex $v \in V \setminus S$ is an endvertex of an edge in the matching $\phi(S, f_S)$. The second endvertex is $f_S(v)$ (in the case $f_S(v) \notin U$) or a copy of $f_S(v)$ in G' (in the case $f_S(v) \in U$). Note that by condition 1, every vertex $w \in S \setminus U$ is an endvertex of at most one chosen edge, thus $\phi(S, f_S)$ is indeed a matching. Moreover, every edge in $\phi(S, f_S)$ has exactly one endpoint in $V \setminus S$. Therefore $|\phi(S, f_S)| = |V| - |S|$.   □

Lemma 2. *Let M be a matching in G'. Then one may construct in polynomial time a feasible solution $\psi(M)$ to the* CONSTRAINED CAPACITATED DOMINATING SET *problem with dominating function $f_{\psi(M)}$ satisfying $|\psi(M)| = |V| - |M|$.*

Proof. Consider the following capacitated dominating set $\psi(M)$ with dominating function $f_{\psi(M)}$:

- $U \subseteq \psi(M)$;
- for $u \in U$ and for each i such that $u_i v \in M$, we take $f_{\psi(M)}(v) = u$;
- for any edge $vw \in M$, where $v, w \notin U$ one of the endpoints (say v) has to satisfy $c(v) > 0$, we add v to $\psi(M)$ and set $f_{\psi(M)}(w) = v$;
- for any $v \notin U$ which is not an endpoint of any edge in M we add v to $\psi(M)$.

It is easy to verify that the above procedure does indeed give a feasible solution to CONSTRAINED CAPACITATED DOMINATING SET. We have $|\psi(M)| = |V| - |M|$ since for each edge in M, exactly one of its endpoints does not belong to $\psi(M)$.   □

We conclude this section with the following theorem.

Theorem 1. *The* CONSTRAINED CAPACITATED DOMINATING SET *problem can be solved in polynomial time.*

Proof. By Lemmas 1 and 2, to find the solution of the CONSTRAINED CAPACITATED DOMINATING SET problem it is enough to find any maximum matching in G', which can be done in polynomial time (see e.g. [20]).

2.2   From CONSTRAINED CAPACITATED DOMINATING SET to CAPACITATED DOMINATING SET

Let us start with the following simple observation. Let S be any capacitated dominating set and let f_S be a dominating function for S. Let

$$U_S = \{v \in S : |f_S^{-1}(v)| \geq 2\}.$$

We have

$$\sum_{v \in S} 1 + |f_S^{-1}(v)| = |S| + \sum_{v \in S} |f_S^{-1}(v)| = n,$$

thus in particular $|U_S| \leq n/3$. Moreover, S with the function f_S is a feasible solution for the CONSTRAINED CAPACITATED DOMINATING SET instance with the graph G and the set U_S. Therefore the following algorithm solves CAPACITATED DOMINATING SET:

1. For each $U \subseteq V$ satisfying $|U| \leq n/3$ solve the CONSTRAINED CAPACITATED DOMINATING SET instance with graph G and subset U.
2. Return the smallest capacitated dominating set from the constructed CONSTRAINED CAPACITATED DOMINATING SET instances.

The CONSTRAINED CAPACITATED DOMINATING SET problem can be solved in polynomial time and there are

$$\sum_{k=0}^{\lceil n/3 \rceil} \binom{n}{k} = O^* \left(\binom{n}{\lceil n/3 \rceil} \right) = O(1.89^n)$$

possible sets U (i.e. sets of cardinality at most $n/3$), thus the whole algorithm works in $O(1.89^n)$ time.

2.3   Approximating CAPACITATED DOMINATING SET

It is known that DOMINATING SET is as hard to approximate as SET COVER and since CAPACITATED DOMINATING SET generalizes DOMINATING SET it is hard to approximate it as well. Thus there probably does not exists a polynomial time algorithm solving CAPACITATED DOMINATING SET with a constant approximation ratio. If we do not have enough time to obtain an exact solution for the CAPACITATED DOMINATING SET problem we can use the following constant approximation scheme. Instead of investigating all subsets $U \subseteq V$ satisfying $|U| \leq n/3$ we can check only smaller sets, namely $|U| \leq cn$ for some constant $c \in (0, \frac{1}{3})$. Thus the approximation algorithm has the following form:

1. For each $U \subseteq V$ satisfying $|U| \leq cn$ solve the CONSTRAINED CAPACITATED DOMINATING SET instance with graph G and subset U.

2. Return the smallest capacitated dominating set from the constructed CONSTRAINED CAPACITATED DOMINATING SET instances.

Theorem 2. *For any fixed constant $c \in (0, \frac{1}{3})$ the described algorithm runs in*

$$O^* \left(\binom{n}{cn} \right) = O^* \left(\left(1 / \left(c^c (1 - c)^{1-c} \right) \right)^n \right)$$

time and polynomial space. For $c \leq 1/4$ the approximation ratio is at most $(\frac{1}{4c} + c)$ and for $c \geq 1/4$ the approximation ratio is at most $2 - 3c$.

Proof. For each subset $U \subseteq V$ the algorithm uses polynomial time only, thus the time bound follows directly from the Stirling formula which can be used to bound the number of subsets $\binom{n}{cn}$.

To calculate the approximation ratio let us consider some optimal solution $OPT \subseteq V$ together with a function $f_{OPT} : V \setminus OPT \to OPT$. By OPT_0, OPT_1 and OPT_2 let us denote subsets of OPT containing vertices which dominate exactly zero, exactly one and at least two vertices from $V \setminus OPT$, according to f_{OPT}, respectively. By m, m_0, m_1, m_2 we denote the cardinalities of sets OPT, OPT_0, OPT_1 and OPT_2 respectively.

We may assume that $m_2 > cn$ since otherwise our algorithm finds the optimal solution. By s let us denote the average number of vertices from $V \setminus OPT$ which a vertex from OPT_2 dominates, i.e. $s = (n - m - m_1)/m_2$. Since our algorithm checks all subsets $U \subseteq V$ satisfying $|U| \leq cn$ it obviously considers the subset $U_0 \subseteq OPT_2$, $|U_0| \leq cn$ containing vertices which dominate the largest number of vertices from $V \setminus OPT$. For this particular subset U_0 note that there exists a feasible solution to CONSTRAINED CAPACITATED DOMINATING SET of size $m + (m_2 - cn)(s - 1)$: we take OPT and for each vertex $v \in OPT_2 \setminus U_0$ we take all but one vertices from $f_{OPT}^{-1}(v)$ since then each vertex outside U_0 dominates at most one other vertex.

Thus the approximation ratio can be bounded by $\alpha = 1 + (m_2 - cn)(s - 1)/m$. Note that if we keep m_2 fixed and increase m_0 and m_1, the approximation ratio decreases — we increase m and decrease s — therefore w.l.o.g. we may assume $m_0 = m_1 = 0$. Denoting $x = n/m_2$, we obtain $\alpha \leq 1 + (1 - cx)(x - 2)$ for $3 \leq x \leq \frac{1}{c}$. The bound for α is a concave function of x with maximum at $x_0 = \frac{1}{2c} + 1$. However when $c \geq \frac{1}{4}$ we have $x_0 \leq 3$. This gives $\alpha \leq \frac{1}{4c} + c$ for $c \leq \frac{1}{4}$ and $\alpha \leq 2 - 3c$ for $\frac{1}{4} \leq c \leq \frac{1}{3}$. □

In Table 1 we gather a few examples of obtained approximation ratios and running times. We compare the running times of our approximation scheme with the running

Table 1. Sample c values, approximation ratios and running times of the algorithm described in Theorem 2 and running times of the trivial approximation algorithm with the same approximation ratio

c	1/3	1/4	1/6	1/10	$0.02506\ldots$
ratio	1 (exact)	5/4	5/3	13/5	10
time	$O(1.89^n)$	$O(1.76^n)$	$O(1.57^n)$	$O(1.39^n)$	$O(1.13^n)$
trivial time	$O^*(2^n)$	$O(1.99^n)$	$O(1.94^n)$	$O(1.81^n)$	$O(1.36^n)$

times of the trivial approximation algorithm needed to obtain the same approximation ratio. This algorithm iterates over all subsets of V that have size at most cn or at least $(1 - c)n$ and returns the smallest feasible solution found.

References

1. Woeginger, G.J.: Exact algorithms for NP-hard problems: A survey. In: Combinatorial Optimization, pp. 185–208 (2001)
2. Fomin, F.V., Grandoni, F., Kratsch, D.: A measure & conquer approach for the analysis of exact algorithms. J. ACM 56(5), 1–32 (2009)
3. van Rooij, J.M.M., Nederlof, J., van Dijk, T.C.: Inclusion/exclusion meets measure and conquer: Exact algorithms for counting dominating sets. In: Fiat, A., Sanders, P. (eds.) ESA 2009. LNCS, vol. 5757, pp. 554–565. Springer, Heidelberg (2009)
4. Björklund, A., Husfeldt, T.: Inclusion–exclusion algorithms for counting set partitions. In: Proc. FOCS'06, pp. 575–582 (2006)
5. Cygan, M., Pilipczuk, M.: Faster exact bandwidth. In: Broersma, H., Erlebach, T., Friedetzky, T., Paulusma, D. (eds.) WG 2008. LNCS, vol. 5344, pp. 101–109. Springer, Heidelberg (2008)
6. Feige, U.: Coping with the NP-hardness of the graph bandwidth problem. In: Halldórsson, M.M. (ed.) SWAT 2000. LNCS, vol. 1851, pp. 10–19. Springer, Heidelberg (2000)
7. Björklund, A., Husfeldt, T., Kaski, P., Koivisto, M.: Fourier meets möbius: fast subset convolution. In: Proc. STOC'07, pp. 67–74 (2007)
8. Nederlof, J.: Fast polynomial-space algorithms using möbius inversion: Improving on steiner tree and related problems. In: Proc. ICALP'09, pp. 713–725 (2009)
9. Amini, O., Fomin, F.V., Saurabh, S.: Counting subgraphs via homomorphisms. In: Albers, S., Marchetti-Spaccamela, A., Matias, Y., Nikoletseas, S., Thomas, W. (eds.) ICALP 2009. LNCS, vol. 5555, pp. 71–82. Springer, Heidelberg (2009)
10. Fomin, F.V., Iwama, K., Kratsch, D.: Moderately exponential time algorithms. Dagstuhl seminar (2008)
11. Niedermeier, R.: Invitation to Fixed-Parameter Algorithms (2006)
12. Dom, M., Lokshtanov, D., Saurabh, S., Villanger, Y.: Capacitated domination and covering: A parameterized perspective. In: Grohe, M., Niedermeier, R. (eds.) IWPEC 2008. LNCS, vol. 5018, pp. 78–90. Springer, Heidelberg (2008)
13. Bodlaender, H., Lokshtanov, D., Penninkx, E.: Planar capacitated dominating set is $W[1]$-hard. In: Proc. IWPEC'09 (to appear, 2009)
14. Feige, U., Talwar, K.: Approximating the bandwidth of caterpillars. In: APPROX-RANDOM, pp. 62–73 (2005)
15. Bourgeois, N., Escoffier, B., Paschos, V.T.: Efficient approximation by "low-complexity" exponential algorithms. Cahier du LAMSADE 271, LAMSADE, Universite Paris-Dauphine (2007)
16. Cygan, M., Kowalik, L., Wykurz, M.: Exponential-time approximation of weighted set cover. Inf. Process. Lett. 109(16), 957–961 (2009)
17. Cygan, M., Pilipczuk, M.: Exact and approximate bandwidth. In: Proc. ICALP'09, pp. 304–315 (2009)
18. Fürer, M., Gaspers, S., Kasiviswanathan, S.P.: An exponential time 2-approximation algorithm for bandwidth. In: Proc. IWPEC'09 (to appear, 2009)
19. Schiermeyer, I.: Efficiency in exponential time for domination-type problems. Discrete Applied Mathematics 156(17), 3291–3297 (2008)
20. Mucha, M., Sankowski, P.: Maximum matchings via gaussian elimination. In: Proc. FOCS'04, pp. 248–255 (2004)

Isomorphism for Graphs of Bounded Feedback Vertex Set Number

Stefan Kratsch and Pascal Schweitzer

Max-Planck-Institut für Informatik, 66123 Saarbrücken, Germany

Abstract. This paper presents an $\mathcal{O}(n^2)$ algorithm for deciding isomorphism of graphs that have bounded feedback vertex set number. This number is defined as the minimum number of vertex deletions required to obtain a forest. Our result implies that GRAPH ISOMORPHISM is fixed-parameter tractable with respect to the feedback vertex set number. Central to the algorithm is a new technique consisting of an application of reduction rules that produce an isomorphism-invariant outcome, interleaved with the creation of increasingly large partial isomorphisms.

1 Introduction

The GRAPH ISOMORPHISM problem is among the few problems in NP for which the complexity is still unknown: Up to now, neither an NP-hardness proof nor an algorithm with provably polynomial running time has appeared. Given two finite graphs G_1 and G_2, the GRAPH ISOMORPHISM problem (GI) asks whether these graphs are structurally equivalent, i.e., whether there exists a bijection from $V(G_1)$, the vertices of G_1, to $V(G_2)$, the vertices of G_2, that preserves the adjacency relationship. Being one of the open problems from Garey and Johnson's list of problems with yet unsettled complexity status [14], the GRAPH ISOMORPHISM problem has been studied extensively throughout the last three decades. During that time, a subexponential-time algorithm for the general problem has been developed by Babai [2]. His algorithm uses a degree reduction method by Zemlyachenko (see [2]) as well as Luks' polynomial-time algorithm for graphs of bounded degree [18]. Schöning's lowness proof [23] showed that GRAPH ISOMORPHISM is not NP-hard, unless the polynomial hierarchy collapses.

Research on GRAPH ISOMORPHISM for restricted graph classes has led to a number of polynomial-time algorithms as well as hardness results. Let us review the known results for classes defined by bounded values of some graph parameter, e.g., graphs of degree bounded by k, from a parameterized point of view. Depending on the parameter GI becomes polynomial-time solvable or it remains GI-complete (i.e., polynomial-time equivalent to GI) even when the parameter is bounded by a constant. The latter is known for bounded chromatic number and bounded chordal deletion number (i.e., number of vertex deletions needed to obtain a chordal graph), since GRAPH ISOMORPHISM is GI-complete for the class of bipartite graphs and the class of chordal graphs (see [27]).

The polynomial results can be split into runtimes of the form $\mathcal{O}(f(k)n^c)$ and $\mathcal{O}(n^{f(k)})$; both are polynomial for bounded k but for the latter the degree

H. Kaplan (Ed.): SWAT 2010, LNCS 6139, pp. 81–92, 2010.
© Springer-Verlag Berlin Heidelberg 2010

Table 1. Upper bounds on the time required to compute some graph parameters as well as running times of the parameterized GRAPH ISOMORPHISM problem. (∗) FPT if the colored $\mathcal{H}$-free isomorphism problem can be solved in polynomial time.

Parameter	Comp. of the parameter		Upper bound for GI	
Chromatic number	3-col. is NP-hard	[14]	GI-hard for $\chi(G) = 2$	
Chordal deletion number	$\mathcal{O}(f(k)n^c)$	[19]	GI-hard for $\mathrm{cvd}(G) = 0$	
Max degree	$\mathcal{O}(m)$		$\mathcal{O}(n^{ck})$	[4]
Genus	$\mathcal{O}(f(k)m)$	[16]	$\mathcal{O}(n^{ck})$	[12,20]
Treewidth	$\mathcal{O}(f(k)n)$	[6]	$\mathcal{O}(n^{k+4.5})$	[5]
Rooted tree distance width	$\mathcal{O}(kn^2)$	[28]	$\mathcal{O}(f(k)n^3)$	[28]
$\mathcal{H}$-free deletion number	$\mathcal{O}(d^k n^d)$	[7]	FPT if (∗)	[Sec. 2]
Feedback vertex set number	$\mathcal{O}(5^k kn^2)$	[8]	$\mathcal{O}((2k + 4k \log k)^k kn^2)$	[Sec. 3]

of the polynomial grows with k. Parameterized complexity (see [9]) studies these function classes in a multivariate analysis of algorithms, motivated by the much better scalability of $\mathcal{O}(f(k)n^c)$ algorithms, so-called *fixed-parameter tractable* algorithms. In the case of GRAPH ISOMORPHISM for a large number of parameters only $\mathcal{O}(n^{f(k)})$ algorithms are known. Such algorithms exist for the parameters degree [18], eigenvalue multiplicity [3], color class size [13], and treewidth [5]. Furthermore, this running time has been shown for the parameter genus [12,20] (extending polynomial-time algorithms for planar graphs [15,24]) and, more general, for the size of an excluded minor [21]. Algorithms of runtime $\mathcal{O}(f(k)n^c)$ are known for the parameters color multiplicity [13], eigenvalue multiplicity [11], rooted tree distance width [28]. For chordal graphs, there is an fpt-algorithm with respect to the size of the simplicial components [26]. The fixed-parameter tractable algorithm for the parameter color multiplicity has recently been extended to hypergraphs [1]. Table 1 summarizes some results for parameterized GRAPH ISOMORPHISM as well as the complexity of computing the parameters.

We develop an $\mathcal{O}(f(k)n^c)$ algorithm for GRAPH ISOMORPHISM parameterized by the feedback vertex set number. The *feedback vertex set number* of a graph G, denoted by $\mathrm{fvs}(G)$, is the size of a smallest subset of vertices S, whose removal leads to a graph that does not contain cycles, i.e., for which $G - S$, the graph induced by the set of vertices $V(G) \setminus S$, is a forest. Our result, a fixed-parameter tractable algorithm, has a running time of $\mathcal{O}(f(k)n^2)$, i.e., it runs in $\mathcal{O}(n^2)$ for graphs of bounded feedback vertex set number.

For a selection of graph parameters, Figure 1 shows the partial order given by the relation stating that a parameter k' is larger than another parameter k, if k can be bounded by a function g of k'. From this it is immediate that if a problem is fixed-parameter tractable (FPT) with respect to some parameter then it is also FPT with respect to any larger (in Figure 1 higher) parameter: time $\mathcal{O}(f(k)n^c)$ with respect to k implies time $\mathcal{O}(f(g(k'))n^c)$ with respect to k' (likewise for runtimes of the form $\mathcal{O}(n^{f(k)})$). The feedback vertex set number, which has been extensively studied in various contexts [8,10,14,22,25], lies above other interesting

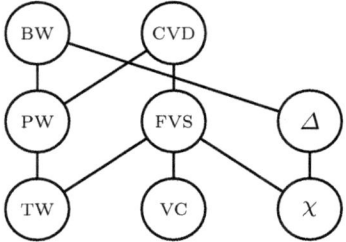

Fig. 1. For various graph parameters, the figure depicts the partial order given by the relation that defines a parameter to be lower than another parameter, if the former can be bounded by a function of the latter. The parameters are: bandwidth (BW), pathwidth (PW), treewidth (TW), size of a minimum vertex cover (VC), size of a minimum feedback vertex set (FVS), vertex deletion distance from chordal graphs (CVD), maximum degree (Δ), and chromatic number (χ).

parameters: As mentioned GI remains hard on graphs of bounded chromatic number, while being polynomially solvable for bounded treewidth. As the rooted tree distance width the feedback vertex set number is a measure for how far a graph is from being a forest. However, these two parameters are incomparable, i.e., neither is bounded by a function of the other.

Our contribution is based on two new techniques: The first makes use of the interplay between deletion sets and small forbidden structures. This is illustrated in Section 2 on the simplified situation where the parameter is the vertex deletion distance to a class of graphs that is characterized by finitely many forbidden induced subgraphs. When we consider the feedback vertex set number in Section 3, the forbidden substructures are cycles, which may be of arbitrary length. The second technique addresses this obstacle by using reduction rules that guarantee short cycles. For the choice of these rules, however, it is crucial that they are compatible with isomorphisms.

2 $\mathcal{H}$-Free Deletion Number

In this section, illustrating the usefulness of deletion sets in the context of GRAPH ISOMORPHISM, we briefly consider the parameter $\mathcal{H}$-free deletion number. For a class $\mathcal{C}$ of graphs we say that a graph G has *vertex deletion distance at most k from $\mathcal{C}$* if there is a *deletion set S* of at most k vertices, for which $G - S \in \mathcal{C}$, i.e., by deleting a most k vertices we obtain a graph in $\mathcal{C}$.

Definition 1. *A class $\mathcal{C}$ of graphs is characterized by finitely many forbidden induced subgraphs, if there is a finite set of graphs $\mathcal{H} = \{H_1, \ldots, H_\ell\}$, such that a graph G is in $\mathcal{C}$ if and only if G does not contain H_i as an induced subgraph for any $i \in \{1, \ldots, \ell\}$. The class $\mathcal{C}$ is called the class of $\mathcal{H}$-free graphs.*

It is known that computing the $\mathcal{H}$-free deletion number k and a corresponding set S of vertices to be removed is FPT with respect to k: There is an algorithm

---

Algorithm 1. IsomorphismIND$_\mathcal{C}$

Input: (G_1, G_2, k): An integer k and two colored graphs G_1, G_2 of distance at most k
      to a fixed class $\mathcal{C}$ that is characterized by a finite set of forbidden induced subgraphs.
Output: An isomorphism ϕ of G_1 and G_2 or **false** if no such isomorphism exists.

> **if** exactly one of G_1 and G_2 contains a forbidden subgraph **return false**
> **if** G_1 and G_2 contain no forbidden subgraphs **then**
>   use an algorithm for colored graph isomorphism for the class $\mathcal{C}$ on G_1 and G_2
>   **return** an isomorphism or **false** if none exists
> 5: **end if**
> **if** $k = 0$ **return false**
> find a forbidden induced subgraph H in G_2
> find a set S of at most k vertices such that $G_1 - S \in \mathcal{C}$
> **for all** $(v_1, v_2) \in S \times V(H)$ **do**
> 10:   **if** v_1 and v_2 are colored with the same color **then**
>   result $\leftarrow$ IsomorphismIND$_\mathcal{C}(G_1 \triangleright v_1, G_2 \triangleright v_2, k - 1)$
>   **if** result $\neq$ **false return** result
>   **end if**
> **end for**
> 15: **return false**

---

with runtime $\mathcal{O}(d^k n^d)$ where d is the number of vertices of the largest forbidden induced subgraph, following a more general result due to Cai [7].

For an FPT-algorithm that solves GRAPH ISOMORPHISM parameterized by the $\mathcal{H}$-free deletion number, we require a method of consistently removing vertices from the graph: Let G be a colored graph, c a vertex coloring of G (not necessarily a proper coloring), and v a vertex of G. We define $G \triangleright v$ to be the colored graph induced by the vertex set $V(G) \setminus \{v\}$ with the coloring given by $(\tau(v, v'), c(v'))$ for all $v' \in V(G) \setminus \{v\}$, where the edge characteristic function $\tau(v, v')$ is 1 if v and v' are adjacent and 0 otherwise. Intuitively, the new coloring encodes at the same time whether in the original graph a vertex is adjacent to v as well as its previous color. In particular, we get the following observation: Suppose that G_1 and G_2 are colored graphs and that v_1 and v_2 are equally colored vertices of G_1 and G_2 respectively, then there is an isomorphism ϕ with $\phi(v_1) = v_2$ if and only if $G_1 \triangleright v_1$ and $G_2 \triangleright v_2$ are isomorphic as colored graphs (where isomorphisms must respect colors).

Theorem 1. *Let some graph class $\mathcal{C}$ be characterized by forbidden induced subgraphs $H_1, \ldots, H_\ell$. If the colored graph isomorphism problem for graphs from $\mathcal{C}$ is in P, then the colored graph isomorphism problem, parameterized by the vertex deletion distance from $\mathcal{C}$, is fixed-parameter tractable.*

Proof. W.l.o.g., both input graphs G_1 and G_2 have distance of at most k from $\mathcal{C}$. Algorithm 1 repeatedly generates a set of candidate pairs of vertices $P = S \times V(H) \subseteq V(G_1) \times V(G_2)$, where S is a minimum deletion set of G_1 and $V(H)$ is the vertex set of a forbidden induced subgraph H in G_2. For any isomorphism ϕ, this ensures that P contains a pair (v_1, v_2) with $\phi(v_1) = v_2$. Indeed, the image $\phi(S)$ is a deletion set of G_2, thereby also intersecting H. The algorithm then makes a recursive call for each choice of $(v_1, v_2) \in P$, removing the vertices v_1 and v_2 from the graphs and correctly coloring the remaining vertices. In the base case, when $G_1, G_2 \in \mathcal{C}$, isomorphism is decided using the polynomial-time algorithm for the class $\mathcal{C}$.

Observe that P has size at most dk where d is the size of the largest graph in $\{H_1, \ldots, H_\ell\}$. Thus there are $\mathcal{O}((dk)^k)$ recursive calls, since k decreases with each call and $k = 0$ terminates a branch. Together with the polynomial-time algorithm for the base case and the FPT-algorithm for distance from $\mathcal{C}$ [7] this gives an $\mathcal{O}(f(k)n^c)$ runtime, proving fixed-parameter tractability.   □

3   Feedback Vertex Set Number

In this section we consider the GRAPH ISOMORPHISM problem parameterized by the feedback vertex set number. Similarly to Section 2 we compute a set that intersects all forbidden structures of the first graph (in our case a feedback vertex set). The image of that set under any isomorphism must intersect every forbidden structure of the second graph (i.e., it must intersect every cycle). To efficiently use this fact, we choose a shortest cycle in the second graph. However, since in general shortest cycles may be of logarithmic size, we perform a sequence of reductions to shorten the cycles. The reduction rules delete all vertices of degree at most one as well as those in a specified set S, and contract vertices of degree two; these are standard reductions for computing feedback vertex sets. Additionally, there is a new rule resulting in the deletion of all components containing at most one cycle. This rule allows us to prove the crucial fact, that exhaustive reduction of graphs behaves well with respect to isomorphism. In order to make this precise, we first show that the result of exhaustively applying the reduction rules is independent of the order in which they are applied.

Lemma 1. *Let G be a graph and let S be a set. Exhaustive application of the following reduction rules in any order has a well-defined result $R_S(G)$, which is a specific graph on a subset of $V(G)$.*

1. *Delete a vertex of degree at most one.*
2. *Delete a vertex in a connected component containing at most one cycle.*
3. *Delete a vertex that is contained in S.*
4. *Contract a vertex of degree two that is not contained in S, i.e., replace the vertex by an edge between its former neighbors; this may create multi-edges and loops.*

Proof. For any graph G and any set S let $L_S(G)$ denote the maximum number of reduction steps that can be applied to G (using S for Rules 3 and 4).

We assume for contradiction that there is a counterexample consisting of a graph G and a set S with minimum value of $L_S(G)$. Let R_1 and R_2 be two maximal sequences of reduction steps for G that yield different results. For $i \in \{1, 2\}$ let v_i be the first vertex reduced by R_i and let G_i be the result of that first step. Observe that $L_S(G_1) < L_S(G)$ and $L_S(G_2) < L_S(G)$, implying that $R_S(G_1)$ and $R_S(G_2)$ are well-defined, by our choice of G.

It suffices for us to show that we can reduce v_2 in G_1 and v_1 in G_2 such that we obtain the same graph G' on the same subset of $V(G)$: Indeed since any further exhaustive reduction has the same outcome, this implies $R_S(G_1) = R_S(G') = R_S(G_2)$ since the result of any maximal sequence of reductions on either G_1 and G_2 is well-defined, and yields the desired contradiction.

The deletion rules (Rules 1–3) are such that a vertex that may be deleted in a graph G may also be deleted in any subgraph of G. Therefore, if *both vertices are deleted from* G, then v_2 can be deleted from G_1 and v_1 can be deleted from G_2; we obtain $G' = G - \{v_1, v_2\}$.

Otherwise *w.l.o.g.* v_1 *is contracted in* G; there are three cases:

1. If v_2 *is not adjacent to* v_1 then the reductions are independent and reducing v_1 in G_2 and v_2 in G_1 yields the same graph G' on the vertex set $V(G) \setminus \{v_1, v_2\}$.
2. If v_2 *is contracted and adjacent to* v_1, then there is a path (u, v_1, v_2, w) and contracting v_1 and v_2 in any order is equivalent to replacing the path by an edge $\{u, w\}$, reducing both graphs to the same graph G' on the vertex set $V(G) \setminus \{v_1, v_2\}$.
3. If v_2 *is deleted and adjacent to* v_1, then the degree of v_2 in G_1 is the same as in G, therefore it can still be deleted. In G_2 the vertex v_1 has degree at most 1, implying that it can be deleted by Rule 1. Both reductions lead to the same graph $G' = G - \{v_1, v_2\}$.   □

We observe that, for any graph G and any set S, the graph $R_S(G)$ has minimum degree at least three and $\mathrm{fvs}(R_S(G)) \leq \mathrm{fvs}(G)$. Concerning the latter, it suffices to observe that vertex deletions do not increase the feedback vertex set number, and that any degree-2-vertex of a feedback vertex set may be replaced by either neighbor while preserving the property of being a feedback vertex set. We denote by $R(G) := R_{\{\}}(G)$ the special case that S is the empty set. Since the result is independent of the order in which the rules are applied, the vertices from the set S may be removed first, i.e., $R_S(G) = R(G - S)$.

As a corollary we conclude that the reduction R maintains isomorphisms.

Corollary 1. *Let ϕ be an isomorphism of graphs G_1 and G_2 and let $S \subseteq V(G_1)$. Then $R_S(G_1)$ and $R_{\phi(S)}(G_2)$ are isomorphic and ϕ restricted to $V(R_S(G_1))$ is an isomorphism from $R_S(G_1)$ to $R_{\phi(S)}(G_2)$.*

The reduction rule that allows vertex deletion in unicyclic components, which is necessary to obtain Corollary 1, has the effect that the set S does not need to be a feedback vertex set, in order for the reduced graph $R_S(G)$ to become empty. As a consequence, removal of such a set S does not necessarily leave a forest, but a graph that may contain a cycle in every component.

Definition 2. *An OC graph is a graph in which every component contains at most one cycle.*

The OC graphs are precisely the graphs which are reduced to the empty graph by repeated application of the reduction rules:

Lemma 2. *A graph G is an OC graph if and only if $R(G)$ is the empty graph.*

Proof. To show that for an arbitrary OC graph G the reduced graph $R(G)$ is the empty graph we assume, w.l.o.g., that the graph is connected. Thus G contains at most one cycle. We claim that after repeatedly removing all vertices of degree at most 1, the graph is empty, or is a cycle: Indeed, suppose v is a vertex in $V(G)$ not contained in a cycle, and v is not removed by the reductions, then in the reduced graph v has degree at least 2, and the longest path through v in the reduced graph ends on one side with a vertex of degree 1, a contradiction. Finally by induction every cycle reduces to the empty graph.

Conversely, to show that a graph which reduces to the empty graph is an OC graph, it suffices to show that any graph which contains two cycles in one component does not reduce to the empty graph. The minimal connected graphs that contain two cycles are the dumbells (i.e., two cycles joined by a path) and the Theta graphs (i.e., two vertices connected by three vertex disjoint paths). By induction they do not reduce to the empty graph, and the lemma follows.   □

For reduced graphs, we can use a nice structural result by Raman et al. [22], stating that graphs of minimum degree at least three must have a large feedback vertex set number or a cycle of length at most six. Thus, in contrast to the general bound of $\log n$ on the girth of a graph, there are few choices for the image of any feedback vertex under an isomorphism between two reduced graphs.

Theorem 2 ([22]). *Let G be a graph on n vertices with minimum degree at least three and of feedback vertex set number at most k. If $n > 2k^2$ then G has a cycle of length at most six.*

The algorithm that we present later branches on possible partial isomorphisms; using Theorem 2 and our reductions the number of choices is reasonably small. On termination there are pairs of vertices that the isomorphism shall respect and removal of those vertices followed by reduction yields two empty graphs. Hence, deletion of the vertices yields two OC graphs. This leaves us with the task of deciding isomorphism for OC graphs, with the restriction that adjacencies with the deleted vertices must be correct. For that purpose we first define OC+k graphs and corresponding isomorphisms.

Definition 3. *A graph with at most one cycle per component plus k distin-guished vertices (OC+k graph) consists of a graph and a k-tuple of its vertices with the property that deletion of those distinguished vertices yields an OC graph.*

An isomorphism of two OC+k graphs is an ordinary isomorphism that maps the k distinguished vertices of one graph to those of the other respecting the order. If the graph is (vertex) colored, then as usual the isomorphism has to respect these colors.

The restriction on the mapping of the distinguished vertices allows efficient iso-morphism testing, mainly requiring an isomorphism test of colored OC graphs.

Theorem 3. GRAPH ISOMORPHISM *for colored OC+k graphs can be solved in $\mathcal{O}(n^2)$ time.*

Proof. We first reduce the problem to colored OC graphs. It suffices to reduce the problem for $k \geq 1$ to the isomorphism problem of OC+$(k-1)$ graphs in $\mathcal{O}(n)$ time. Let G and G' be two given colored OC+k graphs with last distinguished vertex v_k and v'_k respectively. If v_k and v'_k are not equally colored, then the graphs are not isomorphic. Otherwise, as argued in the previous section, the graphs $G_1 \triangleright v_k$ and $G_2 \triangleright v'_k$ are isomorphic, if and only if G_1 and G_2 are isomorphic. The recoloring and the vertex deletion of the reduction require $\mathcal{O}(n)$ time.

We are left with determining the isomorphism of colored OC graphs G_1 and G_2. We assign every vertex v, neighboring a leaf and contained in a component with at least 3 vertices, a color that depends on the multiset of colors of leaf neighbors of the vertex v. We then delete all leaves in these components. Again, the obtained graphs are isomorphic if and only if they were isomorphic prior to the reduction. After this, we rename (in both graphs consistently) the new colors with unused integers in $\{1, \ldots, n\}$ by sorting. By repeated application we obtain graphs in which every component is a cycle or contains at most 2 vertices. This step can be performed in an amortized time of $\mathcal{O}(n \log(n))$, charging the sorting to the removed leaves.

Counting for each isomorphism type the number of components with at most two vertices, it suffices now to determine the isomorphism of disjoint unions of colored cycles. There are at most n such cycles in an OC graph. We solve this task using a string matching algorithm: A colored cycle $\langle c_1, c_2, \ldots, c_n \rangle$ is isomorphic to $\langle c'_1, c'_2, \ldots, c'_{n'} \rangle$ if and only if $n = n'$ and the string $c'_1 c'_2 \ldots c'_n$ or its inverse $c'_n c'_{n-1} \ldots c'_1$ is contained as a substring in the string $c_1 c_2 \ldots c_n c_1 c_2 \ldots c_n$. We repeatedly search two color-isomorphic cycles from each graph and remove them. By employing a linear time string matching algorithm, like the Knuth-Morris-Pratt algorithm [17], we obtain a total running time of $\mathcal{O}(n^2)$.   □

Knowing how to efficiently decide isomorphism of OC+k graphs, we work towards an algorithm that creates sets of pairs of distinguished vertices, such that each isomorphism of the given graphs must respect one of the sets. To that end we show that one can easily compute a set of candidate pairs such that for any isomorphism ϕ of G_1 and G_2 one of the pairs (v_1, v_2) satisfies $\phi(v_1) = v_2$.

Lemma 3. *Let G_1 and G_2 be two graphs of feedback vertex set number at most k and minimum degree at least three. In time $\mathcal{O}(5^k k n^2)$ one can compute a set $P \subseteq V(G_1) \times V(G_2)$ of size at most $2k + 4k \log k$ such that (if G_1 and G_2 are isomorphic) for any isomorphism ϕ there is a pair $(v_1, v_2) \in P$ such that $\phi(v_1) = v_2$ and v_1 is contained in a minimum feedback vertex set of G_1.*

Proof. We choose P as $S \times V(C)$ where S is a minimum feedback vertex set of G_1 and C is a shortest cycle of G_2. The time for this computation is dominated by $\mathcal{O}(5^k k n^2)$ for computing a k-feedback vertex set of G_1.

If $|V(C)| \leq 6$ then P has size at most $6k$. Otherwise, by Theorem 2, G_2 has at most $2k^2$ vertices. Since the girth of any graph with minimum degree 3 is at most $2 \log n$, the cycle C has length at most $2 \log(2k^2) = 2 + 4 \log k$. Thus P contains at most $2k + 4k \log k$ pairs. Note that $k \geq 2$ for graphs with minimum degree 3 and that $6k \leq 2k + 4k \log k$ for $k \geq 2$.

Algorithm 2. IsomorphismFVS

Input: (G_1, G_2, k, FP): Two graphs G_1, G_2, a potential partial isomorphism given by pairs of vertices in $\text{FP} \subseteq V(G_1) \times V(G_2)$, and an integer k such that the feedback vertex set number of $G_1 - S_1$ is at most k, where S_1 is the set of first components of the pairs in FP.

Output: An isomorphism ϕ of G_1 and G_2 that respects FP or **false** if none exists.

    let $(v_{11}, v_{21}), \ldots, (v_{1r}, v_{2r})$ denote the pairs in FP

    $G_1' \leftarrow R(G_1 - \{v_{11}, \ldots, v_{1r}\})$

    $G_2' \leftarrow R(G_2 - \{v_{21}, \ldots, v_{2r}\})$

    if if exactly one of G_1' and G_2' is empty **return false**

5: **if** G_1' and G_2' are empty **then**

       use the algorithm from Theorem 3 on $(G_1, (v_{11}, \ldots, v_{1r}))$ and $(G_2, (v_{21}, \ldots, v_{2r}))$

       return an isomorphism ϕ that respects FP or **false** if none exists

    end if

    if $k = 0$ **return false**

10: compute a set P of candidate pairs according to Lemma 3 for G_1', G_2', and k

    for $(v_1, v_2) \in P$ **do**

       result $\leftarrow$ IsomorphismFVS$(G_1, G_2, k - 1, \text{FP} \cup \{(v_1, v_2)\})$

       if result $\neq$ **false return** result

    end for

15: **return false**

For any isomorphism ϕ from G_1 to G_2 the image $\phi(S)$ must intersect C since $\phi(S)$ is a feedback vertex set of G_2. Hence P contains a pair (v_1, v_2) with $\phi(v_1) = v_2$ as claimed. □

We now design an FPT-algorithm that solves GRAPH ISOMORPHISM parameterized with the feedback vertex set number. Algorithm 2 performs this task in the following way: Given two graphs of feedback vertex set number at most k, it recursively computes an increasingly large partial isomorphism, given by a set of pairs of vertices $\text{FP} \subseteq V(G_1) \times V(G_2)$. This set indicates that, should an isomorphism exist, there is an isomorphism that maps the first vertex of each pair in FP to the corresponding second vertex. The first components are chosen as to be part of a minimal feedback vertex set in the first graph. At the latest when the set FP has reached size k, removal of the vertices in each graph will result in an OC graph each. Isomorphism can then be decided with the algorithm described in Theorem 3.

Definition 4. *We say that an an isomorphism $\phi\colon V(G_1) \to V(G_2)$ respects a set of pairs of vertices $\text{FP} \subseteq V(G_1) \times V(G_2)$, if $\phi(v_1) = v_2$ for all $(v_1, v_2) \in \text{FP}$.*

Given two isomorphic graphs, Algorithm 2 computes an isomorphism:

Lemma 4. *Let G_1 and G_2 be two isomorphic graphs. Suppose $\text{FP} \subseteq V(G_1) \times V(G_2)$ with $|\text{FP}| = r$. Further suppose that there is an isomorphism ϕ from G_1 to G_2 that respects FP and that the feedback vertex set number of $R(G_1 - S_1)$ is at most k, where S_1 is the set of first components of the pairs in FP. Then the call IsomorphismFVS(G_1, G_2, k, FP) will compute an isomorphism from G_1 to G_2 that respects FP.*

Proof. With FP $= \{(v_{11}, v_{21}), \ldots, (v_{1r}, v_{2r})\}$ we define $S_1 = \{v_{11}, \ldots, v_{1r}\}$ as well as $S_2 = \{v_{21}, \ldots, v_{2r}\}$. Let $G'_1 = R(G_1 - S_1)$ and let $G'_2 = R(G_2 - S_2)$. Since ϕ respects FP, it can be restricted to an isomorphism from $G_1 - S_1$ to $G_2 - S_2$. These graphs are therefore isomorphic and, by Corollary 1, the reduced graphs G'_1 and G'_2 are isomorphic, under the restriction of ϕ to $V(G'_1)$.

We show the lemma by induction on k: If $k = 0$, the base case, then $G'_1 = R(G_1 - S_1)$ is empty since it has feedback vertex set number $k = 0$. The isomorphic graph G'_2 is also empty. The graphs $G_1 - S_1$ and $G_2 - S_2$ are OC graphs by Lemma 2. Therefore the graphs $(G_1, (v_{11}, \ldots, v_{1r}))$ and $(G_2, (v_{21}, \ldots, v_{2r}))$ are OC+r graphs and ϕ is an isomorphism of OC+r graphs. Thus the call to the algorithm from Theorem 3 will return an isomorphism that respects FP.

If $k > 0$, we distinguish two cases: Either both G'_1 and G'_2 are empty, in which case we argue as in the base case, or the algorithm computes P for G'_1, G'_2, and k according to Lemma 3. In the set P, since G'_1 and G'_2 are isomorphic (and nonempty), by Lemma 3, there must be a pair $(v_1, v_2) \in P$ such that $\phi(v_1) = v_2$. Lemma 3 additionally guarantees that there must be a feedback vertex set of G'_1 of size at most k that contains v_1, implying that the feedback vertex set number of $R(G'_1 - v_1)$ is at most $k - 1$; by Lemma 1 this extends to $R_{S_1 \cup \{v_1\}}(G_1) = R_{\{v_1\}}(G'_1) = R(G'_1 - v_1)$.

Thus the call IsomorphismFVS$(G_1, G_2, k - 1, \text{FP} \cup \{(v_1, v_2)\})$ has the property that the isomorphism ϕ respects FP $\cup \{(v_1, v_2)\}$ and fvs$(R(G_1 - (S_1 \cup \{v_1\}))) \leq k - 1$. Hence, by induction, it returns an isomorphism ϕ' that respects FP $\cup \{(v_1, v_2)\}$. Thus the call IsomorphismFVS(G_1, G_2, k, FP) returns an isomorphism that respects FP, as claimed. $\square$

The fact that the isomorphism tests for the OC+k graphs are performed in the original input graphs ensures that, even though the reduction R may alter non-isomorphic graphs to be isomorphic (e.g., non-isomorphic trees are reduced to empty graphs), false positives are detected. For this purpose, the partial isomorphism map, encoded by the set FP, has to be maintained by the algorithm, for which Corollary 1 guarantees that it can be lifted into the original graphs and extended to an isomorphism, should it initially arise from an isomorphism.

Theorem 4. GRAPH ISOMORPHISM*(fvs) is fixed-parameter tractable.*

Proof. Let G_1 and G_2 be two graphs of feedback vertex set number at most k. We show that IsomorphismFVS$(G_1, G_2, k, \emptyset)$ correctly determines whether G_1 and G_2 are isomorphic and takes $\mathcal{O}((2k + 4k \log k)^k k n^2)$ time.

The algorithm of Theorem 3 will only return valid isomorphisms. Furthermore, IsomorphismFVS will always find an isomorphism if the given graphs are isomorphic, by Lemma 4. It thus suffices to show that the algorithm terminates in the stated time independent of the outcome.

The call IsomorphismFVS$(G_1, G_2, k, \emptyset)$ leads to a recursive computation of depth at most k. The number of recursive calls is limited by the size of P, computed according to Lemma 3, which is bounded by $2k + 4k \log k$. The computation at each internal node of the branching tree is dominated by the time necessary for generating P, i.e., by $\mathcal{O}(5^k k n^2)$. The calls to the algorithm of Theorem 3 take time $\mathcal{O}(n^2)$. This gives the runtime recurrence $T(k) \leq (2k + 4k \log k)T(k - 1) + \mathcal{O}(5^k k n^2)$, which gives a bound of $\mathcal{O}((2k + 4k \log k)^k k n^2)$.

We conclude that IsomorphismFVS, called as IsomorphismFVS$(G_1, G_2, k, \emptyset)$, is an FPT-algorithm that decides whether G_1 and G_2 are isomorphic. $\square$

Algorithm 2 is also an FPT-algorithm for *colored* graphs with parameter the minimum size of a FVS. For this observe that while Lemma 4 is stated for uncolored graphs it guarantees that for any isomorphism the algorithm finds a corresponding set FP. The algorithm of Theorem 3 will then guarantee that the computed isomorphism respects the colors.

4   Conclusion and Open Problems

We have shown that GRAPH ISOMORPHISM is fixed-parameter tractable with respect to the feedback vertex set number. The feedback vertex set number resides above parameters such as the chromatic and the chordal deletion number, with respect to which GRAPH ISOMORPHISM is not fixed-parameter tractable, unless it may be solved in polynomial time in general. The feedback vertex set number also resides above the parameter treewidth, with respect to which fixed-parameter tractability remains a challenging open problem. In that direction the parameters pathwidth or bandwith are possible further steps to show fixed-parameter tractability of GI with respect to treewidth. Note, that a limited bandwidth simultaneously benefits from a limited treewidth and a limited maximum degree. However even with respect to bandwidth GI might not be fixed-parameter tractable. Showing this, may require a notion of hardness that replaces W[1]-hardness for the not necessarily NP-hard problem GRAPH ISOMORPHISM (W[1] is a parameterized analogue of NP). The reason for this is that prevalent lower bounds from parameterized complexity, such as W[1]-hardness of GI with respect to some parameter, imply that GI is not in P unless FPT=W[1].

References

1. Arvind, V., Das, B., Köbler, J., Toda, S.: Colored hypergraph isomorphism is fixed parameter tractable. ECCC 16(093) (2009)
2. Babai, L.: Moderately exponential bound for graph isomorphism. In: FCT, pp. 34–50. Springer, Heidelberg (1981)
3. Babai, L., Grigoryev, D.Y., Mount, D.M.: Isomorphism of graphs with bounded eigenvalue multiplicity. In: STOC, pp. 310–324. ACM, New York (1982)
4. Babai, L., Luks, E.M.: Canonical labeling of graphs. In: STOC, pp. 171–183. ACM, New York (1983)
5. Bodlaender, H.L.: Polynomial algorithms for graph isomorphism and chromatic index on partial k-trees. Journal of Algorithms 11(4), 631–643 (1990)
6. Bodlaender, H.L.: A linear-time algorithm for finding tree-decompositions of small treewidth. SIAM Journal on Computing 25(6), 1305–1317 (1996)
7. Cai, L.: Fixed-parameter tractability of graph modification problems for hereditary properties. Information Processing Letters 58(4), 171–176 (1996)

8. Chen, J., Fomin, F.V., Liu, Y., Lu, S., Villanger, Y.: Improved algorithms for feedback vertex set problems. Journal of Computer and System Sciences 74(7), 1188–1198 (2008)

9. Downey, R.G., Fellows, M.R.: Parameterized Complexity (Monographs in Computer Science). Springer, Heidelberg (1998)

10. Enciso, R., Fellows, M.R., Guo, J., Kanj, I.A., Rosamond, F.A., Suchý, O.: What makes equitable connected partition easy. In: Chen, J., Fomin, F.V. (eds.) IWPEC 2009. LNCS, vol. 5917, pp. 122–133. Springer, Heidelberg (2009)

11. Evdokimov, S., Ponomarenko, I.N.: Isomorphism of coloured graphs with slowly increasing multiplicity of jordan blocks. Combinatorica 19(3), 321–333 (1999)

12. Filotti, I.S., Mayer, J.N.: A polynomial-time algorithm for determining the isomorphism of graphs of fixed genus. In: STOC, pp. 236–243. ACM, New York (1980)

13. Furst, M.L., Hopcroft, J.E., Luks, E.M.: Polynomial-time algorithms for permutation groups. In: FOCS, pp. 36–41. IEEE, Los Alamitos (1980)

14. Garey, M.R., Johnson, D.S.: Computers and Intractability: A Guide to the Theory of NP-Completeness. W. H. Freeman, New York (1979)

15. Hopcroft, J.E., Wong, J.K.: Linear time algorithm for isomorphism of planar graphs. In: STOC, pp. 310–324. ACM, New York (1974)

16. Kawarabayashi, K., Mohar, B., Reed, B.A.: A simpler linear time algorithm for embedding graphs into an arbitrary surface and the genus of graphs of bounded tree-width. In: FOCS, pp. 771–780. IEEE, Los Alamitos (2008)

17. Knuth, D.E., Morris Jr., J.H., Pratt, V.R.: Fast pattern matching in strings. SIAM Journal on Computing 6(2), 323–350 (1977)

18. Luks, E.M.: Isomorphism of graphs of bounded valence can be tested in polynomial time. Journal of Computer and System Sciences 25(1), 42–65 (1982)

19. Marx, D.: Chordal deletion is fixed-parameter tractable. In: Fomin, F.V. (ed.) WG 2006. LNCS, vol. 4271, pp. 37–48. Springer, Heidelberg (2006)

20. Miller, G.L.: Isomorphism testing for graphs of bounded genus. In: STOC, pp. 225–235. ACM, New York (1980)

21. Ponomarenko, I.N.: The isomorphism problem for classes of graphs closed under contraction. Journal of Mathematical Sciences 55(2), 1621–1643 (1991)

22. Raman, V., Saurabh, S., Subramanian, C.R.: Faster fixed parameter tractable algorithms for finding feedback vertex sets. ACM Transactions on Algorithms 2(3), 403–415 (2006)

23. Schöning, U.: Graph isomorphism is in the low hierarchy. Journal of Computer and System Sciences 37(3), 312–323 (1988)

24. Tarjan, R.E.: A V^2 algorithm for determining isomorphism of planar graphs. Information Processing Letters 1(1), 32–34 (1971)

25. Thomassé, S.: A quadratic kernel for feedback vertex set. In: SODA, pp. 115–119. SIAM, Philadelphia (2009)

26. Toda, S.: Computing automorphism groups of chordal graphs whose simplicial components are of small size. IEICE Transactions 89-D(8), 2388–2401 (2006)

27. Uehara, R., Toda, S., Nagoya, T.: Graph isomorphism completeness for chordal bipartite graphs and strongly chordal graphs. Discrete Applied Mathematics 145(3), 479–482 (2005)

28. Yamazaki, K., Bodlaender, H.L., de Fluiter, B., Thilikos, D.M.: Isomorphism for graphs of bounded distance width. Algorithmica 24(2), 105–127 (1999)

On Feedback Vertex Set
New Measure and New Structures[*]

Yixin Cao[1], Jianer Chen[1], and Yang Liu[2]

[1] Department of Computer Science and Engineering
Texas A&M University
{yixin,chen}@cse.tamu.edu
[2] Department of Computer Science
University of Texas - Pan American
yliu@cs.panam.edu

Abstract. We study the parameterized complexity of the FEEDBACK VERTEX SET problem (FVS) on undirected graphs. We approach the problem by considering a variation of it, the DISJOINT FEEDBACK VERTEX SET problem (DISJOINT-FVS), which finds a disjoint feedback vertex set of size k when a feedback vertex set of a graph is given. We show that DISJOINT-FVS admits a small kernel, and can be solved in polynomial time when the graph has a special structure that is closely related to the maximum genus of the graph. We then propose a simple branch-and-search process on DISJOINT-FVS, and introduce a new branch-and-search measure. The branch-and-search process effectively reduces a given graph to a graph with the special structure, and the new measure more precisely evaluates the efficiency of the branch-and-search process. These algorithmic, combinatorial, and topological structural studies enable us to develop an $O(3.83^k kn^2)$ time parameterized algorithm for the general FVS problem, improving the previous best algorithm of time $O(5^k kn^2)$ for the problem.

1 Introduction

All graphs in our discussion are supposed to be undirected. A *feedback vertex set* (FVS) F in G is a set of vertices in G whose removal results in an acyclic graph. The problem of finding a minimum feedback vertex set in a graph is one of the classical NP-complete problems [16]. The history of the problem can be traced back to early '60s. For several decades, many different algorithmic approaches were tried on this problem, including approximation algorithms, linear programming, local search, polyhedral combinatorics, and probabilistic algorithms (see the survey [10]). There are also exact algorithms finding a minimum FVS in a graph of n vertices in time $\mathcal{O}(1.9053^n)$ [21] and in time $\mathcal{O}(1.7548^n)$ [11].

An important application of the FVS problem is *deadlock recovery* in operating systems [23], in which a deadlock is presented by a cycle in a *system resource-allocation graph* G. Thus, to recover from deadlocks, we need to abort a set of

---

[*] Supported in part by the US NSF under the Grants CCF-0830455 and CCF-0917288.

H. Kaplan (Ed.): SWAT 2010, LNCS 6139, pp. 93–104, 2010.

processes in the system, i.e., to remove a set of vertices in the graph G, so that all cycles in G are broken. Equivalently, we need to find an FVS in G.

In a practical system resource-allocation graph G, it can be expected that the size k of the minimum FVS in G, i.e., the number of vertices in the FVS, is fairly small. This motivated the study of the parameterized version of the problem, which we will name FVS: given a graph G and a parameter k, either construct an FVS of size bounded by k in G or report no such an FVS exists. Parameterized algorithms for the FVS problem have been extensively investigated that find an FVS of k vertices in a graph of n vertices in time $f(k)n^{O(1)}$ for a fixed function f (thus, the algorithms become practically efficient when the value k is small). The first group of parameterized algorithms for FVS was given by Bodlaender [2] and by Downey and Fellows [8]. Since then a chain of dramatic improvements was obtained by different researchers (see Figure 1).

Authors	Complexity	Year
Bodlaender[2] Downey and Fellows [8]	$O(17(k^4)!n^{O(1)})$	1994
Downey and Fellows [9]	$O((2k+1)^k n^2)$	1999
Raman et al.[20]	$O(\max\{12^k, (4\log k)^k\}n^{2.376})$	2002
Kanj et al.[15]	$O((2\log k + 2\log\log k + 18)^k n^2)$	2004
Raman et al.[19]	$O((12\log k/\log\log k + 6)^k n^{2.376})$	2006
Guo et al.[14]	$O((37.7)^k n^2)$	2006
Dehne et al.[7]	$O((10.6)^k n^3)$	2005
Chen et al.[5]	$O(5^k kn^2)$	2008
This paper	$O(3.83^k kn^2)$	2010

Fig. 1. The history of parameterized algorithms for the unweighted FVS problem

Randomized parameterized algorithms have also been studied for the problem. The best randomized parameterized algorithm for the problems is due to Becker et al. [1], which runs in time $O(4^k kn^2)$.

The main result of the current paper is an algorithm that solves the FVS problem. The running time of our algorithm is $O(3.83^k kn^2)$. This improves a long chain of results in parameterized algorithms for the problem. We remark that the running time of our (deterministic) algorithm is even faster than that of the previous best randomized algorithm for the problem as given in [1].

Our approach, as some of the previous ones, is to study a variation of the FVS problem, the DISJOINT FEEDBACK VERTEX SET problem (DISJOINT-FVS), which finds a disjoint feedback vertex set of size k in a graph G when a feedback vertex set of G is given. Our significant contribution to this research includes:

1. A new technique that produces a kernel of size $3k$ for the DISJOINT-FVS problem, and improves the previous best kernel of size $4k$ for the problem [7]. The new kernelization technique is based on a branch and search algorithm for the problem, which is, to our best knowledge, the first time used in the literature of kernelization;

2. A polynomial time algorithm that solves the DISJOINT-FVS problem when the input graph has a special structure;
3. A branch and search process that effectively reduces an input instance of DISJOINT-FVS to an instance of the special structure as given in 2;
4. A new measure that more precisely evaluates the efficiency of the branch and search process in 3;
5. A new algorithm for the FVS problem that significantly improves previous algorithms for the problem.

Due to space limitations, we omit some proofs and refer interested readers to the extended version of the current paper [3].

2   DISJOINT-FVS and Its kernel

We start with a precise definition of our problem.

DISJOINT-FVS. Given a graph $G = (V, E)$, an FVS F in G, and a parameter k, either construct an FVS F' of size k in G such that $F' \subseteq V \setminus F$, or report that no such an FVS exists.

Let $V_1 = V \setminus F$. Since F is an FVS, the subgraph induced by V_1 must be a forest. Moreover, if the subgraph induced by F is not a forest, then it is impossible to have an FVS F' in G such that $F' \subseteq V \setminus F$. Therefore, an instance of DISJOINT-FVS can be written as $(G; V_1, V_2; k)$, and consists of a partition (V_1, V_2) of the vertex set of the graph G and a parameter k such that both V_1 and V_2 induce forests (where $V_2 = F$). We will call an FVS entirely contained in V_1 a V_1-FVS. Thus, the instance $(G; V_1, V_2; k)$ of DISJOINT-FVS is looking for a V_1-FVS of size k in the graph G.

Given an instance $(G; V_1, V_2; k)$ of DISJOINT-FVS, we apply the following rules:

Rule 1. Remove all degree-0 vertices; and remove all degree-1 vertices;
Rule 2. For a degree-2 vertex v in V_1,
- if both neighbors of v are in the same connected component of $G[V_2]$, then include v into the objective V_1-FVS, $G = G \setminus v$, and $k = k - 1$;
- otherwise, move v from V_1 to V_2: $V_1 = V_1 \setminus \{v\}$, $V_2 = V_2 \cup \{v\}$.

Our kernelization algorithm is based on an algorithm proposed in [5], which can be described as follows: on a given instance $(G; V_1, V_2; k)$ of DISJOINT-FVS, keep all vertices in V_1 of degree at least 3 (whenever a vertices in V_1 becomes degree less than 3, applying Rules 1-2 on the vertex), and repeatedly branch on a leaf in the induced subgraph $G[V_1]$. In particular, if the graph G has a V_1-FVS of size bounded by k, then at least one $\mathcal{P}$ of the computational paths in the branching program will return a V_1-FVS F of size bounded by k. The computational path $\mathcal{P}$ can be described by the algorithm in Figure 2.

Lemma 1. *If none of Rule 1 and Rule 2 is applicable on an instance $(G; V_1, V_2; k)$ of DISJOINT-FVS, and $|V_1| > 2k + l - \tau$, then there is no V_1-FVS of size bounded by k in G, where l is the number of connected components in $G[V_2]$ and τ is the number of connected components in $G[V_1]$.*

Algorithm FindingFVS(G, V_1, V_2, k)
INPUT: an instance $(G; V_1, V_2; k)$ of DISJOINT-FVS.
OUTPUT: a V_1-FVS F of size bounded by k in G.
1    $F = \emptyset$;
2   **while** $|V_1| > 0$ **do**
3   pick a leaf w in $G[V_1]$;
4   **case 1:** \\ w is in the objective V_1-FVS F.
5   add w to F and remove w from V_1; $k = k - 1$;
6   **if** the neighbor u of w in $G[V_1]$ becomes degree-2
                then apply Rule 2 on u;
7   **case 2:** \\ w is not in the objective V_1-FVS F.
8   move w from V_1 to V_2.

Fig. 2. The computational path $\mathcal{P}$ that finds the V_1-FVS F of size bounded by k

Note that for those DISJOINT-FVS instances we will meet in Section 4, we always have $|V_2| = k + 1$, which is exactly the characteristic of the iterative compression technique. Also by the simple fact that $l \leq |V_2|$ and $\tau > 0$, we have $2k + l - \tau \leq 3k$, so the kernel size is also bounded by $3k$. With more careful analysis, we can further improve the kernel size to $3k - \tau - \rho(V_1)$, where $\rho(V_1)$ is the size of a maximum matching of the subgraph induced by the vertex set V_1' that consists of all vertices in V_1 of degree larger than 3. The detailed analysis for this fact is given in a complete version of the current paper.

3   A Polynomial Time Solvable Case for DISJOINT-FVS

In this section we consider a special class of instances for the DISJOINT-FVS problem. This approach is closely related to the classical study on graph maximum genus embeddings [4,12]. However, the study on graph maximum genus embeddings that is related to our approach is based on general spanning trees of a graph, while our approach must be restricted to only spanning trees that are constrained by the vertex partition (V_1, V_2) of an instance $(G; V_1, V_2; k)$ of DISJOINT-FVS. We start with the following simple lemma.

Lemma 2. *Let G be a connected graph and let S be a subset of vertices in G such that the induced subgraph $G[S]$ is a forest. Then there is a spanning tree in G that contains the entire induced subgraph $G[S]$, and can be constructed in time $O(m\alpha(n))$, where $\alpha(n)$ is the inverse of Ackermann function [6].*

Let $(G; V_1, V_2; k)$ be an instance for the DISJOINT-FVS problem, recall that (V_1, V_2) is a partition of the vertex set of the graph G such that both induced subgraphs $G[V_1]$ and $G[V_2]$ are forests. By Lemma 2, there is a spanning tree T of the graph G that contains the entire induced subgraph $G[V_2]$. Call a spanning tree that contains the induced subgraph $G[V_2]$ a $T_{G[V_2]}$-*tree*.

Let T be a $T_{G[V_2]}$-tree of the graph G. By the construction, every edge in $G - T$ has at least one end in V_1. Two edges in $G - T$ are V_1-adjacent if they have a common end in V_1. A V_1-adjacency matching in $G - T$ is a partition of the edges in $G - T$ into groups of one or two edges, called 1-groups and 2-groups, respectively, such that two edges in the same 2-group are V_1-adjacent. A maximum V_1-adjacency matching in $G - T$ is a V_1-adjacency matching in $G - T$ that maximizes the number of 2-groups.

Definition 1. Let $(G; V_1, V_2; k)$ be an instance of DISJOINT-FVS. The V_1-adjacency matching number $\mu(G, T)$ of a $T_{G[V_2]}$-tree T in G is the number of 2-groups in a maximum V_1-adjacency matching in $G - T$. The V_1-adjacency matching number $\mu(G)$ of the graph G is the largest $\mu(G, T)$ over all $T_{G[V_2]}$-trees T in G.

An instance $(G; V_1, V_2; k)$ of DISJOINT-FVS is 3-regular$_{V_1}$ if every vertex in the vertex set V_1 has degree exactly 3. Let $f_{V_1}(G)$ be the size of a minimum V_1-FVS for G. Let $\beta(G)$ be the *Betti number* of the graph G that is the total number of edges in $G - T$ for any spanning tree T in G (or equivalently, $\beta(G)$ is the number of *fundamental cycles* in G) [12]. The following lemma is a nontrivial generalization of a result in [17] (the result in [17] is a special case for Lemma 3 in which all vertices in the set V_2 have degree 2).

Lemma 3. For any 3-regular$_{V_1}$ instance $(G; V_1, V_2; k)$ of DISJOINT-FVS, $f_{V_1}(G) = \beta(G) - \mu(G)$. Moreover, a minimum V_1-FVS can be constructed in linear time from a $T_{G[V_2]}$-tree whose V_1-adjacency matching number is $\mu(G)$.

By Lemma 3, in order to construct a minimum V_1-FVS for a 3-regular$_{V_1}$ instance $(G; V_1, V_2, k)$ of DISJOINT-FVS, we only need to construct a $T_{G[V_2]}$-tree in the graph G whose V_1-adjacency matching number is $\mu(G)$. The construction of an unconstrained maximum adjacency matching in terms of general spanning trees has been considered by Furst, Gross and McGeoch in their study of graph maximum genus embeddings [12]. We follow a similar approach, based on cographic matroid parity, to construct a $T_{G[V_2]}$-tree in G whose V_1-adjacency matching number is $\mu(G)$. We start with a quick review on the related concepts in matroid theory. Detailed discussion on matroid theory can be found in [18].

A *matroid* is a pair $(E, \Im)$, where E is a finite set and $\Im$ is a collection of subsets of E that satisfies: (1) If $A \in \Im$ and $B \subseteq A$, then $B \in \Im$; (2) If $A, B \in \Im$ and $|A| > |B|$, then there is an element $a \in A - B$ such that $B \cup \{a\} \in \Im$.

The *matroid parity* problem is stated as follows: given a matroid $(E, \Im)$ and a perfect pairing $\{[a_1, \bar{a}_1], [a_2, \bar{a}_2], \ldots, [a_n, \bar{a}_n]\}$ of the elements in the set E, find a largest subset P in $\Im$ such that for all i, $1 \leq i \leq n$, either both a_i and $\bar{a}_i$ are in P, or neither of a_i and $\bar{a}_i$ is in P.

Each connected graph G is associated with a *cographic matroid* $(E_G, \Im_G)$, where E_G is the edge set of G, and an edge set S is in $\Im_G$ if and only if $G - S$ is connected. It is well-known that matroid parity problem for cographic matroids can be solved in polynomial time [18]. The fastest known algorithm for cographic matroid parity problem runs in time $\mathcal{O}(mn \log^6 n)$ [13].

In the following, we explain how to reduce our problem to the cographic matroid parity problem. Let $(G; V_1, V_2; k)$ be a 3-regular$_{V_1}$ instance of the DISJOINT-FVS problem. Without loss of generality, we make the following assumptions: (1) the graph G is connected (otherwise, we simply work on each connected component of G); and (2) for each vertex v in V_1, there is at most one edge from v to a connected component in $G[V_2]$ (otherwise, we can directly include v in the objective V_1-FVS).

Recall that two edges are V_1-*adjacent* if they share a common end in V_1. For an edge e in G, denote by $d_{V_1}(e)$ the number of edges in G that are V_1-adjacent to e (note that an edge can be V_1-adjacent to the edge e from either end of e).

We construct a *labeled subdivision* G_2 of the graph G as follows.

1. shrink each connected component of $G[V_2]$ into a single vertex; let the resulting graph be G_1;
2. assign each edge in G_1 a distinguished label;
3. for each edge labeled e_0 in G_1, suppose that the edges V_1-adjacent to e_0 are labeled by $e_1, e_2, \ldots, e_d$ (the order is arbitrary), where $d = d_{V_1}(e_0)$; subdivide e_0 into d *segment edges* by inserting $d - 1$ degree-2 vertices in e_0, and label the segment edges by $(e_0 e_1), (e_0 e_2), \ldots, (e_0 e_d)$. Let the resulting graph be G_2. The segment edges $(e_0 e_1), (e_0 e_2), \ldots, (e_0 e_d)$ in G_2 are said to be *from* the edge e_0 in G_1.

There are a number of interesting properties for the graphs constructed above. First, each of the edges in the graph G_1 corresponds uniquely to an edge in G that has at least one end in V_1. Thus, without creating any confusion, we will simply say that the edge is in the graph G or in the graph G_1. Second, because of the assumptions we made on the graph G, the graph G_1 is a simple and connected graph. In consequence, the graph G_2 is also a simple and connected graph. Finally, because each edge in G_1 corresponds to an edge in G that has at least one end in V_1, and because each vertex in V_1 has degree 3, every edge in G_1 is subdivided into at least two segment edges in G_2.

Now in the labeled subdivision graph G_2, pair the segment edge labeled $(e_0 e_i)$ with the segment edge labeled $(e_i e_0)$ for all segment edges (note that $(e_0 e_i)$ is a segment edge from the edge e_0 in G_1 and that $(e_i e_0)$ is a segment edge from the edge e_i in G_1). By the above remarks, this is a perfect pairing $\mathcal{P}$ of the edges in G_2. Now with this edge pairing $\mathcal{P}$ in G_2, and with the cographic matroid $(E_{G_2}, \Im_{G_2})$ for the graph G_2, we call Gabow and Stallmann's algorithm [13] for the cographic matroid parity problem. The algorithm produces a maximum edge subset P in $\Im_{G_2}$ that, for each segment edge $(e_0 e_i)$ in G_2, either contains both $(e_0 e_i)$ and $(e_i e_0)$, or contains neither of $(e_0 e_i)$ and $(e_i e_0)$.

Lemma 4. *From the edge subset P in $\Im_{G_2}$ constructed above, a $T_{G[V_2]}$-tree for the graph G whose V_1-adjacency matching number is $\mu(G)$ can be constructed in time $O(m\alpha(n))$, where n and m are the number of vertices and the number of edges, respectively, of the graph G.*

Now we can solve the 3-regular$_{V_1}$ instance as follows: first shrinking each connected component of $G[V_2]$ into a single vertex; then constructing the labeled subdivision graph G_2 of G, and apply Gabow and Stallmann's algorithm [13] on it to get the edge subset P in $\Im_{G_2}$; finally, building the V_1-adjacency matching M from P, and the V_1-FVS from M. This gives our main result in this section.

Theorem 1. *There is an $\mathcal{O}(n^2 \log^6 n)$ time algorithm that on a 3-regular$_{V_1}$ instance $(G; V_1, V_2; k)$ of the* DISJOINT-FVS *problem, either constructs a V_1-FVS of size bounded by k, if such a V_1-FVS exists, or reports correctly that no such a V_1-FVS exists.*

Combining Theorem 1 and Rule 2, we have

Corollary 1. *There is an $\mathcal{O}(n^2 \log^6 n)$ time algorithm that on an instance $(G; V_1, V_2; k)$ of* DISJOINT-FVS *where all vertices in V_1 have degree bounded by 3, either constructs a V_1-FVS of size bounded by k, if such an FVS exists, or reports correctly that no such a V_1-FVS exists.*

4   An Improved Algorithm for DISJOINT-FVS

Now we are ready for the general DISJOINT-FVS problem. Let $(G; V_1, V_2; k)$ be an instance of DISJOINT-FVS, for which we are looking for a V_1-FVS of size k. Observe that certain structures in the input graph G can be easily processed and then removed from G. For example, the graph G cannot contain self-loops (i.e., edges whose both ends are on the same vertices) because by definition, both induced subgraphs $G[V_1]$ and $G[V_2]$ are forests. Moreover, if two vertices v and w are connected by multiple edges, then exactly one of v and w is in V_1 and the other is in V_2 (this is again because the induced subgraphs $G[V_1]$ and $G[V_2]$ are forests). Thus, in this case, we can directly include the vertex in V_1 in the objective V_1-FVS. Therefore, for a given input graph G, we always first apply a preprocessing that applies the above operations and remove all self-loops and multiple edges in the graph G. In consequence, we can assume, without loss of generality., that the input graph G contains neither self-loops nor multiple edges.

A vertex $v \in V_1$ is a *nice V_1-vertex* if v is of degree 3 in G and all its neighbours are in V_2. Let p be the number of nice V_1-vertices in G, and let l be the number of connected components in the induced subgraph $G[V_2]$. The measure $m = k + \frac{l}{2} - p$ will be used in the analysis of our algorithm.

Lemma 5. *If the measure m is bounded by 0, then there is no V_1-FVS of size bounded by k in G. If all vertices in V_1 are nice V_1-vertices, then a minimum V_1-FVS in G can be constructed in polynomial time.*

Proof. Suppose that $m = k + \frac{l}{2} - p \leq 0$, and that there is a V_1-FVS F of size of $k' \leq k$. Let S be the set of any $p - k'$ nice V_1-vertices that are not in F. The subgraph G' induced by $V_2 \cup S$ must be a forest because F is an FVS and is disjoint with $V_2 \cup S$. On the other hand, the subgraph G' can be constructed from the induced subgraph $G[V_2]$ and the $p - k'$ discrete vertices in S, by adding the $3(p - k')$ edges that are incident to the vertices in S. Since $k' \leq k$, we have $p - k' \geq p - k \geq \frac{l}{2}$. This gives $3(p - k') = 2(p - k') + (p - k') \geq l + (p - k')$.

This contradicts the fact that G' is a forest – in order to keep G' a forest, we can add at most $l + (p - k') - 1$ edges to the structure that consists of the induced subgraph $G[V_2]$ of l connected components and the $p - k'$ discrete vertices in S. This contradiction proves the first part of the lemma.

To prove the second part of the lemma, observe that when all vertices in V_1 are nice V_1-vertices, $(G; V_1, V_2; k)$ is a 3-regular$_{V_1}$ instance for DISJOINT-FVS. By Theorem 1, there is a polynomial time algorithm that constructs a minimum V_1-FVS in G for 3-regular$_{V_1}$ instances of DISJOINT-FVS.   □

The algorithm **Feedback**(G, V_1, V_2, k), for the DISJOINT-FVS problem is given in Figure 3. We first discuss the correctness of the algorithm. The correctness of step 1 and step 2 of the algorithm is obvious. By lemma 5, step 3 is correct. Step 4 is correct by Rule 1 in section 2. After step 4, each vertex in V_1 has degree at least 2 in G.

If the vertex w has two neighbors in V_2 that belong to the same tree T in the induced subgraph $G[V_2]$, then the tree T plus the vertex w contains at least one cycle. Since we are searching for a V_1-FVS, the only way to break the cycles in $T \cup \{w\}$ is to include the vertex w in the objective V_1-FVS. Moreover, the objective V_1-FVS of size at most k exists in G if and only if the remaining graph $G - w$ has a V_1-FVS of size at most $k - 1$ in the subset $V_1 \setminus \{w\}$. Therefore, step 5 correctly handles this case. After this step, all vertices in V_1 has at most one neighbor in a tree in $G[V_2]$.

Because of step 5, a degree-2 vertex at step 6 cannot have both its neighbors in the same tree in $G[V_2]$. By Rule 2, step 6 correctly handles this case. After step 6, all vertices in V_1 have degree at least 3.

A vertex $w \in V_1$ is either in or not in the objective V_1-FVS. If w is in the objective V_1-FVS, then we should be able to find a V_1-FVS F_1 in the graph $G - w$ such that $|F_1| \leq k - 1$ and $F_1 \subseteq V_1 \setminus \{w\}$. On the other hand, if w is not in the objective V_1-FVS, then the objective V_1-FVS for G must be contained in the subset $V_1 \setminus \{w\}$. Also note that in this case, the induced subgraph $G[V_2 \cup \{w\}]$ is still a forest since no two neighbors of w in V_2 belong to the same tree in $G[V_2]$. Therefore, step 7 handles this case correctly. After step 7, every leaf w in $G[V_1]$ that is not a nice V_1-vertex has exactly two neighbors in V_2.

The vertex y in step 8 is either in or not in the objective V_1-FVS . If y is in the objective V_1-FVS, then we should be able to find a V_1-FVS F_1 in the graph $G - y$ such that $|F_1| \leq k - 1$ and $F_1 \subseteq V_1 \setminus \{w\}$. After removing y from the graph G, the vertex w becomes degree-2 and both of its neighbors are in V_2 (note that step 7 is not applicable to w). Therefore, by Rule 2, the vertex w can be moved from V_1 to V_2 (again note that $G[V_2 \cup \{w\}]$ is a forest). On the other hand, if y is not in the objective V_1-FVS, then the objective FVS for G must be contained in the subset $V_1 \setminus \{y\}$. Also note that in this case, the subgraph $G[V_2 \cup \{y\}]$ is a forest since no two neighbors of y in V_2 belong to the same tree in $G[V_2]$. Therefore, step 8 handles this case correctly. Thus, the following conditions hold after step 8:

1. $k > 0$ and G is not a forest (by steps 1 and 2);
2. $p \leq k + \frac{l}{2}$ and not all vertices of V_1 are nice vertices (by step 3);

Algorithm Feedback(G, V_1, V_2, k)
INPUT: an instance $(G; V_1, V_2; k)$ of DISJOINT-FVS.
OUTPUT: a V_1-FVS F of size bounded by k in G if such a V_1-FVS exists.
1   **if** $(k < 0)$ or $(k = 0$ and G is not a forest$)$ **then** return 'No';
2   **if** $k \geq 0$ and G is a forest **then** return $\emptyset$;
     let l be the number of connected components in $G[V_2]$,
     and let p be the number of nice V_1-vertices;
3   **if** $p > k + \frac{l}{2}$ **then** return 'No';
     if $p = |V_1|$ **then** solve the problem in polynomial time;
4   **if** a vertex $w \in V_1$ has degree not larger than 1 **then**
        return **Feedback**$(G - w, V_1 \setminus \{w\}, V_2, k)$;
5   **if** a vertex $w \in V_1$ has two neighbors in the same tree in $G[V_2]$ **then**
        $F_1 = $ **Feedback**$(G - w, V_1 \setminus \{w\}, V_2, k - 1)$;
        if $F_1 = $'No' **then** return 'No' **else** return $F_1 \cup \{w\}$
6   **if** a vertex $w \in V_1$ has degree 2 **then**
        return **Feedback**$(G, V_1 \setminus \{w\}, V_2 \cup \{w\}, k)$;
7   **if** a leaf w in $G[V_1]$ is not a nice V_1-vertex and has ≥ 3 neighbors in V_2
        $F_1 = $ **Feedback**$(G - w, V_1 - \{w\}, V_2, k - 1)$;
7.1   **if** $F_1 \neq$ 'No' **then** return $F_1 \cup \{w\}$
7.2   **else** return **Feedback**$(G, V_1 \setminus \{w\}, V_2 \cup \{w\}, k)$;
8   **if** the neighbor $y \in V_1$ of a leaf w in $G[V_1]$ has at least one neighbor in V_2
        $F_1 = $ **Feedback**$(G - y, V_1 \setminus \{w, y\}, V_2 \cup \{w\}, k - 1)$;
8.1   **if** $F_1 \neq$'No' **then** return $F_1 \cup \{y\}$
8.2   **else** return **Feedback**$(G, V_1 \setminus \{y\}, V_2 \cup \{y\}, k)$;
9   pick a lowest leaf w_1 in any tree T in $G[V_1]$;
        let $w_1, \cdots, w_t$ be the children of w in T;
        $F_1 = $ **Feedback**$(G - w, V_1 \setminus \{w, w_1\}], V_2 \cup \{w_1\}, k - 1)$;
9.1   **if** $F_1 \neq$'No' **then** return $F_1 \cup \{w\}$
9.2   **else** return **Feedback**$(G, V_1 \setminus \{w\}, V_2 \cup \{w\}, k)$.

Fig. 3. Algorithm for DISJOINT-FVS

3. any vertex in V_1 has degree at least 3 in G (by steps 4-6);
4. any leaf in $G[V_1]$ is either a nice V_1-vertex, or has exactly two neighbors in V_2 (by step 7); and
5. for any leaf w in $G[V_1]$, the neighbor $y \in V_1$ of w has no neighbors in V_2 (by step 8).

By condition 4, any tree of single vertex in $G[V_1]$ is a nice V_1-vertex. By condition 5, there is no tree of two vertices in $G[V_1]$. For a tree T with at least three vertices in $G[V_1]$, fix any internal vertex of T as the root. Then we can find a *lowest leaf* w_1 of T in polynomial time. Since the tree T has at least three vertices, the vertex w_1 must have a parent w in T which is in $G[V_1]$.

Vertex w is either in or not in the objective V_1-FVS. If w is in the objective V_1-FVS, then we should find a V_1-FVS F_1 in the graph $G - w$ such that $F_1 \subseteq V_1 \setminus \{w\}$ and $|F_1| \leq k - 1$. Note that after removing w, the leaf w_1 becomes degree-2, and by Rule 2, it is valid to move w_1 from V_1 to V_2 since the two neighbors of w_1 in V_2 are not in the same tree in $G[V_2]$. On the other hand, if w is not in the objective V_1-FVS, then the objective V_1-FVS must be in $V_1 \setminus \{w\}$. In summary, step 9 handles this case correctly.

Theorem 2. *The algorithm* **Feedback**(G, V_1, V_2, k) *correctly solves the* DISJOINT-FVS *problem. The running time of the algorithm is* $\mathcal{O}(2^{k+l/2}n^2)$, *where n is the number of vertices in G, and l is the number of connected components in the induced subgraph $G[V_2]$.*

Proof. The correctness of the algorithm has been verified by the above discussion. Now we consider the complexity of the algorithm. The recursive execution of the algorithm can be described as a search tree $\mathcal{T}$. We first count the number of leaves in the search tree $\mathcal{T}$. Note that only steps 7, 8 and 9 of the algorithm correspond to branches in the search tree $\mathcal{T}$. Let $T(m)$ be the number of leaves in the search tree $\mathcal{T}$ for the algorithm **Feedback**(G, V_1, V_2, k) when $m = k+l/2-p$, where l is the number of connected components (i.e., trees) in the forest $G[V_2]$, and p is the number of nice V_1-vertices.

The branch of step 7.1 has that $k' = k - 1$, $l' = l$ and $p' \geq p$. Thus we have $m' = k' + l'/2 - p' \leq k - 1 + l/2 - p = m - 1$. The branch of step 7.2 has that $k'' = k$, $l'' \leq l - 2$ and $p'' = p$. Thus we have $m'' = k'' + l''/2 - p'' \leq m - 1$. Thus, for step 7, the recurrence is $T(m) \leq 2T(m - 1)$.

The branch of step 8.1 has that $k' = k - 1$, $l' = l - 1$ and $p' \geq p$. Thus we have $m' = k' + l'/2 - p' \leq k - 1 + (l - 1)/2 - p = m - 1.5$. The branch of step 8.2 has that $k'' = k$, $l'' = l$ and $p'' = p + 1$. Thus we have $m'' = k'' + l''/2 - p'' = k + l/2 - (p + 1) = m - 1$. Thus, for step 8, the recurrence is $T(m) \leq T(m - 1.5) + T(m - 1)$.

The branch of step 9.1 has that $k' = k-1$, $l' = l-1$ and $p' \geq p$. Thus we have $m' = k'+l'/2-p' \leq k-1+(l-1)/2-p = m-1.5$. the branch of step 9.2 has that $k'' = k$, $l'' = l+1$ because of w, and $p'' \geq p+2$ because w has at least two children which are leaves. Thus we have $m'' = k'' + l''/2 - p'' \leq k + (l+1)/2 - (p+2) = m - 1.5$. Thus, for step 8, the recurrence is $T(m) \leq 2T(m - 1.5)$.

The worst case happens at step 7. From the recurrence of step 7, we have $T(m) \leq 2^m$. Moreover, steps 1-3 just return an answer; step 4 does not increase measure m since vertex w is not a nice vertex; and step 5 also does not increase m since k decreases by 1 and p decreases by at most 1. Step 6 may increase measure m by 0.5 since l may increase by 1. However, we can simply just bypass vertex w in step 6, instead of putting it into V_2. If we bypass w, then measure m does not change. In Rule 2, we did not bypass w because it is easier to analyze the kernel in section 2 by putting w into V_2. Since $m = k+l/2-p \leq k+l/2$, and it is easy to verify that the computation time along each path in the search tree $\mathcal{T}$ is bounded by $O(n^2)$, we conclude that the algorithm **Feedback**(G, V_1, V_2, k) solves the DISJOINT FVS problem in time $O(2^{k+l/2}n^2)$.   □

5 Concluding Result: An Improved Algorithm for FVS

The results presented in previous sections lead to an improved algorithm for the general FVS problem. Following the idea of *iterative compression* proposed by Reed et al. [22], we formulate the following problem:

> FVS REDUCTION: given a graph G and an FVS F of size $k + 1$ for G, either construct an FVS of size at most k for G, or report that no such an FVS exists.

Lemma 6. *The FVS REDUCTION problem on an n-vertex graph G can be solved in time $\mathcal{O}(3.83^k n^2)$.*

Proof. The proof goes similar to that for Lemma 2 in [3]. Let G be a graph and let F_{k+1} be an FVS of size $k + 1$ in G. For each j, $0 \leq j \leq k$, we enumerate each subset F_{k-j} of $k - j$ vertices in F_{k+1}, and assume that F_{k-j} is the intersection of F_{k+1} and the objective FVS F_k. Therefore, constructing the FVS F_k of size k in the graph G is equivalent to constructing the FVS $F_k - F_{k-j}$ of size j in the graph $G - F_{k-j}$, which, by Theorem 2 (note that $l \leq j+1$), can be constructed in time $\mathcal{O}(2^{j+(j+1)/2} n^2) = \mathcal{O}(2.83^j n^2)$. Applying this procedure for every integer j ($0 \leq j \leq k$) and all subsets of size $k - j$ in F_{k+1} will successfully find an FVS of size k in the graph G, if such an FVS exists. This algorithm solves FVS REDUCTION in time $\sum_{j=0}^{k} \binom{k+1}{k-j} \cdot \mathcal{O}(2.83^j n^2) = \mathcal{O}(3.83^k n^2)$.   □

Finally, by combining Lemma 6 with iterative compression [5], we obtain the main result of this paper.

Theorem 3. *The FVS problem on an undirected graph of n vertices is solvable in time $\mathcal{O}(3.83^k k n^2)$.*

The proof of Theorem 3 is exactly similar to that of Theorem 3 in [5], with the complexity $\mathcal{O}(5^k n^2)$ for solving the FVS REDUCTION problem being replaced by $\mathcal{O}(3.83^k n^2)$, as given in Lemma 6.

References

1. Becker, A., Bar-Yehuda, R., Geiger, D.: Randomized algorithms for the loop cutset problem. J. Artif. Intell. Res. 12, 219–234 (2000)
2. Bodlaender, H.: On disjoint cycles. Int. J. Found. Comput. Sci. 5(1), 59–68 (1994)
3. Cao, Y., Chen, J., Liu, Y.: On Feedback Vertex Set New Measure and New Structures (manuscript, 2010)
4. Chen, J.: Minimum and maximum imbeddings. In: Gross, J., Yellen, J. (eds.) The Handbook of Graph Theory, pp. 625–641. CRC Press, Boca Raton (2003)
5. Chen, J., Fomin, F.V., Liu, Y., Lu, S., Villanger, Y.: Improved algorithms for the feedback vertex set problems. Journal of Computer and System Sciences 74, 1188–1198 (2008)
6. Cormen, T., Leiserson, C., Rivest, R., Stein, C.: Introduction to Algorithms, 2nd edn. The MIT Press and McGraw-Hill Book Company (2001)

7. Dehne, F., Fellows, M., Langston, M., Rosamond, F., Stevens, K.: An $O(2^{O(k)}n^3)$ fpt algorithm for the undirected feedback vertex set problem. In: Wang, L. (ed.) COCOON 2005. LNCS, vol. 3595, pp. 859–869. Springer, Heidelberg (2005)
8. Downey, R., Fellows, M.: Fixed parameter tractability and completeness. In: Complexity Theory: Current Research, pp. 191–225. Cambridge University Press, Cambridge (1992)
9. Downey, R., Fellows, M.: Parameterized Complexity. Springer, New York (1999)
10. Festa, P., Pardalos, P., Resende, M.: Feedback set problems. In: Handbook of Combinatorial Optimization, vol. A(suppl.), pp. 209–258. Kluwer Acad. Publ., Dordrecht (1999)
11. Fomin, F., Gaspers, S., Pyatkin, A.: Finding a minimum feedback vertex set in time $O(1.7548^n)$. In: Bodlaender, H.L., Langston, M.A. (eds.) IWPEC 2006. LNCS, vol. 4169, pp. 184–191. Springer, Heidelberg (2006)
12. Furst, M., Gross, J., McGeoch, L.: Finding a maximum-genus graph imbedding. Journal of the ACM 35(3), 523–534 (1988)
13. Gabow, H., Stallmann, M.: Efficient algorithms for graphic matroid intersection and parity. In: Brauer, W. (ed.) ICALP 1985. LNCS, vol. 194, pp. 210–220. Springer, Heidelberg (1985)
14. Guo, J., Gramm, J., Hüffner, F., Niedermeier, R., Wernicke, S.: Compression-based fixed-parameter algorithms for feedback vertex set and edge bipartization. J. Comput. Syst. Sci. 72(8), 1386–1396 (2006)
15. Kanj, I., Pelsmajer, M., Schaefer, M.: Parameterized algorithms for feedback vertex set. In: Downey, R.G., Fellows, M.R., Dehne, F. (eds.) IWPEC 2004. LNCS, vol. 3162, pp. 235–247. Springer, Heidelberg (2004)
16. Karp, R.: Reducibility among combinatorial problems. In: Complexity of Computer Computations, pp. 85–103. Plenum Press, New York (1972)
17. Li, D., Liu, Y.: A polynomial algorithm for finding the minimul feedback vertex set of a 3-regular simple graph. Acta Mathematica Scientia 19(4), 375–381 (1999)
18. Lovász, L.: The matroid matching problem. In: Algebraic Methods in Graph Theory, Colloquia Mathematica Societatis János Bolyai, Szeged, Hungary (1978)
19. Raman, V., Saurabh, S., Subramanian, C.: Faster fixed parameter tractable algorithms for finding feedback vertex sets. ACM Trans. Algorithms 2(3), 403–415 (2006)
20. Raman, V., Saurabh, S., Subramanian, C.: Faster fixed parameter tractable algorithms for undirected feedback vertex set. In: Bose, P., Morin, P. (eds.) ISAAC 2002. LNCS, vol. 2518, pp. 241–248. Springer, Heidelberg (2002)
21. Razgon, I.: Exact computation of maximum induced forest. In: Arge, L., Freivalds, R. (eds.) SWAT 2006. LNCS, vol. 4059, pp. 160–171. Springer, Heidelberg (2006)
22. Reed, B., Smith, K., Vetta, A.: Finding odd cycle transversals. Oper. Res. Lett. 32(4), 299–301 (2004)
23. Silberschatz, A., Galvin, P.: Operating System Concepts, 4th edn. Addison-Wesley, Reading (1994)

Conflict-Free Coloring Made Stronger

Elad Horev[1], Roi Krakovski[1], and Shakhar Smorodinsky[2]

[1] Computer Science department, Ben-Gurion University, Beer Sheva, Israel
{horevel,roikr}@cs.bgu.ac.il
[2] Mathematics department, Ben-Gurion University, Beer Sheva, Israel
shakhar@math.bgu.ac.il
http://www.math.bgu.ac.il/~shakhar/

Abstract. In FOCS 2002, Even et al. showed that any set of n discs in the plane can be Conflict-Free colored with a total of at most $O(\log n)$ colors. That is, it can be colored with $O(\log n)$ colors such that for any (covered) point p there is some disc whose color is distinct from all other colors of discs containing p. They also showed that this bound is asymptotically tight. In this paper we prove the following stronger results:

(i) Any set of n discs in the plane can be colored with a total of at most $O(k \log n)$ colors such that (a) for any point p that is covered by at least k discs, there are at least k distinct discs each of which is colored by a color distinct from all other discs containing p and (b) for any point p covered by at most k discs, all discs covering p are colored distinctively. We call such a coloring a k-*Strong Conflict-Free* coloring. We extend this result to pseudo-discs and arbitrary regions with linear union-complexity.

(ii) More generally, for families of n simple closed Jordan regions with union-complexity bounded by $O(n^{1+\alpha})$, we prove that there exists a k-Strong Conflict-Free coloring with at most $O(kn^\alpha)$ colors.

(iii) We prove that any set of n axis-parallel rectangles can be k-Strong Conflict-Free colored with at most $O(k \log^2 n)$ colors.

(iv) We provide a general framework for k-Strong Conflict-Free coloring arbitrary hypergraphs. This framework relates the notion of k-Strong Conflict-Free coloring and the recently studied notion of k-colorful coloring.

All of our proofs are constructive. That is, there exist polynomial time algorithms for computing such colorings.

KeyWords: Conflict-Free Colorings, Geometric hypergraphs, Wireless networks, Discrete geometry.

1 Introduction and Preliminaries

Motivated by modeling frequency assignment to cellular antennae, Even et al. [17] introduced the notion of Conflict-Free colorings. A *Conflict-Free* coloring (CF in short) of a hypergraph $H = (V, \mathcal{E})$ is a coloring of the vertices V such that for any non-empty hyperedge $e \in \mathcal{E}$ there is some vertex $v \in e$ whose color

H. Kaplan (Ed.): SWAT 2010, LNCS 6139, pp. 105–117, 2010.

is distinct from all other colors of vertices in e. For a hypergraph H, one seeks the least number of colors l such that there exists an l-coloring of H which is Conflict-Free. It is easily seen that CF-coloring of a hypergraph H coincides with the notion of classical graph coloring in the case when H is a graph (i.e., all hyperedges are of cardinality two). Thus it can be viewed as a generalization of graph coloring. There are two well known generalizations of graph coloring to hypergraph coloring in the literature (see, e.g., [9]). The first generalization requires "less" than the CF requirement and this is the *non-monochromatic* requirement where each hyperedge in $\mathcal{E}$ of cardinality at least two should be non-monochromatic: The *chromatic number* of a hypergraph H, denoted $\chi(H)$, is the least number l such that H admits an l-coloring which is a non-monochromatic coloring. The second generalization requires "more" than the CF requirement and this is the *colorful* requirement where each hyperedge should be colorful (i.e., all of its vertices should have distinct colors). For instance, consider the following hypergraph $H = (V, \mathcal{E})$: Let $V = \{1, 2, \ldots, n\}$ and let $\mathcal{E}$ consist of all subsets of V consisting of consecutive numbers of V. That is, $\mathcal{E}$ consists of all discrete intervals of V. It is easily seen that one can color the elements of V with two colors in order to obtain a non-monochromatic coloring of H. Color the elements of V alternately with 'black' and 'white'. On the other extreme, one needs n colors in any colorful coloring of H. Indeed V itself is also a hyperedge in this hypergraph (an 'interval' containing all elements of V) so all colors must be distinct. However, it is an easy exercise to see that there exists a CF-coloring of H with $\lfloor \log n \rfloor + 1$ colors. In fact, for an integer $k > 0$, if V consist of $2^k - 1$ elements then k colors suffice and are necessary for CF-coloring H.

Let $\mathcal{R}$ be a finite collection of regions in $\mathbb{R}^d$, $d \geq 1$. For a point $p \in \mathbb{R}^d$, define $r(p) = \{R \in \mathcal{R} : p \in R\}$. The hypergraph $(\mathcal{R}, \{r(p)\}_{p \in \mathbb{R}^d})$, denoted $H(\mathcal{R})$, is called the hypergraph *induced* by $\mathcal{R}$. Such hypergraphs are referred to as *geometrically induced* hypergraphs. Informally these are the Venn diagrams of the underlying regions.

In general, dealing with CF coloring for arbitrary hypergraphs is not easier than graph coloring. The paper [17] focused on hypergraphs that are induced by geometric objects such as discs, squares etc. Their motivation was a modeling of frequency assignment to cellular antennae in a manner that reduces the spectrum of frequencies used by a network of antennae. Suppose that antennae are represented by discs in the plane and that every client (holding a cell-phone) is represented by a point. Antennae are assigned frequencies (this is the coloring). A client is served provided that there is at least one antenna 'covering' the client for which the assigned frequency is "unique" and therefore has no "conflict" (interference) with other frequencies used by nearby antennae. When $\mathcal{R}$ is a finite family of n discs in the plane $\mathbb{R}^2$, Even et al. [17] proved that finding an optimal CF-coloring for $\mathcal{R}$ is NP-hard. However, they showed that there is always a CF-coloring of $H(\mathcal{R})$ with $O(\log n)$ colors and that this bound is asymptotically tight. That is, for every n there is a family of n discs which requires $\Omega(\log n)$ colors in any CF-coloring. See [17] for further discussion of this model and the motivation.

CF-coloring finds application also in activation protocols for RFID networks. Radio frequency identification (RFID) is a technology where a reader device can sense the presence of a nearby object by reading a tag device attached to the object. To improve coverage, multiple RFID readers can be deployed in the given region. However, two readers trying to access a tagged device simultaneously might cause mutual interference. One may want to design scheduled access of RFID tags in a multiple reader environment. Assume that we have t time slots and we would like to 'color' each reader with a time slot in $\{1, \ldots, t\}$ such that the reader will try to read all nearby tags in its given time slot. In particular, we would like to read all the tags and minimize the total time slots t. It is easily seen that if we CF-color the family $\mathcal{R}$ of readers then in this coloring every possible tag will have a time slot and a single reader trying to access it in that time slot [18]. The notion of CF-coloring has caught much scientific attention in recent years both from the algorithmic and combinatorial point of view [3,4,6,7,8,11,12,13,16,19,20,22,23,26].

Our Contribution: In this paper we study the notion of *k-Strong-Conflict-Free* (abbreviated, *kSCF*) colorings of hypergraphs. This notion extends the notion of *CF*-colorings of hypergraphs. Informally, in the case of coloring discs, rather than having at least one unique color at every covered point p, we require at least k distinct colors to some k discs such that each of these colors is unique among the discs covering p. The motivation for studying $kSCF$-coloring is rather straightforward in the context of wireless antennae. Having, say $k > 1$ unique frequencies in any given location allows us to serve k clients at that location rather than only one client. In the context of RFID networks, a kSCF coloring will correspond to an activation protocol which is fault-tolerant. That is, every tag can be read even if some $k - 1$ readers are broken.

Definition 1 (k-Strong Conflict-Free coloring:). *Let $H = (V, \mathcal{E})$ be a hypergraph and let $k \in \mathbb{N}$ be some fixed integer. A coloring of V is called k-Strong-Conflict-Free for H if*
(i) for every hyperedge $e \in \mathcal{E}$ with $|e| \geq k$ there exists at least k vertices in e, whose colors are unique among the colors assigned to the vertices of e, and
(ii) for each hyperedge $e \in \mathcal{E}$ with $|e| < k$ all vertices in e get distinct colors.
*   Let $f_H(k)$ denote the least integer l such that H admits a $kSCF$-coloring with l colors.*

Note that a CF-coloring of a hypergraph H is $kSCF$-coloring of H for $k = 1$.

Abellanas et al. [2] were the first to study $kSCF$-coloring[1]. They focused on the special case where V is a finite set of points in the plane and $\mathcal{E}$ consist of all subsets of V which can be realized as an intersection of V with a disc. They showed that in this case the hypergraph admits a kSCF-coloring with $O(\frac{\log n}{\log \frac{ck}{ck-1}})$ $(= O(k \log n))$ colors, for some absolute constant c. See also [1].

---

[1] They referred to such a coloring as k-Conflict-Free coloring.

The following notion of k-*colorful* colorings was recently introduced and studied by Aloupis et al. [5] for the special case of hypergraphs induced by discs.

Definition 2. *Let* $H = (V, \mathcal{E})$ *be a hypergraph, and let* φ *be a coloring of* H. *A hyperedge* $e \in \mathcal{E}$ *is said to be* k-colorful *with respect to* φ *if there exist* k *vertices in* e *that are colored distinctively under* φ. *The coloring* φ *is called* k-colorful *if every hyperedge* $e \in \mathcal{E}$ *is* $\min\{|e|, k\}$-*colorful. Let* $c_H(k)$ *denote the least integer* l *such that* H *admits a* k-*colorful coloring with* l *colors.*

Aloupis et al. were motivated by a problem related to battery lifetime in sensor networks. See [5,10,24] for additional details on the motivation and related problems.

Remark: Every $kSCF$-coloring of a hypergraph H is a k-colorful coloring of H. However, the opposite claim is not necessarily true. A k-colorful coloring assures us that every hyperedge of cardinality at least k has at least k distinct colors present in it. However, these k colors are not necessarily unique since each may appear with multiplicity.

A k-colorful coloring can be viewed as a type of coloring which is "in between" non-monochromatic coloring and colorful coloring. A 2-colorful coloring of H is exactly the classical non-monochromatic coloring, so $\chi(H) = c_H(2)$. If H is a hypergraph with n vertices, then an n-colorful coloring of H is the classical colorful coloring of H. Consider the hypergraph H, consisting of all discrete intervals on $V = \{1, \ldots, n\}$ mentioned earlier. It is easily seen that for any i, an i-colorful coloring with i colors is obtained by coloring V in increasing order with $1, 2, \ldots, i, 1, 2, \ldots, i, 1, 2 \ldots$ with repetition.

In this paper, we study a connection between k-colorful coloring and Strong-Conflict-Free coloring of hypergraphs. We show that if a hypergraph H admits a k-colorful coloring with a "small" number of colors (hereditarily) then it also admits a $(k-1)$SCF-coloring with a "small" number of colors. The interrelation between the quoted terms is provided in Theorems 1 and 2 below.

Let $H = (V, \mathcal{E})$ be a hypergraph and let $V' \subset V$. We write $H[V']$ to denote the sub-hypergraph of H induced by V', i.e., $H[V'] = (V', \mathcal{E}')$ and $\mathcal{E}' = \{e \cap V' | e \in \mathcal{E}\}$. We write $n(H)$ to denote the number of vertices of H.

Theorem 1. *Let* $H = (V, \mathcal{E})$ *be a hypergraph with* n *vertices, and let* $k, \ell \in \mathbb{N}$ *be fixed integers,* $k \geq 2$. *If every induced sub-hypergraph* $H' \subseteq H$ *satisfies* $c_{H'}(k) \leq \ell$, *then* $f_H(k-1) \leq \log_{1+\frac{1}{\ell-1}} n = O(l \log n)$.

Theorem 2. *Let* $H = (V, \mathcal{E})$ *be a hypergraph with* n *vertices, let* $k \geq 2$ *be a fixed integer, and let* $0 < \alpha \leq 1$ *be a fixed real. If every induced sub-hypergraph* $H' \subseteq H$ *satisfies* $c_{H'}(k) = O(kn(H')^{\alpha})$, *then* $f_H(k-1) = O(kn(H')^{\alpha})$.

Consider the hypergraph of "discrete intervals" with n vertices. As mentioned earlier, it has a $(k+1)$-colorful coloring with $k+1$ colors and this holds for every induced sub-hypergraph. Thus, Theorem 1 implies that it also admits a

$kSCF$-coloring with at most $\log_{1+\frac{1}{k}} n = O(k \log n)$ colors. In Section 3.1, we provide an upper bound on the number of colors required by $kSCF$-coloring of geometrically induced hypergraphs as a function of the union-complexity of the regions that induce the hypergraphs. Below we describe the relations between the union-complexity of the regions, k-colorful and $(k-1)$SCF coloring of the underlying hypergraph. First, we need to define the notion of union-complexity.

Definition 3. *For a family $\mathcal{R}$ of n simple closed Jordan regions in the plane, let $\partial \mathcal{R}$ denote the boundary of the union of the regions in $\mathcal{R}$. The union-complexity of $\mathcal{R}$ is the number of intersection points, of a pair of boundaries of regions in $\mathcal{R}$, that belong to $\partial \mathcal{R}$.*

For a set $\mathcal{R}$ of n simple closed planar Jordan regions, let $\mathcal{U}_\mathcal{R} : \mathbb{N} \to \mathbb{N}$ be a function such that $\mathcal{U}_\mathcal{R}(m)$ is the maximum union-complexity of any subset of k regions in $\mathcal{R}$ over all $k \leq m$, for $1 \leq m \leq n$. We abuse the definition slightly and assume that the union-complexity of any set of n regions is at least n. When dealing with geometrically induced hypergraphs, we consider k-colorful coloring and $kSCF$-coloring of hypergraphs that are induced by simple closed Jordan regions having union-complexity at most $O(n^{1+\alpha})$, for some fixed parameter $0 \leq \alpha \leq 1$. The value $\alpha = 0$ corresponds to regions with linear union-complexity such as discs or pseudo-discs (see, e.g., [21]). The value $\alpha = 1$ corresponds to regions with quadratic union-complexity. See [14,15] for additional families with sub-quadratic union-complexity.

In the following theorem we provide an upper bound on the number of colors required by a k-colorful coloring of a geometrically induced hypergraph as a function of k and of the union-complexity of the underlying regions inducing the hypergraph:

Theorem 3. *Let $k \geq 2$, let $0 \leq \alpha \leq 1$, and let c be a fixed constant. Let $\mathcal{R}$ be a set of n simple closed Jordan regions such that $\mathcal{U}_\mathcal{R}(m) \leq cm^{1+\alpha}$, for $1 \leq m \leq n$, and let $H = H(\mathcal{R})$. Then $c_H(k) = O(kn^\alpha)$.*

Combining Theorem 1 with Theorem 3 (for $\alpha = 0$) and Theorem 2 with Theorem 3 (for $0 < \alpha < 1$) yields the following result:

Theorem 4. *Let $k \geq 2$, let $0 \leq \alpha \leq 1$, and let c be a constant. Let $\mathcal{R}$ be a set of n simple closed Jordan regions such that $\mathcal{U}_\mathcal{R}(m) = cm^{1+\alpha}$, for $1 \leq m \leq n$. Let $H = H(\mathcal{R})$. Then:*

$$f_H(k-1) = \begin{cases} O(k \log n), & \alpha = 0, \\ O(kn^\alpha), & 0 < \alpha \leq 1. \end{cases}$$

In Section 3.2 we consider $kSCF$-colorings of hypergraphs induced by axis-parallel rectangles in the plane. It is easy to see that axis-parallel rectangles might have quadratic union-complexity, for example, by considering a grid-like construction of $n/2$ disjoint (horizontally narrow) rectangles and $n/2$ disjoint (vertically narrow) rectangles. For a hypergraph H induced by axis-parallel rectangles, Theorem 4 states that $f_H(k-1) = O(kn)$. This bound is meaningless, since the bound $f_H(k-1) \leq n$ is trivial. Nevertheless, we provide a near-optimal upper bound for this case in the following theorem:

Theorem 5. *Let $k \geq 2$. Let $\mathcal{R}$ be a set of n axis-parallel rectangles, and let $H = H(\mathcal{R})$. Then $f_H(k-1) = O(k \log^2 n)$.*

In order to obtain Theorem 5 we prove the following theorem:

Theorem 6. *Let $H = H(\mathcal{R})$ be the hypergraph induced by a family $\mathcal{R}$ of n axis-parallel rectangles in the plane, and let $k \in \mathbb{N}$ be an integer, $k \geq 2$. For every induced sub-hypergraph $H' \subseteq H$ we have: $c_{H'}(k) \leq k \log n$.*

Theorem 5 is therefore an easy corollary of Theorem 6 combined with Theorem 1.

Har-Peled and Smorodinsky [19] proved that any family $\mathcal{R}$ of n axis-parallel rectangles admits a CF-coloring with $O(\log^2 n)$ colors. Their proof uses the probabilistic method. They also provide a randomized algorithm for obtaining CF-coloring with at most $O(\log^2 n)$ colors. Later, Smorodinsky [26] provided a deterministic polynomial-time algorithm that produces a CF-coloring for n axis-parallel rectangles with $O(\log^2 n)$ colors. Theorem 5 thus generalizes the results of [19] and [26].

All of our proofs are constructive. In other words, there exist deterministic polynomial-time algorithms to obtain the required $kSCF$ coloring with the promised bounds. In this paper, we omit the technical details of the underlying algorithms and we do not make an effort to optimize their running time.

The result of Ali-Abam *et al.*[1] implies that the upper bounds provided in Theorem 4 for $\alpha = 0$ and Theorem 5 are optimal. Specifically, they provide matching lower bounds on the number of colors required by any $kSCF$-coloring of hypergraphs induced by (unit) discs and axis-parallel squares in the plane by a simple analysis of such coloring for the discrete intervals hypergraph mentioned earlier.

Organization. In Section 2 we prove Theorems 1 and 2. In Section 3.1 we prove Theorems 3 and 4. Finally, in Section 3.2 we prove Theorems 5 and 6.

2   A Framework for Strong-Conflict-Free Coloring

In this section, we prove Theorems 1 and 2. To that end we devise a framework for obtaining an upper bound on the number of colors required by a Strong-Conflict-Free coloring of a hypergraph. Specifically, we show that if there exist fixed integers k and l such that an n-vertex hypergraph H admits the hereditary property that every vertex-induced sub-hypergraph H' of H admits a k-colorful coloring with at most l colors, then H admits a $(k-1)SCF$-coloring with $O(l \log n)$ colors. For the case when l is replaced with the function $kn(H')^\alpha$ we get a better bound without the $\log n$ factor.

Framework $\mathcal{A}$:
Input: A hypergraph H satisfying the conditions of Theorems 1 and 2.
Output: A $(k-1)SCF$-coloring of H.

1: $i \leftarrow 1$ {i denotes an unused color.}
2: **while** $V \neq \emptyset$ **do**
3:    **Auxiliary Coloring:** Let $\varphi : V \rightarrow [\ell]$ be a k-colorful coloring of $H[V]$
       with at most ℓ colors.
4:    Let V' be a color class of φ of maximum cardinality.
5:    **Color:** Set $\chi(u) = i$ for every vertex $u \in V'$.
6:    **Discard:** $V \leftarrow V \setminus V'$.
7:    **Increment:** $i \leftarrow i + 1$.
8: **end while**
9: **Return** χ.

Proof of Theorems 1 and 2. We show that the coloring produced by Framework $\mathcal{A}$ is a $(k-1)$SCF-coloring of H with a total number of colors as specified in Theorems 1 and 2.

Let χ denote the coloring obtained by the application of framework $\mathcal{A}$ on H. The number of colors used by χ is the number of iterations performed by $\mathcal{A}$. By the pigeon-hole principle, at least $|V|/\ell$ vertices are removed in each iteration (where V is the set of vertices remained after the last iteration). Therefore, the total number of iterations performed by $\mathcal{A}$ is bounded by $\log_{1+\frac{1}{\ell-1}} n$. Thus, the coloring χ uses at most $\log_{1+\frac{1}{\ell-1}} n$ colors. If in step 3 of the framework, l is replaced with the function $k|V|^{\alpha}$ (for a fixed parameter $0 < \alpha < 1$), then by the pigeon-hole principle at least $\frac{|V|^{1-\alpha}}{k}$ vertices of H are discarded in step 6 of that iteration. It is easily seen that the number of iterations performed in this case is bounded by $O(kn^{\alpha})$ where $n = n(H)$.

Next, we prove that the coloring χ is indeed a $(k-1)SCF$-coloring of H. The colors of χ are the indices of iterations of $\mathcal{A}$. Let $e \in \mathcal{E}$ be a hyperedge of H. If $|e| \leq k$ then it is easily seen that all colors of vertices of e are distinct. Indeed, by the property of the auxiliary coloring φ in step 3 of the framework, every vertex of e is colored distinctively and in each such iteration, at most one vertex from e is colored by χ so χ colors all vertices of e in distinct iterations. Next, assume that $|e| > k$. We prove that e contains at least $k-1$ vertices that are assigned unique colors in χ. For an integer r, let $\{\alpha_1, \ldots, \alpha_r\}$ denote the r largest colors in decreasing order that are assigned to some vertices of e. That is, the color α_1 is the largest color assigned to a vertex of e, the color α_2 is the second largest color and so on. In what follows, we prove a stronger assertion that for every $1 \leq j \leq k-1$ the color α_j exists and is unique in e. The proof is by induction on j. α_1 exists in e by definition. For the base of the induction we prove that α_1 is unique in e. Suppose that the color α_1 is assigned to at least two vertices $u, v \in e$, and consider iteration α_1 of $\mathcal{A}$. Let $H' = H[\{x \in V : \chi(x) \geq \alpha_1\}]$, and let φ be the k-colorful coloring obtained for H' in step 3 of iteration α_1. Put $e' = \{x \in e : \chi(x) \geq \alpha_1\}$. $e' \subset e$ is a hyperedge in H'. Since $u, v \in e'$ then $|e| \geq 2$. φ is k-colorful for H' so e' contains at least two vertices that are colored distinctively in φ. In iteration α_1, the vertices of one color class of φ are removed from e'. Since e' contains vertices from two color classes of φ, it follows that after iteration α_1 at least one vertex of e' remains. Thus, at least one vertex of e' is colored in a later iteration than α_1, a contradiction to the maximality of α_1. The induction hypothesis is that in χ the colors $\alpha_1, \ldots, \alpha_{j-1}$, $1 < j \leq k-1$, all exist and are unique in the hyperedge e. Consider the color α_j. There exists a vertex $u \in e$ such that $\chi(u) = \alpha_j$; for otherwise it follows from the induction hypothesis that $|e| < k-1$ since the colors $\alpha_1, \ldots, \alpha_{j-1}$ are all unique in e and $j - 1 < k - 1$. We prove that the color α_j is unique in e. Assume to the contrary that α_j is not unique at e, and that in χ the color α_j is assigned to at least two vertices $u, v \in e$. Put $H'' = H[\{u \in V : \chi(u) \geq \alpha_j\}]$, and let φ'' be the k-colorful coloring obtained for H'' in step 3 of iteration α_j. Put $e'' = \{u \in e : \chi(u) \geq \alpha_j\}$. e'' is a hyperedge of H''. By the induction hypothesis and the definition of the colors $\alpha_1, \ldots, \alpha_{j-1}$, after iteration α_j a set $U \subset e''$ of exactly $j - 1$ vertices of e'' remains. In addition, $u, v \in e''$ and $U \cap \{u, v\} = \emptyset$. Consequently, $|e''| \geq j + 1$. Since φ'' is k-colorful then e'' contains vertices from $\min\{k, j + 1\}$ color classes of φ''. $j \leq k - 1$ so $\min\{k, j + 1\} = j + 1$. Since in iteration α_j the vertices of one color class of φ'' are removed from e'', it follows that after iteration α_j at least j vertices of e'' remain. This is a contradiction to the induction hypothesis. ∎

Remark. Given a k-colorful coloring of H, the framework $\mathcal{A}$ obtains a Strong Conflict-Free coloring of H in a constructive manner. As mentioned above, in this paper, computational efficiency is not of main interest. However, it can be seen that for certain families of geometrically induced hypergraphs, framework $\mathcal{A}$ produces an efficient algorithm. In particular, for hypergraphs induced by discs or axis-parallel rectangles, framework $\mathcal{A}$ produces an algorithm with a low degree polynomial running time. Colorful-colorings of such hypergraphs can be computed once the arrangement of the discs is computed together with the depth of every face (see, e.g., [25]). Due to space limitation we omit the technical details involving the description of these algorithms for computing k-colorful coloring for those hypergraphs.

3 k-Strong-Conflict-Free Coloring of Geometrically Induced Hypergraphs

Theorems 1 and 2 assert that in order to attain upper bounds on $f_H(k)$, for a hypergraph H, one may concentrate on attaining an upper bound on $c_H(k+1)$. In this section we concentrate on colorful colorings.

3.1 k-Strong-Conflict-Free Coloring and Union Complexity

In this section, we prove Theorems 3 and 4. Before proceeding with a proof of Theorem 3, we need several related definitions and theorems. A simple finite graph G is called k-*degenerate* if every vertex-induced sub-graph of G contains a vertex of degree at most k. For a finite set $\mathcal{R}$ of simple closed planar Jordan regions and a fixed integer k, let $G_k(\mathcal{R})$ denote the graph with vertex set $\mathcal{R}$ and two regions

$r, s \in \mathcal{R}$ are adjacent in $G_k(\mathcal{R})$ if there exists a point $p \in \mathbb{R}^2$ such that (i) $p \in r \cap s$, and (ii) there exists at most k regions in $\mathcal{R} \setminus \{r, s\}$ that contain p.

Theorem 7. *Let $\mathcal{R}$ be a finite set of simple closed planar Jordan regions, let $H = H(\mathcal{R})$, and let k be a fixed integer. If $G_k(\mathcal{R})$ is l-degenerate then $c_H(k) \leq l + 1$.*

Theorem 7 can be proved in a manner similar to that of Aloupis et al. (see [5]) who proved Theorem 7 in the special case when $\mathcal{R}$ is a family of discs. Due to space limitations, we omit a proof of this theorem.

In light of Theorem 7, in order to prove Theorem 3 it is sufficient to prove that for a family of regions satisfying the conditions of Theorem 3 and a fixed integer k, the graph $G_k(\mathcal{R})$ is $O(kn^\alpha)$-degenerate, where α is as in Theorem 3.

Lemma 1. *Let $k \geq 0$, let $0 \leq \alpha \leq 1$, and let c be a fixed constant. Let $\mathcal{R}$ be a set of n simple closed Jordan regions such that $\mathcal{U}_\mathcal{R}(m) \leq cm^{1+\alpha}$, for $1 \leq m \leq n$. Then $G_k(\mathcal{R})$ is $O(kn^\alpha)$-degenerate.*

Our approach to proving Lemma 1 requires several steps. These steps are described in the following lemmas. We shall provide an upper bound on the average degree of every vertex-induced subgraph of $G_k(\mathcal{R})$ by providing an upper bound on the number of its edges. We need the following lemma:

Lemma 2. *([26]) Let $\mathcal{R}$ be a set of n simple closed planar Jordan regions and let $\mathcal{U} : \mathbb{N} \to \mathbb{N}$ be a function such that $\mathcal{U}(m)$ is the maximum union-complexity of any k regions in $\mathcal{R}$ over all $k \leq m$. Then the average degree of $G_0(\mathcal{R})$ is $O(\frac{\mathcal{U}(n)}{n})$.*

For a graph G, we write $E(G)$ to denote the set of edges of G. We use Lemma 2 to obtain the following easy lemma.

Lemma 3. *Let $0 \leq \alpha \leq 1$ and let c be a fixed constant. Let $\mathcal{R}$ be a set of n simple closed Jordan regions such that $\mathcal{U}_\mathcal{R}(m) \leq cm^{1+\alpha}$, for $1 \leq m \leq n$. Then there exists a constant d such that $|E(G_0(\mathcal{R}))| \leq \frac{dn^{1+\alpha}}{2}$.*

Proof: By Lemma 2, it follows that there exists a constant d' such that

$$\frac{2|E(G_0(\mathcal{R}))|}{n} = \frac{\sum_{x \in V(G_0(\mathcal{R}))} deg_{G_0(\mathcal{R})}(x)}{n} \leq \frac{d' cn^{1+\alpha}}{n}.$$

Set $d = d'c$ and the claim follows.   ∎

For a set $\mathcal{R}$ of n simple closed planar Jordan regions, define $I(\mathcal{R})$ to denote the graph whose vertex set is $\mathcal{R}$ and two regions $r, s \in \mathcal{R}$ are adjacent if $r \cap s \neq \emptyset$. The graph $I(\mathcal{R})$ is called the *intersection graph* of $\mathcal{R}$. Note that for any integer k $E(G_k(\mathcal{R})) \subseteq E(I(\mathcal{R}))$. Let $E \subseteq E(I(\mathcal{R}))$ be an arbitrary subset of the edges of $I(\mathcal{R})$. For every edge $e = (a, b) \in E$, pick a point $p_e \in a \cap b$. Note that for distinct edges e and e' in E it is possible that $p_e = p_{e'}$. Put $X_{E,\mathcal{R}} = \{(p_e, r) : e = (a, b) \in E$ and $r \in \mathcal{R} \setminus \{a, b\}$ contains $p_e\}$. In the following two lemmas we obtain a lower bound on $|X_{E,\mathcal{R}}|$ in terms of $|E|$ and $|\mathcal{R}|$.

Lemma 4. *Let $0 \leq \alpha \leq 1$ and let c and d be the constants of Lemma 3. Let $\mathcal{R}$ be a set of n simple closed Jordan regions such that $\mathcal{U}_{\mathcal{R}}(m) \leq cm^{1+\alpha}$, for $1 \leq m \leq n$. Let $E \subseteq E(I(\mathcal{R}))$. Then $|X_{E,\mathcal{R}}| \geq |E| - \frac{dn^{1+\alpha}}{2}$.*

Proof: Apply induction on the value $|E| - \frac{dn^{1+\alpha}}{2}$. Let $P_E = \{p_e : e \in E\}$. One may assume that $|E| - dn^{1+\alpha} \geq 0$ for otherwise the claim follows trivially since $|X_{E,\mathcal{R}}| \geq 0$. Suppose $|E| - \frac{dn^{1+\alpha}}{2} = 1$. Since $|E| > \frac{dn^{1+\alpha}}{2}$, then by Lemma 3 there exists an edge $e = (a,b) \in E \setminus E(G_0(\mathcal{R}))$. Since $e \notin E(G_0(\mathcal{R}))$, it follows that for every point $p \in a \cap b$ there exists a region $r \in \mathcal{R} \setminus \{a,b\}$ such that $p \in r$. Consequently, there exists a region $r \in \mathcal{R} \setminus \{a,b\}$ such that $p_e \in r$. Hence, $(p_e, r) \in X_{E,\mathcal{R}}$ and thus $|X_{E,\mathcal{R}}| \geq 1$. Assume that the claim holds for $|E| - \frac{dn^{1+\alpha}}{2} = i$, where $i > 1$, and consider the case that $|E| - \frac{dn^{1+\alpha}}{2} = i+1$. Let $e = (a,b) \in E$ be an edge such that there exists a region $r \in \mathcal{R} \setminus \{a,b\}$ with $p_e \in r$. Define $E' = E \setminus \{e\}$. Note that $P_{E'} \subset P_E$ and $|E'| - \frac{dn^{1+\alpha}}{2} = i$. By the induction hypothesis it follows that $|X_{E',\mathcal{R}}| \geq |E'| - \frac{dn^{1+\alpha}}{2}$. Observe that $X_{E',\mathcal{R}} \subset X_{E,\mathcal{R}}$ and that $|X_{E,\mathcal{R}}| \geq |X_{E',\mathcal{R}}| + 1$. It follows that

$$|X_{E,\mathcal{R}}| \geq |X_{E',\mathcal{R}}| + 1 \geq |E'| - \frac{dn^{1+\alpha}}{2} + 1 = i + 1 = |E| - \frac{dn^{1+\alpha}}{2}.$$

∎

Observation 8. *Let $0 \leq \alpha \leq 1$ and let X be a binomial random variable with parameters n and p. Then*

$$\mathbf{E}[X^{1+\alpha}] \leq \mathbf{E}[Xn^\alpha] = n^\alpha \mathbf{E}[X] = n^{1+\alpha}p.$$

Lemma 5. *Let $0 \leq \alpha \leq 1$ and let c and d be the constants of Lemma 3. Let $\mathcal{R}$ be a set of n simple closed Jordan regions such that $\mathcal{U}_{\mathcal{R}}(m) \leq cm^{1+\alpha}$ for $1 \leq m \leq n$. Let $E \subseteq E(I(\mathcal{R}))$ such that $|E| > dn^{1+\alpha}$ and let $\{p_e | e \in E\}$ and $X_{E,\mathcal{R}}$ be as before. Then $|X_{E,\mathcal{R}}| \geq \frac{|E|^2}{2dn^{1+\alpha}}$.*

Proof: Let $\mathcal{R}' \subseteq \mathcal{R}$ be a subset of regions of $\mathcal{R}$ chosen randomly and independently such that for every region $r \in \mathcal{R}$, $\mathbf{Pr}[r \in \mathcal{R}'] = p$ for $p = \frac{dn^{1+\alpha}}{|E|}$ (note that $p < 1$). Let $E' \subseteq E$ be the subset of edges that is defined by the intersections of regions in $\mathcal{R}'$. Let $P_{E'} = \{p_e : e \in E'\}$. $P_{E'} \subseteq P_E$ and thus $X_{E',\mathcal{R}'} \subseteq X_{E,\mathcal{R}}$. Each of $|\mathcal{R}'|$, $|E'|$, and $|X_{E',\mathcal{R}'}|$ is a random variable.

By Lemma 4 and by linearity of expectation, it follows that $\mathbf{E}[|X_{E',\mathcal{R}'}|] \geq \mathbf{E}[|E'|] - \mathbf{E}[\frac{d}{2}|\mathcal{R}'|^{1+\alpha}]$.

By Observation 8, $\mathbf{E}[\frac{d}{2}|\mathcal{R}'|^{1+\alpha}] \leq \frac{d}{2}n^{1+\alpha}p$. Hence, it follows that $\mathbf{E}[|X_{E',\mathcal{R}'}|] \geq \mathbf{E}[|E'|] - \frac{d}{2}n^{1+\alpha}p$. For an edge $e = (a,b) \in E$, $\mathbf{Pr}[e \in E'] = \mathbf{Pr}[a,b \in \mathcal{R}'] = p^2$ so $\mathbf{E}[|E'|] = p^2|E|$. In addition, for an edge $e = (a,b)$ and a region $r \in \mathcal{R} \setminus \{a,b\}$ $\mathbf{Pr}[(p_e, r) \in X_{E',\mathcal{R}'}] = \mathbf{Pr}[a,b,r \in \mathcal{R}'] = p^3$. Thus, $\mathbf{E}[|X_{E',\mathcal{R}'}|] = p^3|X_{E,\mathcal{R}}|$. It follows that $|X_{E,\mathcal{R}}| \geq \frac{|E|}{p} - \frac{dn^{1+\alpha}}{2p^2}$. Substituting the value of p in the latter inequality completes the proof of the lemma. ∎

Next, a proof of Lemma 1 is presented.

Proof of Lemma 1. Let d be the constant from Lemma 3. Let $V \subseteq V(G_k(\mathcal{R}))$ be a subset of of m vertices and let G be the subgraph of $G_k(\mathcal{R})$ induced by V. Define $E = E(G)$. Observe that $E \subseteq E(I(\mathcal{R}))$. There are two cases: Either $|E| \leq dm^{1+\alpha}$ or $|E| > dm^{1+\alpha}$. In the former case, the average degree of a vertex in G is at most $2dm^\alpha$. In the latter case, it follows from Lemma 5 that $|X_{E,V}| \geq \frac{|E|^2}{2dm^{1+\alpha}}$. On the other hand, since $E \subseteq E(G_k(\mathcal{R}))$ then by definition, for every edge $e \in E$ the chosen point p_e can belong to at most k other regions of $\mathcal{R}$. Thus $|X_{E,V}| \leq k|E|$. Combining these two inequalities we have: $|E| \leq 2dkm^{1+\alpha}$ and thus the average degree of G in this case is at most $4dkm^\alpha$. Hence, in G there exists a vertex whose degree is at most $\max\{2dm^\alpha, 4dkm^\alpha\} = 4dkm^\alpha$.   ∎

As mentioned in the introduction, Theorem 4 is a corollary of a combination of Theorems 1, 2, and Theorem 3.

3.2 k-Strong Conflict-Free Coloring of Axis-Parallel Rectangles

In this section, we consider $kSCF$-colorings of axis-parallel rectangles and prove Theorems 5 and 6. As mentioned in the introduction, a proof of Theorem 5 can be derived from a combination of Theorem 1 and Theorem 6. Consequently, we concentrate on a proof of Theorem 6. To that end we require the following lemma.

Lemma 6. *Let $k \geq 2$. Let $\mathcal{R}$ be a set of n axis-parallel rectangles such that all rectangles in $\mathcal{R}$ intersect a common vertical line ℓ, and let $H = H(\mathcal{R})$. Then $c_H(k) = O(k)$.*

Proof: Assume, without loss of generality, that the rectangles are in general position (that is, no three rectangles' boundaries intersect at a common point). According to Theorem 3, it is sufficient to prove that for every subset of rectangles $\mathcal{R}' \subseteq \mathcal{R}$, the union-complexity of $\mathcal{R}'$ is at most $O(|\mathcal{R}'|)$. Let $\mathcal{R}' \subseteq \mathcal{R}$ and consider the boundary of the union of the rectangles of $\mathcal{R}'$ that is to the right of the line ℓ. Let $\partial_r \mathcal{R}'$ denote this boundary. An intersection point on $\partial_r \mathcal{R}'$ results from the intersection of a horizontal side of a rectangle and a vertical side of another rectangle. Each horizontal side of a rectangle in $\mathcal{R}'$ may contribute at most one intersection point to $\partial_r \mathcal{R}'$. Indeed, let s be a horizontal rectangle side. Let p be the right-most intersection point on s to the right of the line ℓ and let q be any other intersection point on s to the right of ℓ. Let r be the rectangle whose vertical side defines p on s. Since r intersects ℓ, the point p lies on the right vertical side of r. Hence, $q \in r$; for otherwise either r does not intersect ℓ or q is to the left of ℓ, in which case q does not lie on $\partial_r \mathcal{R}'$. It follows that every horizontal side s of some rectangle contributes at most one point to $\partial_r \mathcal{R}'$. As there are $2|\mathcal{R}'|$ such sides then $\partial_r \mathcal{R}'$ contains at most $2|\mathcal{R}'|$ points. A symmetric argument holds for the boundary of $\partial \mathcal{R}'$ that lies to the left of ℓ. Hence, the union-complexity of $\mathcal{R}'$ is at most $O(|\mathcal{R}'|)$. By Theorem 3, the claim follows.   ∎

Next, we prove Theorem 6.

Proof of Theorem 6. Let ℓ be a vertical line such that at most $n/2$ rectangles lie fully to its right and to its left, respectively. Let $\mathcal{R}'$ and $\mathcal{R}''$ be the sets of rectangles that lie entirely to the right and entirely to the left of ℓ, respectively. Let $\mathcal{R}_\ell$ denote the set of rectangles in $\mathcal{R}$ that intersect ℓ, and let $c(n)$ denote the least number of colors required by a colorful coloring of any n axis-parallel rectangles. By Lemma 6, the set of rectangles $\mathcal{R}_\ell$ can be colored using $O(k)$ colors. In order to obtain a k-colorful coloring of $\mathcal{R}$, we color $\mathcal{R}_\ell$ using a set D of $O(k)$ colors. We then color $\mathcal{R}'$ and $\mathcal{R}''$ recursively by using the same set of colors D' such that $D \cap D' = \emptyset$. The function $c(n)$ satisfies the recurrence $c(n) \leq O(k) + c(n/2)$. Thus, $c(n) = O(k \log n)$. Let φ be the resulting coloring of the above coloring procedure. It remains to prove that φ is a valid k-colorful coloring of $\mathcal{R}$. The proof is by induction on the cardinality of $\mathcal{R}$. Suppose $\mathcal{R}'$ and $\mathcal{R}''$ are colored correctly under φ, and consider a point $p \in \bigcup_{r \in \mathcal{R}} r$. If $r(p) \subset \mathcal{R}_\ell$ or $r(p) \subset \mathcal{R}'$ or $r(p) \subset \mathcal{R}''$ then by Lemma 6 and the induction hypothesis, $r(p)$ is colored correctly under φ. It is not possible that both $r(p) \cap \mathcal{R}' \neq \emptyset$ and $r(p) \cap \mathcal{R}'' \neq \emptyset$. Hence, it remains to consider points p for which either $r(p) \subset \mathcal{R}_\ell \cup \mathcal{R}'$ or $r(p) \subset \mathcal{R}_\ell \cup \mathcal{R}''$. Consider a point p which is, w.l.o.g, of the former type. Let $i = |r(p) \cap \mathcal{R}_\ell|$ and $j = |r(p) \cap \mathcal{R}'|$. If either $i \geq k$ or $j \geq k$, then either by Lemma 6 or by the inductive hypothesis the hyperedge $r(p)$ is k-colorful. It remains to consider the case that $i + j \geq k$ and $i, j < k$. Let φ_ℓ and $\varphi_{\mathcal{R}'}$ be the colorings of $\mathcal{R}_\ell$ and $\mathcal{R}'$ induced by φ, respectively. By the inductive hypothesis, the rectangles in the set $r(p) \cap \mathcal{R}'$ are colored distinctively using j colors under $\varphi_{\mathcal{R}'}$. In addition, by Lemma 6, the rectangles in $r(p) \cap \mathcal{R}_\ell$ are colored using i distinct colors under φ_ℓ. Moreover, the colors used in φ_ℓ are distinct from the ones used in $\varphi_{\mathcal{R}'}$. Hence, $r(p)$ is $\min\{|r(p)|, k\}$-colorful. This completes the proof of the lemma. ∎

References

1. Abam, M.A., de Berg, M., Poon, S.-H.: Fault-tolerant conflict-free coloring. In: Proceedings of the 20th Annual Canadian Conference on Computational Geometry, Montreal, Canada, August 13-15 (2008)
2. Abellanas, M., Bose, P., Garcia, J., Hurtado, F., Nicolas, M., Ramos, P.A.: On properties of higher order delaunay graphs with applications. In: EWCG (2005)
3. Ajwani, D., Elbassioni, K., Govindarajan, S., Ray, S.: Conflict-free coloring for rectangle ranges using $\tilde{O}(n^{.382+\epsilon})$ colors. In: SPAA '07: Proc. 19th ACM Symp. on Parallelism in Algorithms and Architectures, pp. 181–187 (2007)
4. Alon, N., Smorodinsky, S.: Conflict-free colorings of shallow discs. In: SoCG '06: Proc. 22nd Annual ACM Symposium on Computational Geometry, pp. 41–43 (2006)
5. Aloupis, G., Cardinal, J., Collette, S., Langerman, S., Smorodinsky, S.: Coloring geometric range spaces. In: Laber, E.S., Bornstein, C., Nogueira, L.T., Faria, L. (eds.) LATIN 2008. LNCS, vol. 4957, pp. 146–157. Springer, Heidelberg (2008)
6. Bar-Noy, A., Cheilaris, P., Olonetsky, S., Smorodinsky, S.: Online conflict-free colorings for hypergraphs. In: Arge, L., Cachin, C., Jurdziński, T., Tarlecki, A. (eds.) ICALP 2007. LNCS, vol. 4596, pp. 219–230. Springer, Heidelberg (2007)

7. Bar-Noy, A., Cheilaris, P., Olonetsky, S., Smorodinsky, S.: Weakening the online adversary just enough to get optimal conflict-free colorings for intervals. In: SPAA, pp. 194–195 (2007)

8. Bar-Noy, A., Cheilaris, P., Smorodinsky, S.: Conflict-free coloring for intervals: from offline to online. In: SPAA '06: Proceedings of The Eighteenth Annual ACM Symposium on Parallelism in Algorithms and Architectures, pp. 128–137. ACM Press, New York (2006)

9. Berge, C.: Graphs and Hypergraphs. Elsevier Science Ltd., Amsterdam (1985)

10. Buchsbaum, A.L., Efrat, A., Jain, S., Venkatasubramanian, S., Yi, K.: Restricted strip covering and the sensor cover problem. In: SODA, pp. 1056–1063 (2007)

11. Chen, K., Fiat, A., Levy, M., Matoušek, J., Mossel, E., Pach, J., Sharir, M., Smorodinsky, S., Wagner, U., Welzl, E.: Online conflict-free coloring for intervals. SIAM J. Comput. 36, 545–554 (2006); See also in Proc. 16th Annual ACM-SIAM Symposium on Discrete Algorithms (2005)

12. Chen, K., Kaplan, H., Sharir, M.: Online conflict-free coloring for halfplanes, congruent disks, and axis-parallel rectangles (manuscript, 2005)

13. Chen, X., Pach, J., Szegedy, M., Tardos, G.: Delaunay graphs of point sets in the plane with respect to axis-parallel rectangles. In: SODA, pp. 94–101 (2008)

14. Efrat, A.: The complexity of the union of (α, β)-covered objects. In: Proc. 15th Annu. ACM Sympos. Comput. Geom., pp. 134–142 (1999)

15. Efrat, A., Sharir, M.: On the complexity of the union of fat convex objects in the plane. Discrete and Comput. Geom. 23, 171–189 (2000)

16. Elbassioni, K., Mustafa, N.: Conflict-free colorings of rectangles ranges. In: Durand, B., Thomas, W. (eds.) STACS 2006. LNCS, vol. 3884, pp. 254–263. Springer, Heidelberg (2006)

17. Even, G., Lotker, Z., Ron, D., Smorodinsky, S.: Conflict-free colorings of simple geometric regions with applications to frequency assignment in cellular networks. SIAM J. Comput. 33, 94–136 (2003); See also in Proc. 43rd Annual Symposium on Foundations of Computer Science (2002)

18. Gupta, H.: Personal Communication (2006)

19. Har-Peled, S., Smorodinsky, S.: On conflict-free coloring of points and simple regions in the plane. Discrete and Comput. Geom., 47–70 (2005); See also in 19th Annual Symposium on Computational Geometry (2003)

20. Katz, M., Lev-Tov, N., Morgenstern, G.: Conflict-free coloring of points on a line with respect to a set of intervals. In: CCCG '07: Proc. 19th Canadian Conference on Computational Geometry (2007)

21. Kedem, K., Livne, R., Pach, J., Sharir, M.: On the union of Jordan regions and collision-free translational motion amidst polygonal obstacles. Discrete Comput. Geom. 1, 59–71 (1986)

22. Pach, J., Tardos, G.: Coloring axis-parallel rectangles. J. Combin. Theory Ser. A (2009)

23. Pach, J., Tóth, G.: Conflict free colorings. Discrete and Computational Geometry, The Goodman-Pollack Festschrift, 665–671 (2003)

24. Pach, J., Tóth, G.: Decomposition of multiple coverings into many parts. In: Symposium on Computational Geometry, pp. 133–137 (2007)

25. Sharir, M., Agarwal, P.K.: Davenport–Schinzel Sequences and Their Geometric Applications. Cambridge University Press, Cambridge (1995)

26. Smorodinsky, S.: On the chromatic number of some geometric hypergraphs. SIAM Journal on Discrete Mathematics 21, 676–687 (2007); See also in Proc. 17th Annual ACM-SIAM Symposium on Discrete Algorithms (2006)

Polychromatic Coloring for Half-Planes

Shakhar Smorodinsky and Yelena Yuditsky*

Ben-Gurion University, Be'er Sheva 84105, Israel
shakhar@math.bgu.ac.il, yuditsky@bgu.ac.il

Abstract. We prove that for every integer k, every finite set of points in the plane can be k-colored so that every half-plane that contains at least $2k - 1$ points, also contains at least one point from every color class. We also show that the bound $2k - 1$ is best possible. This improves the best previously known lower and upper bounds of $\frac{4}{3}k$ and $4k - 1$ respectively. As a corollary, we also show that every finite set of half-planes can be k colored so that if a point p belongs to a subset H_p of at least $4k - 3$ of the half-planes then H_p contains a half-plane from every color class. This improves the best previously known upper bound of $8k-3$. Another corollary of our first result is a new proof of the existence of small size ϵ-nets for points in the plane with respect to half-planes.

Keywords: Geometric Hypergraphs, Discrete Geometry, Polychromatic Coloring, ϵ-Nets.

1   Introduction

In this contribution, we are interested in coloring finite sets of points in $\mathbb{R}^2$ so that any half-plane that contains at least some fixed number of points, also contains at least one point from each of the color classes.

Before stating our result, we introduce the following definitions:

A *range space* (or *hypergraph*) is a pair $(V, \mathcal{E})$ where V is a set (called the *ground set*) and $\mathcal{E}$ is a set of subsets of V.

A *coloring* of a hypergraph is an assignment of colors to the elements of the ground set. A *k-coloring* is a coloring that uses exactly k colors. More formally, a k-coloring is a function $\chi : V \rightarrow \{1, \ldots, k\}$. A hyperedge $S \in \mathcal{E}$ is said to be *polychromatic* with respect to some k-coloring χ if it contains a point from each of the k color classes. That is for every $i \in \{1, \ldots, k\}$ $S \cap \chi^{-1}(i) \neq \emptyset$. We are interested in hypergraphs induced by an infinite family of geometric regions. Let $\mathcal{R}$ be a family of regions in $\mathbb{R}^d$ (such as all balls, all axis-parallel boxes, all half-spaces, etc.)

Consider the following two functions defined for $\mathcal{R}$ (notations are taken from [3]):

* This work was done while the 2nd author was an M.Sc. student at Ben-Gurion University under the supervision of Shakhar Smorodinsky. Partially supported by the Lynn and William Frankel Center for Computer Sciences.

H. Kaplan (Ed.): SWAT 2010, LNCS 6139, pp. 118–126, 2010.

1. Let $f = f_\mathcal{R}(k)$ denote the minimum number such that any finite point set $P \subset \mathbb{R}^d$ can be k-colored so that every range $R \in \mathcal{R}$ containing at least f points of P is polychromatic.
2. Let $\bar{f} = \bar{f}_\mathcal{R}(k)$ denote the minimum number such that any finite sub-family $\mathcal{R}' \subset \mathcal{R}$ can be k-colored so that for every point $p \in \mathbb{R}^d$, for which the subset $\mathcal{R}'_p \subset \mathcal{R}'$ of regions containing p is of size at least $\bar{f}$, $\mathcal{R}'_p$ is polychromatic.

We note that the functions $f_\mathcal{R}(k)$ and $\bar{f}_\mathcal{R}(k)$ need not necessarily exist, that is $f_\mathcal{R}(k)$ might not be bounded even for $k = 2$. Indeed, suppose $\mathcal{R}$ is the family of all convex sets in the plane and P is a set of more than $2f$ points in convex position. Note that any subset of P can be cut-off by some range in $\mathcal{R}$. By the pigeon-hole principle, any 2 coloring of P contains a monochromatic subset of at least f points, thus illustrating that $f_\mathcal{R}(2)$ does not exist in that case. Also note that $f_\mathcal{R}(k)$ and $\bar{f}_\mathcal{R}(k)$ are monotone non-decreasing, since any upper bound for $f_\mathcal{R}(k)$ would imply an upper-bound for $f_\mathcal{R}(k-1)$ by merging color classes. We sometimes abuse the notation and write $f(k)$ when the family of ranges under consideration is clear from the context.

The functions defined above are related to the so-called *cover-decomposable* problems or the decomposition of *c-fold coverings* in the plane. It is a major open problem to classify for which families $\mathcal{R}$ those functions are bounded and in those cases to provide sharp bounds on $f_\mathcal{R}(k)$ and $\bar{f}_\mathcal{R}(k)$. Pach [10] conjectured that $\bar{f}_\mathcal{T}(2)$ exists whenever $\mathcal{T}$ is a family of all translates of some fixed compact convex set. These functions have been the focus of many recent research papers and some special cases are resolved. See, e.g., [1,2,6,11,12,13,15,16,17]. We refer the reader to the introduction of [3] for more details on this and related problems.

Application to Battery Consumption in Sensor Networks. Let $\mathcal{R}$ be a collection of sensors, each of which monitors the area within a given shape A. Assume further that each sensor has a battery life of one time unit. The goal is to monitor the region A for as long as possible. If we activate all sensors in $\mathcal{R}$ simultaneously, A will be monitored for only one time unit. This can be improved if $\mathcal{R}$ can be partitioned into c pairwise disjoint subsets, each of which covers A. Each subset can be used in turn, allowing us to monitor A for c units of time. Obviously if there is a point in A covered by only c sensors then we cannot partition $\mathcal{R}$ into more than c families. Therefore it makes sense to ask the following question: what is the minimum number $\bar{f}(k)$ for which we know that if every point in A is covered by $\bar{f}(k)$ sensors then we can partition $\mathcal{R}$ into k pairwise disjoint covering subsets? This is exactly the type of problem that we described. For more on the relation between these partitioning problems and sensor networks, see the paper of Buchsbaum *et al.* [4].

Our results. For the family $\mathcal{H}$ of all half-planes, Pach and Tóth showed in [13] that $f_\mathcal{H}(k) = O(k^2)$. Aloupis *et al.* [3] showed that $\frac{4k}{3} \le f_\mathcal{H}(k) \le 4k - 1$. In this paper, we settle the case of half-planes by showing that the exact value of $f_\mathcal{H}(k)$ is $2k - 1$. Keszegh [8] showed that $\bar{f}_\mathcal{H}(2) \le 4$ and Fulek [5] showed that $\bar{f}_\mathcal{H}(2) = 3$. Aloupis *et al.* [3] showed that $\bar{f}_\mathcal{H}(k) \le 8k - 3$. As a corollary of our improved bound for $f_\mathcal{H}(k)$ we also get the improved bound $\bar{f}_\mathcal{H}(k) \le 4k - 3$.

An Application to ϵ-Nets for Half-Planes. Let $H = (V, \mathcal{E})$ be a hypergraph where V is a finite set. Let $\epsilon \in [0, 1]$ be a real number. A subset $N \subseteq V$ is called an ϵ-*net* if for every hyperedge $S \in \mathcal{E}$ such that $|S| \geq \epsilon|V|$ we have also $S \cap N \neq \emptyset$. In other words, N is a hitting set for all "large" hyperedges. Haussler and Welzl [7] proved the following fundamental theorem regarding the existence of small ϵ-nets for hypergraphs with a small VC-dimension.

Theorem 1 (ϵ-net theorem). *Let $H = (V, \mathcal{E})$ be a hypergraph with VC-dimension d. For every $\epsilon \in (0, 1]$, there exists an ϵ-net $N \subset V$ with cardinality at most $O\left(\dfrac{d}{\epsilon} \log \dfrac{1}{\epsilon}\right)$.*

The proof of Haussler and Welzl for the ϵ-net theorem uses a clever probabilistic argument, and in fact it can be shown that a random sample of size $O(\frac{d}{\epsilon} \log \frac{1}{\epsilon})$ is an ϵ-net for H with a positive constant probability. The notion of ϵ-nets is central in several mathematical fields, such as computational learning theory, computational geometry, discrete geometry and discrepancy theory.

Most hypergraphs studied in discrete and computational geometry have a finite VC-dimension. Thus, by the above-mentioned theorem, these hypergraphs admit small size ϵ-nets. Kómlos *et al.* [9] proved that the bound $O(\frac{d}{\epsilon} \log \frac{1}{\epsilon})$ on the size of an ϵ-net for hypergraphs with VC-dimension d is best possible, namely, for a constant d they construct a hypergraph H with VC-dimension d such that any ϵ-net for H must have a size of at least $\Omega(\frac{1}{\epsilon} \log \frac{1}{\epsilon})$. However, their construction is random and seems far from being a "nice" geometric hypergraph. It is believed that for most hypergraphs with VC-dimension d that arise in the geometric context, one can improve on the bound $O(\frac{d}{\epsilon} \log \frac{1}{\epsilon})$.

Consider a hypergraph $H = (P, E)$ where P is a set of n points in the plane and

$$E = \{P \cap h : h \text{ is a half-plane}\}.$$

For this special case, Woeginger [18] showed that for any $\epsilon > 0$ there exists an ϵ-net for H of size at most $\frac{2}{\epsilon} - 1$ (see also, [14]).

As a corollary of Theorem 2, we obtain yet another proof for this fact.

2   Half-Planes

Let $\mathcal{H}$ denote the family of all half-planes in $\mathbb{R}^2$. In this section we prove our main result by finding the exact value of $f_{\mathcal{H}}(k)$ for the family $\mathcal{H}$ of all half-planes.

Theorem 2. $f_{\mathcal{H}}(k) = 2k - 1$.

We start by proving the lower bound $f_{\mathcal{H}}(k) \geq 2k - 1$. Our lower bound construction is simple and is inspired by a lower bound construction for ϵ-nets with respect to half-planes given in [18]. We need to show that there exists a finite set P in $\mathbb{R}^2$ such that for every k-coloring of P there is a half-plane that contains $2k - 2$ points and is not polychromatic. In fact, we show a stronger construction. For every $n \geq 2k - 1$ there is such a set P with $|P| = n$. We construct P as follows: We place $2k-1$ points on a concave curve γ (e.g., the parabola $y = x^2$, $-1 < x < 1$). Let $p_1, p_2, .., p_{2k-1}$ be the points ordered from left to right along their x-coordinates. Notice that for every point p_i on γ there is an open positive half-plane h_i that does not contain p_i and contains the rest of the $2k-2$ points that are on γ. Namely, $h_i \cap \{p_1, ..., p_{2k-1}\} = \{p_1, ...p_{i-1}, p_{i+1}, ..., p_{2k-1}\}$. We choose $h_1, h_2, ..., h_{2k-1}$ in such a way that $\cap_{i=1}^{2k-1} \overline{h_i} \neq \emptyset$ where $\overline{h_i}$ is the complement of h_i. We place $n - (2k-1)$ points in $\cap_{i=1}^{2k-1} \overline{h_i}$. Let $\chi : P \to \{1, ..., k\}$ be some k-coloring of P. There exists a color c that appears at most once among the points on γ (for otherwise we would have at least $2k$ points). If no point on γ is colored with c then a (positive) half-plane bounded by a line separating the parabola from the rest of the points is not polychromatic. Let p_j be the point colored with c. As mentioned, the open half-plane h_j contains all the other points on γ (and only them), so h_j contains $2k-2$ points and misses the color c. Hence, it is not polychromatic. Thus $f_{\mathcal{H}}(k) > 2k - 2$ and this completes the lower bound construction. See Figure 1 for an illustration.

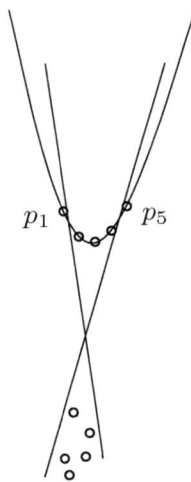

Fig. 1. A construction showing that $f(k) > 2k - 2$ for $n = 10$ and $k = 3$

Next, we prove the upper-bound $f_{\mathcal{H}}(k) \leq 2k - 1$. In what follows, we assume without loss of generality that the set of points P under consideration is in general position, namely, that no three points of P lie on a common line. Indeed, we can slightly perturb the point set to obtain a set P' of points in general position. The perturbation is done in such a way that for any subset of the points of the form $h \cap P$ where h is a half-plane, there is another half-plane h' such that $h \cap P = h' \cap P'$. Thus any valid polychromatic k-coloring for P' also serves as a valid polychromatic k-coloring for P.

For the proof of the upper bound we need the following lemma:

Lemma 1. *Let P be a finite point set in the plane in general position and let $t \geq 3$ be some fixed integer. Let $H' = (P, \mathcal{E}')$ be a hypergraph where $\mathcal{E}' = \{P \cap h :$*

$h \in \mathcal{H}, |P \cap h| = t\}$. Let $P' \subseteq P$ be the set of extreme points of P (i.e., the subset of points in P that lie on the boundary of the convex-hull $CH(P)$ of P). Let $N \subseteq P'$ be a (containment) minimal hitting set for H'. Then for every $E \in \mathcal{E}'$ we have $|N \cap E| \leq 2$.

Proof. First notice that such a hitting set $N \subset P'$ for H' indeed exists since P' is a hitting set.

Assume to the contrary that there exists a hyperedge $E \in \mathcal{E}'$ such that $|N \cap E| \geq 3$. Let h be a half-plane such that $h \cap P = E$ and let l be the line bounding h. Assume, without loss of generality, that l is parallel to the x-axis and that the points of E are below l. If l does not intersect the convex hull $CH(P)$ or is tangent to $CH(P)$ then h contains P and $|P| = t$. Thus any minimal hitting set N contains exactly one point of P, a contradiction. Hence, the line l must intersect the boundary of $CH(P)$ in two points.

Let q, q' be the left and right points of $l \cap \partial CH(P)$ respectively. Let p, r, u be three points in $N \cap E$ ordered according to their counter-clockwise order on $\partial CH(P)$. By the minimality property, there is a half-plane h_r such that $h_r \cap P \in \mathcal{E}'$ and such that $N \cap h_r = \{r\}$, for otherwise, $N \setminus \{r\}$ is also a hitting-set for $\mathcal{E}'$ contradicting the minimality of N. See Figure 2 for an illustration.

Denote the line bounding h_r by l_r and denote by $\bar{h}_r$ the complement half-plane of h_r. Notice that l_r can not intersect the line l in the interior of the segment qq'. Indeed assume to the contrary that l_r intersects the segment qq' in some point x. Then, by convexity, the open segment rx lies in h_r. However, the segment rx must intersect the segment pu. This is impossible since both p and u lie in $\bar{h}_r$ and therefore, by convexity also the segment pu lies in $\bar{h}_r$. Thus the segment pu and the segment rx are disjoint.

Next, suppose without loss of generality that the line l_r intersects l to the right of the segment qq'. Let q'' denote the point $l \cap l_r$. We have that $|h_r \cap P| = t$ and also $|E| = |h \cap P| = t$, therefore there is at least one point r' that is contained in $h_r \cap P$ and is not contained in h, hence it lies above the line l. The segment rr' must intersect the line l to the right of the point q''. Also, by convexity, the segment rr' is contained in $CH(P)$. This implies that the line l must intersect $\partial CH(P)$ to the right of q', i.e intersects $\partial CH(P)$ in three points, a contradiction.

We are ready to prove the second part of Theorem 2: Recall that for a given finite planar set $P \subset \mathbb{R}^2$ and an integer k, we need to show that there is a k-coloring for P such that every half-plane that contains at least $2k - 1$ points is polychromatic.

For $k = 1$ the theorem is obvious. For $k = 2$, put $t = 3$ and let N be a hitting set as in lemma 1. We assign the points of N the color 2 and assign the points of $P \setminus N$ the color 1. Let h be a half-plane such that $|h \cap P| \geq 3$. Assume without loss of generality that h is a negative half-plane. Let l denote the line bounding h. Translate l downwards to obtain a line l', such that for the negative half-plane h' bounded by l', we have $h' \cap P \subseteq h \cap P$ and $|h' \cap P| = 3$. We can assume without loss of generality that no line parallel to l passes through two points of P. Indeed, this can be achieved by rotating l slightly. Obviously $h' \cap N \neq \emptyset$. Moreover, by lemma 1 we have that $h' \cap (P \setminus N) \neq \emptyset$. Hence, h' contains both a

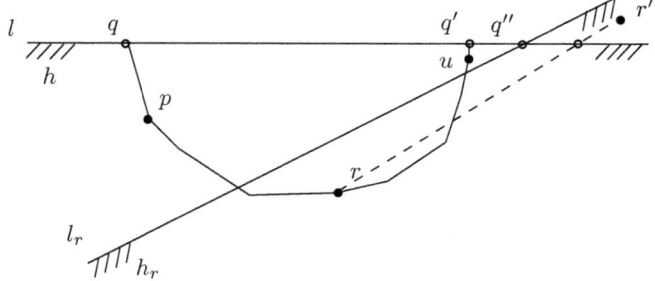

Fig. 2. The line l intersects the boundary of $CH(P)$ in two points

point colored with 1 and a point colored with 2, i.e., h' is polychromatic. Thus h is also polychromatic.

We prove the theorem by induction on the number of colors k. The induction hypothesis is that the theorem holds for all values $i < k$. Let $k > 2$ be an integer. Put $t = 2k - 1$ and let N be a minimal hitting set as in Lemma 1. We assign all points in N the color k. Put $P' = P \setminus N$. By the induction hypothesis, we can color the points of P' with $k - 1$ colors, such that for every half-plane h with $|h \cap P'| \geq 2k-3$, h is polychromatic, i.e., h contains representative points from all the $k-1$ color classes. We claim that this coloring together with the color class N forms a valid k-coloring for P. Consider a half-plane h such that $|h \cap P| \geq 2k - 1$. As before, let h' be a half-plane such that $h' \cap P \subseteq h \cap P$ and $|h' \cap P| = 2k - 1$. It is enough to show that h' is polychromatic. By lemma 1 we know that $1 \leq |h' \cap N| \leq 2$, therefore we can find a half-plane h'' such that $h'' \cap P \subseteq h' \cap P$ and $|h'' \cap (P \setminus N)| = 2k - 3$. By the induction hypothesis, h'' contain representative points from all the initial $k-1$ colors. Thus h' contain a point from N (i.e., colored with k) and a point from each of the initial $k-1$ colors. Hence h' is polychromatic and so is h. This completes the proof of the theorem.

Remark: The above theorem also provides a recursive algorithm to obtain a valid k-coloring for a given finite set P of points. See Algorithm 1. Here, we do not care about the running time of the algorithm. Assume that we have a "black-box" that finds a hitting set N as in lemma 1 in time bounded by some function $f(n, t)$.

Note that a trivial bound on the total running time of the algorithm is $\sum_{i=1}^{k} f(n, 2i - 1)$.

Coloring half-planes. Keszegh [8] investigated the value of $\bar{f}_{\mathcal{H}}(2)$ and proved that $\bar{f}_{\mathcal{H}}(2) \leq 4$. Recently Fulek [5] showed that in fact $\bar{f}_{\mathcal{H}}(2) = 3$. For the general case, Aloupis *et al.* proved in [3] that $\bar{f}_{\mathcal{H}}(k) \leq 8k - 3$. As a corollary of Theorem 2, we also obtain the following improved bound on $\bar{f}_{\mathcal{H}}(k)$:

Theorem 3. $\bar{f}_{\mathcal{H}}(k) \leq 4k - 3$

Theorem 3 is a direct corollary of Theorem 2 and uses the same reduction as in [3]. For the sake of completeness we describe it in detail:

Algorithm 1. Algorithm for polychromatic k-coloring

Input: A finite set $P \subset \mathbb{R}^2$ and an integer $k \geq 1$
Output: A polychromatic k-coloring $\chi : P \to \{1, ..., k\}$
begin
   | **if** $k=1$ **then**
   | | Color all points of P with color 1.
   | **end**
   | **else**
   | | Find a minimal hitting set N as in lemma 1 for all the half-planes of
   | | size $2k - 1$.
   | | Color the points in N with color k.
   | | Set $P = P \setminus N$ and $k = k - 1$. Recursively color P with k colors.
   | **end**
end

Proof. Let $H \subseteq \mathcal{H}$ be a finite set of half-planes. We partition H into two disjoint sets H^+ and H^- where $H^+ \subset H$ (respectively $H^- \subset H$) is the set of all positive half-planes (respectively negative half-planes). It is no loss of generality to assume that all lines bounding the half-planes in H are distinct. Indeed, by a slight perturbation of the lines, one can only obtain a superset of hyperedges in the corresponding hypergraph (i.e, a superset of cells in the arrangement of the bounding lines). Let L^+ (respectively L^-) be the sets of lines bounding the half-planes in H^+ (respectively H^-). Next, we use a standard (incidence-preserving) dualization to transform the set of lines L^+ (respectively L^-) to a set of points L^{+^*} (respectively L^{-^*}). It has the property that a point p is above (respectively incident or below) a line l if and only if the dual line p^* is above (respectively incident or below) the point l^*. See Figure 3 for an illustration.

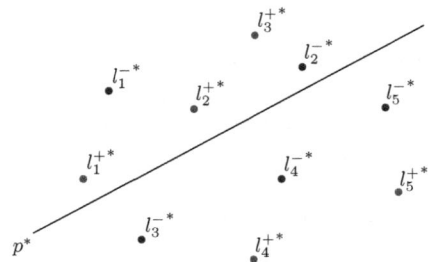

Fig. 3. An illustration of the dualization. In the primal, the point p is contained in the half-planes bounded by the lines l_1^-, l_2^-, l_4^+ and l_5^+.

We then color the sets L^{+^*} and L^{-^*} independently. We color each of them with k-colors so that every half-plane containing $2k - 1$ points of a given set, is also polychromatic. Obviously, by Theorem 2, such a coloring can be found. This coloring induces the final coloring for the set $H = H^+ \cup H^-$. To prove that this coloring is indeed valid, consider a point p in the plane. Let $H' \subseteq H$ be the set of half-planes containing p. We claim that if $|H'| \geq 4k - 3$ then H' is polychromatic. Indeed, if $|H'| \geq 4k - 3$ then, by the pigeon-hole principle, either $|H' \cap H^+| \geq 2k - 1$ or $|H' \cap H^-| \geq 2k - 1$. Suppose without loss of generality that $|H' \cap H^+| \geq 2k - 1$. Let $L_{H'}^+ \subseteq L^+$ be the set of lines bounding the half-planes in $H' \cap H^+$. In the dual, the points in $L_{H'}^+{}^*$ are in the half-plane below the line p^*. Since $|L_{H'}^+{}^*| \geq 2k - 1$, we also have that $L_{H'}^+{}^*$ is polychromatic, thus the set of half-planes $H' \cap H^+$ is polychromatic and so is the set H'. This completes the proof of the theorem.

Remark: Obviously, the result of Fulek [5] implies that the bound $4k - 3$ is not tight already for $k = 2$. It would be interesting to find the exact value of $\bar{f}_{\mathcal{H}}(k)$ for every integer k.

3   Small Epsilon-Nets for Half-Planes

Consider a hypergraph $H = (P, E)$ where P is a set of n points in the plane and $E = \{P \cap h : h \in \mathcal{H}\}$. As mentioned in the introduction Woeginger [18] showed that for any $1 \geq \epsilon > 0$ there exists an ϵ-net for H of size at most $\frac{2}{\epsilon} - 1$.

As a corollary of Theorem 2, we obtain yet another proof for this fact. Recall that for any integer $k \geq 1$ we have $f_{\mathcal{H}}(k) \leq 2k - 1$. Let $\epsilon > 0$ be a fixed real number. Put $k = \lceil \frac{\epsilon n + 1}{2} \rceil$. Let χ be a k-coloring as in Theorem 2. Notice that every half-plane containing at least ϵn points contains at least $2k - 1$ points of P. Indeed such a half-pane must contain at least $\lceil \epsilon n \rceil = \lceil 2(\frac{\epsilon n + 1}{2}) - 1 \rceil = 2k - 1$. Such a half-plane is polychromatic with respect to χ. Thus, every color class of χ is an ϵ-net for H. Moreover, by the pigeon-hole principle one of the color classes has size at most $\frac{n}{k} \leq \frac{2n}{\epsilon n + 1} < \frac{2}{\epsilon}$. Thus such a set contains at most $\frac{2}{\epsilon} - 1$ points as asserted.

The arguments above are general and, in fact, we have the following theorem:

Theorem 4. *Let $\mathcal{R}$ be a family of regions such that $f_{\mathcal{R}}(k) \leq ck$ for some absolute constant c and every integer k. Then for any ϵ and any finite set P there exists an ϵ-net for P with respect to $\mathcal{R}$ of size at most $\frac{c}{\epsilon} - 1$.*

Applying the above theorem for the dual range space defined by a set of n half-planes with respect to points and plugging Theorem 3 we conclude that there exists an ϵ-net for such a range-space of size at most $\frac{4}{\epsilon} - 1$. However, using a more clever analysis one can, in fact, show that there is an ϵ-net of size at most $\frac{2}{\epsilon}$ for such a range-space. We omit the details here.

Acknowledgments. We wish to thank Panagiotis Cheilaris and Ilan Karpas for helpful discussions concerning the problem studied in this paper.

References

1. Aloupis, G., Cardinal, J., Collette, S., Imahori, S., Korman, M., Langerman, S., Schwartz, O., Smorodinsky, S., Taslakian, P.: Colorful strips. In: LATIN (to appear, 2010)

2. Aloupis, G., Cardinal, J., Collette, S., Langerman, S., Orden, D., Ramos, P.: Decomposition of multiple coverings into more parts. In: SODA, pp. 302–310 (2009)
3. Aloupis, G., Cardinal, J., Collette, S., Langerman, S., Smorodinsky, S.: Coloring geometric range spaces. Discrete & Computational Geometry 41(2), 348–362 (2009)
4. Buchsbaum, A.L., Efrat, A., Jain, S., Venkatasubramanian, S., Yi, K.: Restricted strip covering and the sensor cover problem. In: SODA, pp. 1056–1063 (2007)
5. Fulek, R.: Coloring geometric hypergraph defined by an arrangement of half-planes (manuscript), http://dcg.epfl.ch/page74599.html
6. Gibson, M., Varadarajan, K.R.: Decomposing coverings and the planar sensor cover problem. CoRR, abs/0905.1093 (2009)
7. Haussler, D., Welzl, E.: Epsilon-nets and simplex range queries. Discrete & Computational Geometry 2, 127–151 (1987)
8. Keszegh, B.: Weak conflict-free colorings of point sets and simple regions. In: CCCG, pp. 97–100 (2007)
9. Komlós, J., Pach, J., Woeginger, G.J.: Almost tight bounds for epsilon-nets. Discrete & Computational Geometry 7, 163–173 (1992)
10. Pach, J.: Decomposition of multiple packing and covering. In: 2 Kolloq. über Diskrete Geom., pp. 169–178. Inst. Math. Univ. Salzburg (1980)
11. Pach, J., Mani, P.: Decomposition problems for multiple coverings with unit balls. Unpublished manuscript (1987)
12. Pach, J., Tardos, G., Tóth, G.: Indecomposable coverings. In: Akiyama, J., Chen, W.Y.C., Kano, M., Li, X., Yu, Q. (eds.) CJCDGCGT 2005. LNCS, vol. 4381, pp. 135–148. Springer, Heidelberg (2007)
13. Pach, J., Tóth, G.: Decomposition of multiple coverings into many parts. Computational Geometry. Theory and Applications 42(2), 127–133 (2009)
14. Pach, J., Woeginger, G.: Some new bounds for epsilon-nets. In: SCG, pp. 10–15 (1990)
15. Pálvölgyi, D.: Indecomposable coverings with concave polygons. Discrete & Computational Geometry (2009)
16. Pálvölgyi, D., Tóth, G.: Convex polygons are cover-decomposable. Discrete & Computational Geometry 43(3), 483–496 (2010)
17. Tardos, G., Tóth, G.: Multiple coverings of the plane with triangles. Discrete & Computational Geometry 38(2), 443–450 (2007)
18. Woeginger, G.J.: Epsilon-nets for halfplanes. In: van Leeuwen, J. (ed.) WG 1988. LNCS, vol. 344, pp. 243–252. Springer, Heidelberg (1989)

A 3/2-Approximation Algorithm for Multiple Depot Multiple Traveling Salesman Problem

(Extended Abstract)

Zhou Xu[1,*] and Brian Rodrigues[2]

[1] Faculty of Business, The Hong Kong Polytechnic University
lgtzx@polyu.edu.hk
[2] Lee Kong Chian School of Business, Singapore Management University

Abstract. As an important extension of the classical traveling salesman problem (TSP), the multiple depot multiple traveling salesman problem (MDMTSP) is to minimize the total length of a collection of tours for multiple vehicles to serve all the customers, where each vehicle must start or stay at its distinct depot. Due to the gap between the existing best approximation ratios for the TSP and for the MDMTSP in literature, which are 3/2 and 2, respectively, it is an open question whether or not a 3/2-approximation algorithm exists for the MDMTSP. We have partially addressed this question by developing a 3/2-approximation algorithm, which runs in polynomial time when the number of depots is a constant.

Keywords: Approximation algorithm, multiple depots, vehicle routing.

1 Introduction

The multiple depot multiple traveling salesman problem (MDMTSP) is an extension of the classical traveling salesman problem (TSP), where multiple vehicles located at distinct depots are used to serve customers with the total length of their tours minimized. It has many applications in industry. In the liner shipping industry, for example, vessels based at different ports need to be routed to pick up cargos from various origins and deliver them to various destinations.

The problem an be modeled as follows. Given a complete graph $G = (V, E)$ with a vertex set V, and an edge set E, and a depot set $D \subseteq V$, take the vertices which are not depots to be customers given by $I = V \setminus D$, each of which has a unique location. Assuming each vehicle must start or stay at a distinct depot, and since tours for vehicles from the same depot can always be combined to form a tour for one vehicle, the number of depots $k \geq 1$ can be taken to be equal to the number of vehicles, i.e., $|D| = k$. Each edge $e \in E$ has a non-negative length $\ell(e)$ which represents the distance between corresponding locations. We assume distances between locations form a metric, which satisfies

---

* Corresponding author. This work is partially supported by the Internal Competitive Research Grant A-PD0W of the Hong Kong Polytechnic University.

H. Kaplan (Ed.): SWAT 2010, LNCS 6139, pp. 127–138, 2010.

the triangle inequality, since vehicles can always travel the shortest path between any locations. The objective in the model, which we call a multiple depot multiple TSP with k depots or k-MDMTSP for short, seeks to minimize the total length of a collection of tours for vehicles to visit each customer in I exactly once, where each tour must begin at a depot and return to it. In the remainder of this paper, we will use (G, D) to represent an instance of the k-MDMTSP.

Since when $k = 1$ the problem is the TSP, the k-MDMTSP is **NP**-hard for $k \geq 1$ [5]. We are thus motivated to develop approximation algorithms which can give good solutions of practical relevance. A number of researchers have developed approximation algorithms with constant approximation ratios for the TSP [2,15] and other single-depot vehicle routing problems [3,4]. These are typically based on two methods: a tree algorithm [15,9] and a heuristic given by Christofides [2], which achieve approximation ratios of 2 and 3/2 respectively.

In contrast with single-depot vehicle routing problems, most previous work on multiple-depot vehicle routing problems do not provide worst-case analysis, e.g., [6], [14], [7]. Only [12] and [8] provide 2-approximation algorithms, based on the tree algorithm, for the k-MDMTSP and its variant where only $p \leq k$ vehicles can be selected to serve customers. However, since the best approximation ratio 3/2 is obtained using the Christofides heuristic, the best ratio that can be expected for the k-MDMTSP is 3/2 unless the TSP has an approximation algorithm superior to the Christofides heuristic. [13] has recently extended Christofides heuristic to obtain a 3/2-approximation algorithm for a 2-depot Hamiltonian path problem, which is to determine paths instead of tours for salesmen. Analysis of the approximation ratio in this work is more tractable, because it only needs to bound the length of a partial matching for a 2-depot case. Therefore, it is still an open question whether or not such a polynomial time 3/2-approximation algorithm exists for the k-MDMTSP for any $k \geq 2$ [8,12,1][1].

This work contributes to the literature by partially addressing the open question above. We do this by developing a framework in Section 2, which extends Christofides heuristic for the k-MDMTSP. The approximation solution generated by the extended heuristic uses a special spanning forest, to be defined as a depot-rooted spanning forest. By enumerating a number of such spanning forests, and applying the extended heuristic on them to select the shortest solution obtained, we have devised a 3/2-approximation algorithm for the k-MDMTSP in Section 3, which runs in polynomial time when k is a constant. Due to the lack of space, some proofs will omitted from this extended abstract but available in its full article from the authors.

---

[1] Two recent technical reports, "5/3-Approximation Algorithm for a Multiple Depot, Terminal Hamiltonian Path Problem" and "3/2-Approximation Algorithm for a Generalized, Multiple Depot, Hamiltonian Path Problem", authored by S. Rathinam and R. Sengupta [11,10], claim to achieve approximation ratios better than 2 for two multiple depot Hamiltonian path problems. However, we note the proofs for Lemma VI.1 in the first report and for Proposition II.3 in the second are incorrect since they assume that any concave-convex function f of vectors π and x must satisfy $f(\pi^*, x^*) = \max_\pi f(\pi, x^*)$ for all (π^*, x^*) with $f(\pi^*, x^*) = \max_\pi \min_x f(\pi, x)$, which is not always correct, in particular for the (π^*, x^*) constructed in the two reports.

2   The Extended Christofides Heuristic (ECH)

Given an instance (G, D) of the k-MDMTSP, a tour partition of V is defined as a set of routes such that the vertex sets of tours in it forms a partition of V. A tour partition of V is a feasible solution of the k-MDMTSP if each tour in it contains exactly one depot. Thus, a feasible solution with the shortest length is the optimal solution. For a graph H, $V(H)$ and $E(H)$ will denote the vertex and edge set of H, respectively. If $E(H)$ is a multiset, i.e., there is a multiplicity mapping of its elements into the positive integers, then H is a multigraph. For a subgraph H' and an edge subset E of H, we use $H' \setminus E$, $H' \cup E$, and $H' \cap E$ to denote subgraphs on the vertex set $V(H')$ with edge subsets $E(H') \setminus E$, $E(H') \cup E$, and $E(H') \cap E$, respectively.

We next introduce the definition of a depot-rooted spanning forest.

Definition 1. *Given an instance (G, D) of the k-MDMTSP, a depot-rooted spanning forest (DRSF) wrt (G, D) is a set of k trees that are disjoint and include every vertex of G, with each tree in the set containing exactly one depot in D at which it is rooted.*

To adapt the Christofides heuristic to solve the k-MDMTSP for (G, D), it is natural to first compute a DRSF F wrt (G, D). This is since, for every tour in an optimal solution, either deleting an edge from an optimal solution if the edge exists, or otherwise leaving it unchanged, results in a DRSF. Given a DRSF F, the Chrostofides heuristic can be extended in Algorithm 1:

Algorithm 1. ECH(F)
Input: an instance (G, D) of the k-MDMTSP, and a DRSF F wrt (G, D)
Output: a feasible solution to the k-MDMTSP on (G, D)

1: Create an Eulerian multigraph, by adding to F all edges of $M^*(F)$, where $M^*(F)$ is a minimum perfect matching for the subset $\text{Odd}(F)$ of vertices with odd degrees in F. (Notice that the Eulerian multigraph obtained may not be connected but each vertex has an even degree.)

2: Find Eulerian tours for all connected components of the Eulerian multigraph, delete repeated vertices and redundant depots of each tour by shortcuts (using the triangle inequality), and return the collection of resulting tours which constitutes a tour partition of V, and denote this by $\mathcal{C}(F)$.   □

Since the number of vertices with odd degrees in F is always even, there must exist a minimum perfect matching $M^*(F)$ in Step 1. By adding edges of $M^*(F)$ to F, we obtain an Eulerian multigraph, each connected component of which is Eulerian and must contain an Eulerian tour [9]. Notice F contains all vertices and has exactly k trees rooted at distinct depots of D. Thus, the Eulerian multigraph obtained in Step 1 has at most k connected components with each having at least one depot. Hence, $\mathcal{C}(F)$ returned by Step 2 constitutes a tour partition of V. Since each tree of F contains exactly one depot, each tour in $\mathcal{C}(F)$ must also contain exactly one depot after the shortcuts. This implies $\mathcal{C}(F)$ is feasible for the k-MDMTSP. Since the Eulerian tours obtained in Step 2 visit every edge of the Eulerian multigraph exactly once, the total length of these tours equals the total length of F and $M^*(F)$, which is not shorter than $\mathcal{C}(F)$, obtained by shortcuts in Step 2. Thus, the following fact is established:

Theorem 1. *Given a DRSF F wrt (G, D), where $G = (V, E)$ and $D \subseteq V$, the tour partition $\mathcal{C}(F)$ of V obtained by ECH(F) in Algorithm 1 is a feasible solution for the k-MDMTSP on (G, D), and satisfies $\ell(\mathcal{C}(F)) \leq \ell(F) + \ell(M^*(F))$.*

Algorithm 1 is a general framework, because with different F it returns different feasible $\mathcal{C}(F)$. By Theorem 1, to bound $\ell(\mathcal{C}(F))$ it suffices to bound $\ell(F)$ and $\ell(M^*(F))$. Therefore bounding $\ell(M^*(F))$ and choosing F are key to finding good solutions.

In the remainder of this section, we develop a general upper bound on the length of the minimum perfect matching $M^*(F)$ for $\text{Odd}(F)$ of the ECH in Algorithm 1, where F is a given DRSF wrt (G, D), where $G = (V, E)$ and $D \subseteq V$. Using this upper bound on $\ell(M^*(F))$, the length of the heuristic solution $\mathcal{C}(F)$ returned by the ECH can be bounded.

Consider any DRSF F wrt (G, D). For each tour C of G, let $\text{Odd}(F, C)$ denote the subset of the vertices in C with odd degrees in F, so $\text{Odd}(F, C)$ for all C constitutes a partition of $\text{Odd}(F)$. We then introduce the notion of a contracted graph, denoted by $G_S^{[\mathcal{C}]}$, for any edge subset S of G, and for any tour partition $\mathcal{C}$ of V. In $G_S^{[\mathcal{C}]}$, every tour of $\mathcal{C}$ is represented by a unique and distinct vertex, and vertices representing two different tours of $\mathcal{C}$ are incident with an edge if and only they are connected by an edge in S. Thus, $G_S^{[\mathcal{C}]}$ is a multigraph, which must be a subgraph of $G_E^{[\mathcal{C}]}$.

Consider the contracted graph $G_{E(F)}^{[\mathcal{C}]}$. For each vertex ν of $G_{E(F)}^{[\mathcal{C}]}$, we refer to ν as an odd vertex (or even vertex), and define the parity of ν to be odd (or even), iff ν represents a tour $C \in \mathcal{C}$ with an odd (or even) cardinality of $\text{Odd}(F, C)$. With this, we define an auxiliary edge subset of F as follows in Definition 2.

Definition 2. *An edge subset A of F is an auxiliary edge subset of F wrt $\mathcal{C}$, if (i) each edge of A connects a pair of two different tours of $\mathcal{C}$; (ii) different edges of A connect different pairs of tours of $\mathcal{C}$; (iii) the contracted graph $G_A^{[\mathcal{C}]}$ is a spanning forest of $G_{E(F)}^{[\mathcal{C}]}$; and (iv) each tree of $G_A^{[\mathcal{C}]}$ has an even number of odd vertices of $G_{E(F)}^{[\mathcal{C}]}$. An auxiliary edge subset A of F wrt $\mathcal{C}$ is minimal if no proper subset of A is an auxiliary edge subset of F wrt $\mathcal{C}$.*

The following Lemma 1 implies that an auxiliary edge subset A of F wrt $\mathcal{C}$ always exists, and can be obtained from a maximal spanning forest of $G_{E(F)}^{[\mathcal{C}]}$. The proof is omitted due to the lack of space.

Lemma 1. *Given a DRSF F wrt (G, D), and a tour partition $\mathcal{C}$ of V, let A denote an edge subset of F such that every edge of A connects a pair of two different tours of $\mathcal{C}$ and different edges of A connect different pairs of tours of $\mathcal{C}$. If $G_A^{[\mathcal{C}]}$ is a maximal spanning forest of $G_{E(F)}^{[\mathcal{C}]}$, then A is an auxiliary edge subset of F wrt $\mathcal{C}$.*

This now allows us to prove the following lemma, which shows that given any auxiliary edge subset A of a DRSF F wrt a feasible solution $\mathcal{C}$, it is possible to construct another DRSF by replacing an edge in A with an edge in $\mathcal{C}$. This property will be useful in bounding the length of edges in A, by taking F equal to the minimum DRSF, which we do in Section 3. The proof is omitted due to the lack of space.

Lemma 2. *Given any DRSF F wrt (G, D), and any feasible solution $\mathcal{C}$, let A denote an auxiliary edge subset A of F wrt $\mathcal{C}$. Then, for each edge (u, v) of A, where u is the parent of v in F, there exists an edge e in the unique tour $C_v \in \mathcal{C}$ containing v such that such that $(F \setminus \{(u, v)\}) \cup \{e\}$ is a DRSF wrt (G, D).*

Thus, Theorem 2 can be established to provide a general upper bound on $\ell(M^*(F))$ by using an auxiliary edge subset A of F wrt $\mathcal{C}$. The proof is omitted due to the lack of space.

Theorem 2. *Given any DRSF F wrt (G, D), and any tour partition $\mathcal{C}$ of V, let A denote an auxiliary edge subset A of F wrt $\mathcal{C}$. Then, there exists a perfect matching M for $\mathrm{Odd}(F)$, which satisfies:*

$$\ell(M^*(F)) \leq \ell(M) \leq \frac{1}{2}\ell(\mathcal{C}) + \ell(A). \qquad (1)$$

3   3/2-Approximation Algorithm

For the choice of F for the ECH in Algorithm 1, it is natural to compute in Step 1 a DRSF of minimum length, or minimum DRSF wrt (G, D) and denoted by F^*. As shown by [8], finding F^* is computationally tractable, because one can transform it equivalently to the minimum spanning tree problem by contracting multiple depots into a single one. Therefore, $\mathrm{ECH}(F^*)$ can be solved in polynomial time. However, $\mathrm{ECH}(F^*)$ alone cannot achieve the ratio of 3/2.

We thus introduce a variant of the ECH in Algorithm 2, which generates multiple solutions $\mathcal{C}(F)$ by applying the ECH (in Algorithm 1) on multiple F, where each F is a DRSF wrt (G, D) constructed from F^* by replacing edges in E^- of F^* with edges in E^*, where E^- is a subset of $E(F^*)$, and E^+ is a subset of $E \setminus E(F^*)$, with $|E^-| = |E^+| \leq \max\{k - 2, 0\}$. The algorithm then returns the shortest tour from all $\mathcal{C}(F)$.

Algorithm 2. (ECH with multiple DRSF's)
Input: an instance (G, D) of the k-MDMTSP
Output: a solution to the k-MDMTSP on (G, D)
  1: Obtain a minimum DRSF F^* wrt (G, D).
  2: For each pair of edge subsets (E^+, E^-), with $E^- \subseteq E(F^*)$, $E^+ \subseteq E \setminus E(F^*)$, and $|E^-| = |E^+| \leq \max\{k - 2, 0\}$, if replacing edges in E^- of F^* with edges in E^+ constitutes a DRSF F wrt (G, D), apply the ECH in Algorithm 1 on F to obtain a heuristic solution $\mathcal{C}(F)$.
  3: Return the shortest tour from all $\mathcal{C}(F)$ obtained.   □

3.1 Performance Analysis

The following theorem implies that Algorithm 2 achieves an approximation ratio of $3/2$, and has a polynomial time complexity when k is a constant.

Theorem 3. *Given any constant $k \geq 1$, Algorithm 2 achieves an approximation ratio of $3/2$ in polynomial time for the k-MDMTSP.*

To prove the approximation ratio of Algorithm 2, we first need the following Lemma 3, which is proved in Section 3.2.

Lemma 3. *Consider a minimum DRSF F^* wrt (G, D). There always exist disjoint edge subsets, $E^+ \subseteq E \setminus E(F^*)$ and $E^- \subseteq E(F^*)$ with $|E^+| = |E^-| \leq \max\{k - 2, 0\}$, such that $F = (F^* \setminus E^-) \cup E^+$, which replaces edges of E^- in F^* with edges of E^+, constitutes a DRSF wrt (G, D) that satisfies*

(i) $\ell(F) \leq \ell(C^*) - \ell(e_{\max}(C^*))$, and
(ii) There exists an auxiliary edge subset A of F wrt C^* with $\ell(A) \leq \ell(e_{\max}(C^*))$.

We can now prove Theorem 3.

Proof of Theorem 3. First, ECH(F) in Step 2 of Algorithm 2 runs in polynomial time. When k is a constant, since the total number of choices of E^+ and E^- in Step 2 is not greater than $\binom{|F^*|}{\max\{k-2,0\}} \cdot \binom{|E|}{\max\{k-2,0\}}$, which is not greater than a $|V|^{3\max\{k-2,0\}}$, we obtain that Algorithm 2 runs in polynomial time.

From Lemma 3, there must exist disjoint edge subsets, $E^+ \subseteq E \setminus E(F^*)$ and $E^- \subseteq E(F^*)$, such that $F = (F^* \setminus E^-) \cup E^+$ is a DRSF F wrt (G, D), and satisfies conditions (i) and (ii) of Lemma 3, where $|E^+| = |E^-| \leq \max\{k - 2, 0\}$, implying that (E^+, E^-) must be enumerated in Step 2 of Algorithm 2. According to Theorem 2 and Theorem 1, and by conditions (i) and (ii) of Lemma 3, we obtain $\ell(C(F)) \leq \ell(C^*) - \ell(e_{\max}(C^*)) + \frac{1}{2}\ell(C^*) + \ell(e_{\max}(C^*)) \leq \frac{3}{2}\ell(C^*)$. Thus Algorithm 2 achieves an approximation ratio of $3/2$.   □

3.2 The Proof of Lemma 3

We prove the existence of (E^+, E^-) for Lemma 3, by introducing a matroid structure in Lemma 4, and then using an inductive construction in Lemma 5.

First, for any two disjoint edge subsets E^+ and E^- of G, define an (E^+, E^-)-DRSF wrt (G, D) to be a DRSF that contains all edges in E^+ but no edges in E^-. Thus, a minimum (E^+, E^-)-DRSF is the one of the shortest length if it exists. Notice that if $F = (F^* \setminus E^-) \cup E^+$ is a DRSF wrt (G, D), then F must be an (E^+, E^-)-DRSF. We can prove Lemma 3 by using properties of the (E^+, E^-)-DRSF. As shown in the following Lemma 4, under some mild conditions for E^+ and E^-, a minimum (E^+, E^-)-DRSF wrt (G, D) relates to a matroid on graphs and a sufficient and necessary condition of its optimality can be found. The the proof is omitted due to the lack of space.

Lemma 4. *Given a pair of disjoint edge subsets (E^+, E^-) of E, let $E' = E \setminus E^- \setminus E^+$. Consider the family $\mathcal{S}$ of subsets of E' to be such that for any $S \in \mathcal{S}$,* *the subgraph $(V, E^+ \cup S)$ contains no paths between depots of D and contains no cycles. If $\emptyset \in S$ and each connected component of the subgraph $(V, E \setminus E^-)$ contains at least one depot of D, then:*

(i) system (E', S) is a matroid;

(ii) for each maximal independent set X of system (E', S), the subgraph $(V, X \cup E^+)$ is an (E^+, E^-)-DRSF wrt (G, D);

(iii) for each (E^+, E^-)-DRSF F wrt (G, D), the edge set $E(F) \setminus E^+$ is a maximal independent set of system (E', S);

(iv) for each (E^+, E^-)-DRSF F wrt (G, D), F is minimum if and only if there exists no edge pair (e, e'), for $e \in E(F) \setminus E^+$ and $e' \in E'$ with $\ell(e') < \ell(e)$, such that $(F \setminus \{e\}) \cup \{e'\}$ is an (E^+, E^-)-DRSF wrt (G, D).

Consider any minimum DRSF F^* wrt (G, D). Let A^* denote a minimal auxiliary edge subset of F^* wrt C^*, where n_A denotes the number of edges in A^*, and edges in A^* are denoted by $e_1, e_2, ..., e_{n_A}$. For $0 \leq i \leq n_A$, let $A_i = \{e_t : 1 \leq t \leq i\}$ denote the subset of the edges of A^* with i smallest indices, and let $A'_i = A^* \setminus A_i$. Using Lemma 4, we can next establish the following.

Lemma 5. *For each i where $0 \leq i \leq n_A$, there must exist a pair of disjoint subsets (E_i^+, E_i^-) of E which satisfy the following constraints (i)–(v), where $F_i = (F^* \setminus A_i) \cup E_i^+$ replaces edges of A_i in F^* with edges of E_i^+.*

(i) $E_i^+ \subseteq E(C^*)$ with $|E_i^+| = i$ and $E_i^+ \cap E(F^*) = \emptyset$;

(ii) $E_i^- \cap E(C) = \emptyset$ for each tour $C \in C^*$, and $A_i \subseteq E_i^-$;

(iii) F_i is a minimum (E_i^+, E_i^-)-DRSF wrt (G, D);

(iv) $\ell(F_i) \leq [\ell(C^*) - \sum_{C \in C^*} \ell(e_{\max}(C))]$;

(v) A'_i is a minimal auxiliary edge subset of F_i wrt C^*, with $\ell(A'_i) \leq (n_A - i)\ell(e_{\max}(C^*))$.

We prove Lemma 5 by induction on $i = 0, 1, ..., n_A$ as follows. Set $E_0^+ = E_0^- = \emptyset$ and $F_0 = F^*$, which satisfies constraints (i)–(v) for $i = 0$. For each $1 \leq t \leq n_A$, suppose we have obtained (E_{t-1}^+, E_{t-1}^-) which satisfies constraints (i)–(v) of Lemma 5 for $i = t - 1$.

Consider the edge $e_t = (a, b)$ of A^*, where a and b are the endpoints of e_t, and a is the parent of b in F^*. Since $e_t \in A'_{t-1}$ and A'_{t-1} is an auxiliary edge subset of F_{t-1} wrt C^* (due to constraint (v) of Lemma 5 for $i = t-1$), vertices a and b are in the same tree T of F_{t-1} but in different tours of C^*. Thus, $T \setminus \{e_t\}$ has exactly two trees, denoted by T_a and T_b, which contain a and b respectively; and $e_t \notin E(C^*)$, which implies $e_t \notin E_{t-1}^+$ because $E_{t-1}^+ \subseteq E(C^*)$.

Since A'_{t-1} contains e_t and is an auxiliary edge subset of F_{t-1} wrt C^*, Lemma 2 guarantees the existence of such an edge e'_t on tours in C^* that $(F_{t-1} \setminus \{e_t\}) \cup \{e'_t\}$ constitutes a DRSF wrt (G, D). If there are more than one such edges, take the shortest one as e'_t.

Consider the following (E_t^+, E_t^-) and $F_t = (F^* \setminus A_t) \cup E_t^+$. Let $\overline{E}(V(T_b))$ denote the set of edges of G with one endpoint in $V(T_b)$ and one endpoint not in $V(T_b)$. We have $E(F_{t-1}) \cap \overline{E}(V(T_b)) = \{e_t\}$. Since a is the parent of b in F^*, and since $(F_{t-1} \setminus \{e_t\}) \cup \{e_t'\}$ constitutes a DRSF wrt (G, D), edge e_t' must be in $\overline{E}(V(T_b))$. Let us define:

$$E_t^+ = E_{t-1}^+ \cup \{e_t'\}, \tag{2}$$

$$E_t^- = [E_{t-1}^- \cup \overline{E}(V(T_b))] \setminus E(\mathcal{C}^*). \tag{3}$$

Hence, from $e_t \in \overline{E}(V(T_b))$ and $e_t \notin E(\mathcal{C}^*)$ we have $e_t \in E_t^-$. Since $e_t \notin E_{t-1}^+$,

$$F_t = (F^* \setminus A_t) \cup E_t^+ = (F^* \setminus A_{t-1}' \setminus \{e_t\}) \cup E_{t-1}^+ \cup \{e_t'\}$$
$$= (F_{t-1} \setminus \{e_t\}) \cup \{e_t'\}. \tag{4}$$

Thus, F_t is a DRSF wrt (G, D). Moreover, we can show that F_t is an (E_t^+, E_t^-)-DRSF wrt (G, D) by the following Lemma 6.

Lemma 6. F_t is an (E_t^+, E_t^-)-DRSF wrt (G, D), where E_t^+, E_t^- and F_t defined in (2)–(4).

Proof. Since $E(\mathcal{C}^*)$ contains e_t' but not e_t, it can be seen that e_t' and e_t are different. Thus from (2) and (3) we have $e_t \notin E_t^+$ and $e_t' \notin E_t^-$. Since $E(F_{t-1}) \cap \overline{E}(V(T_b)) = \{e_t\}$ and $e_t' \in \overline{E}(V(T_b))$, $e_t' \notin E(F_{t-1})$. By (ii) of Lemma 5 for $i = t - 1$ and $e_t' \in E(\mathcal{C}^*)$, we have $e_t' \notin E_{t-1}^-$ and $e_t' \notin E_t^-$. By (iii) of Lemma 5 for $i = t - 1$ and $e_t' \notin E(F_{t-1})$, we obtain $e_t' \notin E_{t-1}^+$. Thus, due to $e_t \notin E_t^+$, (iii) of Lemma 5 for $i = t - 1$, (4) and (2), F_t includes all edges of E_t^+. Since $E(F_{t-1}) \cap \overline{E}(V(T_b)) = \{e_t\}$, by (iii) of Lemma 5 for $i = t - 1$ and (3), $(F_{t-1} \setminus \{e_t\})$ includes no edges of E_t^-. Due to the fact $e_t' \notin E_t^-$ and (4), F_t contains no edges of E_t^-. Hence, by definition F_t is an (E_t^+, E_t^-)-DRSF wrt (G, D).   □

We are now going to establish Lemmas 7–11 to show that (E_t^+, E_t^-) and $F_t = (F^* \setminus A_t) \cup E_t^+$ satisfy (i)–(v) of Lemma 5, which completes the induction for the proof of Lemma 5.

Lemma 7. E_t^+ defined in (2) satisfies (i) of Lemma 5 for $i = t$, i.e., $E_t^+ \subseteq E(\mathcal{C}^*)$ with $|E_t^+| = t$ and $E_t^+ \cap E(F^*) = \emptyset$.

Proof. Since $e_t' \in E(\mathcal{C}^*)$ and E_{t-1}^+ satisfies (i) of Lemma 5 for $i = t-1$, we know $E_t^+ \subseteq E(\mathcal{C}^*)$. Since $e_t' \notin E_{t-1}^+$, $|E_t^+| = |E_{t-1}^+| + 1 = i$. Since E_{t-1}^- satisfies (ii) of Lemma 5 for $i = t-1$, we know that $(E(F^*) \setminus E_{t-1}^-)$ is a subset of $(E(F^*) \setminus A_{t-1}')$. Since F_{t-1} satisfies (iii) of Lemma 5 for $i = t-1$, we have that $(E(F^*) \setminus E_{t-1}^-)$ is an edge subset of F_{t-1}. Thus, $e_t' \notin E(F^*)$ because $e_t' \notin E_{t-1}^-$ and $e_t' \notin E(F_{t-1})$. By (2), since E_{t-1}^+ is disjoint with $E(F^*)$, we obtain E_t^+ is disjoint with $E(F^*)$. Therefore, E_t^+ satisfies (i) of Lemma 5 for $i = t$.   □

From the proofs of Lemma 6 and Lemma 7, we can summarize that e_t and e_t' satisfy the following properties.

- e_t is in $A_{t-1}' \cap A^* \cap E(F^*) \cap E(F_{t-1}) \cap \overline{E}(V(T_b)) \cap A_t \cap E_t^-$ but not in $E(\mathcal{C}^*) \cup E_{t-1}^+ \cup E_t^+$;
- e_t' is in $E(\mathcal{C}^*) \cap \overline{E}(V(T_b)) \cap E_t^+ \cap F_t$ but not in $E(F_{t-1}) \cup E_{t-1}^+ \cup E_t^- \cup E(F^*)$.

Such properties of e_t and e'_t can help us derive Lemmas 8–11 as follows.

Lemma 8. E_t^- *defined in (3) satisfies (ii) of Lemma 5 for* $i = t$, *i.e.,* $E_t^- \cap E(C) = \emptyset$ *for each tour* $C \in \mathcal{C}^*$, *and* $A_t \subseteq E_t^-$;

Proof. From (3) it is easy to see that E_t^- and $E(\mathcal{C}^*)$ are disjoint. Since $e_t \notin E(\mathcal{C}^*)$ and $e_t \in \overline{E}(V(T_b))$, and E_{t-1}^- satisfies (ii) of Lemma 5 $i = t-1$, by (3) we obtain $A_t^- \subseteq E_t^-$. $\square$

Lemma 9. E_t^+, E_t^- *and* F_t *defined in (2)–(4) satisfy (iii) of Lemma 5 for* $i = t$, *i.e.,* F_t *is a minimum* (E_t^+, E_t^-)*-DRSF wrt* (G, D).

Proof. Since we have already proved that F_t is an (E_t^+, E_t^-)-DRSF wrt (G, D), we only need to show that F_t is minimum. Consider the system $(E', \mathcal{S})$ as defined in Lemma 4 for the pair (E_t^+, E_t^-), where $E' = E \setminus E_t^- \setminus E_t^+$. Since F_t contains all edges of E_t^+ but no edges of E_t^-, subgraph (V, E_t^+) contains no path between depots and contains no cycle, and each connected component of subgraph $(V, E \setminus E_t^-)$ contains at least one depot of D. Therefore, the four statements in Lemma 4 are all true for (E_t^+, E_t^-).

According to statement (iv) of Lemma 4, to prove F_t is minimum, we only need to prove that there exists no edge pair (e, e'), for $e \in E(F_t) \setminus E_t^+$ and $e' \in E \setminus E_t^- \setminus E_t^+$ with $\ell(e') < \ell(e)$, such that F'_t, defined as $(F_t \setminus \{e\}) \cup \{e'\}$, is an (E_t^+, E_t^-)-DRSF wrt (G, D). Supposing there exists such an edge pair (e, e'), we can derive a contradiction as follows.

Let $Q = (F_{t-1} \cup \{e'_t\}) \setminus \{e\}$ and $W = (F_{t-1} \cup \{e'\}) \setminus \{e\}$ denote two subgraphs of G that replace e in F_{t-1} with e'_t and e' respectively. Since $\ell(e') < \ell(e)$, W is always shorter than $\ell(F_{t-1})$, we now show that at least one of Q and W is an (E_{t-1}^+, E_{t-1}^-)-DRSF wrt (G, D) with length shorter than $\ell(F_{t-1})$, which contradicts to the fact that F_{t-1} satisfies (iii) of Lemma 5 for $i = t-1$.

To do this, we first prove edges e, e', e_t, and e'_t are all different from each other. Since $e \notin E_t^+$ and $e'_t \in E_t^+$, e and e'_t are different. Since $e_t \in E_t^-$ and $e' \notin E_t^-$, e' and e_t are different. Since both F_t and $(F_t \setminus \{e\}) \cup \{e'\}$ are DRSFs, and since e is in F_t, e' is not in $E(F_t)$. Notice both e'_t and e are in $E(F_t)$, but neither e_t nor e' is in $E(F_t)$. Edges e, e', e_t, and e'_t are all different. By $e'_t \in E(F_t)$ and $e'_t \notin E(F_{t-1})$, we obtain:

$$Q = (F_{t-1} \cup \{e'_t\}) \setminus \{e\} = [(F_t \setminus \{e'_t\}) \cup \{e_t, e'_t\}] \setminus \{e\}$$
$$= (F_t \setminus \{e\} \cup \{e'\}) \cup \{e_t\} \setminus \{e'\}$$
$$= (F'_t \cup \{e_t\}) \setminus \{e'\} \tag{5}$$
$$W = (F_{t-1} \cup \{e'\}) \setminus \{e\} = [(F_t \setminus \{e'_t\}) \cup \{e_t, e'\}] \setminus \{e\}$$
$$= [(F_t \cup \{e'\}) \setminus \{e, e'_t\}] \cup \{e_t\}$$
$$= (F'_t \setminus \{e'_t\}) \cup \{e_t\} = (F'_t \cup \{e_t\}) \setminus \{e'_t\} \tag{6}$$
$$F'_t = [(F_t \setminus \{e\}) \cup \{e'\}] \tag{7}$$
$$= [(F_{t-1} \cup \{e'_t\}) \setminus \{e_t, e\}] \cup \{e'\} = (F_{t-1} \setminus \{e_t, e\}) \cup \{e'_t, e'\}$$
$$F'_t \setminus \{e'_t\} = (F_{t-1} \setminus \{e_t, e\}) \cup \{e'\} \tag{8}$$

We next prove that both $E(Q)$ and $E(W)$ include E_{t-1}^+ but are disjoint with E_{t-1}^-. On one hand, from $e \in E(F_t) \setminus E_t^+$ and $E_{t-1}^+ \subseteq E_t^+$ we know $e \notin E_{t-1}^+$. From $E_{t-1}^+ \subseteq E(F_{t-1})$ and $Q = (F_{t-1} \cup \{e_t'\}) \setminus \{e\}$ we obtain $E_{t-1}^+ \subseteq E(Q)$. Since $E_{t-1}^+ \subseteq E(F_{t-1})$ and $e \notin E_{t-1}^+$, which implies $E_{t-1}^+ \subseteq E(F_{t-1}) \setminus \{e\}$, we obtain $E_{t-1}^+ \subseteq W$. On the other hand, since $E(F_{t-1})$ is disjoint with E_{t-1}^- and $e_t' \notin E_{t-1}^-$, we have $E(Q)$ is disjoint with E_{t-1}^-. Since E_{t-1}^- is disjoint with $E(\mathcal{C}^*)$, by (3) we have $E_{t-1}^- \subseteq E_t^-$, which implies $e' \notin E_{t-1}^-$ because $e' \notin E_t^-$. Notice $E(F_{t-1})$ is disjoint with E_{t-1}^-. Thus W, which is a subset of $E(F_{t-1}) \cup \{e'\}$, must be disjoint with E_{t-1}^-.

It is now possible to prove that at least one of Q and W is an (E_{t-1}^+, E_{t-1}^-)-DRSF wrt (G, D) with length shorter than $\ell(F_{t-1})$. From (2) and (3), it can be seen that $E_{t-1}^+ \subseteq E_t^+$ and $E_{t-1}^- \subseteq E_t^-$. Thus both F_t and F_t' are (E_{t-1}^+, E_{t-1}^-)-DRSFs wrt (G, D). From (4), since $E(F_{t-1})$ contains e_t but not e_t', we obtain $F_{t-1} = F_t \setminus \{e_t'\} \cup \{e_t\}$. Since both F_t' and F_{t-1} are (E_{t-1}^+, E_{t-1}^-)-DRSFs, by statements (ii) and (iii) of Lemma 4 and by Proposition 2.5 of [16], there must exist an edge $x \in [E(F_t') \setminus (E(F_t) \setminus \{e_t'\})]$, such that $(F_t' \setminus \{x\}) \cup \{e_t\}$ is an (E_{t-1}^+, E_{t-1}^-)-DRSFs. From (7), it is easily verified that $[E(F_t') \setminus (E(F_t) \setminus \{e_t'\})] \subseteq \{e_t', e'\}$. By (5) and (6), we conclude that at least one of Q and W is an (E_{t-1}^+, E_{t-1}^-)-DRSF.

If W is an (E_{t-1}^+, E_{t-1}^-)-DRSF, then we have already shown $\ell(W) < \ell(F_{t-1})$. Consider the case when W is not, but Q is, an (E_{t-1}^+, E_{t-1}^-)-DRSF. We now show $e' \in \overline{E}(V(T_b))$ by contradiction. Suppose e' is not in $\overline{E}(V(T_b))$. Then, since $E(F_{t-1}) \cap \overline{E}(V(T_b)) = \{e_t\}$, by (8) we know $E(F_t') \setminus \{e_t'\}$ has no intersection with $\overline{E}(V(T_b))$. Notice $e_t \in \overline{E}(V(T_b))$ and T_b contains no depot. Since F_t' is an (E_{t-1}^+, E_{t-1}^-)-DRSF, by (6), W must be an (E_{t-1}^+, E_{t-1}^-)-DRSF, leading to a contradiction. Hence, $e' \in \overline{E}(V(T_b))$.

Thus, since $E(F_{t-1}) \setminus \{e_t\}$ has no intersection with $\overline{E}(V(T_b))$ and T_b contains no depot, $(F_{t-1} \setminus \{e_t\}) \cup \{e'\}$ must be an (E_{t-1}^+, E_{t-1}^-)-DRSF wrt (G, D). Moreover, from $e' \notin E_t^-$ and (3) we know $e' \in E(\mathcal{C}^*)$. Since e' is different from e_t, we know $e' \notin E(F_{t-1})$, implying $\ell(e_t') \leq \ell(e')$ by definition of e_t'. Since $\ell(e') < \ell(e)$, we obtain $\ell(e_t') < \ell(e)$. Notice $Q = (F_{t-1} \cup \{e_t'\}) \setminus \{e\}$. Thus $\ell(Q) < \ell(F_{t-1})$.

By the argument above, F_{t-1} cannot be a minimum (E_{t-1}^+, E_{t-1}^-)-DRSF, which contradicts to the fact that F_{t-1} satisfies constraint (iii) of Lemma 5 for $i = t-1$. Therefore, F_t is a minimum (E_t^+, E_t^-)-DRSF wrt (G, D), and satisfies (iii) of Lemma 5 for $i = t$. □

Lemma 10. F_t defined in (4) satisfies (iv) of Lemma 5 for $i = t$, i.e., $\ell(F_t) \leq [\ell(\mathcal{C}^*) - \sum_{C \in \mathcal{C}^*} \ell(e_{\max}(C))]$.

Proof. Notice that $E_t^+ \subseteq E(\mathcal{C}^*)$ and $E_t^- \cap E(\mathcal{C}^*)$ is empty by Lemma 7 and Lemma 8. Thus, deleting $e_{\max}(C)$ from each tour C of $\mathcal{C}^*$ constitutes an (E_t^+, E_t^-)-DRSF wrt (G, D), which has a length not longer than $\ell(F_t)$ due to Lemma 9. Hence, F_t is not longer than $[\ell(\mathcal{C}^*) - \sum_{C \in \mathcal{C}^*} \ell(e_{\max}(C))]$. □

Lemma 11. F_t defined in (4) satisfies (v) of Lemma 5 for $i = t$, i.e., A_t' is a minimal auxiliary edge subset of F_t wrt $\mathcal{C}^*$, with $\ell(A_t') \leq (n_A - t)\ell(e_{\max}(\mathcal{C}^*))$.

Proof. To prove that A'_t is a minimal auxiliary edge subset of F_t wrt $\mathcal{C}^*$, notice that $A'_t = A'_{t-1} \setminus \{e_t\}$ and that A'_{t-1} and F_{t-1} satisfy (v) of Lemma 5 for $i = t-1$. For the edge $e_t = (a, b)$, its endpoints a and b must belong to two different tours of $\mathcal{C}^*$, which are denoted by C_a and C_b containing a and b respectively. Since $e_t \in A'_{t-1}$, vertices representing C_a and C_b must belong to the same connected component denoted by U of $G^{[\mathcal{C}^*]}_{A'_{t-1}}$. Therefore, $G^{[\mathcal{C}^*]}_{A'_{t-1} \setminus \{e_t\}}$ splits U into two connected components U_a and U_b, which contain vertices representing C_a and C_b respectively. Since A'_{t-1} is minimal, e_t is not redundant. Thus, both U_a and U_b contain an odd number of odd vertices of $G^{[\mathcal{C}^*]}_{E(F_{t-1})}$. Since deleting $e_t = (a, b)$ from F_{t-1} changes parities of degrees of vertices a and b only, both U_a and U_b contain an even number of odd vertices of $G^{[\mathcal{C}^*]}_{E(F_{t-1}) \setminus \{e_t\}}$. Therefore, A'_t is an auxiliary edge subset of $F_{t-1} \setminus \{e_t\}$ wrt $\mathcal{C}^*$. Since e'_t belongs to a tour of $\mathcal{C}^*$, vertices of $G^{[\mathcal{C}^*]}$ have the same parities in $G^{[\mathcal{C}^*]}_{E(F_t)}$ as they have in $G^{[\mathcal{C}^*]}_{E(F_{t-1}) \setminus \{e_t\}}$. Thus, A'_t is also an auxiliary edge subset of F_t wrt $\mathcal{C}^*$.

Moreover, by the argument above, vertices representing C_a and C_b are the only vertices that have different parities in $G^{[\mathcal{C}^*]}_{E(F_{t-1})}$ and in $G^{[\mathcal{C}^*]}_{E(F_t)}$. Since they are not connected by any path in $G^{[\mathcal{C}^*]}_{E(F_t)}$, and since A'_{t-1} is minimal, A'_t must be a minimal auxiliary forest of F_t wrt $\mathcal{C}^*$.

To prove $\ell(A'_i) \leq (n_A - i)\ell(e_{\max}(\mathcal{C}^*))$, consider every edge $e = (u, v)$ of A^*, where u is the parent of v in F^*. By Lemma 2, the optimal solution $\mathcal{C}^*$ has a unique tour C_v that contains v, and C_v has an edge e' such that $(F^* \setminus \{(u, v)\}) \cup \{e'\}$ is a DRSF wrt (G, D) with length not shorter than $\ell(F^*)$, which implies $\ell(u, v) \leq \ell(e')$. Therefore,

$$\ell(u, v) \leq \ell(e_{\max}(C_v)) \leq \ell(e_{\max}(\mathcal{C}^*)). \tag{9}$$

Since A'_i is a subset of A^* and $|A'_i| = n_A - i$, for $0 \leq i \leq n_A$, we obtain $\ell(A'_i) \leq (n_A - i)\ell(e_{\max}(\mathcal{C}^*))$. Thus F_t satisfies (v) of Lemma 5 for $i = t$. □

As Lemmas 7–11 complete the induction, Lemma 5 is proved. Using Lemma 5, Lemma 3 now follows:

Proof of Lemma 3. Consider a minimal auxiliary edge subset A^* of F^* wrt $\mathcal{C}^*$. If A^* is empty, F^* satisfies conditions (i) and (ii) in Lemma 3 , and so Lemma 3 is proved. Otherwise, A^* is not empty. By definition of A^*, since $\mathcal{C}^*$ contains exactly k tours, we know $|A^*| \leq k - 1$. Adopting notation defined in Lemma 5, since A^* is not empty, $2 \leq k$ and $1 \leq n_A$. By Lemma 5, there exists a pair of disjoint edge subsets $(E^+_{n_A-1}, E^-_{n_A-1})$ satisfying constraints (i)–(v) of Lemma 5 for $i = n_A-1$. Taking $E^+ = E^+_{n_A-1}$, $E^- = A_{n_A-1}$, $F = F_{n_A-1}$, and $A = A'_{n_A-1}$, we can verify Lemma 3 as follows. By constraint (i) of Lemma 5, $|E^+| = n_A - 1 \leq |A^*| - 1 \leq k - 2$ and $E^+ \subseteq E \setminus F^*$. By definition of A_{n_A-1}, we know $E^- \subseteq F^*$ and $|E^-| \leq k - 1$. Due to constraints (iii) and (iv) of Lemma 5, $F = (F^* \setminus E^-) \cup E^+$ is a DRSF wrt (G, D) with length not longer than $[\ell(\mathcal{C}^*) - \ell(e_{\max}(\mathcal{C}^*))]$. By constraint (v) of Lemma 5, A'_{n_A-1} is a minimal auxiliary edge subset of F_{n_A-1} wrt $\mathcal{C}^*$ with length not longer than $\ell(e_{\max}(\mathcal{C}^*))$. Thus, F satisfies conditions (i) and (ii) in Lemma 3, and so, Lemma 3 is true. $\square$

References

1. Arkin, E.M., Hassin, R., Levin, A.: Approximations for minimum and min-max vehicle routing problems. Journal of Algorithms 59(1), 1–18 (2006)
2. Christofides, N.: Worst-case analysis of a new heuristic for the travelling salesman problem, Technical Report 388, Graduate School of Industrial Administration, Carnegie-Mellon University (1976)
3. Eiselt, H.A., Gendreau, M., Laporte, G.: Arc routing problems, part i: The chinese postman problem. Operations Research 43(2), 231–242 (1995)
4. Eiselt, H.A., Gendreau, M., Laporte, G.: Arc routing problems, part ii: The rural postman problem. Operations Research 43(3), 399–414 (1995)
5. Garey, M.R., Johnson, D.S.: Computers and Intractability: a Guide to the Theory of NP-completeness. Freeman, New York (1979)
6. Giosa, I.D., Tansini, I.L., Viera, I.O.: New assignment algorithms for the multi-depot vehicle routing problem. Journal of the Operational Research Society 53(9), 977–984 (2002)
7. Lim, A., Wang, F.: Multi-depot vehicle routing problem: a one-stage approach. IEEE Transactions on Automation Science and Engineering 2(4), 397–402 (2005)
8. Malik, W., Rathinam, S., Darbha, S.: An approximation algorithm for a symmetric generalized multiple depot, multiple travelling salesman problem. Operations Research Letters 35(6), 747–753 (2007)
9. Papadimitriou, C.H., Steiglitz, K.: Approximation algorithms for the traveling salesman problem. In: Combinatorial Optimization: Algorithms and Complexity, pp. 410–419. Courier Dover Publications (1998)
10. Rathinam, S., Sengupta, R.: 3/2-Approximation Algorithm for a Generalized, Multiple Depot, Hamiltonina Path Problem. Technical report, University of California, Berkeley (2007)
11. Rathinam, S., Sengupta, R.: 5/3-approximation algorithm for a multiple depot, terminal hamiltonian path problem. Technical report, University of California, Berkeley (2007)
12. Rathinam, S., Sengupta, R., Darbha, S.: A resource allocation algorithm for multi-vehicle systems with non holonomic constraints. IEEE Transactions on Automation Sciences and Engineering 4(1), 98–104 (2006)
13. Rathinama, S., Senguptab, R.: 3/2-approximation algorithm for two variants of a 2-depot hamiltonian path problem. Operations Research Letters 38(1), 63–68 (2010)
14. Renaud, J., Laporte, G., Boctor, F.F.: A tabu search heuristic for the multi-depot vehicle routing problem. Computers and Operations Research 23(3), 229–235 (1996)
15. Rosenkrantz, D.J., Stearns, R.E., Lewis, P.M.: An analysis of several heuristics for the traveling salesman problem. SIAM Journal on Computing 6(3), 563–581 (1977)
16. Wolsey, L.A., Nemhauser, G.L.: Integer and Combinatorial Optimization, ch. section III.3, p. 664. Wiley-Interscience, Hoboken (1988)

Minimum and Maximum against k Lies

Michael Hoffmann[1], Jiří Matoušek[2,1],
Yoshio Okamoto[3,*], and Philipp Zumstein[1]

[1] Institute of Theoretical Computer Science, Department of Computer Science,
ETH Zurich, Switzerland
{hoffmann,matousek,zuphilip}@inf.ethz.ch

[2] Department of Applied Mathematics and Institute for Theoretical Computer
Science (ITI), Charles University, Prague, Czech Republic
matousek@kam.mff.cuni.cz

[3] Graduate School of Information Science and Engineering,
Tokyo Institute of Technology, Japan
okamoto@is.titech.ac.jp

Abstract. A neat 1972 result of Pohl asserts that $\lceil 3n/2 \rceil - 2$ comparisons are sufficient, and also necessary in the worst case, for finding both the minimum and the maximum of an n-element totally ordered set. The set is accessed via an oracle for pairwise comparisons. More recently, the problem has been studied in the context of the Rényi–Ulam liar games, where the oracle may give up to k false answers. For large k, an upper bound due to Aigner shows that $(k + \mathrm{O}(\sqrt{k}))n$ comparisons suffice. We improve on this by providing an algorithm with at most $(k+1+C)n + \mathrm{O}(k^3)$ comparisons for some constant C. The known lower bounds are of the form $(k + 1 + c_k)n - D$, for some constant D, where $c_0 = 0.5$, $c_1 = \frac{23}{32} = 0.71875$, and $c_k = \Omega(2^{-5k/4})$ as $k \to \infty$.

1 Introduction

We consider an n-element set X with an unknown total ordering $\leq$. The ordering can be accessed via an oracle that, given two elements $x, y \in X$, tells us whether $x < y$ or $x > y$. It is easily seen that the minimum element of X can be found using $n - 1$ comparisons. This is optimal in the sense that $n - 2$ comparisons are not enough to find the minimum element in the worst case.

One of the nice little surprises in computer science is that if we want to find *both* the minimum and the maximum, we can do significantly *better* than finding the minimum and the maximum separately. Pohl [8] proved that $\lceil 3n/2 \rceil - 2$ is the optimal number of comparisons for this problem ($n \geq 2$). The algorithm first partitions the elements of X into pairs and makes a comparison in each pair. The minimum can then be found among the "losers" of these comparisons, while the maximum is found among the "winners."

Here we consider the problem of determining both the minimum and the maximum in the case where the oracle is not completely reliable: it may sometimes

---

* Supported by Global COE Program "Computationism as a Foundation for the Sciences" and Grant-in-Aid for Scientific Research from Ministry of Education, Science and Culture, Japan, and Japan Society for the Promotion of Science.

H. Kaplan (Ed.): SWAT 2010, LNCS 6139, pp. 139–149, 2010.

give a false answer, but only at most k times during the whole computation, where k is a given parameter.

We refer to this model as *computation against k lies*. Let us stress that we admit repeating the same query to the oracle several times, and each false answer counts as a lie. This seems to be the most sensible definition—if repeated queries were not allowed, or if the oracle could always give the wrong answer to a particular query, then the minimum cannot be determined.

So, for example, if we repeat a given query $2k + 1$ times, we always get the correct answer by majority vote. Thus, we can simulate any algorithm with a reliable oracle, asking every question $2k+1$ times, but for the problems considered here, this is not a very efficient way, as we will see.

The problem of finding both the minimum and the maximum against k lies was investigated by Aigner [1], who proved that $(k + O(\sqrt{k}))n$ comparisons always suffice.[1] We improve on this as follows.

Theorem 1. *There is an algorithm that finds both the minimum and the maximum among n elements against k lies using at most $(k + 1 + C)n + O(k^3)$ comparisons, where C is a constant.*

Our proof yields the constant C reasonably small (below 10, say, at least if k is assumed to be sufficiently large), but we do not try to optimize it.

Lower Bounds. The best known lower bounds for the number of comparisons necessary to determine both the minimum and the maximum against k lies have the form $(k+1+c_k)n - D$, where D is a small constant and the c_k are as follows:

- $c_0 = 0.5$, and this is the best possible. This is the result of Pohl [8] for a truthful oracle mentioned above.
- $c_1 = \frac{23}{32} = 0.71875$, and this is again tight. This follows from a recent work by Gerbner, Pálvölgyi, Patkós, and Wiener [5] who determined the optimum number of comparisons for $k = 1$ up to a small additive constant: it lies between $\lceil \frac{87}{32}n \rceil - 3$ and $\lceil \frac{87}{32}n \rceil + 12$. This proves a conjecture of Aigner [1].
- $c_k = \Omega(2^{-5k/4})$ for all k, as was shown by Aigner [1].

The optimal constant $c_1 = \frac{23}{32}$ indicates that obtaining precise answers for $k > 1$ may be difficult.

Related Work. The problem of determining the minimum alone against k lies was resolved by Ravikumar, Ganesan, and Lakshmanan [9], who proved that finding the minimum against k lies can be performed by using at most $(k + 1)n - 1$ comparisons, and this is optimal in the worst case.

The problem considered in this paper belongs to the area of *searching problems against lies* and, in a wider context, is an example of "computation in the presence of errors." This field has a rich history and beautiful results. A prototype problem, still far from completely solved, is the *Rényi–Ulam liar game* from the 1960s, where one wants to determine an unknown integer x between 1 and n,

---

[1] Here and in the sequel, O(.) and Ω(.) hide only absolute constants, independent of both n and k.

an oracle provides comparisons of x with specified numbers, and it may give at most k false answers. We refer to the surveys by Pelc [7] and by Deppe [2] for more information.

2   A Simple Algorithm

Before proving Theorem 1, we explain a simpler algorithm, which illustrates the main ideas but yields a weaker bound. We begin with formulating a generic algorithm, with some steps left unspecified. Both the simple algorithm in this section and an improved algorithm in the next sections are instances of the generic algorithm.

The generic algorithm

1. For a suitable integer parameter $s = s(k)$, we arbitrarily partition the considered n-element set X into n/s groups $X_1, \ldots, X_{n/s}$ of size s each.[a]
2. In each group X_i, we find the minimum m_i and the maximum M_i. The method for doing this is left unspecified in the generic algorithm.
3. We find the minimum of $\{m_1, \ldots, m_{n/s}\}$ against k lies, and independently, we find the maximum of $\{M_1, M_2, \ldots, M_{n/s}\}$ against k lies.

---

[a] If n is not divisible by s, we can form an extra group smaller than s and treat it separately, say—we will not bore the reader with the details.

The correctness of the generic algorithm is clear, provided that Step 2 is implemented correctly. Eventually, we set $s := k$ in the simple and in the improved algorithm. However, we keep s as a separate parameter, because the choice $s := k$ is in a sense accidental.

In the simple algorithm we implement Step 2 as follows.

Step 2 in the simple algorithm

2.1. (Sorting.) We sort the elements of X_i by an asymptotically optimal sorting algorithm, say mergesort, using $O(s \log s)$ comparisons, and ignoring the possibility of lies. Thus, we obtain an ordering $x_1, x_2, \ldots, x_s$ of the elements of X_i such that *if* all queries during the sorting have been answered correctly, *then* $x_1 < x_2 < \cdots < x_s$. If there was at least one false answer, we make no assumptions, except that the sorting algorithm does not crash and outputs some ordering.
2.2. (Verifying the minimum and maximum.) For each $j = 2, 3, \ldots, s$, we query the oracle $k+1$ times with the pair x_{j-1}, x_j. If any of these queries returns the answer $x_{j-1} > x_j$, we *restart*: We go back to Step 2.1 and repeat the computation for the group X_i from scratch. Otherwise, if all the answers are $x_{j-1} < x_j$, we proceed with the next step.
2.3. We set $m_i := x_1$ and $M_i := x_s$.

Lemma 1 (Correctness). *The simple algorithm always correctly computes the minimum and the maximum against k lies.*

Proof. We note that once the processing of the group X_i in the above algorithm reaches Step 2.3, then $m_i = x_1$ has to be the minimum. Indeed, for every other element x_j, $j \geq 2$, the oracle has answered $k+1$ times that $x_j > x_{j-1}$, and hence x_j cannot be the minimum. Similarly, M_i has to be the maximum, and thus the algorithm is always correct. $\square$

Actually, at Step 2.3 we can be sure that $x_1, \ldots, x_s$ is the sorted order of X_i, but in the improved algorithm in the next section the situation will be more subtle. The next lemma shows, that the simple algorithm already provides an improvement of Aigner's bound of $(k + O(\sqrt{k}))n$.

Lemma 2 (Complexity). *The number of comparisons of the simple algorithm for $s = k$ on an n-element set is $(k + O(\log k))n + O(k^3)$.*

Proof. For processing the group X_i in Step 2, we need $O(s \log s) + (k+1)(s-1) = k^2 + O(k \log k)$ comparisons, provided that no restart is required. But since restarts may occur only if the the oracle lies at least once, and the total number of lies is at most k, there are no more than k restarts for all groups together. These restarts may account for at most $k(k^2 + O(k \log k)) = O(k^3)$ comparisons. Thus, the total number of comparisons in Step 2 is $\frac{n}{s}(k^2 + O(k \log k)) + O(k^3) = (k + O(\log k))n + O(k^3)$.

As we mentioned in the introduction, the minimum (or maximum) of an n-element set against k lies can be found using $(k + 1)n - 1$ comparisons, and so Step 3 needs no more than $2(k+1)(n/s) = O(n)$ comparisons. (We do not really need the optimal algorithm for finding the minimum; any $O((k+1)n)$ algorithm would do.) The claimed bound on the total number of comparisons follows. $\square$

3   The Improved Algorithm: Proof of Theorem 1

In order to certify that x_1 is indeed the minimum of X_i, we want that for every x_j, $j \neq 1$, the oracle declares x_j larger than some other element $k + 1$ times. (In the simple algorithm, these $k + 1$ comparisons were all made with x_{j-1}, but any other smaller elements will do.) This in itself requires $(k + 1)(s - 1)$ queries per group, or $(k + 1)(n - n/s)$ in total, which is already close to our target upper bound in Theorem 1 (we note that s has to be at least of order k, for otherwise, Step 3 of the generic algorithm would be too costly).

Similarly, every x_j, $j \neq s$, should be compared with smaller elements $k + 1$ times, which again needs $(k + 1)(n - n/s)$ comparisons, so all but $O(n)$ comparisons in the whole algorithm should better be used for *both* of these purposes.

In the simple algorithm, the comparisons used for sorting the groups in Step 2.1 are, in this sense, wasted. The remedy is to use most of them also for verifying the minimum and maximum in Step 2.2. For example, if the sorting algorithm has already made comparisons of x_{17} with 23 larger elements, in the verification step it suffices to compare x_{17} with $k + 1 - 23$ larger elements.

One immediate problem with this appears if the sorting algorithm compares x_{17} with some $b > k + 1$ larger elements, the extra $b - (k + 1)$ comparisons are wasted. However, for us, this will not be an issue, because we will have $s = k$, and thus each element can be compared to at most $k - 1$ others (assuming, as we may, that the sorting algorithm does not repeat any comparison).

Another problem is somewhat more subtle. In order to explain it, let us represent the comparisons made in the sorting algorithm by edges of an ordered graph. The vertices are $1, 2, \ldots, s$, representing the elements $x_1, \ldots, x_s$ of X_i in sorted order, and the edges correspond to the comparisons made during the sorting, see the figure below on the left.

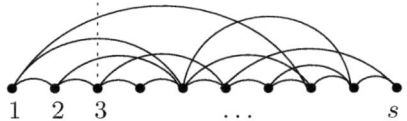
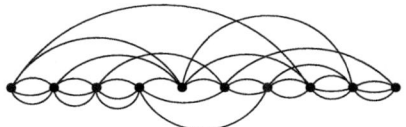

In the verification step, we need to make additional comparisons so that every x_j, $j \neq 1$, has at least $k + 1$ comparisons with smaller elements and every x_j, $j \neq s$, has at least $k + 1$ comparisons with larger elements. This corresponds to adding suitable extra edges in the graph, as in the right drawing above (where $k = 2$, and the added edges are drawn on the bottom side).

As the picture illustrates, sometimes we cannot avoid comparing some element with *more* than $k + 1$ larger ones or $k + 1$ smaller ones (and thus some of the comparisons will be "half-wasted"). For example, no matter how we add the extra edges, the elements x_1, x_2, x_3 together must participate in at least 3 half-wasted comparisons. Indeed, x_2 and x_3 together require 6 comparisons to the left (i.e. with a smaller element). These comparisons can be "provided" only by x_1 and x_2, which together want only 6 comparisons to the right—but 3 of these comparisons to the right were already made with elements larger than x_3 (these are the arcs intersecting the dotted vertical line in the picture).

The next lemma shows that this kind of argument is the only source of wasted comparisons. For an ordered multigraph H on the vertex set $\{1, 2, \ldots, s\}$ as above, let us define $t(H)$, the *thickness* of H, as $\max\{t(j) : j = 2, 3, \ldots, s - 1\}$, where $t(j) := |\{\{a, b\} \in E(H) : a < j < b\}|$ is the number of edges going "over" the vertex j.

Lemma 3. *Let H be an undirected multigraph without loops on $\{1, 2, \ldots, s\}$ such that for every vertex $j = 1, 2, \ldots, s$,*

$$d_H^{\text{left}}(j) := |\{\{i,j\} \in E(H) : i < j\}| \le k+1\,,$$
$$d_H^{\text{right}}(j) := |\{\{i,j\} \in E(H) : i > j\}| \le k+1\,.$$

Then H can be extended to a multigraph $\overline{H}$ by adding edges, so that

(i) *every vertex $j \ne 1$ has at least $k+1$ left neighbors and every vertex $j \ne s$ has at least $k+1$ right neighbors; and*
(ii) *the total number of edges in $\overline{H}$ is at most $(k+1)(s-1)+t(H)$.*

The proof is a network flow argument and therefore constructive. We postpone it to the end of this section.

For a comparison-based sorting algorithm $\mathcal{A}$, we define the *thickness $t_{\mathcal{A}}(s)$* in the natural way: It is the maximum, over all s-element input sequences, of the thickness $t(H)$ of the corresponding ordered graph H (the vertices of H are ordered as in the output of the algorithm and each comparison contributes to an edge between its corresponding vertices). As the above lemma shows, the number of comparisons used for the sorting but not for the verification can be bounded by the thickness of the sorting algorithm.

Lemma 4. *There exists a (deterministic) sorting algorithm $\mathcal{A}$ with thickness $t_{\mathcal{A}}(s) = \mathrm{O}(s)$.*

Proof. The algorithm is based on Quicksort, but in order to control the thickness, we want to partition the elements into two groups of equal size in each recursive step.

We thus begin with computing the median of the given elements. This can be done using $\mathrm{O}(s)$ comparisons (see, e.g., Knuth [6]; the current best deterministic algorithm due to Dor and Zwick [3] uses no more than $2.95s + \mathrm{o}(s)$ comparisons). These algorithms also divide the remaining elements into two groups, those smaller than the median and those larger than the median. To obtain a sorting algorithm, we simply recurse on each of these groups.

The thickness of this algorithm obeys the recursion $t_{\mathcal{A}}(s) \le \mathrm{O}(s)+t_{\mathcal{A}}(\lfloor s/2 \rfloor))$, and thus it is bounded by $\mathrm{O}(s)$. $\qquad\square$

We are going to use the algorithm $\mathcal{A}$ from the lemma in the setting where some of the answers of the oracle may be wrong. Then the median selection algorithm is not guaranteed to partition the current set into two groups of the same size and it is not sure that the running time does not change. However, we can check if the groups have the right size and if the running time does not increase too much. If some test goes wrong, we restart the computation (similar to the simple algorithm).

Now we can describe the improved algorithm, again by specifying Step 2 of the generic algorithm.

Step 2 in the improved algorithm

2.1′. (Sorting.) We sort the elements of X_i by the algorithm $\mathcal{A}$ with thickness $O(s)$ as in Lemma 4. If an inconsistency is detected (as discussed above), we restart the computation for the group X_i from scratch.

2.2′. (Verifying the minimum and maximum.) We create the ordered graph H corresponding to the comparisons made by $\mathcal{A}$, and we extend it to a multigraph $\overline{H}$ according to Lemma 3. We perform the comparisons corresponding to the added edges. If we encounter an inconsistency, then we restart: We go back to Step 2.1′ and repeat the computation for the group X_i from scratch. Otherwise, we proceed with the next step.

2.3′. We set $m_i := x_1$ and $M_i := x_s$.

Proof (of Theorem 1). The correctness of the improved algorithm follows in the same way as for the simple algorithm. In Step 2.2′, the oracle has declared every element x_j, $j \neq 1$, larger than some other element $k+1$ times, and so x_j cannot be the minimum. A similar argument applies for the maximum.

It remains to bound the number of comparisons. From the discussion above, the number of comparisons is at most $((k+1)(s-1)+t_{\mathcal{A}}(s))(\frac{n}{s}+k)+2(k+1)\frac{n}{s}$, with $t_{\mathcal{A}}(s) = O(s)$. For $s = k$, we thus get that the number of comparisons at most $(k+1+C)n + O(k^3)$ for some constant C, as claimed.   □

Proof (of Lemma 3). We will proceed in two steps. First, we construct a multigraph H^* from H by adding a maximum number of (multi)edges such that the left and right degree of every vertex are still bounded above by $k+1$. Second, we extend H^* to $\overline{H}$ by adding an appropriate number of edges to each vertex so that condition (i) holds.

For an ordered multigraph H' on $\{1, 2, \ldots, s\}$ with left and right degrees upper bounded by $k+1$, let us define the *defect* $\Delta(H')$ as

$$\Delta(H') := \sum_{j=1}^{s-1} (k+1 - d_{H'}^{\mathrm{right}}(j)) + \sum_{j=2}^{s} (k+1 - d_{H'}^{\mathrm{left}}(j)) \,.$$

We have $\Delta(H') = 2(k+1)(s-1) - 2e(H')$, where $e(H')$ is the number of edges of H'.

By a network flow argument, we will show that by adding suitable $m^* := (k+1)(s-1) - e(H) - t(H)$ edges to H, one can obtain a multigraph H^* in which all left and right degrees are still bounded by $k+1$ and such that $\Delta(H^*) = 2t(H)$. The desired graph $\overline{H}$ as in the lemma will then be obtained by adding $\Delta(H^*)$ more edges: For example, for every vertex $j \geq 2$ of H^* with $d_{H^*}^{\mathrm{left}}(j) < k+1$, we add $k+1 - d_{H^*}^{\mathrm{left}}(j)$ edges connecting j to 1, and similarly we fix the deficient right degrees by adding edges going to the vertex s.

It remains to construct H^* as above. To this end, we define an auxiliary directed graph G, where each directed edge e is also assigned an integral capacity $c(e)$; see Fig. 1(a).

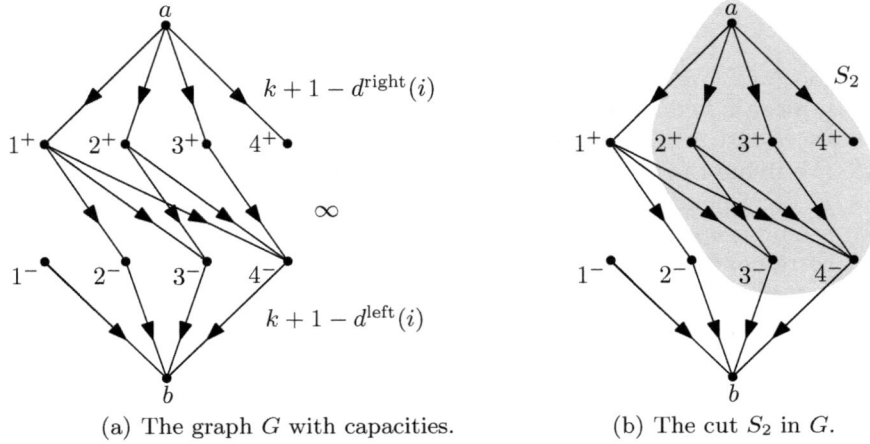

(a) The graph G with capacities.   (b) The cut S_2 in G.

Fig. 1. The directed graph G constructed in the proof of Lemma 3

The vertex set of G consists of a vertex j^- for every $j \in \{1, 2, \ldots, s\}$, a vertex j^+ for every $j \in \{1, 2, \ldots, s\}$, and two special vertices a and b. There is a directed edge in G from a to every vertex j^+ and the capacity of this edge is $k + 1 - d_H^{\text{right}}(j)$. Similarly, there is a directed edge in G from every vertex j^- to b, and the capacity of this edge is $k + 1 - d_H^{\text{left}}(j)$. Moreover, for every i, j with $1 \leq i < j \leq s$, we put the directed edge (i^+, j^-) in G, and the capacity of this edge is ∞ (i.e., a sufficiently large number).

We will check that there is an integral a-b flow in G with value m^* in G. By the max-flow min-cut theorem [4], it suffices to show that every a-b cut in G has capacity at least m^* and there is an a-b cut in G with capacity m^*.

Let $S \subseteq V(G)$ be a minimum a-b cut. Let i be the smallest integer such that $i^+ \in S$. Since the minimum cut cannot use an edge of unbounded capacity, we have $j^- \in S$ for all $j > i$.

We may assume without loss of generality that $j^+ \in S$ for all $j > i$ and $j^- \notin S$ for all $j \leq i$ (the capacity of the cut does not decrease by doing otherwise). Therefore it suffices to consider a-b cuts of the form

$$S_i := \{a\} \cup \{x^+ : x \geq i\} \cup \{x^- : x > i\}$$

for $i = 1, \ldots, s$. The capacity of S_i, see Fig. 1(b), equals

$$\sum_{j<i} c(a, j^+) + \sum_{j>i} c(j^-, b) = (s-1)(k+1) - \sum_{j<i} d_H^{\text{right}}(j) - \sum_{j>i} d_H^{\text{left}}(j).$$

Now let us look at the quantity $\sum_{j<i} d^{\text{right}}(j) + \sum_{j>i} d^{\text{left}}(j)$, and see how much an edge $\{j, j'\}$ $(j < j')$ of H contributes to it: For $j < i < j'$, the contribution is 2, while all other edges contribute 1. Hence the capacity of the cut S_i is $(k+1)(s-1) - e(H) - t(i)$, and the minimum capacity of an a-b-cut is $(k+1)(s-1) - e(H) - t(H) = m^*$ as required.

Thus, there is an integral flow f with value m^* as announced above. We now select the edges to be added to H as follows: For every directed edge (i^+, j^-) of G, we add $f(i^+, j^-)$ copies of the edge $\{i, j\}$, which yields the multigraph H^*. The number of added edges is m^*, the value of the flow f, and the capacity constraints guarantee that all left and right degrees in H^* are bounded by $k+1$. Moreover, the defect of H^* is at most $2t(H)$.   □

4   Concluding Remarks

We can cast the algorithm when $k = 0$ sketched in the introduction into the framework of our generic algorithm. Namely, if we set $s = 2$ and in Step 2 we just compare the two elements in each group, then we obtain that algorithm. The main feature of our algorithm is that every restart only spoils one group. This allows us to keep the effect of lies local.

In order to improve the upper bound of Theorem 1 by the method of this paper, we would need a sorting algorithm with thickness $o(s)$. (Moreover, to make use of the sublinear thickness, we would need to choose s superlinear in k, and thus the sorting algorithm would be allowed to compare every element with only $o(s)$ others.) The following proposition shows, however, that such a sorting algorithm does not exist. Thus, we need a different idea to improve Theorem 1.

Proposition 1. *Every (randomized) algorithm to sort an s-element set has thickness $\Omega(s)$ in expectation.*

Proof. By Yao's principle [10], it is enough to show that every deterministic sorting algorithm $\mathcal{A}$ has expected thickness $\Omega(s)$ for a random input. In our case, we assume that the unknown linear ordering of X is chosen uniformly at random among all the $s!$ possibilities.

In each step, the algorithm $\mathcal{A}$ compares some two elements $x, y \in X$. Let us say that an element $x \in X$ is *virgin* at the beginning of some step if it hasn't been involved in any previous comparison, and elements that are not virgin are *tainted*. A comparison is *fresh* if it involves at least one virgin element.

For notational convenience, we assume that s is divisible by 8. Let $L \subset X$ consist of the first $s/2$ elements in the (random) input order (which is also the order of the output of the algorithm), and let $R := X \setminus L$. Let E_i be the event that the ith fresh comparison is an *LR-comparison*, i.e., a comparison in which one of the two compared elements x, y lies in L and the other in R. We claim that for each $i = 1, 2, \ldots, s/8$, the probability of E_i is at least $\frac{1}{3}$.

To this end, let us fix (arbitrarily) the outcomes of all comparisons made by $\mathcal{A}$ before the ith fresh comparison, which determines the set of tainted elements, and let us also fix the positions of the tainted elements in the input ordering. We now consider the probability of E_i *conditioned* on these choices. The key observation is that the virgin elements in the input ordering are still randomly distributed among the remaining positions (those not occupied by the tainted elements).

Let ℓ be the number of virgin elements in L and r the number of virgin elements in R; we have $s/4 \leq \ell, r \leq s/2$.

We distinguish two cases. First, let only one of the elements x, y compared in the ith fresh comparison be virgin. Say that x is tainted and lies in L. Then the probability of E_i equals $r/(\ell + r) \geq \frac{1}{3}$.

Second, let both of x and y be virgin. Then the probability of E_i is $2\ell r/((\ell + r)(\ell + r - 1))$, and since $s/4 \leq \ell, r \leq s/2$, this probability exceeds $\frac{4}{9}$.

Thus, the probability of E_i conditioned on *every* choice of the outcomes of the initial comparisons and positions of the tainted elements is at least $\frac{1}{3}$, and so the probability of E_i for a random input is at least $\frac{1}{3}$ as claimed. Thus, the expected number of LR-comparisons made by $\mathcal{A}$ is $\Omega(s)$.

Let a be the largest element of L, i.e., the $(s/2)$th element of X, and let b be the smallest element of R, i.e., the $(s/2 + 1)$st element of X. Since we may assume that $\mathcal{A}$ doesn't repeat any comparison, there is at most one comparison of a with b. Every other LR-comparison compares elements that have a or b (or both) between them. Thus, the expected thickness of $\mathcal{A}$ is at least half of the expected number of LR-comparisons, which is $\Omega(s)$. □

Note that the only thing which we needed in the proposition above was that the corresponding ordered graph is simple and has minimum degree at least 1.

Acknowledgments

We thank Döm Pálvölgyi for bringing the problem investigated in this paper to our attention. This work has been started at the 7th Gremo Workshop on Open Problems, Hof de Planis, Stels in Switzerland, July 6–10, 2009. We also thank the participants of the workshop for the inspiring atmosphere.

References

1. Aigner, M.: Finding the maximum and the minimum. Discrete Applied Mathematics 74, 1–12 (1997)
2. Deppe, C.: Coding with feedback and searching with lies. In: Entropy, Search, Complexity. Bolyai Society Mathematical Studies, vol. 16, pp. 27–70 (2007)
3. Dor, D., Zwick, U.: Selecting the median. SIAM Journal on Computing 28, 1722–1758 (1999)
4. Ford Jr., L.R., Fulkerson, D.R.: Maximal flow through a network. Canadian Journal of Mathematics 8, 399–404 (1956)
5. Gerbner, D., Pálvölgyi, D., Patkós, B., Wiener, G.: Finding the maximum and minimum elements with one lie. Discrete Applied Mathematics (to appear)
6. Knuth, D.E.: The Art of Computer Programming. Sorting and Searching, vol. 3. Addison-Wesley Publishing Co., Reading (1973)
7. Pelc, A.: Searching games with errors—fifty years of coping with liars. Theoretical Computer Science 270, 71–109 (2002)

8. Pohl, I.: A sorting problem and its complexity. Communications of the ACM 15, 462–464 (1972)
9. Ravikumar, B., Ganesan, K., Lakshmanan, K.B.: On selecting the largest element in spite of erroneous information. In: Brandenburg, F.J., Wirsing, M., Vidal-Naquet, G. (eds.) STACS 1987. LNCS, vol. 247, pp. 88–99. Springer, Heidelberg (1987)
10. Yao, A.C.-C.: Probabilistic computations: Towards a unified measure of complexity. In: Proceedings of 18th FOCS, pp. 222–227 (1977)

Feasible and Accurate Algorithms for Covering Semidefinite Programs

Garud Iyengar*, David J. Phillips**, and Cliff Stein***

The Department of Industrial Engineering & Operations Research, Columbia University
cliff,garud@ieor.columbia.edu
Mathematics Department, The College of William & Mary
phillips@math.wm.edu

Abstract. In this paper we describe an algorithm to approximately solve a class of semidefinite programs called *covering semidefinite programs*. This class includes many semidefinite programs that arise in the context of developing algorithms for important optimization problems such as Undirected SPARSEST CUT, wireless multicasting, and pattern classification. We give algorithms for covering SDPs whose dependence on ϵ is ϵ^{-1}. These algorithms, therefore, have a better dependence on ϵ than other combinatorial approaches, with a tradeoff of a somewhat worse dependence on the other parameters. For many reasons, including numerical stability and a variety of implementation concerns, the dependence on ϵ is critical, and the algorithms in this paper may be preferable to those of the previous work. Our algorithms exploit the structural similarity between packing and covering semidefinite programs and packing and covering linear programs.

1 Introduction

Semidefinite programming (SDP) is a powerful tool for designing approximation algorithms for NP-hard problems. For example, SDP relaxations are used in the algorithms for the Undirected SPARSEST CUT, c-BALANCED SEPARATOR, and GRAPH CONDUCTANCE [3]. SDP-relaxations are also used in a variety of important applications such as multicast beam-forming [18] and pattern classification [19].

Solving SDPs remains a significant theoretical and practical challenge. Interior point algorithms compute an approximate solution for an SDP with $n \times n$ decision matrices with an absolute error ϵ in $\mathcal{O}(\sqrt{n}(m^3 + n^6)) \cdot \log(\epsilon^{-1}))$ time, where m is the number of constraints [16]. These algorithms have a very low dependence on ϵ, but a high dependence on the other parameters; for example, interior point algorithms require $\mathcal{O}(n^{9.5} \cdot \log(\epsilon^{-1}))$ time to solve the SDPs that arise in Undirected SPARSEST CUT. Thus, interior point algorithms have some drawbacks, and there has been significant work on designing faster algorithms to approximately solve the SDPs that arise in important applications.

---

* Supported in part by NSF grant DMS-06-06712, ONR grant N000140310514 and DOE grant DE-FG02-08ER25856.
** Supported in part by NSF grant DMS-0703532 and a NASA/VSGC New Investigator grant.
*** Supported in part by NSF grants CCF-0728733 and CCF-0915681.

H. Kaplan (Ed.): SWAT 2010, LNCS 6139, pp. 150–162, 2010.

One general class of SDPs with efficient solutions are known as *packing SDPs*. (Packing SDPs are analogous to packing linear programs, see [9] for a precise definition). Examples of packing SDPs include those used to solve MAXCUT, COLORING, Shannon capacity and sparse principal component analysis. For these problems there are essentially three known solution methods, each with its own advantages and drawbacks. The first method is specialized interior point methods which have a good dependence on ϵ, but a poor dependence on the other parameters. A second method, due to Klein and Lu [11], extends ideas of, among others, Plotkin, Shmoys and Tardos [17] for packing linear programs to give algorithms for MAXCUT and coloring that have running times of $\mathcal{O}(nm \log^2(n) \cdot \epsilon^{-2} \log(\epsilon^{-1}))$ and $\mathcal{O}(nm \log^3(n) \cdot \epsilon^{-4})$ respectively on graphs with n nodes and m edges. The drawback of these algorithms is the dependence on ϵ of at least ϵ^{-2}; this bound is inherent in these methods [12] and a significant bottleneck in practice on these types of problems [8,4,9]. A third approach, due to Iyengar, Phillips and Stein[9] also extends ideas from packing linear programs, but starts from the more recent work of Bienstock and Iyengar [5] who build on techniques of Nesterov [14]. Nesterov [15] has also adapted his method to solve the associated saddle point problem with general SDP, although his method does not find feasible solutions. These results for packing SDPs have a dependence of only ϵ^{-1}, but slightly larger dependence on the other parameters than the algorithms of Klein and Lu. The work in [9] is also significant in that it explicitly defines and approximately solves all packing SDPs in a unified manner.

A related approach for a more general class of SDPs is known as the "multiplicative weights method," which appears in several papers [1,2] and generalizes techniques used in packing and covering linear programs, e.g. [6,7]. The work of Arora and Kale [2] gives faster algorithms for solving SDPs (in terms of the problem size), and also extends results to more applications, including Undirected SPARSEST CUT. Their running times achieve a better dependence on n and m than the previous combinatorial algorithms do, but the dependence on ϵ grows to ϵ^{-6} for Undirected SPARSEST CUT. Moreover, they only satisfy each constraint to within an additive error ϵ, and thus, do not necessarily find a feasible solution to the SDP. Their algorithms incorporate a rounding step for each SDP so that finding a feasible solution is not required.

We are not aware of any other work that efficiently finds approximate solutions to large classes of SDPs.

1.1 New Results

In this work, we first define a new class of SDPs that we call *covering SDPs*. Analogous to *covering linear programs* which require that all variables be non-negative and all the constraints are of the form $\mathbf{a}^\top \mathbf{x} \geq 1$, with $\mathbf{a} \geq 0$; in a covering SDP the variable matrix $\mathbf{X}$ is positive semidefinite, and all constraints are of the form $\langle \mathbf{A}, \mathbf{X} \rangle \geq 1$ for positive semidefinite $\mathbf{A}$. We show that several SDPs, including those used in diverse applications such as Undirected SPARSEST CUT, Beamforming, and k-nearest neighbors can be expressed as covering SDPs. We then describe an algorithm for computing a feasible solution to a covering SDP with objective value at most $(1 + \epsilon)$ times the optimal objective value. We call such a solution a relative ϵ-*optimal solution*. (We use the term absolute approximation to refer to an additive approximation bound.) Our

Table 1. Running time comparison. n = matrix dimension, E = number of edges, R = number of receivers, T = number of inputs. We use $\tilde{\mathcal{O}}$ to suppress $\log^k(n)$ factors and all but the highest dependence on ϵ.

Problem	m	This paper	Previous work
Undirected SPARSEST CUT	$\mathcal{O}(n^3)$	$\mathcal{O}(n^4 \cdot \epsilon^{-1})$	$\mathcal{O}(\min\{E^2 \cdot \epsilon^{-4}, n^2 \cdot \epsilon^{-6}\})$ [2]
Beamforming	$\mathcal{O}(R)$	$\mathcal{O}((n^4 + nR) \cdot \epsilon^{-1})$	$\mathcal{O}((R + n^2)^{3.5} \cdot \log(\epsilon^{-1}))$ [18]
k-Nearest Neighbor	$\mathcal{O}(T^2)$	$\mathcal{O}((T^2 n^3 + T^4) \cdot \epsilon^{-1})$	$\mathcal{O}((n^7 + nT^6) \cdot \log(\epsilon^{-1}))$ [19]

algorithm has an ϵ^{-1} dependence, and the dependence on other parameters depends on the specific application. The running times of our algorithm and previous works, applied to three applications, are listed in Table 1. To obtain these results, we first give new SDP formulations of each problem, showing that they are examples of covering SPD. We then apply our main theorem (see Theorem 1 below).

Our algorithm for covering SDP is *not* a simple extension of that for packing SDPs [9]. Several steps that were easy for packing SPDs are not so for covering SDPs and give rise to new technical challenges.

- Computing feasible solutions for covering SDPs is non-trivial; and previous work does *not* compute true feasible solutions to the SDPs. For packing SDPs a feasible solution can be constructed by simply scaling the solution of a Lagrangian relaxation; whereas in covering SDPs, both scaling and shifting by a known strictly feasible point is required.
- In both packing and covering, the SDP is solved by solving a Lagrangian relaxation. In packing, it is sufficient to solve a single Lagrangian relaxation; whereas in covering, we need to solve a sequence of Lagrangian relaxations. The convergence analysis of this scheme is non-trivial.
- Our algorithms use *quadratic* prox functions as opposed to the more usual logarithmic functions [2,9]. Quadratic prox functions avoid the need to compute matrix exponentials and are numerically more stable. The quadratic prox function was motivated by the numerical results in [9].

For Beamforming and k-Nearest Neighbor, our algorithm has an increased dependence on ϵ from $\mathcal{O}(\log(\frac{1}{\epsilon}))$ to $\mathcal{O}(\frac{1}{\epsilon})$ in order to receive a considerable improvement in runtime with respect to the other problem parameters. For Undirected SPARSEST CUT, our algorithm reduces the dependence on ϵ at the expense of an increase in some of the other terms in the running time.

Background and notation. A semidefinite program(SDP) is an optimization problem of the form $\min\{\langle \mathbf{C}, \mathbf{X} \rangle : \langle \mathbf{A}_i, \mathbf{X} \rangle \geq b_i, i = 1, \ldots, m, \mathbf{X} \succeq \mathbf{0}\}$ where $\mathbf{C} \in \mathbb{R}^{n \times n}$ and $\mathbf{A}_i \in \mathbb{R}^{n \times n}, i = 1, \ldots, m$, decision variable $\mathbf{X} \in \mathbb{R}^{n \times n}$ and $\langle ., . \rangle$ denotes the usual *Frobenius* inner product $\langle \mathbf{A}, \mathbf{B} \rangle = \mathbf{Tr}(\mathbf{A}^\top \mathbf{B}) = \sum_{j=1}^{n} \sum_{i=1}^{n} A_{ij} B_{ij}$. The constraint $\mathbf{X} \succeq \mathbf{0}$ indicates that the symmetric matrix $\mathbf{X}$ is *positive semidefinite*, i.e. $\mathbf{X}$ has nonnegative eigenvalues. We use $\mathcal{S}^n$ and $\mathcal{S}_+^n$ to denote the space of $n \times n$ symmetric and positive semidefinite matrices, respectively, and omit the superscript n when the dimension is clear. For a matrix $\mathbf{X} \in \mathcal{S}^n$, we let $\lambda_{\max}(\mathbf{X})$ denote the largest eigenvalue of $\mathbf{X}$.

2 The Covering SDP

We define the *covering SDP* as follows:

$$\nu^* \quad = \quad \min \langle \mathbf{C}, \mathbf{X} \rangle \tag{1}$$
$$\text{s.t.} \ \langle \mathbf{A}_i, \mathbf{X} \rangle \geq 1, \quad i = 1, \ldots, m$$
$$\mathbf{X} \subseteq \mathcal{X} := \{\mathbf{X} : \mathbf{X} \succeq 0, \mathrm{Tr}(\mathbf{X}) \leq \tau\}$$

where $\mathbf{A}_i \succeq 0, i = 1, \ldots, m$, $\mathbf{C} \succeq 0$ and the $\mathcal{X} \subset \mathcal{S}_+$ is a set over which linear optimization is "easy". We refer to the constraints of the form $\langle \mathbf{A}, \mathbf{X} \rangle \geq 1$ for $\mathbf{A} \succeq 0$ as *cover constraints*. We assume the following about the covering SDP (1):

(a) We can compute $\mathbf{Y} \in \mathcal{X}$ such that $\min_{1 \leq i \leq m}\{\langle \mathbf{A}_i, \mathbf{Y} \rangle\} \geq 1 + \frac{1}{q(n,m)}$ for some positive function $q(n, m)$, i.e. $\mathbf{Y}$ is *strictly* feasible with the *margin of feasibility* at least $\frac{1}{q(n,m)}$.
(b) We can compute a lower bound $\nu_L \in \mathbb{R}$ for (1) such that $\nu_U := \langle \mathbf{C}, \mathbf{Y} \rangle \leq \nu_L \cdot p(n, m)$ for some positive function $p(n, m)$, i.e. the *relative error* of $\mathbf{Y}$ can be bounded above by $p(n, m)$.

Each appliction we consider satisfies these assumptions. Our main result is that we can find ϵ-optimal solutions to covering SDPs efficiently.

Theorem 1. *Suppose a covering SDP (1) satisfies Assumptions (a) and (b). Then a relative ϵ-optimal solution can be found in $\mathcal{O}\left(\tau q(n, m)(T_G + \kappa(m)) \|A\| \rho \cdot \frac{1}{\epsilon}\right)$ time where $\|A\| = \max_{1 \leq i \leq m} \lambda_{\max}(\mathbf{A}_i)$, $\rho = \log(p(n, m)\frac{1}{\epsilon})\sqrt{\ln(m)}$ are the logarithmic factors, T_G denotes the running time for computing the negative eigenvalues and the corresponding eigenvectors for a matrix of the form $\mathbf{C} - \sum_{i=1}^m v_i \mathbf{A}_i$, $\mathbf{v} \in \mathbb{R}^n$, and $\kappa(m)$ the running time to calculate $\langle \mathbf{A}_i, \mathbf{Z} \rangle$ for all i and any $\mathbf{Z} \in \mathcal{X}$.*

We prove this theorem to Section 3.3. For the results in Table 1 we set $T_G = n^3$ and calculate $\kappa(m)$ as described in their individual sections. In the remainder of this section, we describe how to formulate our applications as covering SDPs.

2.1 Undirected SPARSEST CUT

Let $\mathcal{G} = (\mathcal{V}, \mathcal{E})$ denote a connected undirected graph with $n = |\mathcal{V}|$ nodes and $E = |\mathcal{E}|$ edges. Let $\mathbf{L}$ denote the Laplacian of $\mathcal{G}$, $\mathbf{K} = n\mathbf{I} - \mathbf{J}$ denote the Laplacian of the complete graph with unit edge weights and $\mathbf{K}_{ij}$ denote the Laplacian of the graph with a single undirected edge (i, j).

As in [3], we assume that $\mathcal{G}$ is unweighted and connected. In [10] we show that the ARV formulation of [3] is equivalent to the following new covering SDP formulation.

USC ARV Formulation	USC Covering SDP Formulation
min $\langle \mathbf{L}, \mathbf{X} \rangle$	min $\langle \mathbf{L}, \mathbf{X} \rangle$
s.t. $\langle \mathbf{K}, \mathbf{X} \rangle = 1$	s.t. $\langle \mathbf{K}, \mathbf{X} \rangle \geq 1$
$\forall i, j, k \ \langle \mathbf{K}_{ij} + \mathbf{K}_{jk} - \mathbf{K}_{ik}, \mathbf{X} \rangle \geq 0$	$\frac{n}{4} \langle \mathbf{K}_{ij} + \mathbf{K}_{jk} - \mathbf{K}_{ik} + 2\mathbf{I}, \mathbf{X} \rangle \geq 1, \ \forall i, j, k$
$\mathbf{X} \succeq 0.$	$\mathbf{X} \in \mathcal{X} = \{\mathbf{Y} : \mathbf{Y} \succeq 0, \mathrm{Tr}(\mathbf{Y}) \leq \frac{2}{n}\}$

Both formulations have $m = \mathcal{O}(n^3)$ constraints. To obtain the covering SDP formulation we relax the first set of constraints, add a trace constraint $\mathbf{Tr}(\mathbf{X}) \leq \frac{2}{n}$, and then shift the second set of linear constraints. Although a trace bound of $\frac{1}{n}$ is sufficient for equivalence, we need the bound to be $\frac{2}{n}$ in order to construct a feasible solution that satisfies Assumption (a).

We now show that Assumption (a) is satisfied. Let $\mathbf{Y} = \frac{2}{n^2(n-1)}\mathbf{K} \succeq 0$. Then $\mathbf{Tr}(\mathbf{Y}) = \frac{2}{n^2(n-1)}(n(n-1)) = \frac{2}{n}$, thus, $\mathbf{Y} \in \mathcal{X}$. For all i, j, k,

$$\frac{n}{4}\left\langle \mathbf{K}ij + \mathbf{K}_{jk} - \mathbf{K}_{ik} + 2\mathbf{I}, \mathbf{Y} \right\rangle = \frac{1}{2n(n-1)}\left\langle \mathbf{K}_{ij} + \mathbf{K}_{jk} - \mathbf{K}_{ik}, \mathbf{K} \right\rangle + 1 = \frac{1}{n-1} + 1.$$

Also, $\langle \mathbf{K}, \mathbf{Y} \rangle = 2 > 1 + \frac{1}{n}$. Thus, $\mathbf{Y}$ satisfies Assumption (a) with $q(n, m) = n$.

Since $\mathcal{G}$ is unweighted, the upper bound $\nu_U = \langle \mathbf{L}, \mathbf{Y} \rangle = \langle \mathbf{L}, \mathbf{K} \rangle = 2$. It was shown in [3] that the ARV formulation has a lower bound of $\frac{1}{\sqrt{\log(n)}}$, i.e., $\nu_L = \frac{1}{\sqrt{\log(n)}}$; thus, $p(m, n) = 2\sqrt{\log(n)}$. Finally, note that $\|A\| = \lambda_{\max}(\mathbf{K}) = n$. Then, since $\tau = \frac{2}{n}$, Theorem 1 implies the following result.

Corollary 1. *An ϵ-optimal solution to the Undirected* SPARSEST CUT *SDP can be found in* $\mathcal{O}\left(n^4\sqrt{\ln(n)} \cdot \frac{1}{\epsilon}\right)$ *time.*

2.2   Beamforming

Sidiropoulos et al [18] consider a wireless multicast scenario with a single transmitter with n antenna elements and R receivers each with a single antenna. Let $\mathbf{h}_i \in \mathbb{C}^n$ denote the complex vector that models the propagation loss and the phase shift from each transmit antenna to the receiving antenna of user $i = 1, \ldots, R$. Let $\mathbf{w}^* \in \mathbb{C}^n$ denote the beamforming weight vector applied to the n antenna elements. Suppose the transmit signal is zero-mean white noise with unit power and the noise at receiver i is zero-mean with variance σ_i^2. Then the received signal to noise ratio (SNR) at receiver i is $|\mathbf{w}^*\mathbf{h}_i|^2 / \sigma_i^2$. Let ρ_i denote the minimum SNR required at receiver i. Then the problem of designing a beamforming vector $\mathbf{w}^*$ that minimizes the transmit power subject to constraints on the received SNR of each user can be formulated as the optimization problem, $\min\{\|\mathbf{w}\|_2^2 : |\mathbf{w}^*\mathbf{h}_i|^2 \geq \rho_i\sigma_i^2, i = 1, \ldots, R\}$. In [18], the authors show that this optimization problem is NP-hard and formulate the following SDP relaxation and its dual.

$$\begin{array}{ll} & \min \langle \mathbf{I}, \mathbf{X} \rangle, \\ \text{(B)} & \text{s.t. } \langle \mathbf{Q}_i, \mathbf{X} \rangle \geq 1, i = 1, \ldots, R, \\ & \mathbf{X} \succeq 0, \end{array} \qquad \begin{array}{l} \max \sum_{i=1}^{R} v_i, \\ \text{s.t. } \sum_{i=1}^{R} v_i \mathbf{Q}_i \preceq \mathbf{I}, \\ \mathbf{v} \geq 0. \end{array}$$

where $\mathbf{Q}_i := \frac{\gamma}{\rho_i\sigma_i^2}\left(\mathbf{g}_i\mathbf{g}_i^\top + \bar{\mathbf{g}}_i\bar{\mathbf{g}}_i^\top\right)$, $\mathbf{g}_i = (\mathbf{Re}(\mathbf{h}_i)^\top \quad \mathbf{Im}(\mathbf{h}_i)^\top)^\top$, $\bar{\mathbf{g}}_i = (\mathbf{Im}(\mathbf{h}_i)^\top \quad \mathbf{Re}(-\mathbf{h}_i)^\top)^\top$, $\mathbf{X} \in \mathcal{S}^{2n}$, and γ is such that $\min_{1 \leq i \leq R} \mathbf{Tr}(\mathbf{Q}_i) = 1$.

Arguments similar to Undirected SPARSEST CUT (see [10]) show that $\tau = 2n$, $q(R, n) = \mathcal{O}(1)$, $\nu_L = 2n$, $\kappa(R) = R$, and $p(n, R) = n$. Thus, Theorem 1 implies the following result.

Corollary 2. *A relative ϵ-optimal feasible solution for the beamforming SDP (B) can be computed in $\tilde{\mathcal{O}}\left(n(n^3 + R) \cdot \frac{1}{\epsilon}\right)$ time.*

2.3 k-Nearest Neighbor Classification

Weinberger et al [19] consider the following model for pattern classification. Let $\mathcal{G} = (\mathcal{V}, \mathcal{E})$ be a graph where $T = |\mathcal{V}|$ and each node $i \in \mathcal{V}$ has an *input*, $(\mathbf{v}_i, y_i), i = 1, \ldots, T$, where $\mathbf{v}_i \in \mathbb{R}^n$ and y_i are labels from a finite discrete set. For each i, there are a constant k adjacent nodes in $\mathcal{G}$, i.e., $\mathcal{G}$ is k-regular, and if $(i, j) \in \mathcal{E}$, then i and j are "near" with respect to some distance function on $\mathbf{v}_i$ and $\mathbf{v}_j$. The goal is to use the T inputs to derive a linear transformation, $\mathbf{H} \in \mathbb{R}^{n \times n}$, so that for input i, $\mathbf{H}\mathbf{v}_i$ is still near its k nearest neighbors but "far" away from inputs j that do not share the same label, i.e., $y_i \neq y_j$. Let $\mathcal{F} = \{(i, j, \ell) : (i, j) \in \mathcal{E}, y_i \neq y_\ell\}$. Let $\mathbf{L}$ denote the Laplacian associated with $\mathcal{G}$ and $\mathbf{C} = \begin{pmatrix} \mathbf{L} & \mathbf{0} \\ \mathbf{0} & c\mathbf{I} \end{pmatrix}$ where $\mathbf{0}$ denotes an appropriately sized matrix of all zeros and $c > 0$ is a given constant. Also, let $\hat{\mathbf{A}}_{ij\ell}$ denote that diagonal block matrix with $\mathbf{K}_{i\ell} - \mathbf{K}_{ij}$ (the edge Laplacians to (i, j) and (i, ℓ) respectively). Finally, let $\mathbf{A}_{ij\ell} = \hat{\mathbf{A}}_{ij\ell} + \hat{\mathbf{I}}$ where $\hat{\mathbf{I}}$ denotes the block diagonal matrix with $\mathbf{I}$ in the upper $n \times n$ block and zeros everywhere else. In [10], we show that the following two formulations are equivalent.

kNN WBS Formulation	kNN Covering SDP Formulation
min $\langle \mathbf{C}, \mathbf{X} \rangle$	min $\langle \mathbf{C}, \mathbf{X} \rangle$
s.t. $\left\langle \hat{\mathbf{A}}_{ij\ell}, \mathbf{X} \right\rangle \geq 1 \ (i, j, \ell) \in \mathcal{E}_D$	s.t. $\langle \mathbf{A}_{ij\ell}, \mathbf{X} \rangle \geq 1 \ (i, j, \ell) \in \mathcal{E}_D$
$\mathbf{X} \succeq \mathbf{0}$	$\mathrm{Tr}(\mathbf{X}) \leq kn$
	$\mathbf{X} \succeq \mathbf{0}.$

The WBS formulation is due to Weinberger, Blitzer and Saul [19]. To obtain the covering SDP formulation, we add the trace constraint trace constraint $\mathbf{Tr}(\mathbf{X}) \leq kT$ and shift the second set of constraints as we did in Undirected SPARSEST CUT. Note the number of constraints is $m = kT^2 = \mathcal{O}(T^2)$ Arguments similar to those used to construct covering SDP formulations for Undirected SPARSEST CUT show that $\nu_L = 1$, $\nu_U = p(n, m) = q(n, m) = \mathcal{O}(T)$ (see [10]). Thus, we have the following corollary to Theorem 1.

Corollary 3. *An ϵ-optimal solution to the k-Nearest Neighbors covering SDP can be found in $\mathcal{O}(T^2(n^3 + T^2)\sqrt{\log(T)} \cdot \frac{1}{\epsilon})$ time.*

3 Computing a Relative ϵ-Optimal Solution for a Covering SDP

In this section we describe the steps of our solution algorithm SOLVECOVERSDP (See Figure 1).

– We start with the Lagrangian relaxation of (1)

$$\nu_\omega^* = \min_{\mathbf{X} \in \mathcal{X}} \max_{v \in \mathcal{V}} \left\{ \left\langle \mathbf{C} - \omega \sum_{i=1}^m v_i \mathbf{A}_i, \mathbf{X} \right\rangle + \omega \sum_{i=1}^m v_i \right\}, \tag{2}$$

where $\mathcal{V} = \{\mathbf{v} : \mathbf{v} \geq \mathbf{0}, \sum_{i=1}^{n} v_i \leq 1\}$ and penalty multiplier ω controls the magnitude of the dual variables. Lagrange duality implies that we need $\omega \rightarrow \infty$ to ensure strict feasibility.

- In Section 3.2 we show that an adaptation of an algorithm due to Nesterov [14] allows us to compute an ϵ-saddle-point, i.e. $(\hat{\mathbf{v}}, \hat{\mathbf{X}})$ such that

$$\max_{\mathbf{v} \in \mathcal{V}} \left\{ \left\langle \mathbf{C} - \omega \sum_{i=1}^{m} v_i \mathbf{A}_i, \hat{\mathbf{X}} \right\rangle + \omega \sum_{i=1}^{m} v_i \right\} - \min_{\mathbf{X} \in \mathcal{X}} \left\{ \left\langle \mathbf{C} - \omega \sum_{i=1}^{m} \hat{v}_i \mathbf{A}_i, \mathbf{X} \right\rangle + \omega \sum_{i=1}^{m} \hat{v}_i \right\} \leq \epsilon,$$

  in $\tilde{\mathcal{O}}(\frac{\|\mathbf{A}\|\omega\tau}{\epsilon})$ iterations of a Nesterov non-smooth optimization algorithm [14], where $\|\mathbf{A}\| = \max_{1 \leq i \leq m} \lambda_{\max}(\mathbf{A}_i)$; thus, large ω leads to larger running times.
- In Section 3.1 we show that an ϵ-saddle-point can be converted into a relative ϵ-optimal solution provided $\omega \geq \frac{1}{g(\mathbf{Y})} \cdot \left(\frac{\nu_U - \nu_L}{\nu_L}\right)$, where ν_U (resp. ν_L) is an upper (resp. lower) bound on the optimal value ν^* of the covering SDP (1), and $g(\mathbf{Y}) = \min_{1 \leq i \leq m} \langle \mathbf{A}_i, \mathbf{Y} \rangle - 1$ denotes the feasibility margin for any strictly feasible point $\mathbf{Y}$. Assumptions (a) and (b) guarantee that $\frac{\nu_U - \nu_L}{\nu_L} \leq p(n, m)$ and that there exists a feasible $\mathbf{Y}$ with $g(\mathbf{Y}) \geq \frac{1}{q(n,m)}$. Thus, it follows that one can compute an ϵ-optimal solution in $\tilde{\mathcal{O}}\left(\frac{p(n,m)q(n,m)\|\mathbf{A}\|}{\epsilon}\right)$ iterations.
- In Section 3.3 we show that by solving a sequence of covering SDPs we can reduce the overall number of iterations to $\tilde{\mathcal{O}}\left(\frac{q(n,m)\|\mathbf{A}\|}{\epsilon}\right)$, i.e. reduce it by a factor $p(n, m)$.

The running time per iteration is dominated by an eigenvalue computation and the solution of an optimization problem via an active set method.

3.1   Rounding the Lagrangian Saddle Point Problem

Let $g(\mathbf{Y}) = \min_{i=1,\ldots,m} \{\langle \mathbf{A}_i, \mathbf{Y} \rangle - b_i\}$ denote the feasibility margin with respect to the covering constraint in (1). Then $\mathbf{Y} \in \mathcal{X}$ is *strictly* feasible for (1) if, and only if, $g(\mathbf{Y}) > 0$.

Lemma 1. *Let* $\mathbf{Y} \in \mathcal{X}$ *be a strictly feasible solution to (1), so* $g(\mathbf{Y}) > 0$. *Define* $\bar{\omega} := \frac{\langle \mathbf{C}, \mathbf{Y} \rangle - \nu^*}{g(\mathbf{Y})}$, *and assume* $\bar{\omega} > 0$. *Choose* $\omega \geq \bar{\omega}$, *and suppose* $(\hat{\mathbf{X}}, \hat{\mathbf{v}})$ *is a δ-saddle-point for (2) i.e.*

$$\max_{\mathbf{v} \in \mathcal{V}} \left\{ \left\langle \hat{\mathbf{C}} - \omega \sum_{i=1}^{m} v_i \mathbf{A}_i, \hat{\mathbf{X}} \right\rangle + \omega \sum_{i=1}^{m} v_i \right\} - \min_{\mathbf{X} \in \mathcal{X}} \left\{ \left\langle \hat{\mathbf{C}} - \omega \sum_{i=1}^{m} v_i \mathbf{A}_i, \mathbf{X} \right\rangle + \omega \sum_{i=1}^{m} \hat{v}_i \right\} \leq \delta.$$

Then $\overline{\mathbf{X}} = \frac{\hat{\mathbf{X}} + \beta(\hat{\mathbf{X}})\mathbf{Y}}{1 + \beta(\hat{\mathbf{X}})}$, *where* $\beta(\hat{\mathbf{X}}) = g(\hat{\mathbf{X}})^- / g(\mathbf{Y})$, *is feasible and absolute δ-optimal for (1).*

Proof. We first show that $\overline{\mathbf{X}}$ is feasible to (1). Since $\overline{\mathbf{X}}$ is a convex combination of $\hat{\mathbf{X}}$ and $\mathbf{Y}$, $\overline{\mathbf{X}} \in \mathcal{X}$. The definition of $\beta(\hat{\mathbf{X}})$ together with the concavity of g implies

$$g(\overline{\mathbf{X}}) \geq \frac{1}{1 + \beta(\hat{\mathbf{X}})} \cdot g(\hat{\mathbf{X}}) + \frac{\beta(\hat{\mathbf{X}})}{1 + \beta(\hat{\mathbf{X}})} \cdot g(\mathbf{Y}) = \frac{1}{1 + \beta(\hat{\mathbf{X}})} \left(g(\hat{\mathbf{X}}) + g(\hat{\mathbf{X}})^- \right) \geq 0.$$

Thus, $\overline{\mathbf{X}}$ is feasible for (1). All that remains is to show that $\left\langle \hat{\mathbf{C}}, \overline{\mathbf{X}} \right\rangle \leq \nu^* + \delta$. First, we show that if the penalty $\omega \geq \bar{\omega}$, then $\nu_\omega^* = \nu^*$. Since $g(\mathbf{X})^- = 0$ for all $\mathbf{X}$ feasible for (1) it follows that $\nu_\omega^* \leq \nu^*$; therefore, we must show that $\nu_\omega^* \geq \nu^*$ when $\omega \geq \bar{\omega}$. Fix $\mathbf{X} \in \mathcal{X}$ and let $\mathbf{X}^\beta = (\mathbf{X} + \beta(\mathbf{X})\mathbf{Y})/(1 + \beta(\mathbf{X}))$, which means $\mathbf{X} = (1 + \beta(\mathbf{X}))\mathbf{X}^\beta - \beta(\mathbf{X})\mathbf{Y}$. Then, by the previous argument, $g(\mathbf{X}^\beta) \geq 0$ so $\left\langle \hat{\mathbf{C}}, \mathbf{X}^\beta \right\rangle \geq \nu^*$. Also,

$$\left\langle \hat{\mathbf{C}}, \mathbf{X} \right\rangle + \omega g(\mathbf{X})^- - \nu^* = (1 + \beta(\mathbf{X}))\underbrace{\left(\left\langle \hat{\mathbf{C}}, \mathbf{X}^\beta \right\rangle - \nu^* \right)}_{\geq 0} + \underbrace{(\omega - \bar{\omega})g(\mathbf{X})^-}_{\geq 0}. \tag{3}$$

Thus, if $\omega \geq \bar{\omega}$ then $\nu_\omega^* = \nu^*$.

We can now show that an $\overline{\mathbf{X}}$ is δ-optimal for (1) when $\omega \geq \bar{\omega}$. Since $\hat{\mathbf{X}}$ is a δ-saddle-point, it follows that

$$\max_{\mathbf{v} \in \mathcal{V}} \left\{ \left\langle \hat{\mathbf{C}} - \omega \sum_{i=1}^m v_i \mathbf{A}_i, \hat{\mathbf{X}} \right\rangle + \omega \sum_{i=1}^m v_i \right\} = \left\langle \hat{\mathbf{C}}, \hat{\mathbf{X}} \right\rangle + \omega g(\hat{\mathbf{X}})^- \leq \nu_\omega^* + \delta = \nu^* + \delta.$$

Thus, the same argument used in (3) indicates that

$$\delta \geq \left\langle \hat{\mathbf{C}}, \hat{\mathbf{X}} \right\rangle + \omega g(\hat{\mathbf{X}})^- - \nu^* = (1 + \beta(\hat{\mathbf{X}}))\left(\left\langle \hat{\mathbf{C}}, \overline{\mathbf{X}} \right\rangle - \nu^* \right) + (\omega - \bar{\omega})g(\hat{\mathbf{X}})^-$$

Since $\omega \geq \bar{\omega}$ and $g(\hat{\mathbf{X}})^- \geq 0$, it follows that $\left\langle \hat{\mathbf{C}}, \overline{\mathbf{X}} \right\rangle - \nu^* \leq \frac{\delta}{1 + \beta(\hat{\mathbf{X}})} \leq \delta$. □

A version of Lemma 1 was established independently by Lu and Monteiro [13].

3.2   Solving the Saddle Point Problems

In this section we describe how to use a variant of the Nesterov non-smooth optimization algorithm [14] to compute δ-optimal saddle-points for the minimax problem (2). We assume some familiarity with the Nesterov algorithm [14]. Let $f(\mathbf{v})$ denote the dual function (or, equivalently the objective function of the $\mathbf{v}$-player): $f(\mathbf{v}) = \sum_{i=1}^m v_i + \omega\tau \min_{\mathbf{X} \in \bar{\mathcal{X}}} \left\{ \left\langle \mathbf{C} - \sum_{i=1}^m v_i \mathbf{A}_i, \mathbf{X} \right\rangle \right\}$, where $\bar{\mathcal{X}} = \{ \mathbf{X} \in \mathcal{S}_+^n : \mathbf{Tr}(\mathbf{X}) \leq 1 \}$. We wish to compute an approximate solution for $\max_{\mathbf{v} \in \mathcal{V}} f(\mathbf{v})$.

In order to use the Nesterov algorithm we need to smooth the non-smooth function f using a strongly convex *prox* function. We smooth f using the spectral quadratic prox function $\sum_{i=1}^n \lambda_i^2(\mathbf{X})$, where $\{\lambda_i(\mathbf{X}) : i = 1, \ldots, n\}$ denotes the eigenvalues of $\mathbf{X}$. Let

$$f_\alpha(\mathbf{v}) := \sum_{i=1}^m v_i + \omega\tau \min_{\mathbf{X} \in \bar{\mathcal{X}}} \left\{ \left\langle \mathbf{C} - \sum_{i=1}^m v_i \mathbf{A}_i, \mathbf{X} \right\rangle + \frac{\alpha}{2} \sum_{i=1}^n \lambda_i^2(\mathbf{X}) \right\}, \tag{4}$$

The Nesterov algorithm requires that $\alpha = \frac{\delta}{D_x}$, where $D_x = \max_{\mathbf{X} \in \bar{\mathcal{X}}} \frac{1}{2}\lambda_i^2(\mathbf{X}) = \frac{1}{2}$.

To optimize $f_\alpha(\mathbf{v})$ the Nesterov algorithm uses a particular regularized gradient descent method where each step involves solving a problem of the form

$$\max_{\mathbf{v} \in \mathcal{V}} \left\{ \bar{\mathbf{g}}^\top \mathbf{v} - \frac{L}{\epsilon} \cdot d(\mathbf{v}, \bar{\mathbf{v}}) \right\}, \tag{5}$$

where $\bar{\mathbf{g}}$ is either a gradient computed at the current iterate ($\mathbf{y}^{(k)}$ computation (see [10]) or a convex combination of gradients of f_α computed at all the previous iterates ($\mathbf{z}^{(k)}$ computation, and $d(\mathbf{v}, \bar{\mathbf{v}})$ denotes the Bregman distance associated with a strongly convex function $\phi(\mathbf{v})$. We use $d(\mathbf{v}, \bar{\mathbf{v}}) = \sum_{i=1}^{n} v_i \ln(v_i/\bar{v}_i)$. Using results from [14], it is easy to show that for this choice of the Bregman distance, the constant $L = \omega^2 \tau^2 \|\mathbf{A}\|^2$, where $\|\mathbf{A}\| = \max_{1 \le i \le m} \lambda_{\max}(\mathbf{A}_i)$ and (5) can be solved in closed form in $\mathcal{O}(m)$ operations.

Let $\mathbf{X}^* = \mathrm{argmin}_{\mathbf{X} \in \bar{\mathcal{X}}} \left\{ \left\langle \hat{\mathbf{C}} - \sum_{i=1}^{m} v_i \mathbf{A}_i, \mathbf{X} \right\rangle + \frac{\alpha}{2} \lambda_i^2(\mathbf{X}) \right\}$. Then $\nabla f_\alpha(\mathbf{v})^\top = \mathbf{1}^\top + \tau\omega \left(\langle \mathbf{A}_1, \mathbf{X}^* \rangle \dots \langle \mathbf{A}_m, \mathbf{X}^* \rangle \right)^\top$ by the envelope theorem. We show below how to compute $\mathbf{X}^*$.

Let $\mathbf{X} = \sum_{i=1}^{n} \lambda_i \mathbf{u}_i \mathbf{u}_i^\top$ and $\hat{\mathbf{C}} - \sum_{i=1}^{m} v_i \mathbf{A}_i = \sum_{i=1}^{n} \theta_i \mathbf{w}_i \mathbf{w}_i^\top$ denote the eigendecompositions of $\mathbf{X}$ and $\mathbf{C} - \sum_{i=1}^{m} v_i \mathbf{A}_i$, respectively. Suppose we fix the eigenvalues $\{\lambda_i\}$ of $\mathbf{X}$ and optimize over the choice of eigenvectors $\{\mathbf{u}_i\}$. Then it is easy to show that $\min_{\mathbf{U}: \mathbf{U}^\top \mathbf{U} = \mathbf{I}} \left\{ \left\langle \hat{\mathbf{C}} - \sum_{i=1}^{m} v_i \mathbf{A}_i, \mathbf{X} \right\rangle \right\} = \sum_{i=1}^{n} \lambda_{(i)} \theta_{[i]}$, where $\lambda_{(i)}$ denotes the i-th *smallest* eigenvalue of $\mathbf{X}$ and $\theta_{[i]}$ denotes the i-th *largest* eigenvalue of $\hat{\mathbf{C}} - \sum_{i=1}^{m} v_i \mathbf{A}_i$, and the minimum is achieved by setting eigenvector $\mathbf{u}_{(i)}$ corresponding to $\lambda_{(i)}$ equal to the eigenvector $\mathbf{w}_{[i]}$ corresponding to $\theta_{[i]}$. Since the prox-function $\frac{1}{2} \sum_{i=1}^{n} \lambda_i^2(\mathbf{X})$ is invariant with respect to the eigenvectors of $\mathbf{X}$, it follows that, by suitably relabeling indices, the expression for the function f_α simplifies to

$$f_\alpha(\mathbf{v}) = \sum_{i=1}^{m} v_i + \omega\tau \min \left\{ \sum_{i=1}^{n} \theta_i \lambda_i + \frac{\alpha}{2} \sum_{i=1}^{n} \lambda_i^2 : \sum_{i=1}^{n} \lambda_i \le 1, \lambda_i \ge 0 \right\}, \qquad (6)$$

and $\mathbf{X}^* = \sum_{i=1}^{n} \lambda_i^* \mathbf{w}_i \mathbf{w}_i^T$, where l^* achieves the minimum in (6). As shown in [10], there is an active set method that solves (6) in $\mathcal{O}(n \log n)$ time. Thus, the computational effort in calculating ∇f_α is the dominated by the effort required to compute the eigendecomposition $(\nu_i, \mathbf{u}_i)$. However, we only have to compute the negative eigenvalues the corresponding eigenvectors. Since $\mathbf{Tr}(\mathbf{A}_k^\top \mathbf{X}^*) = \sum_{i=1}^{n} \lambda_i^* (\mathbf{u}_i^*)^\top \mathbf{A}_k \mathbf{u}_i^*$, it follows that in order to compute the gradient, we only need to compute the product $(\mathbf{u}_i^*)^\top \mathbf{A}_k \mathbf{u}_i^*$. If the constraint matrices $\mathbf{A}_k$ are sparse we can compute the product $\mathbf{A}_k \mathbf{u}_i^*$ in $\mathcal{O}(s_k)$ time, where s_k denotes the number of non-zero elements in $\mathbf{A}_k$ and the product $(\mathbf{u}_i^*)^\top \mathbf{A}_k \mathbf{u}_i^*$ can be computed in $\mathcal{O}(n + s_k)$ time. Thus, the k-th component of the gradient can be computed in $\mathcal{O}(n(s_k + n))$ time, and all the m terms in $\mathcal{O}(m(n + n^2) + n \sum_{k=1}^{m} s_k)$ time.

When a constraint matrix is the Laplacian of a completely connected graph, i.e. $\mathbf{A}_k = \mathbf{K}$, as in Undirected SPARSEST CUT, although there is no sparsity to exploit, we can exploit the fact that $\mathbf{K} = n\mathbf{I} - \mathbf{J}$, where $\mathbf{J}$ is the all ones matrix. We then have that $\mathbf{Tr}(\mathbf{K}^\top \mathbf{X}^*) = \sum_{i=1}^{n} l_i^* \mathbf{Tr}((n\mathbf{I} - \mathbf{J})\mathbf{u}_i^*)(\mathbf{u}_i^*)^\top)$. Since $\mathbf{J}\mathbf{u}_i^*$ is a vector with identical entries (each the sum of the components of $\mathbf{u}_i^*$), the computational cost is $\mathcal{O}(n)$ additions. Also, $\mathbf{I}\mathbf{u}_i^* = \mathbf{u}_i^*$, so the total cost of computing each of the n summand's trace is $\mathcal{O}(n)$ additions and multiplies (by the term n) plus $\mathcal{O}(n)$ additional multiplies and additions to scale and add the main diagonal terms. Thus, the total cost is $\mathcal{O}(n^2)$.

Theorem 2 (Theorem 3 in [14]). *The Nesterov non-smooth optimization procedure computes an δ-saddle-point in $\mathcal{O}\left(\omega\tau(T_G + \kappa(m))\,\|A\|\,\sqrt{\ln(m)}\cdot\frac{1}{\delta}\right)$ time, where ω the saddle-point penalty multiplier, $\|A\| = \max_{1\le i\le m}\lambda_{\max}(\mathbf{A}_i)$, T_G denotes the running time for of computing the gradient ∇f_α and $\kappa(m)$ denotes the running time to compute $\langle \mathbf{A}_i, \mathbf{Z}\rangle$ for $i = 1,\dots,m$ and $\mathbf{Z} \in \mathcal{X}$.*

Note that T_G is typically dominated by the complexity of computing the negative eigenvalues and the corresponding eigenvectors, and is, in the worst case, $\mathcal{O}(n^3)$.

Since $\nu_L \le \nu^* \le \nu_U = \langle \mathbf{C}, \mathbf{Y}\rangle$, Lemma 1 implies that $\omega = \frac{\langle \mathbf{C},\mathbf{Y}\rangle - \nu_L}{g(\mathbf{Y})}$ suffices. Since an absolute $(\epsilon\nu_L)$-optimal solution is a relative ϵ-optimal solution, Assumptions (a) and (b) imply that $\frac{\omega}{\nu_L} = \left(\frac{\nu_U - \nu_L}{\nu_L}\right)\cdot\frac{1}{g(\mathbf{Y})} \le p(n,m)q(n,m)$. Thus, Theorem 2 implies that a relative ϵ-optimal solution can be computed in $\mathcal{O}\left(\tau q(n,m)p(n,m)(T_G + m)\,\|A\|\,\sqrt{\ln(m)}\cdot\frac{1}{\epsilon}\right)$ time. In the next section, we show that bisection can help us improve the overall running time by a factor $p(n,m)$.

3.3 Bisection on the Gap Improves Running Time

Let $\mathbf{Y} \in \mathcal{X}$ denote the strictly point specified in Assumption (a), i.e. $g(\mathbf{Y}) \ge 1 + \frac{1}{q(n,m)}$. In this section, we will assume that $q(n,m) \ge 8$. Let $\Delta = \nu_U - \nu_L$. We initialize the algorithm with $\nu_L^0 = \nu_L$, and $\nu_U^0 = \nu_U$. Let $\nu_L^{(t)}$ and $\nu_U^{(t)}$ denote lower and upper bounds on the optimal value ν^* of (1) and let $\epsilon^{(t)} = \frac{\Delta^{(t)}}{\nu_L^{(t)}}$ denote the relative error at the beginning of the iteration t of SOLVECOVERSDP. Note that Assumption (b) implies $\epsilon^{(0)} = \mathcal{O}(p(n,m))$. In iteration t we approximately solve the SDP

$$\nu^{(t)} = \min \left\langle \frac{1}{\nu_L^{(t)}}\mathbf{C}, \mathbf{X}\right\rangle$$
$$\text{s.t. } \langle \bar{\mathbf{A}}_i, \mathbf{X}\rangle \ge q(n,m) + \gamma^{(t)}, \tag{7}$$
$$\mathbf{X} \in \mathcal{X}$$

where $\gamma^{(t)} = \min\{\alpha^t, \epsilon^{(t)}\}$ for some $\frac{5}{6} < \alpha < 1$ and $\bar{\mathbf{A}}_i = q(n,m)\mathbf{A}_i$. Thus, $\|\bar{\mathbf{A}}\| = q(n,m)\|\mathbf{A}\|$.

Since the right hand sides of the cover constraints in (7) are not all equal to 1, it is *not* a covering SDP. However, since (7) can be converted into a covering SDP by rescaling, we refer to (7) as a covering SDP. In in each iteration we solve (7) with slightly different right hand sides, therefore, we find it more convenient to work with the unscaled version of the problem. Let $g^{(t)}(\mathbf{X}) \triangleq \min_{i=1,\dots,m}\{\langle \bar{\mathbf{A}}_i, \mathbf{X}\rangle - q(n,m)\} - \gamma^{(t)}$ denote the margin of feasibility of $\mathbf{X}$ with respect to the cover constraints in (7). In this section, we let $g(\mathbf{X}) = \min_{i=1,\dots,m}\{\langle \bar{\mathbf{A}}_i, \mathbf{X}\rangle - q(n,m)\}$; thus, $g^{(t)}(\mathbf{X}) = g(\mathbf{X}) - \gamma^{(t)}$.

Lemma 2 ([10]). *For $k \ge 0$, let $\hat{\mathbf{X}}^{(t)} \in \mathcal{X}$ denote an absolute $\frac{\epsilon^{(t)}}{3}$-optimal solution, i.e. $\left\langle \mathbf{C}, \hat{\mathbf{X}}^{(t)}\right\rangle \le \nu^{(t)} + \frac{1}{3}\Delta^{(t)}$. Update $(\nu_L^{(t+1)}, \nu_U^{(t+1)})$ as follows:*

$$(\nu_L^{(t+1)}, \nu_U^{(t+1)}) = \begin{cases} \left(\nu_L^{(t)}, \nu_L^{(t)} + \frac{2}{3}\Delta^{(t)}\right), & \text{if } \left\langle \mathbf{C}, \hat{\mathbf{X}}^{(t)}\right\rangle \le \nu_L^{(t)} + \frac{2}{3}\Delta^{(t)}, \\ \left(\frac{\nu_L^{(t)} + \frac{1}{3}\Delta^{(t)}}{1 + \gamma^{(t)}\frac{\Delta^{(t)}}{\nu_L^{(t)}}}, \nu_U^{(t)}\right), & \text{otherwise.} \end{cases}$$

$\textsc{SolveCoverSDP}$

    Inputs: $\mathbf{C}, \mathbf{A}_K, \mathbf{X}^{(0)}, \epsilon, \nu_U^{(0)}, \nu_L^{(0)}, \delta_t$
    Outputs: $\mathbf{X}^*$

    Set $t \leftarrow 0$ $\epsilon^{(t)} \leftarrow \frac{\nu_U^{(t)} - \nu_L^{(t)}}{\nu_L^{(t)}}$

    while $(\epsilon^{(t)} > \epsilon)$
        do
            Compute $\mathbf{Y}^{(t)}$, $\left(\frac{\epsilon^{(t)}}{3}\right)$-optimal solution to $\nu^{(t)}$ using $\textsc{Nesterov Procedure}$

            Set $(\nu_L^{(t+1)}, \nu_U^{(t+1)}) \leftarrow \begin{cases} (\nu_L^{(t)}, \nu_L^{(t)} + \frac{2}{3}\Delta^{(t)}), & \text{if } \left\langle \mathbf{C}, \mathbf{Y}^{(t)} \right\rangle \le \nu_L^{(t)} + \frac{2}{3}\Delta^{(t)} \\ (\frac{\nu_L^{(t)} + \Delta^{(t)}/3}{1 + \epsilon^{(t)}\nu_L^{(t)}/\Delta^{(t)}}, \nu_U^{(t)}) & \text{otherwise} \end{cases}$

            $t \leftarrow t + 1,$ $\epsilon^{(t)} = \frac{\nu_U^{(t)} - \nu_L^{(t)}}{\nu_L^{(t)}}$

    return $\mathbf{X}^{(t)}$

Fig. 1. Our algorithm for solving the covering SDP

Then, for all $k \ge 0$,

(i) $\frac{\Delta^{(t+1)}}{\nu_L^{(t+1)}} \le \left(\frac{5}{6}\right)^{t+1} \left(\frac{\Delta^0}{\nu_L^0}\right)$, *i.e. the gap converges geometrically to 0.*

(ii) $\hat{\mathbf{X}}^{(t)}$ *is feasible to* $\nu^{(t+1)}$, *and* $g^{(t+1)}(\hat{\mathbf{X}}^{(t)}) \ge (1 - \alpha)\gamma^{(t)} \ge (1 - \alpha)\alpha^t$.

We can now prove our main result.

Theorem 3. *For $\epsilon \in (0,1)$, $\textsc{SolveCoverSDP}$ computes a relative ϵ-optimal solution for the Cover SDP (1) in $\mathcal{O}\left(\tau q(n, m) \log(p(n, m)\frac{1}{\epsilon})(T_G + m) \|A\| \sqrt{\ln(m)} \cdot \frac{1}{\epsilon}\right)$ time, where $q(n, m)$ is the polynomial that satisfies Assumption (a), and T_G denotes the running time for computing the gradient ∇f_α in the Nesterov procedure.*

Proof. From Lemma 2 (i) it follows that $\textsc{SolveCoverSDP}$ terminates after at most $T = \left\lceil \frac{\log(\frac{\epsilon^{(0)}}{\epsilon})}{\log(\frac{6}{5})} \right\rceil = \mathcal{O}(\log(p(n, m)\frac{1}{\epsilon}))$ iterations. From the analysis in the previous sections, we know that the run time for computing an absolute $\frac{1}{3}\epsilon^{(t)}$-optimal solution is $\tilde{\mathcal{O}}\left(\frac{\epsilon^{(t)}}{g^{(t)}(\hat{\mathbf{X}}^{(t)})} \cdot \frac{3}{\epsilon^{(t)}}\right) = \tilde{\mathcal{O}}\left(\frac{1}{\gamma^{(t-1)}}\right)$, where we have ignored polynomial factors. Since $1 > \alpha > \frac{5}{6}$, Lemma 2 implies $\gamma^{(t+1)} \le \gamma^{(t)}$, so the overall running time of $\textsc{SolveCoverSDP}$ is $\tilde{\mathcal{O}}(T \cdot \frac{1}{\gamma^{(T)}})$. Thus, all that remains to show is that $\mathcal{O}(1/\gamma^{(T)}) = \mathcal{O}(1/\epsilon)$.

Recall $\gamma^{(t)} = \min\{\epsilon^{(t)}, \alpha^t\}$ and let $T_\gamma = \inf\{t : \epsilon^{(0)}(\frac{5}{6})^t < \alpha^t\}$, i.e., the first iteration where $\gamma^{(t)} = \epsilon^{(t)}$. Then, the runtime is bound by $\mathcal{O}(\max\{T, T_\gamma\}\frac{1}{\epsilon})$. Then the theorem follows since

$$T_\gamma = \left\lceil \frac{\log(\epsilon^{(0)})}{\log(6\alpha/5)} \right\rceil = \mathcal{O}(\log(p(n, m))). \qquad \square$$

Acknowledgements. We would like to thank Satyen Kale for his insightful comments.

References

1. Arora, S., Hazan, E., Kale, S.: Fast algorithms for approximate semidefinite programming using the multiplicative weights update method. In: Proceedings of the 46th Annual Symposium on Foundations of Computer Science (2005)
2. Arora, S., Kale, S.: A combinatorial, primal-dual approach to semidefinite programs. In: Proceedings of the 39th Annual ACM Symposium on Theory of Computing (2007)
3. Arora, S., Rao, S., Vazirani, U.: Expander flows, geometric embeddings, and graph partitionings. In: Proceedings of the 36th Annual ACM Symposium on Theory of Computing, pp. 222–231 (2004)
4. Bienstock, D.: Potential function methods for approximately solving linear programming problems: theory and practice, Boston, MA. International Series in Operations Research & Management Science, vol. 53 (2002)
5. Bienstock, D., Iyengar, G.: Solving fractional packing problems in $O^*(\frac{1}{\epsilon})$ iterations. In: Proceedings of the 36th Annual ACM Symposium on Theory of Computing, pp. 146–155 (2004)
6. Fleischer, L.: Fast approximation algorithms for fractional covering problems with box constraint. In: Proceedings of the 15th ACM-SIAM Symposium on Discrete Algorithms (2004)
7. Garg, N., Konemann, J.: Faster and simpler algorithms for multicommodity flow and other fractional packing problems. In: Proceedings of the 39th Annual Symposium on Foundations of Computer Science, pp. 300–309 (1998)
8. Goldberg, A.V., Oldham, J.D., Plotkin, S.A., Stein, C.: An implementation of a combinatorial approximation algorithm for minimum-cost multicommodity flow. In: Bixby, R.E., Boyd, E.A., Ríos-Mercado, R.Z. (eds.) IPCO 1998. LNCS, vol. 1412, pp. 338–352. Springer, Heidelberg (1998)
9. Iyengar, G., Phillips, D.J., Stein, C.: Approximation algorithms for semidefinite packing problems with applications to maxcut and graph coloring. In: Proceedings of the 11th Conference on Integer Programming and Combinatorial Optimization, pp. 152–166 (2005); Submitted to SIAM Journal on Optimization
10. Iyengar, G., Phillips, D.J., Stein, C.: Feasible and accurate algorithms for covering semidefinite programs. Tech. rep., Optimization online (2010)
11. Klein, P., Lu, H.-I.: Efficient approximation algorithms for semidefinite programs arising from MAX CUT and COLORING. In: Proceedings of the Twenty-eighth Annual ACM Symposium on the Theory of Computing, Philadelphia, PA, pp. 338–347. ACM, New York (1996)
12. Klein, P., Young, N.: On the number of iterations for Dantzig-Wolfe optimization and packing-covering approximation algorithms. In: Cornuéjols, G., Burkard, R.E., Woeginger, G.J. (eds.) IPCO 1999. LNCS, vol. 1610, pp. 320–327. Springer, Heidelberg (1999)
13. Lu, Z., Monteiro, R., Yuan, M.: Convex optimization methods for dimension reduction and coefficient estimation in multivariate linear regression, Arxiv preprint arXiv:0904.0691 (2009)
14. Nesterov, Y.: Smooth minimization of nonsmooth functions. Mathematical Programming 103, 127–152 (2005)
15. Nesterov, Y.: Smoothing technique and its applications in semidefinite optimization. Mathematical Programming 110, 245–259 (2007)

16. Nesterov, Y., Nemirovski, A.: Interior-point polynomial algorithms in convex programming. SIAM Studies in Applied Mathematics, vol. 13. Society for Industrial and Applied Mathematics (SIAM), Philadelphia (1994)
17. Plotkin, S., Shmoys, D.B., Tardos, E.: Fast approximation algorithms for fractional packing and covering problems. Mathematics of Operations Research 20, 257–301 (1995)
18. Sidiropoulos, N., Davidson, T., Luo, Z.: Transmit beamforming for physical-layer multicasting. IEEE Transactions on Signal Processing 54, 2239 (2006)
19. Weinberger, K., Saul, L.: Distance metric learning for large margin nearest neighbor classification. The Journal of Machine Learning Research 10, 207–244 (2009)

The Quantitative Analysis of User Behavior Online – Data, Models and Algorithms

Prabhakar Raghavan

Yahoo! research
pragh@yahoo-inc.com

By blending principles from mechanism design, algorithms, machine learning and massive distributed computing, the search industry has become good at optimizing monetization on sound scientific principles. This represents a successful and growing partnership between computer science and microeconomics. When it comes to understanding how online users respond to the content and experiences presented to them, we have more of a lacuna in the collaboration between computer science and certain social sciences. We will use a concrete technical example from image search results presentation, developing in the process some algorithmic and machine learning problems of interest in their own right. We then use this example to motivate the kinds of studies that need to grow between computer science and the social sciences; a critical element of this is the need to blend large-scale data analysis with smaller-scale eye-tracking and "individualized" lab studies.

H. Kaplan (Ed.): SWAT 2010, LNCS 6139, p. 163, 2010.
© Springer-Verlag Berlin Heidelberg 2010

Systems of Linear Equations over $\mathbb{F}_2$ and Problems Parameterized above Average

Robert Crowston[1], Gregory Gutin[1], Mark Jones[1],
Eun Jung Kim[1], and Imre Z. Ruzsa[2]

[1] Department of Computer Science
Royal Holloway, University of London
Egham, Surrey TW20 0EX, UK
{robert,gutin,markj,eunjung}@cs.rhul.ac.uk
[2] Alfréd Rényi Institute of Mathematics, Hungarian Academy of Sciences
H-1053, Budapest, Hungary
ruzsa@renyi.hu

Abstract. In the problem Max Lin, we are given a system $Az = b$ of m linear equations with n variables over $\mathbb{F}_2$ in which each equation is assigned a positive weight and we wish to find an assignment of values to the variables that maximizes the excess, which is the total weight of satisfied equations minus the total weight of falsified equations. Using an algebraic approach, we obtain a lower bound for the maximum excess.

Max Lin Above Average (Max Lin AA) is a parameterized version of Max Lin introduced by Mahajan et al. (Proc. IWPEC'06 and J. Comput. Syst. Sci. 75, 2009). In Max Lin AA all weights are integral and we are to decide whether the maximum excess is at least k, where k is the parameter.

It is not hard to see that we may assume that no two equations in $Az = b$ have the same left-hand side and $n = \text{rank}A$. Using our maximum excess results, we prove that, under these assumptions, Max Lin AA is fixed-parameter tractable for a wide special case: $m \leq 2^{p(n)}$ for an arbitrary fixed function $p(n) = o(n)$. This result generalizes earlier results by Crowston et al. (arXiv:0911.5384) and Gutin et al. (Proc. IW-PEC'09). We also prove that Max Lin AA is polynomial-time solvable for every fixed k and, moreover, Max Lin AA is in the parameterized complexity class W[P].

Max r-Lin AA is a special case of Max Lin AA, where each equation has at most r variables. In Max Exact r-SAT AA we are given a multiset of m clauses on n variables such that each clause has r variables and asked whether there is a truth assignment to the n variables that satisfies at least $(1 - 2^{-r})m + k2^{-r}$ clauses. Using our maximum excess results, we prove that for each fixed $r \geq 2$, Max r-Lin AA and Max Exact r-SAT AA can be solved in time $2^{O(k \log k)} + m^{O(1)}$. This improves $2^{O(k^2)} + m^{O(1)}$-time algorithms for the two problems obtained by Gutin et al. (IWPEC 2009) and Alon et al. (SODA 2010), respectively.

It is easy to see that maximization of arbitrary pseudo-boolean functions, i.e., functions $f : \{-1, +1\}^n \to \mathbb{R}$, represented by their Fourier expansions is equivalent to solving Max Lin. Using our main maximum excess result, we obtain a tight lower bound on the maxima of pseudo-boolean functions.

H. Kaplan (Ed.): SWAT 2010, LNCS 6139, pp. 164–175, 2010.

1   Introduction

In the problem MAX LIN, we are given a system $Az = b$ of m linear equations in n variables over $\mathbb{F}_2$ in which each equation is assigned a positive weight and we wish to find an assignment of values to the variables in order to maximize the total weight of satisfied equations. A special case of MAX LIN when each equation has at most r variables is called MAX r-LIN.

Various algorithmic aspects of MAX LIN have been well-studied (cf. [2,10,11]). Perhaps, the best known result on MAX LIN is the following inapproximability theorem of Håstad [10]: unless P=NP, for each $\epsilon > 0$ there is no polynomial time algorithm for distinguishing instances of MAX 3-LIN in which at least $(1 - \epsilon)m$ equations can be simultaneously satisfied from instances in which less than $(1/2 + \epsilon)m$ equations can be simultaneously satisfied.

Notice that maximizing the total weight of satisfied equations is equivalent to maximizing the *excess*, which is the total weight of satisfied equations minus the total weight of falsified equations. In Section 2, we investigate lower bounds for the maximum excess. Using an algebraic approach, we prove the following main result: Let $Az = b$ be a MAX LIN system such that rank$A = n$ and no pair of equations has the same left-hand side, let $w_{\min}$ be the minimum weight of an equation in $Az = b$, and let $k \geq 2$. If $k \leq m \leq 2^{n/(k-1)} - 2$, then the maximum excess of $Az = b$ is at least $k \cdot w_{\min}$. Moreover, we can find an assignment that achieves an excess of at least $k \cdot w_{\min}$ in time $m^{O(1)}$.

Using this and other results of Section 2 we prove parameterized complexity results of Section 3. To describe these results we need the following notions, most of which can be found in monographs [6,7,15].

A *parameterized problem* is a subset $L \subseteq \Sigma^* \times \mathbb{N}$ over a finite alphabet Σ. L is *fixed-parameter tractable (FPT)* if the membership of an instance (x, k) in $\Sigma^* \times \mathbb{N}$ can be decided in time $f(k)|x|^{O(1)}$, where f is a computable function of the parameter k. When the decision time is replaced by the much more powerful $|x|^{O(f(k))}$, we obtain the class XP, where each problem is polynomial-time solvable for any fixed value of k. There is an infinite number of parameterized complexity classes between FPT and XP (for each integer $t \geq 1$, there is a class $W[t]$) and they form the following tower: $FPT \subseteq W[1] \subseteq W[2] \subseteq \cdots \subseteq W[P] \subseteq XP$. Here $W[P]$ is the class of all parameterized problems (x, k) that can be decided in $f(k)|x|^{O(1)}$ time by a nondeterministic Turing machine that makes at most $f(k) \log |x|$ nondeterministic steps for some computable function f. For the definition of the classes $W[t]$, see, e.g., [7] (we do not use these classes in the rest of the paper).

Given a pair L, L' of parameterized problems, a *bikernelization from L to L'* is a polynomial-time algorithm that maps an instance (x, k) to an instance (x', k') (the *bikernel*) such that (i) $(x, k) \in L$ if and only if $(x', k') \in L'$, (ii) $k' \leq f(k)$, and (iii) $|x'| \leq g(k)$ for some functions f and g. The function $g(k)$ is called the *size* of the bikernel. The notion of a bikernelization was introduced in [1], where it was observed that a parameterized problem L is fixed-parameter tractable if and only if it is decidable and admits a bikernelization from itself to a parameterized problem L'. A *kernelization* of a parameterized problem L

is simply a bikernelization from L to itself; the bikernel is the *kernel*, and $g(k)$ is the *size* of the kernel. Due to applications, low degree polynomial size kernels are of main interest.

Note that $W/2$ is a tight lower bound on the maximum weight of satisfiable equations in a MAX LIN system $Az = b$. Indeed, $W/2$ is the average weight of satisfied equations (as the probability of each equation to be satisfied is $1/2$) and, thus, is a lower bound; to see the tightness consider a system of pairs of equations of the form $\sum_{i \in I} z_i = 0$, $\sum_{i \in I} z_i = 1$ of weight 1. Mahajan et al. [13,14] parameterized MAX LIN as follows: given a MAX LIN system $Az = b$, decide whether the total weight of satisfied equations minus $W/2$ is at least k', where W is the total weight of all equations and k' is the parameter. This is equivalent to asking whether the maximum excess is at least k, where $k = 2k'$ is the parameter. (Note that since $k = 2k'$, these two questions are equivalent from the complexity point of view.) Since $W/2$ is the average weight of satisfied equations, we will call the parameterized MAX LIN problem MAX LIN ABOVE AVERAGE or MAX LIN AA. Since the parameter k is more convenient for us to use, in what follows we use the version of MAX LIN AA parameterized by k.

Mahajan et al. [13,14] raised the question of determining the parameterized complexity of MAX LIN AA. It is not hard to see (we explain it in detail in Section 2) that we may assume that no two equations in $Az = b$ have the same left-hand side and $n = \mathrm{rank} A$. Using our maximum excess results, we prove that, under these assumptions, (a) MAX LIN AA is fixed-parameter tractable if $m \leq 2^{p(n)}$ for an arbitrary fixed function $p(n) = o(n)$, and (b) MAX LIN AA has a polynomial-size kernel if $m \leq 2^{n^a}$ for an arbitrary $a < 1$. We conjecture that under the two assumptions if $m < 2^{an}$ for some constant $a > 0$, then MAX LIN AA is W[1]-hard, i.e., result (a) is best possible in a sense. In addition, we prove that MAX LIN AA is in XP (thus, MAX LIN AA is polynomial-time solvable for every fixed k), and, moreover, it is in W[P].

Recall that MAX r-LIN AA is a special case of MAX LIN AA, where each equation has at most r variables. In MAX EXACT r-SAT AA we are given a multiset of m clauses on n variables such that each clause has r variables and asked whether there is a truth assignment to the n variables that satisfies at least $(1 - 2^{-r})m + k2^{-r}$ clauses. Using our maximum excess results, we prove that for each fixed $r \geq 2$ Max r-Lin AA has a kernel with $O(k \log k)$ variables and, thus, it can be solved in time $2^{O(k \log k)} + m^{O(1)}$. This improves a kernel with $O(k^2)$ variables for Max r-Lin AA obtained by Gutin et al. [8]. Similarly, we prove that for each $r \geq 2$ MAX EXACT r-SAT AA has a kernel with $O(k \log k)$ variables and it can be solved in time $2^{O(k \log k)} + m^{O(1)}$ improving a kernel with $O(k^2)$ variables for MAX EXACT r-SAT AA obtained by Alon et al. [1]. Note that while the kernels with $O(k^2)$ variables were obtained using a probabilistic approach, our results are obtained using an algebraic approach. Using a graph-theoretical approach Alon et al. [1] obtained a kernel of MAX EXACT 2-SAT AA with $O(k)$ variables, but it is unlikely that their approach can be extended beyond $r = 2$.

Fourier analysis of pseudo-boolean functions, i.e., functions $f : \{-1, +1\}^n \to \mathbb{R}$, has been used in many areas of computer science(cf. [1,16,17]). In Fourier analysis, the Boolean domain is often assumed to be $\{-1, +1\}^n$ rather than more usual $\{0, 1\}^n$ and we will follow this assumption in our paper. Here we use the following well-known and easy to prove fact [16] that each function $f : \{-1, +1\}^n \to \mathbb{R}$ can be uniquely written as

$$f(x) = \sum_{S \subseteq [n]} c_S \prod_{i \in S} x_i, \tag{1}$$

where $[n] = \{1, 2, \ldots, n\}$ and each c_S is a real. Formula (1) is the Fourier expansion f, c_S are the Fourier coefficients of f, and the monomials $\prod_{i \in S} x_i$ form an orthogonal basis of (1) (thus, the monomials are often written as $\chi_S(x)$ but we will use only $\prod_{i \in S} x_i$ as it is more transparent).

Optimization of pseudo-boolean functions is useful in many areas including computer science, discrete mathematics, operations research, statistical mechanics and manufacturing; for many results and applications of pseudo-boolean function optimization, see a well-cited survey [3]. In classical analysis, there is a large number of lower bounds on the maxima of trigonometric Fourier expansions, cf. [4]. In Section 3, we prove a sharp lower bound on the maximum of a pseudo-boolean function using its Fourier expansion. The bound can be used in algorithmics, e.g., for approximation algorithms.

2   Results on Maximum Excess

Consider two reduction rules for MAX LIN introduced in [8] for MAX LIN AA. These rules are of interest due to Lemma 1.

Reduction Rule 1. *Let $t = \text{rank} A$ and let columns $a^{i_1}, \ldots, a^{i_t}$ of A be linearly independent. Then delete all variables not in $\{z_{i_1}, \ldots, z_{i_t}\}$ from the equations of $Az = b$.*

Reduction Rule 2. *If we have, for a subset S of $[n]$, an equation $\sum_{i \in S} z_i = b'$ with weight w', and an equation $\sum_{i \in S} z_i = b''$ with weight w'', then we replace this pair by one of these equations with weight $w' + w''$ if $b' = b''$ and, otherwise, by the equation whose weight is bigger, modifying its new weight to be the difference of the two old ones. If the resulting weight is 0, we delete the equation from the system.*

Lemma 1. *Let $A'z' = b'$ be obtained from $Az = b$ by Rule 1 or 2. Then the maximum excess of $A'z' = b'$ is equal to the maximum excess of $Az = b$. Moreover, $A'z' = b'$ can be obtained from $Az = b$ in time polynomial in n and m.*

To see the validity of Rule 1, consider an independent set I of columns of A of cardinality $\text{rank} A$ and a column $a^j \notin I$. Observe that $a^j = \sum_{i \in I'} a^i$, where $I' \subseteq I$. Consider an assignment $z = z^0$. If $z_j^0 = 1$ then for each $i \in I' \cup \{j\}$ replace z_i^0 by $z_i^0 + 1$. The new assignment satisfies exactly the same equations as the initial assignment. Thus, we may assume that $z_j = 0$ and remove z_j from the system. For a different proof, see [8]. If we cannot change a weighted system $Az = b$ using Rules 1 and 2, we call it *irreducible*.

Consider the following algorithm that tries to maximize the total weight of satisfied equations of $Az = b$. We assume that, in the beginning, no equation or variable in $Az = b$ is marked.

ALGORITHM $\mathcal{H}$

While the system $Az = b$ is nonempty do the following:

1. Choose an arbitrary equation $\sum_{i \in S} z_i = b$ and mark z_l, where $l = \min\{i : i \in S\}$.
2. Mark this equation and delete it from the system.
3. Replace every equation $\sum_{i \in S'} z_i = b'$ in the system containing z_l by $\sum_{i \in S} z_i + \sum_{i \in S'} z_i = b + b'$. (The weight of the equation is unchanged.)
4. Apply Reduction Rule 2 to the system.

Note that algorithm $\mathcal{H}$ replaces $Az = b$ with an *equivalent* system under the assumption that the marked equations are satisfied; that is, for every assignment of values to the variables $z_1, \ldots, z_n$ that satisfies the marked equations, both systems have the same excess.

The *maximum $\mathcal{H}$-excess* of $Az = b$ is the maximum possible total weight of equations marked by $\mathcal{H}$ for $Az = b$ taken over all possible choices in Step 1 of $\mathcal{H}$.

Lemma 2. *The maximum excess of $Az = b$ equals its maximum $\mathcal{H}$-excess.*

Proof. We first prove that the maximum excess of $Az = b$ is not smaller than its maximum $\mathcal{H}$-excess.

Let K be the set of equations marked by $\mathcal{H}$. A method first described in [5] can find an assignment of values to the variables such that the equations in K are satisfied and, in the remainder of the system, the total weight of satisfied equations is not smaller than the total weight of falsified equations.

For the sake of completeness, we repeat the description here. By construction, for any assignment that satisfies all the marked equations, exactly half of the non-marked equations are satisfied. Therefore it suffices to find an assignment to the variables such that all marked equations are satisfied. This is possible if we find an assignment that satisfies the last marked equation, then find an assignment satisfying the equation marked before the last, etc. Indeed, the equation marked before the last contains a (marked) variable z_l not appearing in the last equation, etc. This proves the first part of our lemma.

Now we prove that the maximum $\mathcal{H}$-excess of $Az = b$ is not smaller than its maximum excess. Let $z = (z_1, \ldots, z_n)$ be an assignment that achieves the maximum excess, t. Observe that if at each iteration of $\mathcal{H}$ we mark an equation that is satisfied by z, then $\mathcal{H}$ will mark equations of total weight t. $\square$

Remark 1. It follows from Lemma 2 that the maximum excess of a (nonempty) irreducible system $Az = b$ with smallest weight $w_{\min}$ is at least $w_{\min}$. If all weights are integral, then the maximum excess of $Az = b$ is at least 1.

Clearly, the total weight of equations marked by $\mathcal{H}$ depends on the choice of equations to mark in Step 1. Below we consider one such choice based on the following theorem. The theorem allows us to find a set of equations such that we can mark each equation in the set in successive iterations of $\mathcal{H}$. This means we can run $\mathcal{H}$ a guaranteed number of times, which we can use to get a lower bound on the $\mathcal{H}$-excess.

Theorem 1. *Let M be a set in $\mathbb{F}_2^n$ such that M contains a basis of $\mathbb{F}_2^n$, the zero vector is in M and $|M| < 2^n$. If k is a positive integer and $k + 1 \leq |M| \leq 2^{n/k}$ then, in time $|M|^{O(1)}$, we can find a subset K of M of $k + 1$ vectors such that no sum of two or more vectors of K is in M.*

Proof. We first consider the case when $k = 1$. Since $|M| < 2^n$ and the zero vector is in M, there is a non-zero vector $v \notin M$. Since M contains a basis for $\mathbb{F}_2^n$, v can be written as a sum of vectors in M and consider such a sum with the minimum number of summands: $v = u_1 + \cdots + u_\ell$, $\ell \geq 2$. Since $u_1 + u_2 \notin M$, we may set $K = \{u_1, u_2\}$. We can find such a set K in polynomial time by looking at every pair in $M \times M$.

We now assume that $k > 1$. Since $k + 1 \leq |M| \leq 2^{n/k}$ we have $n \geq k + 1$.

We proceed with a greedy algorithm that tries to find K. Suppose we have a set $L = \{a_1, \ldots, a_l\}$ of vectors in M, $l \leq k$, such that no sum of two or more elements of L is in M. We can extend this set to a basis, so $a_1 = (1, 0, 0, \ldots, 0)$, $a_2 = (0, 1, 0, \ldots, 0)$ and so on. For every $a \in M \backslash L$ we check whether $M \backslash \{a_1, \ldots, a_l, a\}$ has an element that agrees with a in all co-ordinates $l + 1, \ldots, n$. If no such element exists, then we add a to the set L, as no element in M can be expressed as a sum of a and a subset of L.

If our greedy algorithm finds a set L of size at least $k + 1$, we are done and L is our set K. Otherwise, we have stopped at $l \leq k$. In this case, we do the next iteration as follows. Recall that L is part of a basis of M such that $a_1 = (1, 0, 0, \ldots, 0)$, $a_2 = (0, 1, 0, \ldots, 0), \ldots$. We create a new set M' in $\mathbb{F}_2^{n'}$, where $n' = n - l$. We do this[1] by removing the first l co-ordinates from M, and then identifying together any vectors that agree in the remaining n' co-ordinates. We are in effect identifying together any vectors that only differ by a sum of some elements in L. It follows that every element of M' was created by identifying together at least two elements of M, since otherwise we would have had an element in $M \backslash L$ that should have been added to L by our greedy algorithm. Therefore it follows that $|M'| \leq |M|/2 \leq 2^{n/k-1}$. ¿From this inequality and the fact that $n' \geq n - k$, we get that $|M'| \leq 2^{n'/k}$. It also follows by construction of M' that M' has a basis for $\mathbb{F}_2^{n'}$, and that the zero vector is in M'. (Thus, we have

---

[1] For the reader familiar with vector space terminology: $\mathbb{F}_2^{n'}$ is $\mathbb{F}_2^n$ modulo $\mathrm{span}(L)$, the subspace of $\mathbb{F}_2^n$ spanned by L, and M' is the image of M in $\mathbb{F}_2^{n'}$.

$|M'| \geq n' + 1$.) If $n' \geq k + 1$ we complete this iteration by running the algorithm on the set M' as in the first iteration. Otherwise ($n' \leq k$), the algorithm stops.

Since each iteration of the algorithm decreases n', the algorithm terminates. Now we prove that at some iteration, the algorithm will actually find a set K of $k + 1$ vectors. To show this it suffices to prove that we will never reach the point when $n' \leq k$. Suppose this is not true and we obtained $n' \leq k$. Observe that $n' \geq 1$ (before that we had $n' \geq k + 1$ and we decreased n' by at most k) and $|M'| \geq n' + 1$. Since $|M'| \leq 2^{n'/k}$, we have $n' + 1 \leq 2^{n'/k}$, which is impossible due to $n' \leq k$ unless $n' = 1$ and $k = 1$, a contradiction with the assumption that $k > 1$.

It is easy to check that the running time of the algorithm is polynomial in $|M|$.

$\square$

Remark 2. It is much easier to prove a non-constructive version of the above result. In fact we can give a non-constructive proof that $k + 1 \leq |M| \leq 2^{n/k}$ can be replaced by $2k < |M| < 2^{n/k}((k-1)!)^{1/k}$. We will extend our proof above for the case $k = 1$. We may assume that $k \geq 2$. Observe that the number of vectors of $\mathbb{F}_2^n$ that can be expressed as the sum of at most k vectors of M is at most

$$\binom{|M|}{k} + \binom{|M|}{k-1} + \cdots + \binom{|M|}{1} + 1 \leq |M|^k/(k-1)! \text{ for } |M| > 2k.$$

Since $|M| < 2^{n/k}((k-1)!)^{1/k}$ we have $|\mathbb{F}_2^n| > |M|^k/(k-1)!$ and, thus, at least for one vector a of $\mathbb{F}_2^n$ we have $a = m_1 + \cdots + m_\ell$, where ℓ is minimum and $\ell > k$. Note that, by the minimality of ℓ, no sum of two or more summands of the sum for a is in M and all summands are distinct. Thus, we can set $K = \{m_1, \ldots, m_{k+1}\}$.

Theorem 2. *Let $Az = b$ be an irreducible system, let $w_{\min}$ be the minimum weight of an equation in $Az = b$, and let $k \geq 2$. If $k \leq m \leq 2^{n/(k-1)} - 2$, then the maximum excess of $Az = b$ is at least $k \cdot w_{\min}$. Moreover, we can find an assignment that achieves an excess of at least $k \cdot w_{\min}$ in time $m^{O(1)}$.*

Proof. Consider a set M of vectors in $\mathbb{F}_2^n$ corresponding to equations in $Az = b$ as follows: for each $\sum_{i \in S} z_i = b_S$ in $Az = b$, the vector $v = (v_1, \ldots, v_n) \in M$, where $v_i = 1$ if $i \in S$ and $v_i = 0$, otherwise. Add the zero vector to M. As $Az = b$ is reduced by Rule 1 and $k \leq m \leq 2^{n/(k-1)} - 2$, we have that M contains a basis for $\mathbb{F}_2^n$ and $k \leq |M| \leq 2^{n/(k-1)} - 1$. Therefore, using Theorem 1 we can find a set K of k vectors such that no sum of two or more vectors in K belongs to M.

Now run Algorithm $\mathcal{H}$ choosing at each Step 1 an equation of $Az = b$ corresponding to a member of K, then equations picked at random until the algorithm terminates. Algorithm $\mathcal{H}$ will run at least k iterations as no equation corresponding to a vector in K will be deleted before it has been marked. Indeed, suppose that this is not true. Then there are vectors $w \in K$ and $v \in M$ and a pair of nonintersecting subsets K' and K'' of $K \setminus \{v, w\}$ such that $w + \sum_{u \in K'} u = v + \sum_{u \in K''} u$. Thus, $v = w + \sum_{u \in K' \cup K''} u$, a contradiction with the definition of K.

In fact, the above argument shows that no equation of $Az = b$ corresponding to a member of K will change its weight during the first k iterations of $\mathcal{H}$. Thus, by Lemma 2, the maximum excess of $Az = b$ is at least $k \cdot w_{\min}$. It remains to observe that we can once again use the algorithm given in the proof of Lemma 2 to find an assignment that gives an excess of at least $k \cdot w_{\min}$.   □

We now provide a useful association between weighted systems of linear equations on $\mathbb{F}_2^n$ and Fourier expansions of functions $f : \{-1, +1\} \rightarrow \mathbb{R}$. Let us rewrite (1), the Fourier expansion of such a function, as

$$f(x) = c_\emptyset + \sum_{S \in \mathcal{F}} c_S \prod_{i \in S} x_i, \tag{2}$$

where $\mathcal{F} = \{\emptyset \neq S \subseteq [n] : c_S \neq 0\}$.

Now associate the polynomial $\sum_{S \in \mathcal{F}} c_S \prod_{i \in S} x_i$ in (2) with a weighted system $Az = b$ of linear equations on $\mathbb{F}_2^n$: for each $S \in \mathcal{F}$, we have an equation $\sum_{i \in S} z_i = b_S$ with weight $|c_S|$, where $b_S = 0$ if c_S is positive and $b_S = 1$, otherwise. Conversely, suppose we have a system $Az = b$ of linear equations on $\mathbb{F}_2^n$ in which each equation $\sum_{i \in S} z_i = b_S$ is assigned a weight $w_S > 0$ and no pair of equations have the same left-hand side. This system can be associated with the polynomial $\sum_{S \in \mathcal{F}} c_S \prod_{i \in S} x_i$, where $c_S = w_S$, if $b_S = 0$, and $c_S = -w_S$, otherwise. The above associations provide a bijection between Fourier expansions of functions $f : \{-1, +1\} \rightarrow \mathbb{R}$ with $c_\emptyset = 0$ and weighted systems of linear equations on $\mathbb{F}_2^n$. This bijection is of interest due to the following:

Proposition 1. *An assignment $z^{(0)} = (z_1^{(0)}, \ldots, z_n^{(0)})$ of values to the variables of $Az = b$ maximizes the total weight of satisfied equations of $Az = b$ if and only if $x^{(0)} = ((-1)^{z_1^{(0)}}, \ldots, (-1)^{z_n^{(0)}})$ maximizes $f(x)$. Moreover, $\max_{x \in \{-1,+1\}^n} f(x) - c_\emptyset$ equals the maximum excess of $Az = b$.*

Proof. The claims of this lemma easily follow from the fact that an equation $\sum_{i \in S} z_i = 0$ is satisfied if and only if $\prod_{i \in S} x_i > 0$, where $x_i = (-1)^{z_i}$.   □

3   Corollaries

This section contains a collection of corollaries of Theorem 2 establishing parameterized complexity of special cases of MAX LIN AA, of MAX EXACT r-SAT, and of a wide class of constraint satisfaction problems. In addition, we will prove that MAX LIN AA is in X[P] and obtain a sharp lower bound on the maximum of a pseudo-boolean function.

3.1   Parameterized Complexity of Max Lin AA

Corollary 1. *Let $p(n)$ be a fixed function such that $p(n) = o(n)$. If $m \leq 2^{p(n)}$ then MAX LIN AA is fixed-parameter tractable. Moreover, a satisfying assignment can be found in time $g(k)m^{O(1)}$ for some computable function g.*

Proof. We may assume that $m \geq n > k > 1$. Observe that $m \leq 2^{n/k}$ implies $m \leq 2^{n/(k-1)} - 2$. Thus, by Theorem 2, if $p(n) \leq n/k$, the answer to MAX LIN AA is YES, and there is a polynomial algorithm to find a suitable assignment. Otherwise, $n \leq f(k)$ for some function dependent on k only and MAX LIN AA can be solved in time $2^{f(k)}m^{O(1)}$ by checking every possible assignment.   □

Let ρ_i be the number of equations in $Az = b$ containing z_i, $i = 1, \ldots, n$. Let $\rho = \max_{i \in [n]} \rho_i$ and let r be the maximum number of variables in an equation of $Az = b$. Crowston et al. [5] proved that MAX LIN AA is fixed-parameter tractable if either $r \leq r(n)$ for some fixed function $r(n) = o(n)$ or $\rho \leq \rho(m)$ for some fixed function $\rho(m) = o(m)$.

For a given $r = r(n)$, we have $m \leq \sum_{i=1}^{r} \binom{n}{i}$. By Corollary 23.6 in [9], $m \leq 2^{nH(r/n)}$, where $H(y) = -y \log_2 y - (1 - y) \log_2(1 - y)$, the entropy of y. It is easy to see that if $y = o(n)/n$, then $H(y) = o(n)/n$. Hence, if $r(n) = o(n)$, then $m \leq 2^{o(n)}$. By Corollary 23.5 in [9] (this result was first proved by Kleitman et al. [12]), for a given $\rho = \rho(m)$ we have $m \leq 2^{nH(\rho/m)}$. Therefore, if $\rho(m) = o(m)$ then $m \leq 2^{n \cdot o(m)/m}$ and, thus, $m \leq 2^{o(n)}$ (as $n \leq m$, if $n \to \infty$ then $m \to \infty$ and $o(m)/m \to 0$). Thus, both results of Crowston et al. [5] follow from corollary 1.

Similarly to Corollary 1 it is easy to prove the following:

Corollary 2. *Let $0 < a < 1$ be a constant. If $m < 2^{O(n^a)}$ then MAX LIN AA has a kernel with $O(k^{1/(1-a)})$ variables.*

By Corollary 1 it is easy to show that MAX LIN AA is in XP.

Proposition 2. MAX LIN AA *can be solved in time $O(m^{k+O(1)})$.*

Proof. We may again assume $m \geq n > k > 1$. As in the proof of Corollary 1, if $m \leq 2^{n/k}$ then the answer to MAX LIN AA is YES and a solution can be found in time $m^{O(1)}$. Otherwise, $2^n < m^k$ and MAX LIN AA can be solved in time $O(m^{k+2})$.   □

In fact, it is possible to improve this result, as the next theorem shows. We will not give its proof due to the page limit.

Theorem 3. MAX LIN AA *is in W[P].*

3.2   Max r-Lin AA, Max Exact r-SAT AA and Max r-CSP AA

Using Theorem 2 we can prove the following two results.

Corollary 3. *Let $r \geq 2$ be a fixed integer. Then MAX r-LIN AA has a kernel with $O(k \log k)$ variables and can be solved in time $2^{O(k \log k)} + m^{O(1)}$.*

Proof. Observe that $m \leq n^r$ and $n^r \leq 2^{n/(k-1)} - 2$ if $n \geq c(r)k \log_2 k$ provided $c(r)$ is large enough ($c(r)$ depends only on r). Thus, by Theorem 2, if $n \geq c(r)k \log_2 k$ then the answer to MAX r-LIN AA is YES. Hence, we obtain a problem kernel with at most $c(r)k \log_2 k = O(k \log k)$ variables and, therefore, can solve MAX r-LIN AA in time $2^{O(k \log k)} + m^{O(1)}$.   □

Corollary 4. *Let $r \geq 2$ be a fixed integer. Then there is a bikernel from* MAX EXACT r-SAT *to* MAX r-LIN AA *with $O(k \log k)$ variables. Moreover,* MAX EXACT r-SAT *has a kernel with $O(k \log k)$ variables and can be solved in time $2^{O(k \log k)} + m^{O(1)}$.*

Proof. Let F be an r-CNF formula with clauses $C_1, \ldots, C_m$ in the variables $x_1, x_2, \ldots, x_n$. We may assume that $x_i \in \{-1, 1\}$, where -1 corresponds to TRUE. For F, following [1] consider

$$g(x) = \sum_{C \in F} [1 - \prod_{x_i \in \mathrm{var}(C)} (1 + \epsilon_i x_i)],$$

where $\mathrm{var}(C)$ is the set of variables of C, $\epsilon_i \in \{-1, 1\}$ and $\epsilon_i = 1$ if and only if x_i is in C. It is shown in [1] that the answer to MAX EXACT r-SAT is YES if and only if there is a truth assignment x^0 such that $g(x^0) \geq k$.

Algebraic simplification of $g(x)$ will lead us to Fourier expansion of $g(x)$:

$$g(x) = \sum_{S \in \mathcal{F}} c_S \prod_{i \in S} x_i, \tag{3}$$

where $\mathcal{F} = \{\emptyset \neq S \subseteq [n] : c_S \neq 0, |S| \leq r\}$. Thus, $|\mathcal{F}| \leq n^r$. By Proposition 1, $\sum_{S \in \mathcal{F}} c_S \prod_{i \in S} x_i$ can be viewed as an instance of MAX r-LIN and, thus, we can reduce MAX EXACT r-SAT into MAX r-LIN in polynomial time (the algebraic simplification can be done in polynomial time as r is fixed). By Corollary 3, MAX r-LIN has a kernel with $O(k \log k)$ variables. This kernel is a bikernel from MAX EXACT r-SAT to MAX r-LIN. Using this bikernel, we can solve MAX EXACT r-SAT in time $2^{O(k \log k)} + m^{O(1)}$.

It remains to use the transformation described in [1] of a bikernel from MAX EXACT r-SAT to MAX r-LIN into a kernel of MAX EXACT r-SAT. This transformation gives us a kernel with $O(k \log k)$ variables. □

In the Boolean Max-r-Constraint Satisfaction Problem (MAX-r-CSP), we are given a collection of Boolean functions, each involving at most r variables, and asked to find a truth assignment that satisfies as many functions as possible. We will consider the following parameterized version of MAX-r-CSP. We are given a set Φ of Boolean functions, each involving at most r variables, and a collection $\mathcal{F}$ of m Boolean functions, each $f \in \mathcal{F}$ being a member of Φ, and each acting on some subset of the n Boolean variables $x_1, x_2, \ldots, x_n$ (each $x_i \in \{-1, 1\}$). We are to decide whether there is a truth assignment to the n variables such that the total number of satisfied functions is at least $E + k2^{-r}$, where E is the average value of the number of satisfied functions.

Corollary 5. *Let $r \geq 2$ be a fixed integer. Then there is a bikernel from* MAX r-CSP *to* MAX r-LIN AA *with $O(k \log k)$ variables.* MAX r-CSP *can be solved in time $2^{O(k \log k)} + m^{O(1)}$.*

Proof. Following [2] for a boolean function f of $r(f) \leq r$ boolean variables $x_{i_1}, \ldots, x_{i_{r(f)}}$, introduce a polynomial $h_f(x)$, $x = (x_1, x_2, \ldots, x_n)$ as follows. Let $V_f \subset \{-1, 1\}^{r(f)}$ denote the set of all satisfying assignments of f. Then

$$h_f(x) = 2^{r-r(f)} \sum_{(v_1,\ldots,v_{r(f)}) \in V_f} [\prod_{j=1}^{r(f)} (1 + x_{i_j} v_j) - 1].$$

Let $h(x) = \sum_{f \in \mathcal{F}} h_f(x)$. It is easy to see (cf. [1]) that the value of $h(x)$ at x^0 is precisely $2^r(s - E)$, where s is the number of the functions satisfied by the truth assignment x^0, and E is the average value of the number of satisfied functions. Thus, the answer to MAX-r-CSP is YES if and only if there is a truth assignment x^0 such that $h(x^0) \geq k$. The rest of the proof is similar to that of Corollary 4.   □

3.3   Lower Bound on Maxima of Pseudo-boolean Functions

Corollary 6. $\max_{x \in \{-1,+1\}^n} f(x) \geq c_\emptyset + (1 + \lfloor \frac{\text{rank}A}{\log_2(|\mathcal{F}|+2)} \rfloor) \cdot \min_{S \in \mathcal{F}} |c_S|$.

Proof. Consider the system $Az = b$ associated with the Fourier expansion of f according to the bijection described before Proposition 1. We may assume that the weighted system $Az = b$ has been simplified using Rule 1 and, thus, its number n' of variables equals $\text{rank}A$. Note that $n' \leq m$, where m is the number of equations in $Az = b$. By Theorem 2, Proposition 1 and the fact that $\min_{S \in \mathcal{F}} |c_S| = \min_j w_j$, it follows that if $k \leq m \leq 2^{n'/(k-1)} - 2$ then

$$\max_{x \in \{-1,+1\}^n} f(x) - c_\emptyset \geq k \min_{S \in \mathcal{F}} |c_S|.$$

To complete the proof, recall that $n' = \text{rank}A$, $m = |\mathcal{F}|$ and observe that the maximum possible (integral) value of k satisfying $m \leq 2^{n'/(k-1)} - 2$ is $1 + \lfloor \frac{\text{rank}A}{\log_2(|\mathcal{F}|+2)} \rfloor$ and thus, the above inequality remains valid after substituting k with $1 + \lfloor \frac{\text{rank}A}{\log_2(|\mathcal{F}|+2)} \rfloor$.   □

This bound is tight. Indeed, consider the function $f(x) = -\sum_{\emptyset \neq S \subseteq [n]} \prod_{i \in S} x_i$. Observe that $n = \text{rank}A$, $|\mathcal{F}| = 2^n - 1$ and, thus, $\max_{x \in \{-1,+1\}^n} f(x) \geq 1 + \lfloor \frac{\text{rank}A}{\log_2(|\mathcal{F}|+2)} \rfloor = 1$. If $x = (1, 1, \ldots, 1)$ then $f(x) = -|\mathcal{F}|$ and if we set some $x_i = -1$ then after canceling out of monomials we see that $f(x) = 1$. Therefore, $\max_{x \in \{-1,+1\}^n} f(x) = 1$, and, thus, the bound of corollary 6 is tight. It is easy to see that the bound remains tight if we delete one monomial from $f(x)$. A sightly more complicated function showing that the bound is tight is as follows: $g(x) = -\sum_{\emptyset \neq S \subseteq [n_1]} \prod_{i \in S} x_i - \sum_{S \in \mathcal{G}} \prod_{i \in S} x_i$, where $n_1 < n$ and $\mathcal{G} = \{S : \emptyset \neq S \subseteq [n], [n_1] \cap S = \emptyset\}$.

Acknowledgments. Gutin is thankful to Ilia Krasikov and Daniel Marx for discussions on the topic of the paper. Research of Gutin, Jones and Kim was supported in part by an EPSRC grant. Research of Gutin was also supported in part by the IST Programme of the European Community, under the PASCAL 2 Network of Excellence. Research of Ruzsa was supported by ERC–AdG Grant No. 228005 and Hungarian National Foundation for Scientific Research (OTKA), Grants No. 61908.

References

1. Alon, N., Gutin, G., Kim, E.J., Szeider, S., Yeo, A.: Solving MAX-r-SAT above a tight lower bound. Tech. Report arXiv:0907.4573. A priliminary version was published in Proc. ACM-SIAM Symposium on Discrete Algorithms, SODA 2010, pp. 511–517 (2010), http://arxiv.org/abs/0907.4573
2. Alon, N., Gutin, G., Krivelevich, M.: Algorithms with large domination ratio. J. Algorithms 50, 118–131 (2004)
3. Boros, E., Hammer, P.L.: Pseudo-boolean optimization. Discrete Appl. Math. 123, 155–225 (2002)
4. Borwein, P.: Computational excursions in analysis and number theory. Springer, New York (2002)
5. Crowston, R., Gutin, G., Jones, M.: Note on Max Lin-2 above average. Tech. Report, arXiv:0911.5384, http://arxiv.org/abs/0911.5384
6. Downey, R.G., Fellows, M.R.: Parameterized complexity. Springer, Heidelberg (1999)
7. Flum, J., Grohe, M.: Parameterized Complexity Theory. Springer, Heidelberg (2006)
8. Gutin, G., Kim, E.J., Szeider, S., Yeo, A.: A probabilistic approach to problems parameterized above tight lower bound. In: Chen, J., Fomin, F.V. (eds.) Proc. IWPEC'09. LNCS, vol. 5917, pp. 234–245. Springer, Heidelberg (2009)
9. Jukna, S.: Extremal combinatorics: with applications in computer science. Springer, Heidelberg (2001)
10. Håstad, J.: Some optimal inapproximability results. J. ACM 48(4), 798–859 (2001)
11. Håstad, J., Venkatesh, S.: On the advantage over a random assignment. Random Structures Algorithms 25(2), 117–149 (2004)
12. Kleitman, D.J., Shearer, J.B., Sturtevant, D.: Intersection of k-element sets. Combinatorica 1, 381–384 (1981)
13. Mahajan, M., Raman, V., Sikdar, S.: Parameterizing MAX SNP problems above guaranteed values. In: Bodlaender, H.L., Langston, M.A. (eds.) IWPEC 2006. LNCS, vol. 4169, pp. 38–49. Springer, Heidelberg (2006)
14. Mahajan, M., Raman, V., Sikdar, S.: Parameterizing above or below guaranteed values. J. Computer System Sciences 75(2), 137–153 (2009)
15. Niedermeier, R.: Invitation to fixed-parameter algorithms. Oxford University Press, Oxford (2006)
16. O'Donnell, R.: Some topics in analysis of boolean functions. Technical report, ECCC Report TR08-055. Paper for an invited talk at STOC'08 (2008), http://www.eccc.uni-trier.de/eccc-reports/2008/TR08-055/
17. de Wolf, R.: A Brief Introduction to fourier analysis on the boolean cube. Theory Of Computing Library Graduate Surveys 1, 1–20 (2008), http://theoryofcomputing.org

Capacitated max-Batching with Interval Graph Compatibilities

Tim Nonner[*]

Albert Ludwigs University of Freiburg, Germany
nonner@informatik.uni-freiburg.de

Abstract. We consider the problem of partitioning interval graphs into cliques of bounded size. Each interval has a weight, and the weight of a clique is the maximum weight of any interval in the clique. This natural graph problem can be interpreted as a batch scheduling problem. Solving a long-standing open problem, we show NP-hardness, even if the bound on the clique sizes is constant. Moreover, we give a PTAS based on a novel dynamic programming technique for this case.

1 Introduction

We consider the problem of partitioning interval graphs into cliques of bounded size. Each interval has a weight, and the cost of a clique is the maximum weight of any interval in the clique. Specifically, let J denote the intervals/vertices of the graph, where each interval $I \in J$ has a weight $w_I \in \mathbb{R}^+$, and let then $w_C := \max_{I \in C} w_I$ be the weight of a clique $C \subseteq J$. Moreover, let k be the bound on the clique size. The objective is hence to find a partition σ of J into cliques of at most size k such that $\mathrm{cost}(\sigma) := \sum_{C \in \sigma} w_C$ is minimized. We refer to this problem as CB_k.

We can think of an interval as a *job* and of a clique as a *batch* of jobs which satisfy the compatibility constraint implied by the interval graph structure. Thus, CB_k can be interpreted as a capacitated batch scheduling problem, where the maximum weight of a job in a batch is the time needed to process this batch [8,5], called *max-batching* [10], and the objective function given above is hence the completion time of a partition/schedule σ. This problem can be generalized to arbitrary graphs instead of interval graphs, as done in [9,8,6]. In this case, the problem is clearly NP-hard, since it contains graph coloring [12]. We will mostly use the terms batch and schedule instead of clique and partition throughout this paper, respectively.

Previous work. Finke et al. [8] showed that CB_k can be solved via dynamic programming in polynomial time for $k = \infty$. A similar result was independently obtained by Becchetti et al. [3] in the context of data aggregation. Moreover, this result was extended by Gijswijt, Jost, and Queyranne [9] to value-polymatroidal

---

[*] Supported by DFG research program No 1103 *Embedded Microsystems*. Parts of this work were done while the author was visiting the IBM T.J. Watson Research Center.

H. Kaplan (Ed.): SWAT 2010, LNCS 6139, pp. 176–187, 2010.

cost functions, a subset of the well-known submodular cost functions. Using this result as a relaxation, Correa et al. [6] recently presented a 2-approximation algorithm for CB_k for arbitrary k. However, it was raised as an open problem in [8,5,6] whether the case $k < \infty$ is NP-hard or not.

If all interval weights are identical, then CB_k simplifies to finding a clique partition of minimum cardinality [4]. We can also think of this problem as a hitting set problem with uniform capacity k, where we want to stab intervals with vertical lines corresponding the batches. Since the natural greedy algorithm solves this problem in polynomial time [8], Even et al. [7] addressed the more complicated case of non-uniform capacities. They presented a polynomial time algorithm based on a general dynamic programming approach introduced by Baptiste [2] for the problem of scheduling jobs such that the number of gaps is minimized.

Contributions. First, we settle the complexity of CB_k for $k < \infty$ with the following theorem.

Theorem 1. CB_k *is strongly NP-hard, even for $k = 3$ and two different interval lengths.*

Note that this hardness result is tight, since CB_k can be solved in polynomial time for $k = 2$ by using an algorithm for weighted matching. Unfortunately, due to space limitations, we have to defer a proof of this theorem to the full version of this paper. On the other hand, we obtain the following positive result.

Theorem 2. *There is a PTAS for CB_k for any constant k.*

This paper mostly deals with a high-level description of this PTAS in Section 3, which is based on a quite sophisticated dynamic program. It is worth mentioning that this dynamic program differs significantly from the dynamic programs introduced before for the related problems discussed above [3,6,7,2]. Instead, the way we decompose an instance is more related to the hierarchical quadtree decomposition used by Arora [1] to obtain a PTAS for the travelling salesman problem in the plane. Another important ingredient is the observation in Section 2 that we may assume that the number of different interval weights is constant.

Related work. Also the complementary problem, called *max-coloring*, where we want to partition a graph into independent sets instead of cliques, has raised a considerable amount of attention [14,13]. Pemmaraju, Raman, and Varadarajan [14] showed that this complementary problem is NP-hard for $k = \infty$, even for interval graphs. Moreover, Pemmaraju and Raman [13] showed that a graph class admits a $4c$-approximation algorithm if there is a c-approximation algorithm for the simpler coloring problem. Finally, note that our results for CB_k can be applied to this complementary problem for the graph class of co-interval graphs and constant k.

2   Preliminaries

To distinguish the intervals in J from other intervals, we refer to these intervals as *due intervals*. Let then n be the number of due intervals, and assume that all endpoints of due intervals are elements in the range $\{1, \dots, T\}$ for $T := 2n$. Motivated by the interpretation of a due interval as a temporal constraint in [3], we refer to an element in $\{1, \dots, T\}$ as a *period*. Assume then w.l.o.g. that all endpoints of due intervals are distinct, which is possible since $T = 2n$. Moreover, assume w.l.o.g. that n and T are powers of 2, and that all due intervals contain at least two periods. Finally, let $\text{OPT} := \text{cost}(\sigma^*)$ denote the cost of an optimal schedule σ^* for all due intervals.

Consider a schedule σ. For each batch $C \in \sigma$, there is a period t_C such that $t_C \in I$ for each due interval $I \in C$. If t_C is non-unique, simply choose t_C to be the smallest such period. Thus, we can also think of a schedule σ as a function $\sigma : J \to \{1, \dots, T\}$ that *assigns* each due interval I to the period $\sigma(I) = t_C$, where $C \in \sigma$ is the batch with $I \in C$.

Let $J_1, J_2, \dots, J_m$ with $m \le n$ be a partition of the due intervals such that (1) $w_I = w_{I'}$ for each due interval pair $I, I' \in J_i$, and (2) $w_I < w_{I'}$ for $I \in J_i$ and $I' \in J_{i'}$ with $i < i'$. Motivated by the application as a data aggregation problem in [3], we refer to the indices $1, 2, \dots, m$ as *nodes*, and we call $i_I := i$ the *release node* of a due interval $I \in J_i$. Analogously to w_C, define then $i_C := \max_{I \in C} i_I$ for a batch C.

2.1   Geometric Interpretation

If we interpret the release node i_I of each due interval I as a vertical height, then we can also think of a batch C as a vertical line with x-coordinate t_C starting at y-coordinate i_C and ending at y-coordinate 0. Thus, we say that each due interval contained in C is *stabbed* by this batch. We illustrate this geometric interpretation with an example instance containing five due intervals in Figure 1. For $k = 3$, the vertical dashed lines represent of schedule σ containing two batches C and C' with $t_C = 3$, $i_C = 3$, $t_{C'} = 7$, and $i_{C'} = 3$. The fat dots indicate which due intervals are stabbed by which batches. Note that batch C' does not fully exploit its capacity.

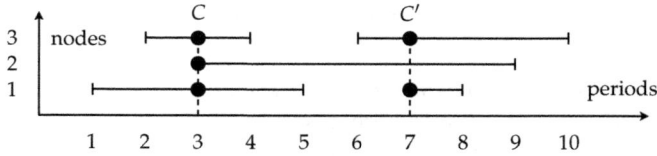

Fig. 1. Geometric interpretation

This geometric nature makes CB_k a natural target for dynamic programming techniques. Indeed, many algorithms for related problems are dynamic programming based [7,3,6,8,2]. However, none of these approaches seems to apply to CB_k.

2.2   Trading Nodes for Accuracy

The following lemma allows us to trade the number of nodes for accuracy such that we may even assume that m is constant. However, even if we additionally assume that k is constant, this does not seem to straightforward result in a PTAS. Instead, we require a sequence of additional arguments in Section 3.

Lemma 1. *For any $\epsilon > 0$, by losing an $(1+\epsilon)$-factor in the approximation ratio, we may assume that the number of nodes m is constant.*

Proof. We only sketch the proof. First, for some arbitrary small $\alpha > 1$, do a geometric rounding step by rounding each due interval weight up to the next power of α. Afterwards, use the shifting technique of Hochbaum and Maas [11] to decompose the input instance J into a sequence of subinstances, each requiring only a constant number of nodes. Use then properties of the geometric series with base α to show that combining schedules for these subinstances to a schedule for J adds only an $(1 + \epsilon)$-factor to the approximation ratio.   □

3   A PTAS for Constant k and m

In this section, we present a PTAS for constant k and m. To this end, we need several independent ingredients. First, we introduce the notion of an easy instance in Subsection 3.1. Such that instance provides some additional information about an optimal schedule that allows us to apply a hierarchical decomposition. Moreover, we introduce the notion of a normal schedule in Subsection 3.2, which allows us to restrict the search space by using a simple swapping argument about the assignment of due intervals to periods. Using this, we show then in Subsection 3.3 that there is a dynamic programming based quasipolynomial time algorithm for easy instances, and we extend this dynamic program to a QPTAS for general instances in Subsection 3.4. We obtain this extension by trading the size of the search space for accuracy. On the other hand, we extend this dynamic program to a PTAS for easy instances in Subsection 3.5. This extension is again obtained by trading the search space for accuracy. Combining both extensions allows us to finally prove Theorem 2 in Subsection 3.5.

3.1   Easy Instances

Let R be an ordered complete binary tree with T leaves. In what follows, a *vertex* will always be a vertex in R (this contrast to a *node*, which represents a class of due interval weight as defined in Section 2). Let $\overleftarrow{u}$ and $\overrightarrow{u}$ denote the left and right child of a non-leaf vertex u, respectively. Moreover, let $\hat{u}$ denote the parent of a non-root vertex u. For a vertex pair v, u, we write $v \overleftarrow{\succ} u$ and $v \overrightarrow{\succ} u$ if and only if $\overleftarrow{v}$ and $\overrightarrow{v}$ is an ancestor of u, respectively. We use here that u is an ancestor of itself, and hence $v \overleftarrow{\succ} \overleftarrow{v}$ and $v \overrightarrow{\succ} \overrightarrow{v}$. Additionally, we write $v \succ u$ if and only if $v \overleftarrow{\succ} u$ or $v \overrightarrow{\succ} u$. Thus, $v \succ u$ implies that $v \neq u$. Finally, let d_u denote the depth of a vertex u in R, where the root of R has depth 0.

Each vertex u corresponds naturally to the leaves of the subtree rooted at u, which form a subrange of periods $P_u \subseteq \{1, \ldots, T\}$ with $|P_u| = T/2^{d_u}$. More specifically, let $u_1, u_2, \ldots, u_T$ be the natural ordering of the leaves of R. For each period t, define then $P_{u_t} := \{t\}$, and, for each non-leaf vertex u, inductively define $P_u := P_{\overleftarrow{u}} \cup P_{\overrightarrow{u}}$. Note that $P_u = \{1, \ldots, T\}$ for the root u.

For each non-leaf vertex u, let p_u be an arbitrary element with $\max P_{\overleftarrow{u}} < p_u < \min P_{\overrightarrow{u}}$, and, for each node i, let then

$$R_u^i := \{I \mid i_I = i \text{ and } p_u \in I \subseteq [\min P_u, \max P_u]\}$$

Additionally, for each node i and leaf u, define $R_u^i := \emptyset$. Since we assume that each due interval contains at least two periods, we have that the due interval sets R_u^i are a pairwise disjoint partition of all due intervals.

Let A always denote a tuple of due interval sets with $A_u^i \subseteq R_u^i$ for each vertex u and node i, and if we say that such a tuple is u-based, then A_v^i is only defined for all vertices $v \succ u$. Since m is constant and $T = 2n$, note that the dimension of a tuple A is $m|R| = m(2T-1) = \mathcal{O}(n)$, but if A is u-based, then its dimension is at most $m \log T = \mathcal{O}(\log n)$, where the *dimension* of a tuple is the size of its index set. In what follows, we will sometimes require that a tuple is u-based to ensure that its dimension is logarithmic.

Moreover, we say that a schedule σ *satisfies* a tuple A if, for each vertex u and node i, it holds for each due interval $I \in R_u^i$ assigned by σ that

$$\sigma(I) \begin{cases} < p_u & \text{if } I \in A_u^i, \\ > p_u & \text{otherwise.} \end{cases}$$

Now we are ready to define the central notion of an easy instance. We say that the input instance J is *easy* if we additionally know a tuple A such that there is an optimal schedule σ^* which satisfies A. Clearly, in general, we cannot assume that we know such a tuple A. However, for each vertex u and node i, knowing such a tuple allows us to 'decompose' J at p_u such that all intervals in A_u^i can be assigned 'to the left', and all due intervals in $R_u^i \backslash A_u^i$ can be assigned 'to the right'.

3.2   Normal Schedules

Consider a fixed tuple A as described in Subsection 3.1, vertex pair $v \overleftarrow{\succ} u$, and node i, and let $I_1, I_2, \ldots, I_t$ be an ordering of the due intervals A_u^i according to their left endpoints such that, for each $1 \leq z \leq t-1$, the left endpoint of I_z is strictly smaller than the left endpoint of I_{z+1}. Recall that we assume that all endpoints are distinct. Let then

$$A_{vu}^i := \{I_r, I_{r+1}, \ldots, I_s\} \tag{1}$$

be the the consecutive subsequence of due intervals in A_v^i whose left endpoints are in P_u, and define

$$
A_{vu}^i(z) := \begin{cases} \emptyset & \text{for } 0 \le z < r, \\ \{I_r, I_{r+1}, \ldots, I_z\} & \text{for } r \le z \le s, \\ A_{vu}^i & \text{for } s < z \le n. \end{cases}
$$

Note that we can analogously define A_{vu}^i and $A_{vu}^i(z)$ for $v \overrightarrow{\succ} u$, but in this case, we use $R_v^i \setminus A_v^i$ instead of A_v^i and switch sides, i.e., we use an ordering $I_1, I_2, \ldots, I_t$ of the due intervals $R_v^i \setminus A_v^i$ according to their right endpoints such that, for each $1 \le z \le t - 1$, the right endpoint of I_z is strictly larger than the right endpoint of I_{z+1}. Define then $A_{vu}^i := \{I_r, I_{r+1}, \ldots, I_s\}$ to be the consecutive subsequence of due intervals in $R_v^i \setminus A_v^i$ whose right endpoints are in P_u, and finally $A_{vu}^i(z)$ as above. Note that the definition of $A_{vu}^i(z)$ hence includes both cases, $v \overleftarrow{\succ} u$ and $v \overrightarrow{\succ} u$.

Let a always denote an integer tuple with $0 \le a_{vu}^i \le n$ for each vertex pair $v \succ u$ and node i. Given such an integer tuple a and a tuple A, we abbreviate $A_{vu}^{ai} = A_{vu}^i(a_{vu}^i)$. We will always have that $a_{vu}^i = n$ for $v = \hat{u}$, which implies that

$$
A_{vu}^{ai} = \begin{cases} A_v^i & \text{if } u = \overleftarrow{v}, \\ R_v^i \setminus A_v^i & \text{if } u = \overrightarrow{v}. \end{cases} \tag{2}
$$

Moreover, if we say that a is *u-based*, then a_{vu}^i is only defined for all vertices $v \succ u$.

Consider now a schedule σ, and note that there is obviously some tuple A that is satisfied by σ. Specifically, for each vertex v and node i, simply define $A_v^i := \{I \in R_v^i \mid \sigma(I) < p_v\}$. Using this, we say that σ is *normal* if there is an integer tuple a such that, for each vertex pair $v \succ u$ and node i, σ assigns a due interval $I \in A_{vu}^i$ to a period in P_u if and only if $I \in A_{vu}^{ai}$. We then also say that σ *satisfies* a. The following lemma, which can be proven via a simple swapping argument, motivates this definition, since it says that we may restrict our attention to normal schedules.

Lemma 2. *For any schedule σ', there is a normal schedule σ assigning the same due intervals with $\mathrm{cost}(\sigma) = \mathrm{cost}(\sigma')$.*

We finally need the following simple structural lemma.

Lemma 3. *Let σ be a normal schedule that satisfies a tuple a. Then, for each vertex pair $v \succ u$ where u is not a leaf and node i, we may assume that*

$$
A_{v\overleftarrow{u}}^{ai} \begin{cases} \subseteq A_{vu}^{ai} \cap A_{v\overleftarrow{u}}^i & \text{if } v \overleftarrow{\succ} u, \\ = A_{vu}^{ai} \cap A_{v\overleftarrow{u}}^i & \text{if } v \overrightarrow{\succ} u, \end{cases}
$$

$$
A_{v\overrightarrow{u}}^{ai} \begin{cases} \subseteq A_{vu}^{ai} \cap A_{v\overrightarrow{u}}^i & \text{if } v \overleftarrow{\succ} u, \\ = A_{vu}^{ai} \cap A_{v\overrightarrow{u}}^i & \text{if } v \overrightarrow{\succ} u, \end{cases}
$$

and it is possible that the inclusions are not tight.

3.3   A Quasipolynomial Time Algorithm for Easy Instances

In this subsection, we present a dynamic programming based quasipolynomial time algorithm for easy instances. Recall that if the input instance J is easy, then we know a tuple A such that there is an optimal schedule σ^* that satisfies A. Hence, we can think of A as a given global variable. We will show in Subsection 3.4 how to get rid of this assumption. Let x always denote an integer tuple with $0 \leq x_i \leq n$ for each node i.

Let Π be the dynamic programming array, which, for each vertex u, integer tuple x, and u-based integer tuple a includes an entry $\Pi(u, x, a, A)$ that contains the cost of an optimal normal schedule σ assigning the due intervals

$$\Phi(u, x, a, A) := J[P_u] \cup P_u^x \cup \bigcup_{i=1}^{m} \bigcup_{v \in R: v \succ u} A_{vu}^{ai}$$

to periods in P_u subject to the constraint that σ satisfies A, where the first part $J[P_u] := \{I \mid I \subseteq [\min P_u, \max P_u]\}$ is the set of all due intervals that are contained in $[\min P_u, \max P_u]$. Moreover, the second part P_u^x stands for a set of due intervals that, for each node i, contains exactly x_i many due intervals I with release node i and $P_u \subseteq I$. Note that such a due interval can be assigned to any period in P_u. Hence, since we consider an optimal schedule σ assigning due intervals only to periods in P_u, we do not need to further specify these due intervals. Indeed, we will only count the number of such intervals with x when filling array Π. Finally, the due intervals in the third part have the property that exactly one endpoint is contained in P_u. Recall that Lemma 2 implies that taking a normal schedule σ is no restriction.

Since m is constant, R has $2T - 1 = \mathcal{O}(n)$ many vertices, and a has at most $m \log T = \mathcal{O}(\log n)$ many dimensions, we conclude that the size of Π is at most

$$(2T - 1) \cdot n^m \cdot n^{m \log T} = n^{\mathcal{O}(\log n)}, \tag{3}$$

which gives quasipolynomial running time if we can somehow inductively fill this array.

To fill array Π, for each non-leaf vertex u, use the recurrence relation

$$\Pi(u, x, a, A) = \min_{\overleftarrow{x}, \overrightarrow{x}, \overleftarrow{a}, \overrightarrow{a}} \{\Pi(\overleftarrow{u}, \overleftarrow{x}, \overleftarrow{a}, A) + \Pi(\overrightarrow{u}, \overrightarrow{x}, \overrightarrow{a}, A)\}, \tag{4}$$

where, for each node i, we have the constraints

$$\overleftarrow{a}^i_{v\overleftarrow{u}} \in \{0, \ldots, n\} \text{ for } v \succ u \text{ s.t. } A_{v\overleftarrow{u}}^{\overleftarrow{a}i} \begin{cases} \subseteq A_{vu}^{ai} \cap A_{v\overleftarrow{u}}^i & \text{if } v \overleftarrow{\succ} u, \\ = A_{vu}^{ai} \cap A_{v\overleftarrow{u}}^i & \text{if } v \overrightarrow{\succ} u, \end{cases} \tag{5}$$

$$\overrightarrow{a}^i_{v\overrightarrow{u}} \in \{0, \ldots, n\} \text{ for } v \succ u \text{ s.t. } A_{v\overrightarrow{u}}^{\overrightarrow{a}i} \begin{cases} \subseteq A_{vu}^{ai} \cap A_{v\overrightarrow{u}}^i & \text{if } v \overrightarrow{\succ} u, \\ = A_{vu}^{ai} \cap A_{v\overrightarrow{u}}^i & \text{if } v \overleftarrow{\succ} u, \end{cases}$$

and

$$\overleftarrow{y}_i \in \{0, \ldots, n\} \text{ and } \overrightarrow{y}_i \in \{0, \ldots, n\} \text{ s.t. } x_i = \overleftarrow{y}_i + \overrightarrow{y}_i, \tag{6}$$

and

$$\overleftarrow{x}_i = \sum_{v \in R : v \succ u} |(A_{vu}^{ai} \cap A_{v\overrightarrow{u}}^i) \setminus A_{v\overrightarrow{u}}^{\overrightarrow{a}i}| + \overleftarrow{y}_i, \tag{7}$$

$$\overrightarrow{x}_i = \sum_{v \in R : v \succ u} |(A_{vu}^{ai} \cap A_{v\overleftarrow{u}}^i) \setminus A_{v\overleftarrow{u}}^{\overleftarrow{a}i}| + \overrightarrow{y}_i.$$

Consider an optimal normal schedule σ assigning the due intervals $\Phi(u, x, a, A)$ to periods in P_u subject to the constraint that σ satisfies A. To show correctness, we need to argue that Recurrence Relation (4) defines a valid decomposition of σ. Since a is u-based, this integer tuple is only defined for all vertex pairs $v \succ u$. However, by the definition of $\Phi(u, x, a, A)$, we have that σ satisfies a at these vertex pairs, i.e., for each vertex $v \succ u$ and node i, σ assigns a due interval $I \in A_{vu}^i$ to a period in P_u if and only if $I \in A_{vu}^{ai}$. Moreover, since σ is normal, we can extend a to all vertex pairs $v \succ \overleftarrow{u}$ and $v \succ \overrightarrow{u}$ such that σ also satisfies a at these vertex pairs.

Each due interval in P_u^x needs to be assigned to either a period in $P_{\overleftarrow{u}}$ or $P_{\overrightarrow{u}}$. However, since such a due interval may be assigned to any period in $P_u = P_{\overleftarrow{u}} \cup P_{\overrightarrow{u}}$, using Constraints (6), we simply split the number of such due intervals x_i with release node i into two parts, namely $\overleftarrow{y}_i$ and $\overrightarrow{y}_i$. These parts are then added to $\overleftarrow{x}_i$ and $\overrightarrow{x}_i$ in Constraints (7), respectively. However, we further increase $\overleftarrow{x}_i$ and $\overrightarrow{x}_i$ with a sum, which is due to the fact that the inclusions in Constraints (5) are not necessarily tight. We discuss this in the next paragraph.

We need to argue that we can find two integer tuples $\overleftarrow{a}$ and $\overrightarrow{a}$ such that Constraints (5) are satisfied. However, we only consider $\overleftarrow{a}$, since the same arguments work for $\overrightarrow{a}$ as well. Recall that $\overleftarrow{a}$ is $\overleftarrow{u}$-based, i.e., $\overleftarrow{a}$ is only defined for all vertex pairs $v \succ \overleftarrow{u}$. However, recall that we require $\overleftarrow{a}_{u\overleftarrow{u}}^i = n$ for each node i per definition of such tuples. This is indeed necessary, since σ satisfies A, and hence, for each node i, all due intervals A_u^i are assigned to a period in $P_{\overleftarrow{u}}$. But since $\overleftarrow{a}_{u\overleftarrow{u}}^i = n$, we obtain with Equation (2) that $A_{u\overleftarrow{u}}^{\overleftarrow{a}i} = A_u^i$. Thus, we only need to consider the vertex pairs $v \succ \overleftarrow{u}$ with $v \succ u$. To this end, recall that we extended a to all vertex pairs $v \succ \overleftarrow{u}$ above. Hence, simply set $\overleftarrow{a}_{v\overleftarrow{u}}^i := a_{v\overleftarrow{u}}^i$ for each vertex $v \succ u$ and node i. Because of Lemma 3, this setting of $\overleftarrow{a}$ satisfies Constraints (5). However, for example, consider now a fixed vertex pair $v \succ u$. If we have that $A_{v\overleftarrow{u}}^{\overleftarrow{a}i} \subset A_{vu}^{ai} \cap A_{v\overleftarrow{u}}^i$, i.e., this inclusion is not tight, then there are some due intervals in $A_{vu}^{ai} \cap A_{v\overleftarrow{u}}^i$ which are assigned by σ to a period in $P_{\overrightarrow{u}}$. However, it is easy to see that any such due interval contains all periods in $P_{\overrightarrow{u}}$, and hence we only need to increase $\overrightarrow{x}_i$ by one with the additional sum in Constraints (7) to represent each such due interval. This shows that Recurrence Relation (4) defines a valid decomposition of σ into two parts. For the sake of exposition, we avoid a formal induction, since the provided arguments are sufficient to verify correctness.

Finally, to initiate array Π, recall that P_u contains only a single period if u is a leaf, say t. Hence, in this case, it is easy to see how an optimal schedule σ assigning the due intervals $\Phi(u, x, a, A)$ to t looks like, and what its cost is.

This can be done by using a similar top-down decomposition as used by Correa et al. [6]. Therefore, we can easily fill all entries $\Pi(u, x, a, A)$ where u is a leaf.

This completes the definition of the dynamic program. Note that, in contrast to the initialization of Π, we treat all nodes independently during Recurrence Relation (4). Specifically, we have a separate set of constraints for each node.

Combining all arguments in this subsection gives the correctness of this approach, since it implies that, if u is the root and x the integer tuple that contains only 0's, then $\Pi(u, x, a, A)$ contains the cost of an optimal schedule σ subject to the constraint that σ satisfies A. Recall here that an u-based integer tuple a is empty if u is the root. Moreover, recall the assumption that there is an optimal schedule σ^* satisfying A, and hence $\text{cost}(\sigma) = \text{cost}(\sigma^*) = \text{OPT}$.

3.4   A QPTAS

In this subsection, we extend the dynamic program for easy instances from Subsection 3.3 to general instances. We first need some additional definitions.

Let K always denote a tuple of sets of due interval sets with $K_v^i \subseteq \mathcal{P}(R_v^i)$ for each vertex v and node i, where $\mathcal{P}(R_v^i)$ denotes the power set of R_v^i. Define $|K| := \max_{v,i} |K_v^i|$. For a tuple A, we write $A \in K$ if $A_v^i \in K_v^i$ for each vertex v and node i. Analogously, for a u-based tuple A, we write $A \in K$ if this property holds for each vertex $v \succ u$, and we say that a schedule σ *satisfies* a K-extension of A if we can extend A to all other vertices $v \not\succ u$ such that still $A \in K$ and moreover σ satisfies A.

Now recall that the dynamic program in Subsection 3.3 requires that the input instance J is easy, which implies that we know a tuple A such that there is an optimal schedule σ^* that satisfies A. However, in general, we do not know such a tuple A. We could enumerate all such tuples, but this is not feasible in general, since there are too many ways to select a subset $A_v^i \subseteq R_v^i$, even for a single vertex v and node i. However, the following lemma, whose proof is deferred to the full version of this paper, allows us to trade the number of tuples A for accuracy. This lemma requires that k is constant.

Lemma 4. *For any $\epsilon > 0$, we can compute a tuple K with $|K| \leq c$ for some constant c in polynomial time such that there is a tuple $A \in K$ and a schedule σ that satisfies A with $\text{cost}(\sigma) \leq (1 + \epsilon)\text{OPT}$.*

Lemma 4 basically says that by losing an $(1 + \epsilon)$-factor we only have to consider a constant number of subsets $A_v^i \subseteq R_v^i$ for each vertex v and node i, namely only the at most c many subsets in K_v^i. In combination with the fact that R has logarithmic depth $\log T = \mathcal{O}(\log n)$, this allows us to incorporate an enumeration of sufficiently many tuples A in the dynamic program.

More specifically, assume that we are given a tuple K as described in Lemma 4, and hence, instead of a global variable A as in Subsection 3.3, we have a global variable K. We extend array Π such that, for each vertex u, integer tuple x, u-based integer tuple a, and u-based tuple $A \in K$, $\Pi(u, x, a, A)$ contains the cost of an optimal normal schedule σ assigning the due intervals $\Phi(u, x, a, A)$

to periods in P_u subject to the constraint that σ satisfies a K-extension of A. Because $|K| \leq c$ and A has at most $m \log T = \mathcal{O}(\log n)$ dimensions, we find that the size of Π listed in (3) only increases by the polynomial factor $c^{m \log T} = n^{\mathcal{O}(1)}$, and hence array Π has still quasipolynomial size.

To fill the extended array Π, instead of Recurrence Relation (4), we use the extended recurrence relation

$$\Pi(u, x, a, A) = \min_{\overleftarrow{x}, \overrightarrow{x}, \overleftarrow{a}, \overrightarrow{a}, \overline{A}} \left\{ \Pi(\overleftarrow{u}, \overleftarrow{x}, \overleftarrow{a}, \overline{A}) + \Pi(\overrightarrow{u}, \overrightarrow{x}, \overrightarrow{a}, \overline{A}) \right\}$$

subject to Constraints (5), (6) and (7), but we have the additional constraints that, for each node i,

$$\overline{A}_v^i \begin{cases} \in K_u^i & \text{for } v = u, \\ = A_v^i & \text{for } v \succ u. \end{cases} \tag{8}$$

Note that A is now a parameter in the array Π instead of a global variable. The new Constraints (8) are used to restrict the parameter $\overline{A}$ with respect to A and K. More specifically, $\overline{A}$ extends A at node u using K. Note that $\overline{A}$ is a $\overleftarrow{u}$- and $\overrightarrow{u}$-based tuple.

The correctness of this extension can be straightforward proven by extending the arguments in Subsection 3.3. Therefore, if u is the root and x the tuple that contains only 0's, then $\Pi(u, x, a, A)$ contains now the cost of an optimal schedule σ subject to the constraint that σ satisfies a K-extension of A. Note that a u-based tuple A is empty if u is the root, and hence this K-extension can be any tuple $A \in K$. Because we choose K according to Lemma 4, this shows that the extension of the dynamic program introduced in this subsection yields a QPTAS.

3.5   A PTAS for Easy Instances

In this subsection, we extend the dynamic program for easy instances described in Subsection 3.3 to a PTAS for easy instances. Recall that if the input instance J is easy, then we know a tuple A such that there is an optimal schedule σ^* that satisfies A. However, we have already shown in Subsection 3.4 how to get rid of this assumption, but, for the sake of exposition, we assume in this subsection again that J is easy.

Observe that the base n of the quasipolynomial size listed in (3) is due to the fact that we only have the bound $0 \leq a_{vu}^i \leq n$ on the entries of an integer tuple a. But if we even had a bound $0 \leq a_{vu}^i \leq c$ for some constant c, then we would immediately obtain an array Π of polynomial size, and hence polynomial running time. However, restricting the search space in this way does clearly not yield an approximation scheme. A more general way to restrict the search space is to only restrict the number of different values each entry of a may take. To this end, we say that an integer tuple a is *c-restricted* for some positive integer c if, for each vertex pair $v \succ u$ and node i, it satisfies

$$a_{vu}^i \in B_{vu}^{ci} := \{ y \in z\mathbb{N} \mid r - z \leq y \leq s + z \},$$

where z is the smallest power of 2 such that $|A^i_{vu}|/z \leq 2c$, and the integers $0 \leq r \leq s \leq n$ are such that $\{I_r, I_{r+1}, \ldots, I_s\} = A^i_{vu}$ as given in (1). We additionally require that $a^i_{vu} = \max B^{ci}_{vu}$ if $v = \hat{u}$, which ensures that Equation (2) holds as well for c-restricted integer tuples. On the other hand, note that if $a^i_{vu} = \min B^{ci}_{vu}$, then $A^{ai}_{vu} = \emptyset$. The following lemma, whose proof is deferred to the full version of this paper, states that c-restricted integer tuples yield an arbitrary good approximation. Hence, it allows us to trade the number of integer tuples a for accuracy. This lemma also requires that k is constant.

Lemma 5. *For any $\epsilon > 0$, there is a constant c such that there is a c-restricted integer tuple a and a schedule σ satisfying A and a with $\mathrm{cost}(\sigma) \leq (1 + \epsilon)\mathrm{OPT}$.*

We can restrict the dynamic program to c-restricted integer tuples by adding the constraints that, for each node i,

$$\overleftarrow{a}^i_{v\overleftarrow{u}} \in B^{ci}_{v\overleftarrow{u}} \text{ for } v \succ u, \tag{9}$$
$$\overrightarrow{a}^i_{v\overrightarrow{u}} \in B^{ci}_{v\overrightarrow{u}} \text{ for } v \succ u.$$

In this case, since each entry a^i_{vu} of a c-restricted integer tuple a may take at most $2c + 2$ many different values, in contrast to (3), we obtain an array Π of polynomial size

$$(2T - 1) \cdot n^m \cdot (2c + 2)^{m \log T} = n^{\mathcal{O}(1)}.$$

However, we need to argue that adding the new Constraints (9) is indeed correct. To this end, consider some fixed vertex pair $v \succ u$ and node i. Since $|A^i_{vu}| \geq |A^i_{v\overleftarrow{u}}|$ and $|A^i_{vu}| \geq |A^i_{v\overrightarrow{u}}|$, we have that $B^{ci}_{vu} \subseteq B^{ci}_{v\overleftarrow{u}} \cup B^{ci}_{v\overrightarrow{u}}$. Using this, it is easy to see that Lemma 3 holds as well for a schedule σ satisfying a c-restricted integer tuple a. Therefore, by extending the arguments in Subsection 3.3, we can straightforward prove that, if u is the root and x the tuple that contains only 0's, then $\Pi(u, x, a, A)$ contains now the cost of an optimal schedule σ subject to the constraint that σ satisfies A and a c-restricted integer tuple. Together with Lemma 5, this shows that the extension of the dynamic program introduced in this subsection yields a PTAS for easy instances. Moreover, in combination with the extension introduced in Subsection 3.4, this proves Theorem 2.

4 Conclusion

It is worth mentioning that the presented dynamic program is quite flexible with respect to the used objective function, e.g. it is also possible to incorporate the penalization of gaps as in [2]. Indeed, we think that it is a promising general technique to deal with interval stabbing type problems. To obtain the PTAS, we first showed that we may assume that the number of nodes m is constant, and then we gave a PTAS for this special case. However, it is not clear whether this special case is still NP-hard or admits a polynomial time algorithm. Finally, our PTAS requires that k is constant. On the other hand, a 2-approximation algorithm for arbitrary k is known [6], but it is an open problem whether this case admits a PTAS as well. We think that the methods developed in this paper are limited to constant k.

Acknowledgements

The author would like to thank Maxim Sviridenko and Nicole Megow for valuable discussions.

References

1. Arora, S.: Polynomial time approximation schemes for euclidean traveling salesman and other geometric problems. Journal of the ACM 45(5), 753–782 (1998)
2. Baptiste, P.: Scheduling unit tasks to minimize the number of idle periods: a polynomial time algorithm for offline dynamic power management. In: Proceedings of the 17th Annual ACM-SIAM Symposium on Discrete Algorithms (SODA'06), pp. 364–367 (2006)
3. Becchetti, L., Korteweg, P., Marchetti-Spaccamela, A., Skutella, M., Stougie, L., Vitaletti, A.: Latency constrained aggregation in sensor networks. In: Proceedings of the 14th Annual European Symposium on Algorithms (ESA'06), pp. 88–99 (2006)
4. Bodlaender, H.L., Jansen, K.: Restrictions of graph partition problems. part I. Theoretical Computer Science 148(1), 93–109 (1995)
5. Boudhar, M.: Dynamic scheduling on a single batch processing machine with split compatibility graphs. Journal of Mathematical Modelling and Algorithms 2(1), 17–35 (2003)
6. Correa, J., Megow, N., Raman, R., Suchan, K.: Cardinality constrained graph partitioning into cliques with submodular costs. In: Proceedings of the 8th Cologne-Twente Workshop on Graphs and Combinatorial Optimization (CTW'09), pp. 347–350 (2009)
7. Even, G., Levi, R., Rawitz, D., Schieber, B., Shahar, S., Sviridenko, M.: Algorithms for capacitated rectangle stabbing and lot sizing with joint set-up costs. ACM Transactions on Algorithms 4(3) (2008)
8. Finke, G., Jost, V., Queyranne, M., Sebö, A.: Batch processing with interval graph compatibilities between tasks. Discrete Applied Mathematics 156(5), 556–568 (2008)
9. Gijswijt, D., Jost, V., Queyranne, M.: Clique partitioning of interval graphs with submodular costs on the cliques. RAIRO - Operations Research 41(3), 275–287 (2007)
10. Hochbaum, D.S., Landy, D.: Scheduling semiconductor burn-in operations to minimize total flowtime. Operations Research 45, 874–885 (1997)
11. Hochbaum, D.S., Maass, W.: Approximation schemes for covering and packing problems in image processing and VLSI. Journal of the ACM 32(1), 130–136 (1985)
12. Garey, M.R., Johnson, D.S.: Computers and Intracability, A Guide to the Theory of NP-Completeness. W.H. Freeman and Company, New York (1979)
13. Pemmaraju, S.V., Raman, R.: Approximation algorithms for the max-coloring problem. In: Proceedings of the 32nd International Colloquium on Automata, Languages and Programming (ICALP'05), pp. 1064–1075 (2005)
14. Pemmaraju, S.V., Raman, R., Varadarajan, K.: Buffer minimization using max-coloring. In: Proceedings of the 15th annual ACM-SIAM Symposium on Discrete Algorithms (SODA'04), pp. 562–571 (2004)

A Weakly Robust PTAS for Minimum Clique Partition in Unit Disk Graphs*

(Extended Abstract)**

Imran A. Pirwani and Mohammad R. Salavatipour

Dept. of Computing Science, University of Alberta,
Edmonton, Alberta T6G 2E8, Canada
{pirwani,mreza}@cs.ualberta.ca

Abstract. We consider the problem of partitioning the set of vertices of a given unit disk graph (UDG) into a minimum number of cliques. The problem is NP-hard and various constant factor approximations are known, with the best known ratio of 3. Our main result is a *weakly robust* polynomial time approximation scheme (PTAS) for UDGs expressed with edge-lengths and $\varepsilon > 0$ that either (i) computes a clique partition, or (ii) produces a certificate proving that the graph is not a UDG; if the graph is a UDG, then our partition is guaranteed to be within $(1 + \varepsilon)$ ratio of the optimum; however, if the graph is not a UDG, it either computes a clique partition, or detects that the graph is not a UDG. Noting that recognition of UDG's is NP-hard even with edge lengths, this is a significant weakening of the input model.

We consider a weighted version of the problem on vertex weighted UDGs that generalizes the problem. We note some key distinctions with the unweighted version, where ideas crucial in obtaining a PTAS breakdown. Nevertheless, the weighted version admits a $(2+\varepsilon)$-approximation algorithm even when the graph is expressed, say, as an adjacency matrix. This is an improvement on the best known 8-approximation for the *unweighted* case for UDGs expressed in standard form.

1 Introduction

A standard network model for homogeneous networks is the *unit disk graph* (UDG). A graph $G = (V, E)$ is a UDG if there is a mapping $f : V \mapsto \mathbb{R}^2$ such that $\|f(u) - f(v)\|_2 \leq 1 \Leftrightarrow \{u, v\} \in E$; $f(u)^1$ models the position of the node u while the unit disk centered at $f(u)$ models the range of radio communication. Two nodes u and v are said to be able to directly communicate if they lie in the unit disks placed at each others' centers. There is a vast collection of literature on algorithmic problems studied on UDGs. See the survey [2].

---

* The authors were supported by Alberta Ingenuity. Work of the second author was additionally supported by a grant from NSERC.

** Several details are left out and proofs omitted due to space constraints. For a full version of the paper, please see [11].

[1] $f(.)$ is called a realization of G. Note that G may not come with a realization.

H. Kaplan (Ed.): SWAT 2010, LNCS 6139, pp. 188–199, 2010.
© Springer-Verlag Berlin Heidelberg 2010

Clustering of a set of points is an important subroutine in many algorithmic and practical applications and there are many kinds of clusterings depending upon the application. A typical objective in clustering is to minimize the number of "groups" such that each "group" (cluster) satisfies a set of criteria. Mutual proximity of points in a cluster is one such criterion, where points in a cluster forming a clique in the underlying network is an extreme form of mutual proximity. We study an optimization problem related to clustering, called the *minimum clique partition* problem on UDGs.

Minimum clique partition on unit disk graphs (MCP): Given a unit disk graph, $G = (V, E)$, partition V into a smallest number of cliques.

Besides being theoretically interesting, MCP has been used as a black-box for a number of recent papers, for example [9].

On general graphs, the clique-partition problem is equivalent to the minimum graph coloring on the complement graph which is not approximable within $n^{1-\varepsilon}$, for any $\varepsilon > 0$, unless P=NP [12]. MCP has been studied for special graph classes. It is shown to be MaxSNP-hard for cubic graphs and NP-complete for planar cubic graphs [4]; they also give a 5/4-approximation algorithm for graphs with maximum degree at most 3. MCP is NP-hard for a subclass of UDGs, called *unit coin graphs*, where the interiors of the associated disks are pairwise disjoint [5]. Good approximations, however, are possible on UDGs. The best known approximation is due to [5] who give a 3-approximation via partitioning the vertices into co-comparability graphs, and solving the problem exactly on them. They give a 2-approximation algorithm for coin graphs. MCP has also been studied on UDGs expressed in standard form. For UDGs expressed in standard form [10] give an 8-approximation algorithm.

Our Results and Techniques: Our main result is a weakly-robust (defined below) PTAS for MCP on a given UDG. For ease of exposition, first we show this (in Section 2.1) when the UDG is given with a realization, $f(.)$. The holy-grail is a PTAS when the UDG is expressed in standard form, say, as an adjacency matrix. However, falling short of proving this, we show (in Section 2.2) how to get a PTAS when the input UDG is expressed in standard form along with associated edge-lengths corresponding to some (unknown) realization. Note that an algorithm is called *robust* if it either computes an answer or declares that the input is not from the restricted domain; if the algorithm computes an answer then it is correct and has the guaranteed properties. Our algorithm is weakly-robust in the sense that it either (i) computes a clique partition of the input graph or (ii) gives a certificate that the input graph is not a UDG. If the input is indeed a UDG then the algorithm returns a clique partition (case (i)) which is a $(1 + \varepsilon)$-approximation (for a given $\epsilon > 0$). However, if the input is not a UDG, the algorithm either computes a clique partition but with no guarantee on the quality of the solution or returns that it is not a UDG. Therefore, this algorithm should be seen as a weakly-robust PTAS.

The generation of a polynomial-sized certificate which proves why the input graph is not a UDG should be seen in the context of the negative result of [1] which says that even if edge lengths are given, UDG recognition is NP-hard. Due to space constraints, we assume that the input graph is a-priori known to be a UDG. The full version of the paper does not make this assumption [11].

In Section 3 we explore a weighted version of MCP where we are given a vertex weighted UDG. In this formulation, the weight of a clique is the weight of a heaviest vertex in it, and the weight of a clique partition is the sum of the weights of the cliques in it. We note some key distinctions between the weighted and the unweighted versions of the problem and show that the ideas that help in obtaining a PTAS do not help in the weighted case. Nevertheless, we show that the problem admits a $(2 + \varepsilon)$-approximation algorithm for the weighted case *using only adjacency*. This result should be contrasted with the unweighted case where it is not clear as to how to remove the dependence on the use of edge-lengths, which was crucially exploited in deriving a PTAS.

Throughout, we use OPT to denote an optimum clique partition and opt to denote its size (or, in Section 3, weight); n and m denote the number of points (i.e. nodes of $G = (V, E)$) and the number of edges, respectively. Due to space constraints, all proofs appear in a full version of this paper [11].

2   A PTAS for Unweighted Unit Disk Graphs

For simplicity, we first describe an algorithm when the input is given with a geometric realization, and a parameter $\varepsilon > 0$.

2.1   A PTAS for UDGs with a Geometric Realization

We assume the input UDG is expressed with geometry. Using a randomly shifted grid whose cell size is $k \times k$ (for $k = k(\varepsilon)$) we partition the plane. Since the diameter of the convex hull of each clique is at most 1, for large values of k, a fixed clique is cut by this grid with probability at most $\frac{2}{k}$. Therefore, if we could efficiently compute an optimal clique partition in each $k \times k$ cell, then taking the union of these cliques yields a solution whose expected size is at most $(1+\varepsilon)$opt. We call the algorithm `MinCP1`. Therefore, the main portion of the algorithm is to solve the problem optimally for a bounded diameter region (i.e. a $k \times k$ cell).

Optimal Clique Partition of a UDG in a $k \times k$ Square. Unlike optimization problems such as maximum (weighted) independent set and minimum dominating set, where one can "guess" only a small-sized subset of points to obtain an optimal solution, the combinatorial complexity of any single clique in an optimal solution can be high. Therefore, it is unclear as to how to "guess" even few cliques, each of which may be large. A result of Capoyleas et al. [3] comes to our aid; a version of their result says that there exists an optimal clique partition where the convex hulls of the cliques are pair-wise non-overlapping. This phenomenon of *separability* of an optimal partition, coupled with the fact that the size of an optimal partition in a small region is small, allows us to circumvent the above difficulty. The following simple lemma bounds the size of an optimal solution of an instance of bounded diameter.

Lemma 1. *Any set of UDG vertices P in a $k \times k$ square has a clique partition of size $O(k^2)$.*

We state a variant of a result by Capoyleas et al. [3] which says that there is an optimal clique partition with cliques having non-overlapping convex hulls.[2]

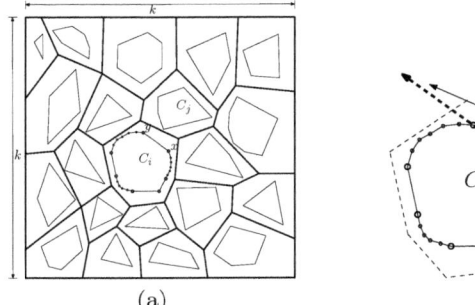

(a)   (b)

Fig. 1. (a) An optimal clique partition of UDG points in a bounded region; each light convex shape corresponds to a clique in the clique partition. The heavy line-segments represent segments of the corresponding separators. (b) A close-up view of C_i and C_j. A separator line, l_{ij} is shown which separates C_i and C_j, corresponding to the segment in (a). Note that l'_{ij} is also a separator for C_i and C_j and l'_{ij} is passing through points x and y in C_i.

Theorem 1 (Separation Theorem [3]). *There is an optimal clique partition in which the convex hulls of the cliques are pairwise non-overlapping, i.e., there is a straight line l_{ij} that separates a pair of cliques C_i, C_j such that all vertices of C_i are on one side of l_{ij}, and all the vertices of C_j are on the other side of l_{ij}. (see Figure 1). Such a separable clique partition can be computed efficiently.*

The general structure of the algorithm for computing optimal solution of a $k \times k$ cell is as follows. In order to reduce the search space for separator lines, one can find a characterization of the separator lines with some extra properties. Let C_i, C_j be a pair of cliques each having at least two points. Let L_{ij} be the (infinite) set of distinct separator lines. Since C_i and C_j are convex, there exists at least one line in L_{ij} that goes through two points of C_i (or C_j) (see Figure 1(b)). Therefore, given two cliques C_i and C_j in a clique partition (with pairwise non-overlapping parts) there is a separator line l_{ij} that goes through two vertices of one of them, say $u, v \in C_i$ such that all the vertices of C_j are on one side of this line and all the vertices of C_i are on the other side or on the line. Since there are $O(k^2)$ cliques in an optimal partition of $k \times k$ cell, there are $O(k^4)$ pairs of cliques in the partition and their convex hulls are pairwise non-overlapping. In fact, a more careful analysis shows that the dual graph of the regions is planar (see

---

[2] We gave a proof of this theorem [11] before it was brought to our attention that Capoyleas, Rote, and Woeginger [3] proved this much earlier in a different context.

Figure 1(a)); thus there are $O(k^2)$ distinct straight lines, each of which separate a pair of cliques in our optimal solution[3]. For every clique C_i, the separator lines l_{ij} (for all values of j) define a convex region that contains clique C_i. So once we guess this set of $O(k^2)$ lines, these convex regions define the cliques. We will try all possible (non-equivalent) sets of $O(k^2)$ separator lines and check if each of the convex regions indeed defines a clique and if we obtain a clique partition. This can be performed in $n^{O(k^2)}$ time (see [3] for more details). Thus,

Theorem 2. *MinCP1 returns a clique partition of size at most* $(1+\varepsilon)opt$ *w.h.p.*

2.2   A PTAS for UDGs with Edge-Lengths Only

Here, we assume only edge-lengths are known with respect to a feasible (unknown) realization of the UDG. We prove that,

Theorem 3. *Given a UDG G with associated (rational) edge-lengths and $\varepsilon > 0$, there is a polynomial time algorithm which computes a clique partition of G whose size is a $(1 + \varepsilon)$-approximation of the optimum clique partition.*

The high level idea of the algorithm is as follows. As in the geometric case, we first decompose the graph into bounded diameter regions and show that if we can compute the optimum clique partition of each region then the union of these clique partitions is within $(1 + \varepsilon)$ fraction of the optimum. There are two main difficulties here for which we need new ideas. The first major difference is that we cannot use the random shift argument as in the geometric case. To overcome this, we use a ball growing technique that yields bounded diameter regions. This is inspired by [8] who give local PTAS for weighted independent set, and minimum dominating set for UDGs without geometry. The second major difference is that, even if we have the set of points belonging to a bounded region (a ball) it is unclear as to how to use the separation theorem to obtain an optimal solution for this instance. We show how to compute an optimal clique partition when the diameter of the input is bounded.

Let $B_r(v) = \{u : d(u, v) \leq r\}$, where by $d(u, v)$ we mean the number of edges on a shortest path from u to v. So, $B_r(v)$ can be computed using a breadth-first search (BFS) tree rooted at v. We describe our decomposition algorithm which partitions the graph into bounded diameter subgraphs in Algorithm 1. We will describe a procedure, called OPT-CP which, given a graph induced by the vertices of $B_r(v)$ and a parameter $\ell = \text{poly}(r)$, runs in time $|B_r(v)|^{O(\ell^2)} \leq n^{O(\ell^2)}$ and computes an optimal clique partition of $B_r(v)$. We only call this procedure for "small" values of $r = r(\varepsilon)$.

Clearly, $\mathcal{C}$ on "Step 8" is a clique partition. Let us assume that each ball $B_r(v)$ we consider, OPT-CP returns an optimal clique partition $C_r(v)$. We show that then, $|\mathcal{C}| \leq (1 + \varepsilon)opt$. We also show that for any iteration of the outer "while–loop", "Step 4" of MinCP2 is executed in time polynomial in n, by using edge-lengths instead of Euclidean coordinates.

---

[3] The fact that there are $O(k^2)$ separable lines was independently observed by [6].

Algorithm 1. MinCP2(G, ε)

1: $\mathcal{C} \leftarrow \emptyset$; $\beta \leftarrow \lceil c_0 \frac{1}{\varepsilon} \log \frac{1}{\varepsilon} \rceil$; $\ell \leftarrow c_1 \beta^2$.
   {where c_0 is the constant in Lemma 5, and c_1 is the constant in inequality (1).}
2: **while** $V \neq \emptyset$ **do**
3:     Pick an arbitrary vertex $v \in V$; $r \leftarrow 0$
       {Let $C_r(v)$ denote a clique partition of $B_r(v)$ computed by calling OPT-CP}
4:     **while** $|C_{r+2}(v)| > (1 + \varepsilon) \cdot |C_r(v)|$ **do**
5:         $r \leftarrow r + 1$
6:     $\mathcal{C} \leftarrow \mathcal{C} \cup C_{r+2}(v)$
7:     $V \leftarrow V \setminus B_{r+2}(v)$
8: **return** $\mathcal{C}$ as our clique partition

For an iteration i of the outer loop, let v_i be the vertex chosen in "Step 3" and let r_i^* be the value of r for which the "while-loop" on "Step 4" terminates, that is, $|C_{r_i^*+2}(v_i)| \leq (1+\varepsilon) \cdot |C_{r_i^*}(v_i)|$. Let k be the maximum number of iterations of the outer loop. The following lemma states that two distinct balls grown around vertices are far from each other.

Lemma 2. *Every two vertices $v \in B_{r_i^*}(v_i)$ and $u \in B_{r_j^*}(v_j)$ are non-adjacent.*

Using the above lemma, one can bound opt from below in the following,

Lemma 3. $opt \geq \sum_{i=1}^{k} |C_{r_i^*}(v_i)|$

We can now relate the cost of our solution to opt as follows,

Lemma 4. *If for each i, $|C_{r_i^*+2}(v_i)| \leq (1 + \varepsilon) \cdot |C_{r_i^*}(v_i)|$, then $\sum_{i=1}^{k} |C_{r_i^*+2}(v_i)| \leq (1 + \varepsilon) \cdot opt$*

Finally, we show that the inner "while-loop" terminates in $\tilde{O}(\frac{1}{\varepsilon})$, so $r_i^* \in \tilde{O}(\frac{1}{\varepsilon})$. Obviously, the "while-loop" on "Step 4" terminates eventually, so r_i^* exists. By definition of r_i^*, for all smaller values of $r < r_i^*$: $|C_r(v_i)| > (1 + \varepsilon) \cdot |C_{r-2}(v_i)|$. Since the diameter of $B_r(v_i)$ is $O(r)$, if $B_r(v)$ is a UDG, there is a realization of it in which all the points fit into a $2r \times 2r$ square. Thus, $|C_r(v_i)| \in O(r^2)$. So for some $\alpha \in O(1)$:

$$\alpha \cdot r^2 > |C_r(v_i)| > (1 + \varepsilon) \cdot |C_{r-2}(v_i)| > \ldots > (1 + \varepsilon)^{\frac{r}{2}} \cdot |C_0(v_i)| = \Theta((\sqrt{1 + \varepsilon})^r),$$

when r is even (for odd values of r we obtain $|C_r(v_i)| > (1 + \varepsilon)^{\frac{r-1}{2}} \cdot |C_1(v_i)| \geq \Theta((\sqrt{1 + \varepsilon})^{r-1})$. Therefore we have:

Lemma 5. *There is a constant $c_0 > 0$ such that for each i: $r_i^* \leq c_0/\varepsilon \cdot \log 1/\varepsilon$.*

In the next subsection, we show that the algorithm OPT-CP, given $B_r(v)$ and an upper bound ℓ on $|C_r(v)|$, computes an optimal clique partition. The algorithm runs in time $n^{O(\ell^2)}$. By the above arguments, there is a constant $c_1 > 0$ such that:

$$|C_r(v)| = O(r_i^{*2}) \leq c_1 \cdot \frac{c_0^2}{\varepsilon^2} \log^2 \frac{1}{\varepsilon}. \tag{1}$$

Let $\ell = \lceil c_1 \frac{c_0^2}{\varepsilon^2} \log^2 \frac{1}{\varepsilon} \rceil$ for any call to OPT-CP as an upper bound, where c_1 is the constant in $O((r_i^*)^2)$. So, the running time of the algorithm is $n^{\tilde{O}(1/\varepsilon^4)}$.

2.3   An Optimal Clique Partition for $B_r(v)$

Here we present the algorithm OPT-CP that given $B_r(v)$ (henceforth referred to as G') and an upper bound ℓ on the size of an optimal solution for G', computes an optimal clique partition of it. The algorithm runs in time $n^{O(\ell^2)}$. Since, by Lemma 5, ℓ is independent of n in each call to this algorithm, the running time of OPT-CP is polynomial in n. Our algorithm is based on the separation theorem [3]. Even though we do not have a realization of the nodes on the plane, we show how to apply the separation theorem [3] as in the geometric setting. We use node/point to refer to a vertex of G' and/or its corresponding point on the plane for some realization of G'.

Lemma 6. *Suppose we have four mutually adjacent nodes p, a, b, r and their pairwise distances with respect to some realization on the Euclidean plane. Then there is a poly-time procedure that can decide if p and r are on the same side of the line that goes through a and b or are on different sides.*

Say that G' has an optimum clique partition of size $\alpha \leq \ell$. The cliques fall in two categories: small (with at most $2\alpha - 2$ points), and large (with at least $2\alpha - 1$ points). We focus only on finding the large cliques since we can guess all the small cliques. Suppose for each pair $C_i, C_j \in$ OPT of large cliques, we guess their respective representative nodes, c_i and c_j, through which no separating lines pass. Also, suppose that we guess a separating line l_{ij} correctly which goes through points u_{ij} and v_{ij}. For a point p adjacent to c_i or c_j we want to efficiently test if p is on the the same side of line l_{ij} as c_i (the *positive* side), or on c_j's side (the *negative* side), using only edge-lengths. Without loss of generality, let both u_{ij} and v_{ij} belong to clique C_i. For each p different from the representatives:

- Suppose p is adjacent to all of c_i, u_{ij}, v_{ij}, c_j. Observe that we also have the edges $c_i u_{ij}$ and $c_i v_{ij}$. Given the edge-lengths of all the six edges among the four vertices c_i, u_{ij}, v_{ij}, p using Lemma 6 we can decide if in a realization of these four points, the line going through u_{ij}, v_{ij} separates the two points p and c_i or not. If p and c_i are on the same side, we say p is on the positive side of l_{ij} for C_i. Else, it is on the positive side of l_{ij} for C_j.
- Suppose p is adjacent to c_i (and also to u_{ij} and v_{ij}) but not to c_j. Given the edge-lengths of all the six edges among the four vertices c_i, u_{ij}, v_{ij}, p using Lemma 6 we can decide if in a realization of these four points, the line going through u_{ij}, v_{ij} separates the two points p and c_i or not. If p and c_i are on the same side, we say p is on the positive side of l_{ij} for C_i. Else, it is on the positive side of l_{ij} for C_j.

For each C_i and all the lines l_{ij}, consider the set of nodes that are on the positive side of all these lines with respect to C_i; we place these nodes in C_i. After obtaining the large and the small cliques, we obtain sets $C_1, \ldots, C_\alpha$. At the end we check if each C_i forms a clique and if their union covers all the points. The number of guesses for representatives is $n^{O(\alpha)}$ and the number of guesses for the separator lines is $n^{O(\alpha^2)}$. So there are a total of $n^{O(\alpha^2)}$ configurations that we consider. Clearly, some set of guesses is a correct one, allowing us to obtain an optimum clique partition.

3   Weighted Clique Partition Using Only Adjacency

We now consider a generalization of the minimum clique partition on UDGs, which we call *minimum weighted clique partition* (MWCP). Given a node-weighted graph $G = (V, E)$ with vertex weight $\mathbf{wt}\,(v)$, the weight of a clique C is defined as the weight of the heaviest vertex in it. For a clique partition $\mathcal{C} = \{C_1, C_2, \ldots, C_t\}$, the weight of $\mathcal{C}$ is defined as sum of the weights of the cliques in $\mathcal{C}$, i.e. $\mathbf{wt}\,(\mathcal{C}) = \sum_{i=1}^t \mathbf{wt}\,(C_i)$. The problem is, given G in standard form, say, as an adjacency matrix, construct a clique partition $\mathcal{C} = \{C_1, C_2, \ldots, C_t\}$ while minimizing $\mathbf{wt}\,(\mathcal{C})$. The weighted version of the problem as it is defined above has also been studied in different contexts. See [7], for example.

MWCP distinguishes itself from MCP in two important ways: (i) The *separability* property which was crucially used earlier to devise a PTAS does not hold in the weighted case (Figure 2(a)), and (ii) the number of cliques in an optimal solution for a UDG in a region of bounded radius is not bounded by the diameter of the region (Figure 2(b)). To the best of our knowledge, MWCP has not been investigated before on UDGs. We, however, note that a modification to the algorithm by [10] also yields a factor-8 approximation to the weighted case, a generalization which they do not consider. Here, we give an algorithm which runs in time $O(n^{\text{poly}(1/\varepsilon)})$ for a given $\varepsilon > 0$ and computes a $(2+\varepsilon)$-approximation to MWCP for UDGs expressed in standard form, for example, as an adjacency matrix. The algorithm returns a clique partition which is a $(2+\varepsilon)$-approximation.

Theorem 4. *Given a UDG G expressed in standard form, and $\varepsilon > 0$, there is a polynomial time algorithm which computes a clique partition whose weight is a $(2 + \varepsilon)$-approximation of the minimum weighted clique partition.*

We will employ a similar ball growing technique (as in Section 2) that will give us bounded diameter regions. We then show that we can compute a clique partition whose weight is within a factor $(2 + \varepsilon)$ of the optimal. For the case of bounded diameter region, although the optimum solution may have a large number of cliques, we can show that there is a clique partition with few cliques whose cost is within $(1 + \varepsilon)$-factor of optimum. First we describe the main algorithm. Then in Subsection 3.1 we show that for each subgraph $B_r(v)$ (of bounded diameter) there is a near optimal clique partition with $\tilde{O}(r^2)$ cliques. Then in Subsection 3.2 we discuss a near 2-approx for such a near optimal clique partition.

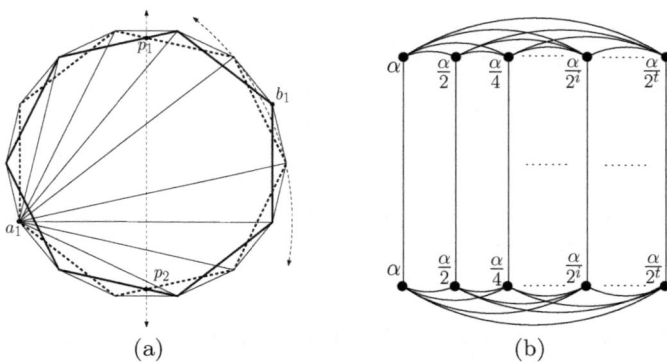

(a)   (b)

Fig. 2. (a) Two overlapping weighted cliques, $A = \{a_1, \ldots, a_k\}$ and $B = \{b_1, \ldots, b_k\}$ are shown, a_i, b_i are independent for all i. The heavy polygon has vertices weighted k while the dashed ones are weighted 1. $opt = k + 1$ while any separable partition must pay a cost of at least $2k$. (b) A UDG for which OPT contains t cliques. The weight is less than $2 \cdot \alpha$.

Our decomposition algorithm (Algorithm 2) is similar to Algorithm 1 and partitions the graph into bounded diameter subgraphs. The procedure CP, given a UDG induced by the vertices of $B_r(v)$ and a parameter $\ell = \mathrm{poly}(r)$, runs in time $n^{O(\ell^2)}$ and computes a clique partition of $B_r(v)$ whose weight is within a factor $(2 + \varepsilon)$ of the optimum. We only call this procedure for constant values of r. In the following, let $0 < \gamma \le \frac{\sqrt{9+4\varepsilon}-3}{2}$ be a rational number. Let k be the maximum number of iterations of the outer "while-loop". The following lemma states that distinct balls are far from each other.

Algorithm 2. MinCP3(G, γ)

1: $\mathcal{C} \leftarrow \emptyset;\ \beta \leftarrow \lceil c_0 \frac{1}{\gamma} \log \frac{1}{\gamma} \rceil;\ \ell \leftarrow c_1 \beta^2$.
   {where c_0 is the constant in Lemma 10, and c_1 is the constant in inequality (2).}
2: **while** $V \neq \emptyset$ **do**
3:     $v \leftarrow \arg\max_u \{\mathbf{wt}(u)\};\ r \leftarrow 0$
       {Let $C_r(v)$ denote a factor-$(2+\gamma)$ partition of $B_r(v)$ computed by calling CP}
4:     **while** $\mathbf{wt}(C_{r+2}(v)) > (1 + \gamma) \cdot \mathbf{wt}(C_r(v))$ **do**
5:         $r \leftarrow r + 1$
6:     $\mathcal{C} \leftarrow \mathcal{C} \cup C_{r+2}(v)$
7:     $V \leftarrow V \setminus B_{r+2}(v)$
8: **return** $\mathcal{C}$ as our clique partition

Lemma 7. *Every two vertices $v \in B_{r_i^*}(v_i)$ and $u \in B_{r_j^*}(v_j)$ are non-adjacent.*

The above lemma can be used to bound opt from below, in the following,

Lemma 8. $(2 + \gamma) \cdot opt \ge \sum_{i=1}^{k} \mathbf{wt}\left(C_{r_i^*}(v_i)\right)$

We can now relate the cost of our clique partition to opt as follows.

Lemma 9. *If for each i, $\boldsymbol{wt}\left(C_{r_i^*+2}(v_i)\right) \leq (1+\gamma)\boldsymbol{wt}\left(C_{r_i^*}(v_i)\right)$, then $\sum_{i=1}^{k} \boldsymbol{wt}\left(C_{r_i^*+2}(v_i)\right) \leq (2+\varepsilon)opt$*

Next, we show that the inner "while-loop" terminates in $\tilde{O}(\frac{1}{\gamma})$, that is each r_i^* is bounded by $\tilde{O}(\frac{1}{\gamma})$. This is similar to Lemma 5. Since the while loop terminates, r_i^* exists and by definition of r_i^*, it must be the case that for all smaller values of $r < r_i^*$, $\boldsymbol{wt}\left(C_r(v_i)\right) > (1+\gamma) \cdot \boldsymbol{wt}\left(C_{r-2}(v_i)\right)$. Because the diameter of $B_r(v_i)$ is $O(r)$, there is a realization of it in which all the points fit into a $r \times r$ grid. Also, since v_i is a heaviest vertex in the (residual) graph, there is a clique partition whose weight is at most $\alpha \cdot \boldsymbol{wt}(v_i) \cdot r^2$, for some constant α. Therefore, $\boldsymbol{wt}\left(C_r(v_i)\right) < \alpha \cdot \boldsymbol{wt}(v_i) \cdot r^2$ So:

$$\alpha \cdot \boldsymbol{wt}(v_i) \cdot r^2 > \boldsymbol{wt}\left(C_r(v_i)\right) > (1+\gamma) \cdot \boldsymbol{wt}\left(C_{r-2}(v_i)\right) > \dots$$
$$> (1+\gamma)^{\frac{r}{2}} \cdot \boldsymbol{wt}\left(C_0(v_i)\right) = \boldsymbol{wt}(v_i) \cdot \left(\sqrt{1+\gamma}\right)^r,$$

which implies $\alpha \cdot r^2 > \left(\sqrt{1+\gamma}\right)^r$, for the case that r is even. If r is odd we obtain $\alpha \cdot r^2 > \left(\sqrt{1+\gamma}\right)^{r-1}$. Thus, the following lemma easily follows:

Lemma 10. *There is a constant $c_0 > 0$ such that for each i: $r_i^* \leq c_0/\gamma \cdot \log 1/\gamma$.*

In Subsection 3.2, we discuss the algorithm CP that given $B_r(v)$ and an upper bound ℓ on $|C_r(v)|$, computes a clique partition having weight within a factor $(2+\gamma)$ of opt; the algorithm runs in time $n^{O(\ell^2)}$. By the above arguments, there is a constant $c_1 > 0$ such that:

$$|C_r(v)| = O(r_i^{*2}) \leq c_1 \cdot \frac{c_0^2}{\gamma^2} \log^2 \frac{1}{\gamma} \tag{2}$$

Let $\ell = \lceil c_1 \frac{c_0^2}{\gamma^2} \log^2 \frac{1}{\gamma} \rceil$ for any call to CP as an upper bound, where c_1 is the constant in $O(r_i^{*2})$. So, the running time of the algorithm is $n^{\tilde{O}(1/\varepsilon^4)}$.

3.1 Existence of a Small Clique Partition of $B_r(v)$ Having Near-Optimal Weight

Unlike the unweighted case, an optimal weighted clique partition in a small region may contain a large number of cliques. Yet, there exists a partition whose weight is within a factor $(1+\frac{\gamma}{2})$ of the minimum weight which contains few cliques (where by "few" we mean ℓ as in Algorithm 2). The existence of a light and small partition allows us to enumerate them in the same manner in the algorithm of subsection 2.3, yielding a $(2+\gamma)$-approximation for the problem instance in a ball of small radius. In the following, let $r \in \tilde{O}(\frac{1}{\gamma})$; we focus on the subproblem that lies in some $B_r(v)$. Recall that any ball of radius r can be partitioned into $O(r^2)$ cliques (Lemma 1). We begin with a simple lemma which says that for any clique partition $\mathcal{C}$, if the vertices can be be covered by another clique partition $\mathcal{C}'$ containing x cliques then the sum of the weights of the x cliques in $\mathcal{C}'$ is not significantly more than the weight of the heaviest clique in $\mathcal{C}$.

Lemma 11. *For any collection of disjoint cliques $\mathcal{C} = \{C_1, C_2, \ldots, C_t\}$ having weights such that $\mathbf{wt}(C_1) \geq \mathbf{wt}(C_2) \geq \ldots \geq \mathbf{wt}(C_t)$ suppose the vertices of $\mathcal{C}$ can be partitioned into x cliques $\mathcal{C}' = \{C'_1, C'_2, \ldots, C'_x\}$. Then $\mathbf{wt}(\mathcal{C}') \leq x \cdot \mathbf{wt}(C_1)$*

Also, for any clique, C_i, in an optimal partition of a ball of radius r, the sum of the weights of the lighter cliques is not significantly more than $\mathbf{wt}(C_i)$.

Lemma 12. *Let $\mathcal{C} = \{C_1, C_2, \ldots, C_t\}$ be OPT and let $\mathbf{wt}(C_1) \geq \mathbf{wt}(C_2) \geq \ldots \geq \mathbf{wt}(C_t)$. Suppose there is another clique partition $\mathcal{C}' = \{C'_1, \ldots, C'_x\}$ of the vertices of $\mathcal{C}$. Then, for every $1 \leq i < t$: $(x-1) \cdot \mathbf{wt}(C_i) \geq \sum_{l=i+1}^{t} \mathbf{wt}(C_l)$.*

We now are ready to state the main result of this section which says that for any optimal weighted clique partition of a ball of radius r, there exists another clique partition whose weight is arbitrarily close to the weight of the optimal partition, but has $O(r^2)$ cliques in it. Since the radius of the ball within which the subproblem lies is small, $r \in \tilde{O}(\frac{1}{\gamma})$, this means that if we were to enumerate all the clique partitions of the subproblem up to $O(r^2)$, we will see one whose weight is arbitrarily close to the weight of an optimal clique. Choosing a lightest one from amongst all such cliques guarantees that we will choose a one whose weight is arbitrarily close to the optimal weight.

Lemma 13. *Let $\gamma > 0$ and $r \in \tilde{O}(1/\gamma)$ be two constants. Let $\mathcal{C} = \{C_1, C_2, \ldots, C_t\}$ be an optimal weighted clique partition of $B_r(v)$ and let $\mathcal{C}' = \{C'_1, \ldots, C'_x\}$ be another clique partition of vertices of $\mathcal{C}$ with $x \in O(r^2)$. Let $\mathbf{wt}(C_1) \geq \mathbf{wt}(C_2) \geq \ldots \geq \mathbf{wt}(C_t)$. Then, there is a partition of vertices of $\mathcal{C}$ into at most $j+x$ cliques for some constant $j = j(\gamma)$, with cost at most $(1 + \frac{\gamma}{2})opt$.*

3.2 $(2 + \gamma)$-Approximation for MWCP in $B_r(v)$

Lemma 13 shows that there exists a clique partition of $B_r(v)$ that has $\tilde{O}(1/\gamma)$ cliques, whose weight is at most $(1 + \frac{\gamma}{2})$ of an optimal clique partition of $B_r(v)$. Thus, if an efficient algorithm can cover each of the $\tilde{O}(1/\gamma)$ cliques with at most two cliques, then this will yield a $(2+\gamma)$-approximation for MWCP in $B_r(v)$. We employ ideas from [10] to efficiently find such a cover. Due to space constraints, we leave the details out, but refer the reader to a full version of the paper [11].

4   Concluding Remarks

Our PTAS applies to the planar case; it does not extend even to three dimensions. It is an open question as to how to obtain a PTAS in higher, but fixed, dimensions. It is also unclear how to obtain a PTAS for the weighted, planar case with geometry. The main technical challenges are to show some exploitable properties, as we exploited separability in Section 2, that allows us to find the few cliques efficiently. It will be interesting to see how to obtain a PTAS when the (unweighted) UDG is expressed in standard form. We leave these as interesting open questions for the future.

Acknowledgments. We thank Sriram Pemmaraju, Lorna Stewart, and Zoya Svitkina for helpful discussions. Our thanks to an anonymous source for pointing out the result of Capoyleas et al. [3].

References

1. Aspnes, J., Goldenberg, D.K., Yang, Y.R.: On the computational complexity of sensor network localization. In: Nikoletseas, S.E., Rolim, J.D.P. (eds.) ALGOSEN-SORS 2004. LNCS, vol. 3121, pp. 32–44. Springer, Heidelberg (2004)
2. Balasundaram, B., Butenko, S.: Optimization problems in unit-disk graphs. In: Floudas, C.A., Pardalos, P.M. (eds.) Encyclopedia of Optimization, pp. 2832–2844. Springer, Heidelberg (2009)
3. Capoyleas, V., Rote, G., Woeginger, G.J.: Geometric clusterings. J. Algorithms 12(2), 341–356 (1991)
4. Cerioli, M.R., Faria, L., Ferreira, T.O., Martinhon, C.A.J., Protti, F., Reed, B.: Partition into cliques for cubic graphs: Planar case, complexity and approximation. Discrete Applied Mathematics 156(12), 2270–2278 (2008)
5. Cerioli, M.R., Faria, L., Ferreira, T.O., Protti, F.: On minimum clique partition and maximum independent set on unit disk graphs and penny graphs: complexity and approximation. Electronic Notes in Discrete Mathematics 18, 73–79 (2004)
6. Dumitrescu, A., Pach, J.: Minimum clique partition in unit disk graphs. CoRR, abs/0909.1552 (2009)
7. Finke, G., Jost, V., Queyranne, M., Sebö, A.: Batch processing with interval graph compatibilities between tasks. Discrete Applied Mathematics 156(5), 556–568 (2008)
8. Nieberg, T., Hurink, J., Kern, W.: Approximation schemes for wireless networks. ACM Transactions on Algorithms 4(4), 1–17 (2008)
9. Pandit, S., Pemmaraju, S., Varadarajan, K.: Approximation algorithms for domatic partitions of unit disk graphs. In: APPROX-RANDOM, pp. 312–325 (2009)
10. Pemmaraju, S.V., Pirwani, I.A.: Good quality virtual realization of unit ball graphs. In: Arge, L., Hoffmann, M., Welzl, E. (eds.) ESA 2007. LNCS, vol. 4698, pp. 311–322. Springer, Heidelberg (2007)
11. Pirwani, I.A., Salavatipour, M.R.: A weakly robust PTAS for minimum clique partition in unit disk graphs. CoRR, abs/0904.2203 (2009)
12. Zuckerman, D.: Linear degree extractors and the inapproximability of max clique and chromatic number. Theory of Computing 3(1), 103–128 (2007)

Representing a Functional Curve by Curves with Fewer Peaks*

Danny Z. Chen, Chao Wang, and Haitao Wang**

Department of Computer Science and Engineering
University of Notre Dame, Notre Dame, IN 46556, USA
{dchen,cwang1,hwang6}@nd.edu

Abstract. We study the problems of (approximately) representing a functional curve in 2-D by a set of curves with fewer peaks. Let $\mathbf{f}$ be an input nonnegative piecewise linear functional curve of size n. We consider the following problems. (1) Uphill-downhill pair representation (UDPR): Find two nonnegative piecewise linear curves, one nondecreasing and one nonincreasing, such that their sum approximately represents $\mathbf{f}$. (2) Unimodal representation (UR): Find a set of k nonnegative unimodal (single-peak) curves such that their sum approximately represents $\mathbf{f}$. (3) Fewer-peak representation (FPR): Find a nonnegative piecewise linear curve with at most k peaks that approximately represents $\mathbf{f}$. For each problem, we consider two versions. For UDPR, we study the *feasibility* version and the *min-ϵ* version. For each of the UR and FPR problems, we study the *min-k* version and the *min-ϵ* version. Little work has been done previously on these problems. We solve all problems (except the UR *min-ϵ*) in optimal $O(n)$ time, and the UR *min-ϵ* version in $O(n+m \log m)$ time, where $m < n$ is the number of peaks of $\mathbf{f}$. Our algorithms are based on new geometric observations and interesting techniques.

1 Introduction

In this paper, we study the problems of approximately representing a 2-D functional curve by a set of curves with fewer peaks. Let $\mathbf{f}$ be an arbitrary input piecewise linear functional curve of size n. In general, when representing $\mathbf{f}$ by one or more structurally simpler curves, $\mathbf{g}^{(1)}, \mathbf{g}^{(2)}, \ldots, \mathbf{g}^{(k)}$ ($k \geq 1$), we are interested in the following aspects of the representation: (1) the *representation mode*, which defines the types of and constraints on the simpler curves used, (2) the *representation complexity*, which is the number of simpler curves involved in the representation, and (3) the *representation error*, which is the vertical distance between $\mathbf{f}$ and the sum of the simpler curves in the representation, i.e., $\sum_{i=1}^{k} \mathbf{g}^{(i)}$.

---

* This research was supported in part by NSF under Grants CCF-0515203 and CCF-0916606.
** Corresponding author. The work of this author was also supported in part by a graduate fellowship from the Center for Applied Mathematics, University of Notre Dame.

H. Kaplan (Ed.): SWAT 2010, LNCS 6139, pp. 200–211, 2010.

For simplicity, we describe the input piecewise linear curve $\mathbf{f}$ by $(f_1, f_2, \ldots, f_n)$, where $f_i = \mathbf{f}(x_i)$ is the value of $\mathbf{f}$ at the i-th x-coordinate x_i ($x_i < x_{i+1}$ for each i). Without loss of generality (WLOG), the x_i's are all omitted in our discussion. For the consistency of our algorithmic manipulation and analysis, we need to define carefully the *peaks* of $\mathbf{f} = (f_1, f_2, \ldots, f_n)$, with a little subtlety. Clearly, a peak is at a local maximal height. If multiple consecutive vertices of $\mathbf{f}$ all have the same local maximal height and if this group of vertices does not include the last vertex of $\mathbf{f}$, then we define the *peak* for this group of vertices as only the *first* vertex of the group. However, if the group includes the last vertex of $\mathbf{f}$, then we define the *peak* as the *last* vertex of the group (and of $\mathbf{f}$). Fig. 1 (a) shows an example. The precise definition of peaks is as follows: we call f_i a *peak* of $\mathbf{f}$ if (1) $i = 1$ and there is a j with $1 < j \leq n$ such that $f_1 = \cdots = f_{j-1} > f_j$, or (2) $1 < i < n$, $f_{i-1} < f_i$ and there is a j with $i < j \leq n$ such that $f_i = \cdots = f_{j-1} > f_j$, or (3) $i = n$ and there is a j with $1 \leq j \leq n - 1$ such that $f_j < f_{j+1} = \cdots = f_n$.

Specifically, we consider three modes of representation in this paper. (1) **Uphill-downhill pair representation (UDPR)**: Represent $\mathbf{f}$ by two curves, one nondecreasing (uphill) and one nonincreasing (downhill). (2) **Unimodal representation (UR)**: Represent $\mathbf{f}$ by a set of unimodal curves. A functional curve $\mathbf{g}$ is *unimodal* (or *single-peak*) if there is only one peak on $\mathbf{g}$. (3) **Fewer-peak representation (FPR)**: Represent $\mathbf{f}$ by a functional curve with at most a given number k of peaks. It is interesting to note that a nondecreasing curve and a nonincreasing curve of size n each can sum up to form a functional curve $\mathbf{f}$ with $O(n)$ peaks (e.g., see Fig. 1 (b)). The error measure we use is the *uniform error metric*, also known as the L_∞ metric.

We consider several problem versions. For UDPR, its representation complexity (i.e., the number of curves in the representation) is always 2. We consider: (1) the *feasibility* version, which seeks to decide whether an uphill-downhill pair representation is feasible subject to a given bound ϵ on the representation error, and (2) the *min-ϵ* version, which aims to minimize the representation error ϵ^* among all feasible uphill-downhill pair representations. For UR, the representation complexity is the number of unimodal curves in the representation. For FPR, the representation complexity is the number of peaks on the sought curve. For each of the UR and FPR problems, we consider: (1) the *min-k* version, which minimizes the representation complexity k^* subject to a given error bound ϵ, and (2) the *min-ϵ* version, which minimizes the representation error ϵ^* subject to a given bound k on the representation complexity. For these problems, we require that $\mathbf{f}$ and all the simpler functional curves involved be *nonnegative* (i.e., on or above the x-axis), which is justified by real applications discussed later. Note that this actually makes the problems more theoretically interesting. Without the nonnegativeness constraint, some problems become much easier to solve.

Based on nontrivial and interesting geometric observations, we develop efficient algorithms for these problems. For the UDPR problem, we give $O(n)$ time algorithms for both its feasibility version and *min-ϵ* version. For the UR problem, we present an $O(n)$ time algorithm for its *min-k* version, and an $O(n + m \log m)$

time algorithm for its *min-ε* version, where $m < n$ is the number of peaks on **f**. For the FPR problem, our *min-k* and *min-ε* algorithms both take $O(n)$ time.

1.1   Motivations and Related Work

Motivated by applications in data mining [5,6], Chun *et al.* gave a linear time algorithm [2] for approximating a piecewise linear curve by a single unimodal curve, under the L_2 error measure. In [3], Chun *et al.* studied an extended case in which the approximating function has k peaks, for a given number k, under the L_p error measure. This problem is similar to our FPR *min-ε* problem except that our error measure is different. The algorithm in [3] computes an optimal solution in $O(km^2 + nm \log n)$ time, where m is the number of peaks on the input curve. In addition, an $O(n \log n)$ time algorithm for computing an optimal unimodal function to approximate a piecewise linear curve under the L_p error measure is also given [3]. As shown in [2,3], the algorithms above are applicable to certain data mining problems. Motivated by applications in statistics, Stout [10] considered the unimodal regression problem, aiming to approximate a set of n points by a unimodal step function. He gave three algorithms with time bounds $O(n \log n)$, $O(n)$, and $O(n)$ for the problem under the L_1, L_2, and L_∞ error measures, respectively.

Our studies are also motivated by a dose decomposition problem in intensity-modulated radiation therapy (IMRT). IMRT is a modern cancer treatment technique aiming to deliver a prescribed conformal radiation dose to a target tumor while sparing the surrounding normal tissue and critical structures [13]. A prescribed dose function **f** is always nonnegative and normally takes a piecewise linear form (defined on a grid domain). In the rotational delivery approach [12] (also called dynamic IMRT), a prescribed dose function **f** is delivered by repeatedly rotating the radiation source around the patient. In each rotation (called a *path*), a portion of the prescribed dose **f** is delivered in a continuous manner. It has been observed that a unimodal dose function can be delivered by a path more smoothly and accurately than an arbitrary dose function. Thus, for a fast and accurate delivery of the prescribed dose, it is desirable to represent exactly or approximately a nonnegative (arbitrary) dose curve **f** by the sum of a minimum set of unimodal curves.

Additionally, various curve approximation problems have been studied extensively. Most of them seek to simplify a given curve by other "simpler" curves (e.g., with fewer line segments) under certain error criteria (e.g., see [1,8,9,11] and their references). For the general curve representation problems studied in this paper, we are not aware of any previous efficient algorithms.

Due to the space limit, in the following paper the proofs of some lemmas and the algorithms for FPR problem are omitted but can be found in our full paper.

2   The Uphill-Downhill Pair Representation (UDPR)

The UDPR problem is defined as follows. Given $\mathbf{f} = (f_1, f_2, \ldots, f_n)$ $(n \geq 2)$ and $\epsilon \geq 0$, find a pair of piecewise linear curves $\mathbf{y} = (y_1, y_2, \ldots, y_n)$ and $\mathbf{z} =$

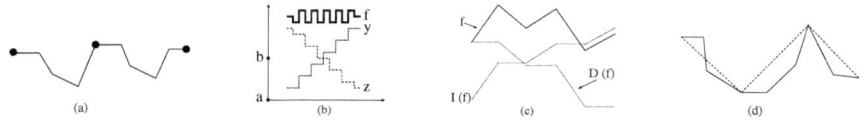

Fig. 1. (a) The peaks (the black points) on a piecewise linear curve; (b) an uphill curve **y** and a downhill curve **z** of size n each can sum up to form a functional curve **f** with $O(n)$ peaks; (c) the profile curves $I(\mathbf{f})$ and $D(\mathbf{f})$ of the curve **f**; (d) the skeleton curve $SK(\mathbf{f})$ (the dashed curve) of **f** (the solid curve)

$(z_1, z_2, \ldots, z_n)$, such that (1) $|y_i + z_i - f_i| \leq \epsilon$ for every $1 \leq i \leq n$, (2) $y_1 \leq y_2 \leq \cdots \leq y_n$, (3) $z_1 \geq z_2 \geq \cdots \geq z_n$, and (4) **y** and **z** are both nonnegative. If constraint (4) is removed, i.e., the sought curves need not be nonnegative, then we call it the *relaxed UDPR problem*. Interestingly, our solutions for the UDPR problems are used as a subroutine for solving the UR problems in Section 3.

We first define some notations. Given $\mathbf{f} = (f_1, f_2, \ldots, f_n)$, we define a nondecreasing (uphill) piecewise linear functional curve $I(\mathbf{f}) = (I(f_1), I(f_2), \ldots, I(f_n))$ and a nonincreasing (downhill) curve $D(\mathbf{f}) = (D(f_1), D(f_2), \ldots, D(f_n))$ as follows: $I(f_1) = 0$, $I(f_i) = I(f_{i-1}) + \max\{f_i - f_{i-1}, 0\}$ for $2 \leq i \leq n$; $D(f_1) = f_1$, $D(f_i) = D(f_{i-1}) - \max\{f_{i-1} - f_i, 0\}$ for $2 \leq i \leq n$. Fig. 1 (c) shows an example. We call these two curves $I(\mathbf{f})$ and $D(\mathbf{f})$ the *profile curves* of **f**. Observe that since $f_i = I(f_i) + D(f_i)$ for each i, the profile curves of **f** form a solution for the relaxed UDPR problem of **f** with any error $\epsilon \geq 0$. In Fig. 1 (b), the curves **y** and **z** form a feasible solution for the *relaxed* UDPR problem on **f** and $\epsilon = 0$, but not for the UDPR problem if the x-axis passes the point b instead of a.

For $\mathbf{f} = (f_1, f_2, \ldots, f_n)$, similar to the peak definition, we call f_i a *valley* if (1) $i = 1$ and there is a j with $1 < j \leq n$ such that $f_1 = \cdots = f_{j-1} < f_j$, or (2) $1 < i < n$, $f_{i-1} > f_i$ and there is a j with $i < j \leq n$ such that $f_i = \cdots = f_{j-1} < f_j$, or (3) $i = n$ and there is a j with $1 \leq j \leq n - 1$ such that $f_j > f_{j+1} = \cdots = f_n$. Clearly, there is exactly one valley (resp., peak) between any two consecutive peaks (resp., valleys) on a curve. Given **f**, define its *skeleton* $SK(\mathbf{f})$ by connecting each peak (resp., valley) to its right side consecutive valley (resp., peak) with a line segment (see Fig. 1 (d)). A curve $\mathbf{f}'$ is called a *skeleton curve* if each f_i' is either a peak or a valley and a *general curve* otherwise.

Given a skeleton curve $\mathbf{f} = (f_1, f_2, \ldots, f_n)$ and $\epsilon \geq 0$, the *characteristic curve* of **f** and ϵ, denoted by $R(\mathbf{f}, \epsilon)$, is defined as $R(\mathbf{f}, \epsilon) = (R_1, R_2, \ldots, R_n)$, where $R_1 = f_1 + \epsilon$, R_i is equal to $f_i - \epsilon$ if $R_{i-1} < f_i - \epsilon$, R_i is $f_i + \epsilon$ if $R_{i-1} > f_i + \epsilon$, and $R_i = R_{i-1}$ otherwise (see Fig. 2 (a)).

2.1   The Feasibility of the UDPR Problem

For $\epsilon \geq 0$, we say a curve $\mathbf{f} = (f_1, f_2, \ldots, f_n)$ is ϵ-*UDP-representable* if the UDPR problem on **f** and ϵ is feasible. We first discuss the UDPR feasibility algorithm for a skeleton curve and then deal with the general curve.

Fig. 2. (a) The characteristic curve $R(\mathbf{f}, \epsilon)$ (the dashed one); (b) an allied pair (R_i, R_j) (black points) on $R(\mathbf{f}, \epsilon)$ (the dashed curve); (c) $\mathcal{M}(\epsilon)$ is a step function; (d) $\epsilon'_i = (f_i - f_{k'})/2$ for the peak f_i

Lemma 1. *Given a skeleton curve $\mathbf{f} = (f_1, f_2, \ldots, f_n)$ and $\epsilon > 0$, suppose $R(\mathbf{f}, \epsilon)$ is its characteristic curve. Then $\mathbf{f}$ is ϵ-UDP-representable if and only if $D(R_n) \geq 0$. Moreover, if $\mathbf{f}$ is ϵ-UDP-representable, then the profile curves of $R(\mathbf{f}, \epsilon)$ form a UDPR solution.*

Given a skeleton curve $\mathbf{f}$ and $\epsilon \geq 0$, since $R(\mathbf{f}, \epsilon)$ and its profile curves can all be computed in linear time, by Lemma 1, we have the following result.

Lemma 2. *The UDPR feasibility problem on a skeleton curve $\mathbf{f}$ and $\epsilon \geq 0$ is solvable in $O(n)$ time.*

The problem on a general curve can be handled by the next lemma.

Lemma 3. *Given $\epsilon \geq 0$, $\mathbf{f}$ is ϵ-UDP-representable if and only if $SK(\mathbf{f})$ is ϵ-UDP-representable. Furthermore, given a feasible solution for $SK(\mathbf{f})$ (resp., $\mathbf{f}$), a feasible solution for $\mathbf{f}$ (resp., $SK(\mathbf{f})$) can be obtained in $O(n)$ time.*

In light of Lemmas 2 and 3, we have the following theorem.

Theorem 1. *The UDPR feasibility problem on a general piecewise linear functional curve $\mathbf{f}$ of size n and $\epsilon \geq 0$ is solvable in $O(n)$ time.*

2.2 The *min-ε* Version of the UDPR Problem

In this section, we consider the *min-ε* version, seeking the minimum possible error ϵ^* for $\mathbf{f}$ to be ϵ^*-UDP-representable. By Lemma 3, we only need to develop an algorithm for the skeleton of $\mathbf{f}$.

Before giving the algorithm, we first discuss some geometric observations. Given a skeleton curve $\mathbf{f} = (f_1, f_2, \ldots, f_n)$ and $\epsilon \geq 0$, by Lemma 1, $\mathbf{f}$ is ϵ-UDP-representable if and only if $D(R_n) \geq 0$. By the definition of the profile curves, we have $D(R_n) = R_1 - \sum_{i=2}^{n} \max\{0, R_{i-1} - R_i\}$ and $R_1 = f_1 + \epsilon$. For a general functional curve $\mathbf{h} = (h_1, h_2, \ldots, h_n)$, we define $\mathcal{H}(\mathbf{h})$ to be $\sum_{i=2}^{n} \max\{0, h_{i-1} - h_i\}$. Geometrically, the value of $\mathcal{H}(\mathbf{h})$ is the sum of the "height drops" of all the "downhill" portions of the curve $\mathbf{h}$. Then we have $D(R_n) = f_1 + \epsilon - \mathcal{H}(R(\mathbf{f}, \epsilon))$.

On $R(\mathbf{f}, \epsilon) = (R_1, R_2, \ldots, R_n)$, we call R_i an *R-peak* if R_i is a peak on $R(\mathbf{f}, \epsilon)$ with $1 < i < n$. Thus R_1 and R_n cannot be R-peaks. For each R-peak R_i, we define its *allied R-valley* to be R_j, where R_j is the first valley on $R(\mathbf{f}, \epsilon)$ to the right of R_i, i.e., $j = \min\{t \mid t > i$ and R_t is a valley$\}$. An R-peak R_i and its allied R-valley R_j form an *allied pair* (R_i, R_j) (see Fig. 2 (b)).

Observation 1. *For any $\epsilon \geq 0$, if R_i is an R-peak on $R(\mathbf{f}, \epsilon)$, then $R_i = f_i - \epsilon$ and f_i is a peak on $\mathbf{f}$; if R_j is the allied R-valley of the R-peak R_i on $R(\mathbf{f}, \epsilon)$, then $R_j = f_j + \epsilon$ and f_j is a valley on $\mathbf{f}$.*

We name the sequence of the allied pairs of $R(\mathbf{f}, \epsilon)$ (from left to right) the *topology* of $R(\mathbf{f}, \epsilon)$.

Lemma 4. *Given a skeleton curve $\mathbf{f}$ and an error $\epsilon \geq 0$, if both $R(\mathbf{f}, \epsilon)$ and $R(\mathbf{f}, \epsilon + \Delta\epsilon)$ has the same topology for a value $\Delta\epsilon$, then $\mathcal{H}(R(\mathbf{f}, \epsilon + \Delta\epsilon)) = \mathcal{H}(R(\mathbf{f}, \epsilon)) - 2\Delta\epsilon \cdot \alpha$, where α is the number of allied pairs on $R(\mathbf{f}, \epsilon)$.*

The above lemma implies that if the topology of $R(\mathbf{f}, \epsilon)$ does not change for $\epsilon \in [\epsilon_1, \epsilon_2]$, then $\mathcal{H}(R(\mathbf{f}, \epsilon))$ is a continuous decreasing linear function in that interval. Denote by $\mathcal{M}(\epsilon)$ the number of allied pairs on $R(\mathbf{f}, \epsilon)$. Thus $\mathcal{M}(0)$ is the number of allied pairs on $\mathbf{f}$ (when $\epsilon = 0$, $R(\mathbf{f}, \epsilon) = \mathbf{f}$). Note that as ϵ increases from 0 to ∞, at some values of ϵ, the topology of $R(\mathbf{f}, \epsilon)$ will change and the value of $\mathcal{M}(\epsilon)$ will decrease by some integer $t \geq 1$. When ϵ is large enough, $\mathcal{M}(\epsilon)$ becomes zero and never decreases any more. Thus, $\mathcal{M}(\epsilon)$ is a nonincreasing step function (see Fig. 2 (c)), and the number of steps is at most $\mathcal{M}(0)$. Suppose the i-th "step" of $\mathcal{M}(\epsilon)$ is defined on the interval $[\epsilon_i, \epsilon_{i+1})$; then we call ϵ_i a *critical error* if $i \geq 1$ ($\epsilon_1 = 0$ is not considered to be a critical error). Formally, ϵ' is a critical error if and only if $\mathcal{M}(L(\epsilon')) - \mathcal{M}(\epsilon') > 0$, where $L(\epsilon')$ is a value less than ϵ' but infinitely close to it. We use a multi-set E to denote the set of all critical errors: For each critical error ϵ', if $\mathcal{M}(L(\epsilon')) - \mathcal{M}(\epsilon') = t \geq 1$, then E contains t copies of ϵ'. Thus $|E|$ is exactly equal to $\mathcal{M}(0)$.

From a geometric point of view, $R(\mathbf{f}, \epsilon)$ changes its topology only when a peak of the curve $\mathbf{f} - \epsilon$ "touches" some point of a horizontal segment of $R(\mathbf{f}, \epsilon)$ starting at a valley of $\mathbf{f} + \epsilon$. When a peak $f_i - \epsilon$ of $\mathbf{f} - \epsilon$ touches a horizontal segment starting at a valley $f_j + \epsilon$ of $\mathbf{f} + \epsilon$, we have $f_i - \epsilon = f_j + \epsilon$, implying $\epsilon = \frac{|f_i - f_j|}{2}$. Let $E' = \{|f_i - f_j|/2 \mid$ for any peak f_i and valley f_j on $\mathbf{f}\}$. Clearly, the critical error set E is a subset of E'. Thus we have the following lemma.

Lemma 5. *Given a skeleton curve $\mathbf{f}$, the function $\mathcal{G}(\epsilon) = f_1 + \epsilon - \mathcal{H}(R(\mathbf{f}, \epsilon))$ (i.e., $\mathcal{G}(\epsilon) = D(R_n)$) is a continuous increasing piecewise linear function for $\epsilon \geq 0$. More specifically, the interval $[0, +\infty)$ for ϵ can be partitioned into $|E'| + 1$ sub-intervals by the elements in E', such that in each such sub-interval, $\mathcal{G}(\epsilon)$ is an increasing linear function of ϵ.*

The Algorithm. The idea of our linear-time algorithm is to first determine the multi-set E explicitly and then compute ϵ^*. Suppose we already have the set E explicitly; then ϵ^* can be computed by the following lemma.

Lemma 6. *After E is obtained, ϵ^* can be computed in $O(|E|)$ time.*

Proof. Assume that the elements in E are $\epsilon_1 \leq \epsilon_2 \leq \cdots \leq \epsilon_M$, where $M = |E| = \mathcal{M}(0)$ (this assumption is only for analysis since we do not sort them in the algorithm). By Lemma 5, the function $\mathcal{G}(\epsilon) = f_1 + \epsilon - \mathcal{H}(R(\mathbf{f}, \epsilon))$ is increasing, and thus ϵ^* is the unique value with $\mathcal{G}(\epsilon^*) = 0$. By Lemma 4, $\mathcal{G}(0) = f_1 - \mathcal{H}(R(\mathbf{f}, 0))$, $\mathcal{G}(\epsilon_1) = \mathcal{G}(0) + \epsilon_1 + 2M \cdot \epsilon_1$, and $\mathcal{G}(\epsilon_2) = \mathcal{G}(0) + \epsilon_2 + 2M \cdot \epsilon_1 + 2(M-1) \cdot (\epsilon_2 - \epsilon_1)$. Generally, if we let $\epsilon_0 = 0$, then for $1 \leq i \leq M$, $\mathcal{G}(\epsilon_i) = \mathcal{G}(0) + \epsilon_i + 2\sum_{t=0}^{i-1}(M - t)(\epsilon_{t+1} - \epsilon_t)$. Thus, geometrically, $\mathcal{G}(\epsilon)$ is a piecewise linear concave increasing function whose slope, when $\epsilon \in [\epsilon_i, \epsilon_{i+1})$, is $1 + 2(M - i)$ for any $0 \leq i \leq M$ (let ϵ_{M+1} be ∞). Note that if the elements in E are already sorted, then it is easy to compute ϵ^* in linear time since each $\mathcal{G}(\epsilon_i)$ can be obtained from $\mathcal{G}(\epsilon_{i-1})$ in $O(1)$ time and $\mathcal{G}(\epsilon)$ is an increasing function. However, as we show below, we can still compute ϵ^* in linear time without sorting the elements in E. Define $h(i, j) = \sum_{t=i}^{j-1}(M - t)(\epsilon_{t+1} - \epsilon_t)$. Then $\mathcal{G}(\epsilon_i) = \mathcal{G}(0) + \epsilon_i + 2h(0, i)$. By a simple deduction, we can get $h(i, j) = \sum_{t=i+1}^{j-1} \epsilon_t + (M - j + 1)\epsilon_j - (M - i)\epsilon_i$. Thus, we can compute the value of $h(i, j)$ in $O(j - i)$ time if we know all the values $\epsilon_i, \epsilon_{i+1}, \ldots, \epsilon_j$. Further, $\mathcal{G}(0)$ can be easily computed in linear time.

To obtain ϵ^*, we do the following: (1) Search in E for the two elements ϵ' and ϵ'' such that ϵ' is the largest element in E with $\mathcal{G}(\epsilon') \leq 0$ and ϵ'' is the smallest one with $\mathcal{G}(\epsilon'') > 0$; (2) compute the smallest value $\epsilon^* \in [\epsilon', \epsilon'']$ such that $\mathcal{G}(\epsilon^*) = 0$. In step (1), to find ϵ', a straightforward way is to first sort all elements in E, and then from the smallest element to the largest one, check the value of $\mathcal{G}(\epsilon_i)$ for each ϵ_i. But that takes $O(M \log M)$ time. An $O(M)$ time algorithm, based on prune and search, works as follows. We first use the selection algorithm [4] to find the median $\epsilon_{M/2}$ in E and compute $\mathcal{G}(\epsilon_{M/2})$, for which we need to spend $O(\frac{M}{2})$ time to compute $h(0, \frac{M}{2})$. If $\mathcal{G}(\epsilon_{M/2}) = 0$, then the algorithm stops with $\epsilon^* = \epsilon_{M/2}$. Otherwise, let $E_1 = \{\epsilon_i \mid i < \frac{M}{2}\}$ and $E_2 = \{\epsilon_i \mid i > \frac{M}{2}\}$. If $\mathcal{G}(\epsilon_{M/2}) < 0$, then we continue the same procedure on E_2. Since we already have the value of $h(0, \frac{M}{2})$, when computing $h(0, j)$ for $j > \frac{M}{2}$, we only need to compute $h(\frac{M}{2} + 1, j)$ because $h(0, j) = h(0, \frac{M}{2}) + h(\frac{M}{2} + 1, j)$, which takes $O(j - \frac{M}{2})$ time. If $\mathcal{G}(\epsilon_{M/2}) > 0$, then we continue the same procedure on E_1. Thus the total time for computing ϵ' is $O(M)$. To obtain ϵ'', note that ϵ'' is the smallest element in E that is larger than ϵ', and thus ϵ'' can be found in linear time. Step (2) takes $O(1)$ time since when $\epsilon \in [\epsilon', \epsilon'']$, $\mathcal{G}(\epsilon)$ is a linear function.

It remains to compute the multi-set E. Let $P(\mathbf{f})$ denote the set of indices of all peaks on $\mathbf{f}$ except f_1 and f_n. When $\epsilon = 0$, since $R(\mathbf{f}, \epsilon)$ is the same as $\mathbf{f}$, R_i is an R-peak on $R(\mathbf{f}, \epsilon)$ if and only if $i \in P(\mathbf{f})$. Thus $|P(\mathbf{f})| = \mathcal{M}(0)$. For each $i \in P(\mathbf{f})$, let $i' = \min\{t \mid i < t \leq n + 1, f_t > f_i\}$ (with $f_{n+1} = +\infty$); in other words, $f_{i'}$ is the leftmost peak to the right of f_i that is *larger* than f_i, or $i' = n + 1$ if there is no such peak on $\mathbf{f}$. Let $i'' = \max\{t \mid 0 \leq t < i, f_t \geq f_i\}$ (with $f_0 = +\infty$), i.e., $f_{i''}$ is the rightmost peak to the left of f_i that is *larger than or equal to* f_i, or $i'' = 0$ if there is no such peak (see Fig. 2 (d)). For each $i \in P(\mathbf{f})$, let $f_{k'} = \min\{f_t \mid i < t < i'\}$, $f_{k''} = \min\{f_t \mid i'' < t < i\}$, and $\epsilon'_i = (f_i - \max\{f_{k'}, f_{k''}\})/2$. Fig. 2 (d) shows an example. The next lemma is crucial for computing E.

Lemma 7. *For any $\epsilon \geq 0$, $1 < i < n$, R_i is an R-peak on $R(\mathbf{f}, \epsilon)$ if and only if $i \in P(\mathbf{f})$ and $\epsilon < \epsilon'_i$.*

Consequently, the multi-set E can be obtained by the following lemma.

Lemma 8. $E = \{\epsilon'_i \mid$ *for each* $i \in P(\mathbf{f})\}$ *and* E *can be computed in* $O(n)$ *time.*

Theorem 2. *The UDPR min-ϵ problem on* $\mathbf{f}$ *can be solved in* $O(n)$ *time.*

3   The Unimodal Representation Problem (UR)

In this section, we study the UR problem. Since we require all unimodal curves be nonnegative, in the following, a unimodal curve means a nonnegative one.

For two integers $i' < i''$, denote by $[i' \ldots i'']$ the sequence of integers between i' and i'', i.e., $[i' \ldots i''] = \{i', i'+1, \ldots, i''\}$. For a curve $\mathbf{f} = (f_1, f_2, \ldots, f_n)$, denote by $\mathbf{f}[i' \ldots i'']$ the portion of $\mathbf{f}$ restricted to the indices in $\{i', i'+1, \ldots, i''\}$. The two lemmas below outlines the underlying geometric structures.

Lemma 9. *Let* $\mathbf{h}^{(1)}, \mathbf{h}^{(2)}, \ldots, \mathbf{h}^{(k)}$ *be* $k \geq 1$ *unimodal functional curves defined on* $[1 \ldots n]$*. Assume that for each* j*,* $\mathbf{h}^{(j)}$ *peaks at* i^*_j*, with* $1 \leq i^*_1 \leq i^*_2 \leq \cdots \leq i^*_k \leq n$*. Then the curve* $\mathbf{h} = \sum_{j=1}^{k} \mathbf{h}^{(j)}$ *satisfies: (1)* $\mathbf{h}$ *is nonnegative and nondecreasing on* $[1 \ldots i^*_1]$*, (2)* $\mathbf{h}$ *is 0-UDP-representable on* $[i^*_j \ldots i^*_{j+1}]$ *for each* $j = 1, 2, \ldots, k-1$*, and (3)* $\mathbf{h}$ *is nonnegative and nonincreasing on* $[i^*_k \ldots n]$*.*

Proof. Since every $\mathbf{h}^{(j)}$ is nondecreasing on $[1 \ldots i^*_1]$ and nonincreasing on $[i^*_1 \ldots n]$, (1) and (3) of the lemma follow. (2) of the lemma holds due to the fact that on $[i^*_j \ldots i^*_{j+1}]$, $\mathbf{h}^{(1)}, \mathbf{h}^{(2)}, \ldots, \mathbf{h}^{(j)}$ are all nonincreasing, and $\mathbf{h}^{(j+1)}, \mathbf{h}^{(j+2)}, \ldots, \mathbf{h}^{(k)}$ are all nondecreasing. Thus for each j, the portion of the curve $\mathbf{h}$ on $[i^*_j \ldots i^*_{j+1}]$ is equal to the sum of a nondecreasing curve $\mathbf{y}^{(j)} = \sum_{t=j+1}^{k} \mathbf{h}^{(t)}$ and a nonincreasing curve $\mathbf{z}^{(j)} = \sum_{t=1}^{j} \mathbf{h}^{(t)}$.

Lemma 10. *Given a curve* $\mathbf{h}$ *defined on* $[1 \ldots n]$*, if there exist* $k \geq 1$ *integers* $i^*_1 \leq i^*_2 \leq \cdots \leq i^*_k$ *in* $[1 \ldots n]$ *such that (1)* $\mathbf{h}$ *is nonnegative and nondecreasing on* $[1 \ldots i^*_1]$*, (2)* $\mathbf{h}$ *is 0-UDP-representable on* $[i^*_j \ldots i^*_{j+1}]$ *for each* $j = 1, 2, \ldots, k-1$*, and (3)* $\mathbf{h}$ *is nonnegative and nonincreasing on* $[i^*_k \ldots n]$*, then there exist* k *unimodal curves* $\mathbf{h}^{(1)}, \mathbf{h}^{(2)}, \ldots, \mathbf{h}^{(k)}$ *defined on* $[1 \ldots n]$ *such that* $\mathbf{h} = \sum_{j=1}^{k} \mathbf{h}^{(j)}$*.*

3.1   The *min-k* Version of the Unimodal Representation Problem

Lemmas 9 and 10 imply that the *min-k* version of the UR problem on $\mathbf{f}$ and ϵ is equivalent to finding the minimum number of intermediate points $i^*_1 \leq i^*_2 \leq \cdots \leq i^*_k$ in $[1 \ldots n]$, such that (1) $\mathbf{f}[1 \ldots i^*_1]$ (resp., $\mathbf{f}[i^*_k \ldots n]$) can be represented by a nonnegative nondecreasing (resp., nonincreasing) curve with an error no more than ϵ, (2) for each j with $1 \leq j \leq k-1$, $\mathbf{f}[i^*_j \ldots i^*_{j+1}]$ is ϵ-UDP-representable.

The following lemma follows a similar spirit as the UDPR feasibility problem.

Lemma 11. *Given* $\mathbf{f} = (f_1, f_2, \ldots, f_n)$ *and* $\epsilon \geq 0$*,* $\mathbf{f}$ *can be represented by a nonnegative nondecreasing (resp., nonincreasing) curve with an error no bigger than* ϵ *if and only if* $f_j - \epsilon \leq f_i + \epsilon$ *(resp.,* $f_j + \epsilon \geq f_i - \epsilon$*) holds for all* $1 \leq j < i \leq n$*. Moreover, if the problem is feasible, then it always has a solution* $\mathbf{y}$ *defined by* $y_i = \max\{0, \max_{j=1}^{i}\{f_j - \epsilon\}\}$ *(resp.,* $y_i = \min_{j=1}^{i}\{f_j + \epsilon\}$*), which can be computed in* $O(n)$ *time.*

Given $\mathbf{f} = (f_1, f_2, \ldots, f_n)$ and $\epsilon \geq 0$, our UR *min-k* algorithm works in a greedy fashion: (1) Find the largest index i_1^* such that $\mathbf{f}[1 \ldots i_1^*]$ can be represented by a nonnegative nondecreasing curve with an error $\leq \epsilon$; (2) find the smallest index c, such that $\mathbf{f}$ can be represented by a nonnegative nonincreasing curve on $[c \ldots n]$ with an error $\leq \epsilon$; (3) if $i_1^* \geq c$, then we are done; otherwise, by a linear scan from i_1^*, find the largest index i_2^* such that $\mathbf{f}[i_1^* \ldots i_2^*]$ is ϵ-UDP-representable; the same procedure continues until $i_k^* \geq c$. When the algorithm stops, k is the minimum number of unimodal curves needed to represent $\mathbf{f}$. By Theorem 1 and Lemma 11, the above *min-k* algorithm takes $O(n)$ time.

Theorem 3. *The UR min-k problem on $\mathbf{f}$ and $\epsilon \geq 0$ is solvable in $O(n)$ time.*

Additionally, we have the following result which will be useful for our UR *min-ϵ* algorithm given in the next section.

Lemma 12. *Given a curve $\mathbf{f} = (f_1, f_2, \ldots, f_n)$ and $\epsilon \geq 0$, $\mathbf{f}$ can be represented by k unimodal curves if and only if $SK(\mathbf{f})$ can be represented by k unimodal curves. Furthermore, given a feasible solution for $SK(\mathbf{f})$ (resp., $\mathbf{f}$), a feasible solution for $\mathbf{f}$ (resp., $SK(\mathbf{f})$) can be obtained in $O(n)$ time.*

3.2   The *min-ϵ* Version of the Unimodal Representation Problem

The UR *min-ϵ* problem is: Given $\mathbf{f} = (f_1, f_2, \ldots, f_n)$ and $k > 0$, find the smallest error ϵ^* such that $\mathbf{f}$ can be represented by at most k unimodal curves.

Given a curve $\mathbf{f}$, denote by $\mathcal{K}(\epsilon)$ the minimum number of unimodal curves for representing $\mathbf{f}$ with an error $\leq \epsilon$. Clearly, $\mathcal{K}(\epsilon)$ changes in a monotone fashion with respect to ϵ. To solve the *min-ϵ* we use our *min-k* algorithm as a procedure, and perform a search for the optimal error ϵ^*. The structures of the unimodal representations specified in Lemmas 9 and 10 imply that we only need to consider those ϵ values that may cause a feasibility change to one of the following representations: (1) representing $\mathbf{f}[i' \ldots i'']$ ($1 \leq i' < i'' \leq n$) by a pair of nondecreasing and nonincreasing curves with an error $\leq \epsilon$, (2) representing $\mathbf{f}[1 \ldots i]$ ($1 \leq i \leq n$) by a nondecreasing curve with an error $\leq \epsilon$, or (3) representing $\mathbf{f}[j \ldots n]$ ($1 \leq j \leq n$) by a nonincreasing curve with an error $\leq \epsilon$.

Given a curve $\mathbf{f} = (f_1, f_2, \ldots, f_n)$, by Lemma 12, it suffices to consider the UR *min-ϵ* algorithm for its $SK(\mathbf{f})$ curve. After obtaining the minimum error ϵ^* for $SK(\mathbf{f})$, we need $O(n)$ additional time to produce the solution curves for $\mathbf{f}$. The next algorithm focuses on $SK(\mathbf{f})$ although it works for any general curve. In the following, we assume $SK(\mathbf{f}) = \mathbf{g} = (g_1, g_2, \ldots, g_m)$ (i.e., $|SK(\mathbf{f})| = m$).

Given $k > 0$, our UR *min-ϵ* algorithm has two steps. (1) Search in $S = \{0\} \cup \{|g_i - g_j|/2 \mid 1 \leq i, j \leq m\}$ for $\epsilon', \epsilon'' \in S$, such that ϵ' is the largest element in S with $\mathcal{K}(\epsilon') > k$ and ϵ'' is the smallest element in S with $\mathcal{K}(\epsilon'') \leq k$. (2) With ϵ' and ϵ'', find the smallest value $\epsilon^* \in [\epsilon', \epsilon'']$ with $\mathcal{K}(\epsilon^*) \leq k$. In the following, we focus on showing an efficient implementation of these two steps.

Our algorithm involves a search technique, called *binary search on sorted arrays*, which is shown in the following lemma.

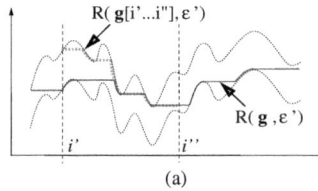 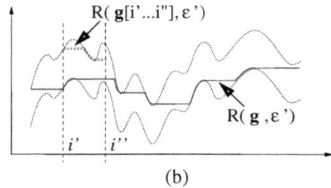

(a)   (b)

Fig. 3. The two cases for computing the representability of $R(\mathbf{g}[i' \ldots i''], \epsilon')$: (a) $R(\mathbf{g}[i' \ldots i''], \epsilon')$ merges into $R(\mathbf{g}, \epsilon')$; (b) $R(\mathbf{g}[i' \ldots i''], \epsilon')$ does not merge into $R(\mathbf{g}, \epsilon')$

Lemma 13. *Given M arrays A_i, $1 \leq i \leq M$, each containing $O(N)$ elements in sorted order, a sought element δ in $A = \cup_{i=1}^{M} A_i$ can be determined in $O((M + T)\log(NM))$ time, where $O(T)$ is the time of a procedure to report whether $a \leq \delta$ or $a > \delta$ for any given value a.*

For Step (1), note that $\mathcal{K}(\epsilon)$ is monotone with respect to ϵ; further, the set S can be represented implicitly as $O(m)$ sorted arrays of size $O(m)$ each. Thus, this step clearly takes $O(m \log m)$ time based on Lemma 13, using our UR *min-k* algorithm as a procedure.

For Step (2), note that ϵ' and ϵ'' are two consecutive elements in S in the sense that for any $\hat{\epsilon} \in S$, either $\hat{\epsilon} \leq \epsilon'$ or $\hat{\epsilon} \geq \epsilon''$. Thus, by Lemma 11, changing the error ϵ from ϵ' to ϵ'' does not cause a feasibility change on representing $\mathbf{g}[1 \ldots i]$ (resp., $\mathbf{g}[j \ldots n]$) by an uphill (resp., downhill) curve. Therefore, when ϵ changes from ϵ' to ϵ'', the decreasing of the function $\mathcal{K}(\epsilon)$ is due to the feasibility change of the uphill-downhill pair representations of some $\mathbf{g}[i' \ldots i'']$'s, for $1 \leq i' < i'' \leq m$. Denote by $\epsilon[i', i'']$ the minimum error ϵ such that $\mathbf{g}[i' \ldots i'']$ is ϵ-UDP-representable, and define $S' = \{\epsilon[i', i''] \mid 1 \leq i' < i'' \leq m\}$. Thus, S' must contain ϵ^*. Step (2) can be carried out by performing a similar search as in Step (1) on S' for $\epsilon^* \in [\epsilon', \epsilon'']$. Further, since $\epsilon^* \in [\epsilon', \epsilon'']$, we only need to consider those elements of S' which are in $[\epsilon', \epsilon'']$. The key to this to compute efficiently, for any $1 \leq i' < i'' \leq m$, the value $\epsilon[i', i'']$ (if it is in $[\epsilon', \epsilon'']$).

We design a data structure such that, after $O(m)$ time preprocessing, for any query $q(i', i'')$, $1 \leq i' < i'' \leq m$, the following can be determined in $O(1)$ time: Whether $\epsilon[i', i''] \in [\epsilon', \epsilon'']$; if it is, report the value of $\epsilon[i', i'']$; otherwise, report whether $\epsilon[i', i''] < \epsilon'$ or $\epsilon[i', i''] > \epsilon''$. Define $\mathcal{G}(\epsilon, \mathbf{g}[i' \ldots i''])$ to be $g_{i'} + \epsilon - \mathcal{H}(R(\mathbf{g}[i' \ldots i''], \epsilon))$. If we replace $\mathcal{G}(\epsilon)$ in Lemma 5 by $\mathcal{G}(\epsilon, \mathbf{g}[i' \ldots i''])$, by the definition of ϵ' and ϵ'', when $\epsilon \in [\epsilon', \epsilon'']$, $\mathcal{G}(\epsilon, \mathbf{g}[i' \ldots i''])$ is a linear function and $\epsilon[i', i'']$ is the unique error $\hat{\epsilon}$ such that $\mathcal{G}(\hat{\epsilon}, \mathbf{g}[i' \ldots i'']) = 0$. If $\epsilon[i', i''] \in [\epsilon', \epsilon'']$, then once $\mathcal{G}(\epsilon', \mathbf{g}[i' \ldots i''])$ and $\mathcal{G}(\epsilon'', \mathbf{g}[i' \ldots i''])$ are available, $\epsilon[i', i'']$ can be obtained in $O(1)$ time. Further, $\epsilon[i', i''] < \epsilon'$ if and only if $\mathcal{G}(\epsilon', \mathbf{g}[i' \ldots i'']) > 0$, and $\epsilon[i', i''] > \epsilon''$ if and only if $\mathcal{G}(\epsilon'', \mathbf{g}[i' \ldots i'']) < 0$. Thus, to answer each query $q(i', i'')$ in $O(1)$ time, it suffices to compute the two values $\mathcal{H}(R(\mathbf{g}[i' \ldots i''], \epsilon'))$ and $\mathcal{H}(R(\mathbf{g}[i' \ldots i''], \epsilon''))$ (and consequently, $\mathcal{G}(\epsilon', \mathbf{g}[i' \ldots i''])$ and $\mathcal{G}(\epsilon'', \mathbf{g}[i' \ldots i''])$) in $O(1)$ time. This is made possible by our $O(m)$ time preprocessing algorithm given below. We only show the preprocessing algorithm for $\mathcal{H}(R(\mathbf{g}[i' \ldots i''], \epsilon'))$ (the case for $\mathcal{H}(R(\mathbf{g}[i' \ldots i''], \epsilon''))$ is handled similarly).

The main idea is to use the geometric relations between the two characteristic curves $R(\mathbf{g}, \epsilon')$ and $R(\mathbf{g}[i' \ldots i''], \epsilon')$. Specifically, we use the value $\mathcal{H}(R(\mathbf{g}, \epsilon'))$ to help compute $\mathcal{H}(R(\mathbf{g}[i' \ldots i''], \epsilon'))$. As part of the preprocessing, we compute, in $O(m)$ time, the value $\mathcal{H}(R(\mathbf{g}, \epsilon'))$, and further, keep all the *prefix values* $\mathcal{H}(R(\mathbf{g}[1 \ldots i], \epsilon'))$ for $1 \leq i \leq n$. Considering the relations between $R(\mathbf{g}, \epsilon')$ and $R(\mathbf{g}[i', i''], \epsilon')$, there are two possible cases: (I) $R(\mathbf{g}[i' \ldots i''], \epsilon')$ "merges" into $R(\mathbf{g}, \epsilon')$ (Fig. 3(a)); (II) $R(\mathbf{g}[i' \ldots i''], \epsilon')$ does not merge into $R(\mathbf{g}, \epsilon')$ (Fig. 3(b)). To deal with Case (I), after $R(\mathbf{g}, \epsilon')$ is computed, with $O(m)$ time preprocessing (in Lemma 14), we can store the merge point $\overline{i'}$ for every i' in $[1 \ldots m]$ (this merge point does not depend on i''), as well as the total amount of "downhill drops" from i' to its $\overline{i'}$ (denote this amount by $C(i')$). In this way, the value of $\mathcal{H}(R(\mathbf{g}[i' \ldots i''], \epsilon'))$ is equal to $C(i') + \mathcal{H}(R(\mathbf{g}[1 \ldots i''], \epsilon')) - \mathcal{H}(R(\mathbf{g}[1 \ldots \overline{i'}], \epsilon'))$ (see Fig. 3(a)), which can be obtained in $O(1)$ time from the prefix values $\mathcal{H}(R(\mathbf{g}[1 \ldots i''], \epsilon'))$ and $\mathcal{H}(R(\mathbf{g}[1 \ldots \overline{i'}], \epsilon'))$. Note that the merge point $\overline{i'}$ of i' also allows us to decide in $O(1)$ time which of the two cases holds for a query $q(i', i'')$. For Case (II), the key observation is that the value of $\mathcal{H}(R(\mathbf{g}[i' \ldots i''], \epsilon'))$ is $g_{i'} - h[i', i'']$, where $h[i', i'']$ is the minimum value of $\mathbf{g}$ on $[i' \ldots i'']$. With a range minima data structure [7] (which can be constructed in linear time), we can report $h[i', i'']$, and consequently $\mathcal{H}(R(\mathbf{g}[i' \ldots i''], \epsilon'))$, in $O(1)$ time.

Lemma 14. *All merge points $\overline{i}$'s for $i \in [1 \cdots m]$ can be obtained in $O(m)$ time.*

Proof. We first compute the two curves $\mathbf{g} + \epsilon'$ and $\mathbf{g} - \epsilon'$, and then $R(\mathbf{g}, \epsilon')$. In the region $\mathcal{R}$ bounded between $\mathbf{g} + \epsilon'$ and $R(\mathbf{g}, \epsilon')$, we perform a rightwards horizontal trapezoidal decomposition from the vertices of $\mathbf{g} + \epsilon'$, which can be done in linear time. This produces a set L of horizontal line segments in $\mathcal{R}$. We then connect these segments to $R(\mathbf{g}, \epsilon')$ by following downhill paths along $L \cup (\mathbf{g} + \epsilon')$, until reaching some points on $R(\mathbf{g}, \epsilon')$ (if a point on $R(\mathbf{g}, \epsilon')$ is reachable). Note that for each segment $l \in L$, such a downhill path connecting l to $R(\mathbf{g}, \epsilon')$ (if any) is unique. This process creates a forest, with the whole curve $R(\mathbf{g}, \epsilon')$ being the root of one of the trees, T. For each vertex v of $\mathbf{g} + \epsilon'$ in $\mathcal{R}$, we then find the first point on $R(\mathbf{g}, \epsilon')$ along T, denoted by $\overline{v}$ (if v is not in the tree T containing $R(\mathbf{g}, \epsilon')$, then $\overline{v} = +\infty$). Clearly, these structures can all be built in $O(m)$ time. Thus, in $O(m)$ time, we can compute all merge points. □

Since we are only concerned with those error values in $[\epsilon', \epsilon'']$, for a query $q(i', i'')$, if $\epsilon[i', i''] > \epsilon''$, we simply set $\epsilon[i', i''] = +\infty$. Likewise, if $\epsilon[i', i''] < \epsilon'$, we set $\epsilon[i', i''] = -\infty$. With this value-setting, for any $[j' \ldots j''] \subseteq [i' \ldots i'']$, we have $\epsilon[j', j''] \leq \epsilon[i', i'']$. Thus, the set S' can be viewed as consisting of $O(m)$ sorted arrays of size $O(m)$ each. Precisely, for each $1 \leq i' \leq m - 1$, let array $A_i = \{\epsilon[i', i''] \mid i'' = i' + 1, \ldots, m\}$. Further, S' can be represented implicitly as discussed above and any element of S' can be obtained in $O(1)$ time (after an $O(m)$ time preprocessing). Therefore, by using the searching technique in Lemma 13, we can find the error $\epsilon^* \in S'$ in $O(m \log m)$ time.

Theorem 4. *Given an integer $k > 0$, the UR min-ϵ problem on a curve $\mathbf{f} = (f_1, f_2, \ldots, f_n)$ is solvable in $O(n + m \log m)$ time, where m is the size of $SK(\mathbf{f})$.*

References

1. Barequet, G., Chen, D.Z., Daescu, O., Goodrich, M., Snoeyink, J.: Efficiently approximating polygonal paths in three and higher dimensions. Algorithmica 33(2), 150–167 (2002)
2. Chun, J., Sadakane, K., Tokuyama, T.: Linear time algorithm for approximating a curve by a single-peaked curve. Algorithmica 44(2), 103–115 (2006)
3. Chun, J., Sadakane, K., Tokuyama, T., Yuki, M.: Peak-reducing fitting of a curve under the L_p metric. Interdisciplinary Information Sciences 11(2), 191–198 (2005)
4. Cormen, T., Leiserson, C., Rivest, R., Stein, C.: Introduction to Algorithms, 2nd edn. MIT Press, Cambridge (2001)
5. Fukuda, T., Morimoto, Y., Morishita, S., Tokuyama, T.: Implementation and evaluation of decision trees with range and region splitting. Constraints 2, 401–427 (1997)
6. Fukuda, T., Morimoto, Y., Morishita, S., Tokuyama, T.: Data mining with optimized two-dimensional association. ACM Trans. Database Systems 26, 179–213 (2001)
7. Gabow, H.N., Bentley, J., Tarjan, R.E.: Scaling and related techniques for geometry problems. In: Proc. of the 16th Annual ACM Symposium on Theory of Computing (STOC), pp. 135–143 (1984)
8. Goodrich, M.: Efficient piecewise-linear function approximation using the uniform metric. In: Proc. of the 10th Annual ACM Symposium on Computational Geometry, pp. 322–331 (1994)
9. Guha, S., Shim, K.: A note on linear time algorithms for maximum error histograms. IEEE Transactions on Knowledge and Data Engineering 19(7), 993–997 (2007)
10. Stout, Q.F.: Unimodal regression via prefix isotonic regression. Computational Statistics & Data Analysis 53(2), 289–297 (2008)
11. Varadarajan, K.: Approximating monotone polygonal curves using the uniform metric. In: Proc. of the 12th Annual ACM Symposium on Computational Geometry, pp. 311–318 (1996)
12. Wang, C., Luan, S., Tang, G., Chen, D.Z., Earl, M.A., Yu, C.X.: Arc-modulated radiation therapy (AMRT): A single-arc form of intensity-modulated arc therapy. Physics in Medicine and Biology 53(22), 6291–6303 (2008)
13. Webb, S.: The Physics of Three-Dimensional Radiation Therapy. Institute of Physics Publishing, Bristol (1993)

Bregman Clustering for Separable Instances[*]

Marcel R. Ackermann and Johannes Blömer

Department of Computer Science, University of Paderborn, Germany
{mra,bloemer}@uni-paderborn.de

Abstract. The Bregman k-median problem is defined as follows. Given a Bregman divergence D_ϕ and a finite set $P \subseteq \mathbb{R}^d$ of size n, our goal is to find a set C of size k such that the sum of errors $\mathrm{cost}(P, C) = \sum_{p \in P} \min_{c \in C} D_\phi(p, c)$ is minimized. The Bregman k-median problem plays an important role in many applications, e.g., information theory, statistics, text classification, and speech processing. We study a generalization of the kmeans++ seeding of Arthur and Vassilvitskii (SODA '07). We prove for an almost arbitrary Bregman divergence that if the input set consists of k well separated clusters, then with probability $2^{-\mathcal{O}(k)}$ this seeding step alone finds an $\mathcal{O}(1)$-approximate solution. Thereby, we generalize an earlier result of Ostrovsky et al. (FOCS '06) from the case of the Euclidean k-means problem to the Bregman k-median problem. Additionally, this result leads to a constant factor approximation algorithm for the Bregman k-median problem using at most $2^{\mathcal{O}(k)}n$ arithmetic operations, including evaluations of Bregman divergence D_ϕ.

1 Introduction

Clustering is the problem of grouping a set of objects into subsets (known as clusters) such that similar objects are grouped together. The quality of a clustering is often measured using a well defined cost function involving a dissimilarity measure that bears meaning for the given application. A cost function that has been proved useful in the past decades is the k-median cost function. Here the objective is to partition a set of objects into k clusters, each with a given cluster representative (known as cluster center) such that the sum of distances from each object to their representative is minimized.

Many approximation algorithms and techniques for this minimization problem have been developed when the dissimilarity measure used is a metric such as the Euclidean distance (known as the *Euclidean k-median problem*) [1,2,3,4,5,6], or when the squared Euclidean distance is used (known as the *Euclidean k-means problem*) [7,4,8,9,10,11]. However, until recently, for non-metric dissimilarity measures almost no approximation algorithms were known. This stands in sharp contrast to the fact that many non-metric dissimilarity measures are used in various applications. To name just a few examples, the Mahalanobis divergence is used in statistics, the Itakura-Saito divergence is used in speech processing, and the Kullback-Leibler divergence is used in machine learning, data mining,

---

[*] Research supported by Deutsche Forschungsgemeinschaft (DFG), grant BL-314/6-1.

H. Kaplan (Ed.): SWAT 2010, LNCS 6139, pp. 212–223, 2010.

and information theory. These examples are instance of a broader class of dissimilarity measures that has only recently attained considerable attention: the class of *Bregman divergences*. Although a PTAS has been given for this broader class of problems [12,13], it suffers from impractically huge constants that are involved in the running time. Hence, in this paper, we focus on the statement and analysis of an efficient and practical sampling scheme to obtain a constant factor approximate solution for the *Bregman k-median problem*.

Related work. One particular algorithm for the Euclidean k-means problem that has appealed to practitioners during the past decades is *Lloyd's k-means algorithm* [14]. Starting with an arbitrary set of k center points $c_1, c_2, \ldots, c_k$, Lloyd's local improvement strategy iterates between two steps: First, each input point is assigned to its closest center point to build a partition $P_1, P_2, \ldots, P_k$ of the input points. Then, for each set P_i the center point c_i is recomputed as the centroid (center of mass) of P_i. These steps are repeated until the partition and the center points become stable. It can easily be seen that after a finite number of steps Lloyd's algorithm terminates with a locally optimal clustering. In addition, a single Lloyd iteration can be implemented to run quite fast in practice.

However, it is known that the speed of convergence as well as the quality of the local optimum computed depends considerably on the choice of initial center points. While it has been shown recently that the expected number of iterations is polynomial in n, k, and d in the smoothed complexity model [15], in the worst case, an exponential number of $2^{\Omega(n)}$ iterations are necessary, even in the plane [16]. Furthermore, there are simple examples of input sets such that a poor choice of initial centers leads to arbitrarily bad clusterings (i.e., the approximation ratio is unbounded). To deal with these problems in practice, the initial center points are usually chosen uniformly at random among the input points. However, no non-trivial approximation guarantees are known for uniform seeding. Recently, Arthur and Vassilvitskii proposed a new non-uniform initial seeding procedure for the Euclidean k-means problem [17], known as *kmeans++ seeding*. It has been shown that this seeding step alone computes an $\mathcal{O}(\log k)$-approximate set of centers. Any subsequent Lloyd iteration only improves this solution. Independently, a different analysis of essentially the same seeding procedure has been given by Ostrovsky et al. [18]. In this analysis it has been shown, that for certain well separated input instances with probability $2^{-\mathcal{O}(k)}$ the non-uniform seeding step gives an $\mathcal{O}(1)$-approximate set of centers. Recently, it has been shown that kmeans++ seeding of $\mathcal{O}(k)$ centers gives a constant factor bi-criteria approximation [19].

It has been observed that Lloyd's algorithm is also applicable to whole class of Bregman divergences [20]. Moreover, it has been shown that under some mild continuity assumption the class of Bregman divergences is exactly the the class of dissimilarity measures for which Lloyd's algorithm is applicable [21]. Results on the worst-case and smoothed number of iterations have also been generalized to the Bregman k-median problem [22]. Furthermore, kmeans++ seeding has been adapted to the case of Bregman divergences independently in at least three different publications [23,24,13]. We call this generalization *bregmeans++ seeding*. In detail, it has been shown that in the case of μ-similar Bregman divergences (i.e., Bregman divergences that feature $\mathcal{O}(1/\mu)$-approximate metric properties for some constant $0 < \mu \leq 1$), this seeding step alone computes an $\mathcal{O}(\mu^{-2} \log k)$-approximate set of centers [13].

Our contribution. In this paper, we give an analysis of bregmeans++ seeding when restricted to certain input sets known as *separable input instances*. This restriction is assumed for the following reason. When using solutions of the Bregman k-median problem in real-world applications, we implicitly assume that these center points provide a meaningful representation of the input data. That is, we expect the input set to consist of k well separated clusters, and that each of the k-medians distinctively characterizes one of these clusters. If this is not the case then, obviously, a different number of clusters should be considered.

This motivates the notion of separable input sets: A k-median input instance is called (k, α)-separable if no clustering of a cost within a factor of $1/\alpha$ of the optimal k-median cost can be achieved by using only $k-1$ or fewer centers, where α is some small positive constant. This models the situation where we have agreed on a small number of k such that the centers are a meaningful representation of the input points, while fewer clusters would lead to an unreasonable rise of the cost function. The notion of separable input instances has been used before to analyze clustering algorithms [25,9,18].

Another notion frequently used to describe meaningful k-clustering is the notion of *stable clusterings* [26]. Here a clustering is assumed to be stable if a small perturbation of the input points leads to essentially the same optimal partition of the input set into clusters (i.e., the symmetric difference of perturbed and unperturbed optimal clusters is small). However, in [18] it is shown that the notions of separable inputs and stable clusterings are equivalent. Hence, in this paper, we concentrate on the notion of separable input instances.

We analyse bregmeans++ seeding for a Bregman divergence from the large subclass of μ-similar Bregman divergences. We prove that in case of separable input instances with constant probability the non-uniform seeding of bregmeans++ yields a constant factor approximation. Stated in detail, our main result is as follows.

Theorem 1. *Let $0 < \mu \leq 1$ be constant, let D_ϕ be a μ-similar Bregman divergence on domain $\mathbb{X} \subseteq \mathbb{R}^d$, and let $P \subseteq \mathbb{X}$ be a $(k, \mu/8)$-separable input set. Then with probability at least $2^{-\mathcal{O}(k)}\mu^k$, bregmeans++ seeding computes an $\mathcal{O}(1/\mu)$-approximate solution of the Bregman k-median problem with respect to D_ϕ and input instance P.*

Earlier, it has been shown that a non-uniform seeding very similar to kmeans++ seeding provides a constant factor approximate solution for separable instances in the context of the Euclidean k-means problem [18]. Unfortunately, this result relies crucially on the metric properties of the Euclidean distance. Our result can be seen as a generalization of the result from [18] to the class of Bregman divergences, which includes even asymmetric dissimilarity measures. In particular, we obtain our result by a somewhat simplified proof that focuses on the combinatorial and statistical properties of the Bregman k-median problem.

An immediate consequence of Theorem 1 is that we can compute a constant factor approximation to the k-median problem of separable input instance P of size $|P| = n$ with arbitrary high probability by computing the bregmeans++ seeding $2^{\mathcal{O}(k)}$ times independently and choosing the best set of centers obtained this way. This leads to a constant factor approximation algorithm for the Bregman k-median problem using at most $2^{\mathcal{O}(k)}n$ arithmetic operations, including evaluations of D_ϕ. Note that a small number of Lloyd iterations can still significantly improve the quality of this solution. However, no approximation guarantee for this improvement is known, and the theoretically provable approximation factor already applies to the solution computed by the seeding step.

2   Preliminaries

Bregman k-median clustering. The dissimilarity measures known as Bregman divergences were introduced in 1967 by Lev M. Bregman [27]. Intuitively, a Bregman divergence can be seen as the error when approximating a convex function by a tangent hyperplane. We use the following formal definition.

Definition 2. *Let $\mathbb{X} \subseteq \mathbb{R}^d$ be convex. For any strictly convex, differentiable function $\phi : \mathbb{X} \to \mathbb{R}$ we define the Bregman divergence with respect to ϕ as*

$$D_\phi(p, q) = \phi(p) - \phi(q) - \nabla\phi(q)^\top (p - q)$$

for $p, q \in \mathbb{X}$. Here $\nabla\phi(q)$ denotes the gradient of ϕ at point q.

Bregman divergences include many prominent dissimilarity measures like the squared Euclidean distance $D_{\ell_2^2}(p, q) = \|p - q\|_2^2$ (with $\phi_{\ell_2^2}(t) = \|t\|_2^2$), the generalized Kullback-Leibler divergence $D_{\mathrm{KL}}(p, q) = \sum p_i \ln \frac{p_i}{q_i} - \sum (p_i - q_i)$ (with $\phi_{\mathrm{KL}}(t) = \sum t_i \ln t_i - t_i$), and the Itakura-Saito divergence $D_{\mathrm{IS}}(p, q) = \sum (\frac{p_i}{q_i} - \ln \frac{p_i}{q_i} - 1)$ (with $\phi_{\mathrm{IS}}(t) = -\sum \ln t_i$). We point out that, in general, Bregman divergences are asymmetric and do not satisfy the triangle inequality. Furthermore, D_ϕ may possess singularities, i.e., there may exist points $p, q \in \mathbb{X}$ such that $D_\phi(p, q) = \infty$.

For $p \in \mathbb{X}$ and any set $C \subseteq \mathbb{X}$ we also write $D_\phi(p, C) = \min_{c \in C} D_\phi(p, c)$ to denote the minimal dissimilarity from point p towards any point from set C. For any finite point set $P \subseteq \mathbb{X}$ and $C \subseteq \mathbb{X}$ of size $|C| = k$ we denote the k-median cost of input set P towards the centers from C by $\mathrm{cost}(P, C) = \sum_{p \in P} D_\phi(p, C)$. The goal of the *$k$-median problem* with respect to Bregman divergence D_ϕ and input instance $P \subseteq \mathbb{X}$ is to find a set $C \subseteq X$ of size k such that $\mathrm{cost}(P, C)$ is minimized. We denote the cost of such an optimal solution by $opt_k(P)$. The elements of an optimal C are called *k-medians* of P.

Furthermore, it is an important observation that all Bregman divergences satisfy the following *central identity*.

Lemma 3 ([20], proof of Proposition 1). *Let D_ϕ be a Bregman divergence on domain $\mathbb{X} \subseteq \mathbb{R}^d$, and let $c_P = \frac{1}{|P|} \sum_{p \in P} p$ denote the centroid (center of mass) of point set P. For all $q \in \mathbb{X}$ we have $\mathrm{cost}(P, q) = opt_1(P) + |P| D_\phi(c_P, q)$.*

As an immediate consequence of Lemma 3 and of the non-negativity of D_ϕ, we find that that for all Bregman divergences the centroid c_P is the optimal 1-median of P.

μ-similarity. One particular class of dissimilarity measures stands out among the class of Bregman divergences: *Mahalanobis distances.* For any symmetric positive definite matrix $U \in \mathbb{R}^{d \times d}$ the Mahalanobis distance with respect to U is defined as $D_U(p,q) = (p-q)^\top U (p-q)$ for $p, q \in \mathbb{R}^d$. The Mahalanobis distance was introduced in 1936 by P. C. Mahalanobis [28] based on the inverse of the covariance matrix of two random variables. It is a Bregman divergence given by the generating function $\phi_U(t) = t^\top U t$. Unlike all other Bregman divergences, Mahalanobis distances are symmetric. Furthermore, they satisfy the following *double triangle inequality.*

Lemma 4 ([13], Lemma 2.1). *Let D_U be a Mahalanobis distance on domain $\mathbb{R}^d$. For all $p, q, r \in \mathbb{R}^d$ we have $D_U(p,q) \leq 2\big(D_U(p,r) + D_U(r,q)\big)$.*

To some extent, Mahalanobis distances are prototypical for many Bregman divergences that are used in practice. This observation is formalized in the following notion of μ-similarity.

Definition 5. *A Bregman divergence D_ϕ on domain $\mathbb{X} \subseteq \mathbb{R}^d$ is called μ-similar for constant $0 < \mu \leq 1$ if there exists a symmetric positive definite matrix U such that for Mahalanobis distance D_U and for each $p, q \in \mathbb{X}$ we have*

$$\mu D_U(p,q) \leq D_\phi(p,q) \leq D_U(p,q).$$

It can be shown that all Bregman divergences D_ϕ that are used in practice are μ-similar when restricted to a domain $\mathbb{X}$ that avoids the singularities of D_ϕ (cf. [29], where also an overview of some μ-similar Bregman divergences can be found). It easily follows from Lemma 4 that μ-similar Bergman divergences are approximately symmetric and approximately satisfy the triangle inequality, both within a factor of $\mathcal{O}(1/\mu)$.

3   Sampling Methods for μ-Similar Bregman Divergences

Uniform sampling. It has been shown that in the case of μ-similar Bregman divergences, the 1-median of input set P can be well approximated by taking a sample set $S \subseteq P$ of constant size and solving the 1-median problem S exactly [12]. Considering a single sample point, a straightforward corollary of Lemma 3.4 from [12] can be stated as follows.

Lemma 6. *Let D_ϕ be μ-similar Bregman divergence on domain $\mathbb{X}$ and let $P \subseteq \mathbb{X}$. Then with probability at least $\frac{1}{2}$ a uniformly sampled point $s \in P$ satisfies $\mathrm{cost}(P, s) \leq \big(1 + \frac{2}{\mu}\big) \mathrm{opt}_1(P)$.*

Hence, if a cluster P_i of the optimal k-median clustering of P is known then an $\mathcal{O}(1/\mu)$-approximate center can be obtain by sampling a single point uniformly at random from P_i. Of course, if our goal is to solve the k-median problem with $k \geq 2$ then the optimal clusters of P are not known in advance.

Non-uniform sampling. The non-uniform sampling scheme which we call breg-means++ is a random sampling scheme as follows. The first of k approximate medians is chosen uniformly at random from input instance P. After that, assume that we have already chosen approximate medians $A_j = \{a_1, a_2, \ldots, a_j\}$. The next approximate median a_{j+1} is chosen from P with probability proportional to $\mathrm{D}_\phi(a_{j+1}, A_j)$, that is, for all $p \in P$ we have

$$\Pr\left[p = a_{j+1} \;\middle|\; A_j = \{a_1, a_2, \ldots, a_j\} \text{ already chosen}\right] = \frac{\mathrm{D}_\phi(p, A_j)}{\mathrm{cost}(P, A_j)}.$$

This sampling scheme is repeated until we have chosen k points. In this case, we say the set $A = \{a_1, a_2, \ldots, a_k\}$ is chosen *at random according to* D_ϕ. Assuming that random sampling of a single element according to a given probability distribution can be achieved in constant time, it is easy to see that sampling of k points according to D_ϕ can be achieved using at most $\mathcal{O}(k\,|P|)$ arithmetic operations, including evaluations of Bregman divergence D_ϕ.

This sampling scheme has been originally proposed in the context of Euclidean k-means clustering, as well as for the k-median problem using a t-th power of the Euclidean distance [17]. The following theorem can be proven the same way as Theorem 3.1 from [17], using the $\mathcal{O}(1/\mu)$-approximate metric properties of the μ-similar Bregman divergence D_ϕ.

Theorem 7 ([13], Theorem A.1). *If D_ϕ is a μ-similar Bregman divergence on domain $\mathbb{X}$ and $A \subseteq \mathbb{X}$ with $|A| = k$ is chosen at random according to D_ϕ, then we have* $\mathrm{E}\left[\mathrm{cost}(P, A)\right] \leq \frac{8}{\mu^2}(2 + \ln k)\,opt_k(P)$.

From Markov's inequality it follows that with high probability bregmeans++ seeding yields a factor $\mathcal{O}(\mu^{-2} \log k)$ approximation of $opt_k(P)$.

4 Analysis on Separable Input Instances

In our analysis, we concentrate on the case of input instances for which the optimal k-medians indeed give a meaningful representation of the input set, that is, the case of separable input instances. The motivation behind this notion is discussed in Section 1. We use the following formal definition.

Definition 8. *Let $0 < \alpha < 1$. An input instance $P \subseteq \mathbb{X}$ is called (k, α)-separable if $opt_k(P) \leq \alpha\, opt_{k-1}(P)$.*

We show that for (k, α)-separable instance P, where $\alpha < \mu/8$, with constant probability sampling according to D_ϕ computes a factor $\mathcal{O}(1/\mu)$-approximate solution to the μ-similar Bregman k-median problem. This implies that on such instances, bregmeans++ seeding has in fact a better approximation guarantee than is suggested by the result from Theorem 7. In detail, below we prove that a set A of k points chosen at random according to D_ϕ satisfies

$$\Pr\left[\mathrm{cost}(P, A) \leq \left(1 + \frac{2}{\mu}\right) opt_k(P)\right] \geq \frac{1}{2}\left(\frac{\mu}{20}\right)^{k-1}. \tag{1}$$

Theorem 1 is an immediate consequence of inequality (1).

Proof of Theorem 1. Let $P_1, P_2, \ldots, P_k$ denote the clusters of an optimal k-median clustering of P and let $C = \{c_1, c_2, \ldots, c_k\}$ be the corresponding optimal k-medians, i.e., we have $\text{cost}(P, C) = opt_k(P)$ and $\text{cost}(P_i, c_i) = opt_1(P_i)$ for all $1 \leq i \leq k$. Furthermore, for all $1 \leq i \leq k$ we define

$$X_i = \left\{ x \in P_i \,\middle|\, D_\phi(c_i, x) \leq \frac{2}{\mu|P_i|} \, opt_1(P_i) \right\} ,$$

and $Y_i = P_i \setminus X_i$. Note that in the definition of X_i the optimal median c_i is used as the first argument of Bregman divergence D_ϕ. From the central identity of Lemma 3 we know that the elements $x \in X_i$ are exactly the points from P_i that are $(1 + 2/\mu)$-approximate medians of P_i since

$$\text{cost}(P_i, x) = opt_1(P_i) + |P_i| \, D_\phi(c_i, x) \leq \left(1 + \frac{2}{\mu}\right) opt_1(P_i) .$$

Analogously, we know that the elements $y \in Y_i$ are exactly the points from P_i that fail to be $(1 + 2/\mu)$-approximate medians of P_i.

Let $A = \{a_1, a_2, \ldots, a_k\}$ be the set of points chosen iteratively at random according to D_ϕ. Our strategy to prove Theorem 1 is to show that for separable input instance P with probability at least $2^{-\mathcal{O}(k)}\mu^k$, set A consists of one point from each set $X_1, X_2, \ldots, X_k$ and no point from any set Y_i. In that case, assuming $a_i \in X_i$ for all $1 \leq i \leq k$, we conclude

$$\text{cost}(P, A) \leq \sum_{i=1}^{k} \text{cost}(P_i, a_i) \leq \left(1 + \frac{2}{\mu}\right) \sum_{i=1}^{k} opt_1(P_i) = \left(1 + \frac{2}{\mu}\right) opt_k(P) . \quad (2)$$

We start by noting that each set X_i is indeed a large subset of P_i. This observation is an immediate consequence of Lemma 6.

Lemma 9. *For all $1 \leq i \leq k$ we have $|X_i| \geq \frac{1}{2}|P_i| \geq |Y_i|$.*

Hence, let us consider the first, uniformly chosen point a_1. In the sequel, let $P_{[i,j]}$ denote the union $\bigcup_{t=i}^{j} P_t$, and $X_{[i,j]}$ the union $\bigcup_{t=i}^{j} X_t$. Using Lemma 9 we immediately obtain the following lemma.

Lemma 10. $\Pr\left[a_1 \in X_{[1,k]}\right] \geq \frac{1}{2}$.

Now, let us assume that we have already sampled set $A_j = \{a_1, a_2, \ldots, a_j\}$ with $1 \leq j \leq k - 1$ and $a_i \in X_i$ for all $1 \leq i \leq j$. Our goal is to show that with significant probability the next sampled point a_{j+1} is chosen from $X_{[j+1,k]}$. In a first step towards this goal, the following lemma states that with high probability point a_{j+1} is chosen from $P_{[j+1,k]}$.

Lemma 11. $\Pr\left[a_{j+1} \in P_{[j+1,k]} \,\middle|\, a_1 \in X_1, \ldots, a_j \in X_j\right] \geq 1 - \frac{3\alpha}{\mu}$.

Proof. For (k, α)-separable P we have

$$\text{cost}(P, A_j) \geq opt_j(P) \geq \frac{1}{\alpha} \, opt_k(P) = \frac{1}{\alpha} \sum_{i=1}^{k} opt_1(P_i) . \quad (3)$$

From $a_1 \in X_1, \ldots, a_j \in X_j$ we know that for all $1 \le i \le j$ we find

$$\text{cost}(P_i, a_i) \le \left(1 + \frac{2}{\mu}\right) opt_1(P_i) \le \frac{3}{\mu} opt_1(P_i) \tag{4}$$

since $1 + \frac{2}{\mu} \le \frac{3}{\mu}$ for $\mu \le 1$. Using (3) and (4) we obtain

$$\text{cost}(P, A_j) \ge \frac{1}{\alpha} \sum_{i=1}^{j} opt_1(P_i) \ge \frac{\mu}{3\alpha} \sum_{i=1}^{j} \text{cost}(P_i, a_i) \ge \frac{\mu}{3\alpha} \text{cost}(P_{[1,j]}, A_j) . \tag{5}$$

Hence, using (5) we conclude $\frac{\text{cost}(P_{[1,j]}, A_j)}{\text{cost}(P, A_j)} \le \frac{3\alpha}{\mu}$. The lemma follows. $\qquad\square$

Next, we show that if $a_i \in X_i$ for all $1 \le i \le j$ and we have that $a_{j+1} \in P_{[j+1,k]}$, it follows that with significant probability a_{j+1} is chosen from $X_{[j+1,k]}$.

Lemma 12. $\Pr\left[a_{j+1} \in X_{[j+1,k]} \,\middle|\, a_1 \in X_1, \ldots, a_i \in X_j, a_{j+1} \in P_{[j+1,k]}\right] \ge \frac{\mu}{5}\left(1 - \frac{4\alpha}{\mu}\right)$

Proof. We start by noting that for a separable instance P the set $A_j \subseteq X_{[1,j]}$ is indeed a poor choice as approximate medians for $P_{j+1}, P_{j+2}, \ldots, P_k$. More precisely, for (k, α)-separable P from inequalities (3) and (5) we know

$$\text{cost}(P, A_j) \ge \frac{1}{\alpha} \sum_{i=1}^{j} opt_1(P_i) + \frac{1}{\alpha} \sum_{i=j+1}^{k} opt_1(P_i)$$

$$\ge \frac{\mu}{3\alpha} \sum_{i=1}^{j} \text{cost}(P_i, a_i) + \frac{1}{\alpha} \sum_{i=j+1}^{k} \text{cost}(P_i, c_i) .$$

Using $\alpha \le \frac{\mu}{8}$ we have $\frac{\mu}{3\alpha} > 1$, and we obtain

$$\text{cost}(P, A_j) \ge \sum_{i=1}^{j} \text{cost}(P_i, a_i) + \frac{1}{\alpha} \sum_{i=j+1}^{k} \text{cost}(P_i, c_i) .$$

Hence,

$$\text{cost}(P_{[j+1,k]}, A_j) \ge \frac{1}{\alpha} \sum_{i=j+1}^{k} \text{cost}(P_i, c_i) . \tag{6}$$

Now, we make use of bound (6) to show that the cost of $X_{[j+1,k]}$ towards A_j is at least a significant fraction of the cost of $P_{[j+1,k]}$ towards A_j. To this end, fix an index $i > j$. Let D_U be a Mahalanobis distance such that $\mu D_U(p, q) \le D_\phi(p, q) \le D_U(p, q)$ for all $p, q \in X$, and let $a_x^* = \arg\min_{a \in A_j} D_\phi(x, a)$ for any $x \in X_i$. Obviously, $D_\phi(y, A_j) \le D_\phi(y, a_x^*) \le D_U(y, a_x^*)$. Using the double triangle inequality of D_U (Lemma 4) we deduce that for all $x \in X_i$ and all $y \in Y_i$ we have

$$D_\phi(y, A_j) \le 4\left(D_U(y, c_i) + D_U(x, c_i) + D_U(x, a_x^*)\right)$$

$$\le \frac{4}{\mu}\left(D_\phi(y, c_i) + D_\phi(x, c_i) + D_\phi(x, A_j)\right) . \tag{7}$$

Due to Lemma 9 we know that $|X_i| \geq |Y_i|$. Hence, there exists an injective mapping $\sigma : Y_i \to X_i$ such that inequality (7) can be applied to each $y \in Y_i$ using a different intermediate point $\sigma(y) \in X_i$. Therefore, by summing up over all $y \in Y_i$ we obtain

$$
\begin{aligned}
\text{cost}(Y_i, A_j) &\leq \frac{4}{\mu}\Big(\text{cost}(Y_i, c_i) + \text{cost}\big(\sigma(Y_i), c_i\big) + \text{cost}\big(\sigma(Y_i), A_j\big)\Big) \\
&\leq \frac{4}{\mu}\big(\text{cost}(Y_i, c_i) + \text{cost}(X_i, c_i) + \text{cost}(X_i, A_j)\big) \\
&= \frac{4}{\mu}\,\text{cost}(P_i, c_i) + \frac{4}{\mu}\,\text{cost}(X_i, A_j) \ .
\end{aligned}
$$

Hence,

$$
\begin{aligned}
\text{cost}(P_i, A_j) &\leq \frac{4}{\mu}\,\text{cost}(P_i, c_i) + \left(\frac{4}{\mu} + 1\right)\text{cost}(X_i, A_j) \\
&\leq \frac{4}{\mu}\,\text{cost}(P_i, c_i) + \frac{5}{\mu}\,\text{cost}(X_i, A_j)
\end{aligned}
$$

since $\frac{4}{\mu} + 1 \leq \frac{5}{\mu}$ for $\mu \leq 1$. Summing up over all indices $i > j$ and using (6) leads to

$$
\begin{aligned}
\text{cost}(P_{[j+1,k]}, A_j) &\leq \frac{4}{\mu}\sum_{i=j+1}^{k} \text{cost}(P_i, c_i) + \frac{5}{\mu}\,\text{cost}(X_{[j+1,k]}, A_j) \\
&\leq \frac{4\alpha}{\mu}\,\text{cost}(P_{[j+1,k]}, A_j) + \frac{5}{\mu}\,\text{cost}(X_{[j+1,k]}, A_j) \ .
\end{aligned}
$$

Thus,

$$
\left(1 - \frac{4\alpha}{\mu}\right)\text{cost}(P_{[j+1,k]}, A_j) \leq \frac{5}{\mu}\,\text{cost}(X_{[j+1,k]}, A_j) \ . \tag{8}
$$

Using (8) we conclude $\frac{\text{cost}(X_{[j+1,k]}, A_j)}{\text{cost}(P_{[j+1,k]}, A_j)} \geq \frac{\mu}{5}\left(1 - \frac{4\alpha}{\mu}\right)$, and the lemma follows. $\square$

Finally, we use Lemmas 10–12 to prove that with probability at least $2^{-\mathcal{O}(k)}\mu^k$, set A obtained by sampling according to D_ϕ contains exactly one point from each set $X_1, X_2, \ldots, X_k$. Lemma 13 together with inequality (2) concludes the proof of Theorem 1.

Lemma 13. $\Pr\left[\forall\, 1 \leq i \leq k :\ A \cap X_i \neq \emptyset\right] \geq \frac{1}{2}\left(\frac{\mu}{20}\right)^{k-1}$.

Proof. In the following, let $\nu_j = |\{i \,|\, A_j \cap X_i \neq \emptyset\}|$ denote the number of sets X_i that have been considered by the first j sampled points $A_j = \{a_1, a_2, \ldots, a_j\}$. We prove this lemma inductively by showing that for all $1 \leq j \leq k$ we have

$$
\Pr[\nu_j = j] \geq \frac{1}{2}\left(\frac{\mu}{20}\right)^{j-1} \ . \tag{9}
$$

From Lemma 10 we know that with probability at least $\frac{1}{2}$ we have $a_1 \in X_{[1,k]}$. Since the X_i form a partition of $X_{[1,k]}$, in this case we have $\nu_1 = 1$. This proves the inductive base case of $j = 1$.

Now, assume that inequality (9) holds for j with $1 \leq j < k$. That is, with probability at least $\frac{1}{2}\left(\frac{\mu}{20}\right)^{j-1}$ and without loss of generality we may assume $a_i \in X_i$ for all $1 \leq i \leq j$. In this case, by using Lemma 11 and Lemma 12 we deduce

$$
\begin{aligned}
\Pr[\nu_{j+1} = j+1 \,|\, \nu_j = j] &\geq \Pr[a_{j+1} \in X_{[j+1,k]} \,|\, a_1 \in X_1, \ldots, a_j \in X_j] \\
&\geq \Pr\left[a_{j+1} \in X_{[j+1,k]} \,|\, a_1 \in X_1, \ldots, a_j \in X_j, a_{j+1} \in P_{[j+1,k]}\right] \\
&\quad \cdot \Pr\left[a_{j+1} \in P_{[j+1,k]} \,|\, a_1 \in X_1, \ldots, a_j \in X_j\right] \\
&\geq \frac{\mu}{5}\left(1 - \frac{3\alpha}{\mu}\right)\left(1 - \frac{4\alpha}{\mu}\right) \\
&\geq \frac{\mu}{20} \tag{10}
\end{aligned}
$$

since $\frac{3\alpha}{\mu} \leq \frac{4\alpha}{\mu} \leq \frac{1}{2}$ for $\alpha \leq \frac{\mu}{8}$. Hence, by using induction hypothesis (9) and inequality (10) we conclude

$$
\Pr[\nu_{j+1} = j+1] \geq \Pr\left[\nu_{j+1} = j+1 \,|\, \nu_j = j\right] \cdot \Pr[\nu_j = j] \geq \frac{1}{2}\left(\frac{\mu}{20}\right)^j . \qquad \square
$$

5 Discussion

In this paper we have introduced and analyzed a practical approximation algorithm applicable to the μ-similar Bregman k-median problem. The sampling technique presented in this paper is easy to implement and runs quite fast in practice, i.e., it has an asymptotic running time of $\mathcal{O}(kn)$ and there are no large hidden constants involved in this running time. In [17] some experiments on real-world input instances for the Euclidean k-means problem have been conducted. It turns out that using the kmeans++ seeding outperforms the standard implementation of Lloyd's algorithm using uniform seeding both in terms of speed of convergence and cost of the clustering. In fact, if the data set consists of k well separated clusters, experiments on synthetic data sets indicate that the quality of the clustering is improved by up to several orders of magnitude.

As for Theorem 1, in contrast to the analysis of [18] our analysis emphasizes the combinatorial and statistical structure of the Bregman k-median problem. However, the approximate triangle inequality and the approximate symmetry of the μ-similar Bregman divergence D_ϕ is needed in a single argument in the proof of Lemma 12. It remains an open problem to find a proof of the approximation guarantee that relies purely on the combinatorial properties of the Bregman k-median problem. Moreover, it is still unknown whether it is possible to give high quality clustering algorithms (say, constant factor approximations) for the Bregman k-median problem that do not rely on assumptions such as μ-similarity.

Acknowledgments. The authors are very grateful to the anonymous reviewers for their detailed and helpful comments.

References

1. Arora, S., Raghavan, P., Rao, S.: Approximation schemes for Euclidean k-medians and related problems. In: Proceedings of the 30th Annual ACM Symposium on Theory of Computing (STOC '98), pp. 106–113 (1998)
2. Kolliopoulos, S.G., Rao, S.: A nearly linear-time approximation scheme for the Euclidean κ-median problem. In: Nešetřil, J. (ed.) ESA 1999. LNCS, vol. 1643, pp. 378–389. Springer, Heidelberg (1999)
3. Bǎdoiu, M., Har-Peled, S., Indyk, P.: Approximate clustering via core-sets. In: Proceedings of the 34th Annual ACM Symposium on Theory of Computing (STOC'02), pp. 250–257. Association for Computing Machinery (2002)
4. Har-Peled, S., Mazumdar, S.: On coresets for k-means and k-median clustering. In: Proceedings of the 36th Annual ACM Symposium on Theory of Computing (STOC'04), pp. 291–300. Association for Computing Machinery (2004)
5. Kumar, A., Sabharwal, Y., Sen, S.: Linear time algorithms for clustering problems in any dimensions. In: Caires, L., Italiano, G.F., Monteiro, L., Palamidessi, C., Yung, M. (eds.) ICALP 2005. LNCS, vol. 3580, pp. 1374–1385. Springer, Heidelberg (2005)
6. Chen, K.: On k-median clustering in high dimensions. In: Proceedings of the 17th Annual ACM-SIAM Symposium on Discrete Algorithms (SODA '06), pp. 1177–1185. Society for Industrial and Applied Mathematics (2006)
7. Matoušek, J.: On approximate geometric k-clustering. Discrete and Computational Geometry 24(1), 61–84 (2000)
8. Fernandez de la Vega, W., Karpinski, M., Kenyon, C., Rabani, Y.: Approximation schemes for clustering problems. In: Proceedings of the 35th Annual ACM Symposium on Theory of Computing (STOC'03), pp. 50–58. Association for Computing Machinery (2003)
9. Kumar, A., Sabharwal, Y., Sen, S.: A simple linear time $(1+\epsilon)$-approximation algorithm for k-means clustering in any dimensions. In: Proceedings of the 45th Annual IEEE Symposium on Foundations of Computer Science (FOCS '04), pp. 454–462. IEEE Computer Society, Los Alamitos (2004)
10. Chen, K.: On coresets for k-median and k-means clustering in metric and Euclidean spaces and their applications. SIAM Journal on Computing 39(3), 923–947 (2009)
11. Feldman, D., Monemizadeh, M., Sohler, C.: A PTAS for k-means clustering based on weak coresets. In: Proceedings of the 23rd ACM Symposium on Computational Geometry (SCG '07), pp. 11–18. Association for Computing Machinery (2007)
12. Ackermann, M.R., Blömer, J., Sohler, C.: Clustering for metric and non-metric distance measures. In: Proceedings of the 19th Annual ACM-SIAM Symposium on Discrete Algorithms (SODA '08), pp. 799–808. Society for Industrial and Applied Mathematics (2008); Full version to appear in ACM Transactions on Algorithms (special issue on SODA '08).
13. Ackermann, M.R., Blömer, J.: Coresets and approximate clustering for Bregman divergences. In: Proceedings of the 20th Annual ACM-SIAM Symposium on Discrete Algorithms (SODA'09), pp. 1088–1097. Society for Industrial and Applied Mathematics (2009)
14. Lloyd, S.P.: Least squares quantization in PCM. IEEE Transactions on Information Theory 28(2), 129–137 (1982)
15. Arthur, D., Manthey, B., Röglin, H.: k-means has polynomial smoothed complexity. In: Proceedings of the 50th Symposium on Foundations of Computer Science (FOCS '09). IEEE Computer Society Press, Los Alamitos (2009) (to appear)

16. Vattani, A.: k-means requires exponetially many iterations even in the plane. In: Proceedings of the 25th Annual Symposium on Computational Geometry (SCG '09), pp. 324–332. Association for Computing Machinery (2009)

17. Arthur, D., Vassilvitskii, S.: k-means++: the advantages of careful seeding. In: Proceedings of the 18th Annual ACM-SIAM Symposium on Discrete Algorithms (SODA '07), pp. 1027–1035. Society for Industrial and Applied Mathematics (2007)

18. Ostrovsky, R., Rabani, Y., Schulman, L.J., Swamy, C.: The effectiveness of Lloyd-type methods for the k-means problem. In: Proceedings of the 47th Annual Symposium on Foundations of Computer Science (FOCS '06), pp. 165–176. IEEE Computer Society, Los Alamitos (2006)

19. Aggarwal, A., Deshpande, A., Kannan, R.: Adaptive sampling for k-means clustering. In: Proceedings of the 12th International Workshop on Approximation Algorithms for Combinatorial Optimization Problems (APPROX '09), pp. 15–28. Springer, Heidelberg (2009)

20. Banerjee, A., Merugu, S., Dhillon, I.S., Ghosh, J.: Clustering with Bregman divergences. Journal of Machine Learning Research 6, 1705–1749 (2005)

21. Banerjee, A., Guo, X., Wang, H.: On the optimality of conditional expectation as a Bregman predictor. IEEE Transactions on Information Theory 51(7), 2664–2669 (2005)

22. Manthey, B., Röglin, H.: Worst-case and smoothed analysis of k-means clustering with Bregman divergences. In: Dong, Y., Du, D.-Z., Ibarra, O. (eds.) ISAAC 2009. LNCS, vol. 5878, pp. 1024–1033. Springer, Heidelberg (2009)

23. Nock, R., Luosto, P., Kivinen, J.: Mixed Bregman clustering with approximation guarantees. In: Daelemans, W., Goethals, B., Morik, K. (eds.) ECML PKDD 2008, Part II. LNCS (LNAI), vol. 5212, pp. 154–169. Springer, Heidelberg (2008)

24. Sra, S., Jegelka, S., Banerjee, A.: Approximation algorithms for Bregman clustering, co-clustering and tensor clustering. Technical Report MPIK-TR-177, Max Planck Institure for Biological Cybernetics (2008)

25. Kanungo, T., Mount, D.M., Netanyahu, N.S., Piatko, C.D., Silverman, R., Wu, A.Y.: An efficient k-means clustering algorithm: Analysis and implementation. IEEE Transactions on Pattern Analysis and Machine Intelligence 24(7), 881–892 (2002)

26. Ben-Hur, A., Elisseeff, A., Guyon, I.: A stability based method for discovering structure in clustered data. In: Proceedings of the 7th Pacific Symposium on Biocomputing (PSB '02), pp. 6–17. World Scientific, Singapore (2002)

27. Bregman, L.M.: The relaxation method of finding the common points of convex sets and its application to the solution of problems in convex programming. USSR Computational Mathematics and Mathematical Physics 7, 200–217 (1967)

28. Mahalanobis, P.C.: On the generalized distance in statistics. In: Proceedings of the National Institute of Sciences of India, vol. 2(1), pp. 49–55. Indian National Science Academy (1936)

29. Ackermann, M.R.: Algorithms for the Bregman k-Median Problem. PhD thesis, University of Paderborn, Department of Computer Science (2009)

Improved Methods For Generating Quasi-gray Codes[*]

Prosenjit Bose[1], Paz Carmi[2], Dana Jansens[1], Anil Maheshwari[1],
Pat Morin[1], and Michiel Smid[1]

[1] Carleton University, Ottawa ON, Canada
[2] Ben-Gurion University of the Negev

Abstract. Consider a sequence of bit strings of length d, such that each string differs from the next in a constant number of bits. We call this sequence a quasi-Gray code. We examine the problem of efficiently generating such codes, by considering the number of bits read and written at each generating step, the average number of bits read while generating the entire code, and the number of strings generated in the code. Our results give a trade-off between these constraints, and present algorithms that do less work on average than previous results, and that increase the number of bit strings generated.

Keywords: Gray codes, quasi-Gray codes, decision trees, counting, combinatorial generation.

1 Introduction

We are interested in efficiently generating a sequence of bit strings. The class of bit strings we wish to generate are cyclic quasi-Gray codes. A *Gray code* [3] is a sequence of bit strings, such that any two consecutive strings differ in exactly one bit. We use the term *quasi-Gray code* [2] to refer to a sequence of bit strings where any two consecutive strings differ in at most c bits, where c is a constant defined for the code. A Gray code (quasi-Gray code) is called *cyclic* if the first and last generated bit strings also differ in at most 1 bit (c bits).

We say a bit string that contains d bits has *dimension* d, and are interested in efficient algorithms to generate a sequence of bit strings that form a quasi-Gray code of dimension d. After generating a bit string, we say the algorithm's data structure corresponds exactly to the generated bit string, and it's *state* is the bit string itself. In this way, we restrict an algorithm's data structure to using exactly d bits. At each step, the input to the algorithm will be the previously generated bit string, which is the algorithm's previous state. The output will be a new bit string that corresponds to the state of the algorithm's data structure.

The number of consecutive unique bit strings generated is equal to the number of consecutive unique states for the generating data structure, and we call this value L, the *length* of the generated code. Clearly $L \leq 2^d$. We define the *space*

[*] Research supported by NSERC.

H. Kaplan (Ed.): SWAT 2010, LNCS 6139, pp. 224–235, 2010.
© Springer-Verlag Berlin Heidelberg 2010

efficiency of an algorithm as the ratio $L/2^d$, that is, the fraction of bit strings generated out of all possible bit strings given the dimension of the strings. When the space efficiency is 1 we call the data structure *space-optimal*, as it generates all possible bit strings. When $L < 2^d$ the structure is non-space-optimal.

We are concerned with the efficiency of our algorithms in the following ways. First, we would like to know how many bits the algorithm must read in the worst case in order to make the appropriate changes in the input string and generate the next bit string in the code. Second, we would like to know how many bits must change in the worst case to reach the successor string in the code, which must be a constant value to be considered a quasi-Gray code. Third, we examine the average number of bits read at each generating step. And last, we would like our algorithms to be as space efficient as possible, ideally generating as many bit strings as their dimension allows, with $L = 2^d$. Our results give a trade-off between these different goals.

Our decision to limit the algorithm's data structure to exactly d bits differs from previous work, where the data structure could use more bits than the strings it generated [2,5]. To compare previous results to our own, we consider the extra bits in their data structure to be a part of their generated bit strings. This gives a more precise view of the space efficiency of an algorithm.

Each generated bit string of dimension d has a distinct totally ordered rank in the generated code. Given a string of rank k in a code of length L, where $0 \le k < L$, we want to generate the bit string of rank $(k + 1) \mod L$.

We work within the bit probe model [4,5], counting the average-case and the worst-case number of bits read and written for each bit string generated. We use the Decision Assignment Tree (DAT) model [2] to construct algorithms for generating quasi-Gray codes and describe the algorithms' behaviour, as well as to discuss upper and lower bounds.

We use a notation for the iterated log function of the form $\log^{(c)} n$ where c is a non-negative whole number, and is always surrounded by brackets to differentiate it from an exponent. The value of the function is defined as follows. When $c = 0$, $\log^{(c)} n = n$. If $c > 0$, then $\log^{(c)}(n) = \log^{(c-1)}(\log(n))$. We define the function $\log^* n$ to be equal to the smallest non-negative value of c such that $\log^{(c)} n \le 1$. Throughout, the base of the log function is assumed to be 2 unless stated otherwise.

1.1   Results Summary

Our results, as well as previous results, are summarized in Table 1.

First, we present some space-optimal algorithms. Although our space-optimal algorithms read a small number of bits in the average case, they all read d bits in the worst case. In Section 3.1, we describe the Recursive Partition Gray Code (RPGC) algorithm, which generates a Gray code of dimension d while reading on average no more than $O(\log d)$ bits. This improves the average number of bits read for a space-optimal Gray code from d to $O(\log d)$. In Section 3.2, we use the RPGC to construct a DAT that generates a quasi-Gray code while reducing the average number of bits read. We then apply this technique iteratively in

Table 1. Summary of results. When "Worst-Case Bits Written" is a constant then the resulting code is a quasi-Gray code, and when it is 1, the code is a Gray code. $c \in \mathbb{Z}$ and t are constants greater than 0.

Dimension	Space Efficiency	Bits Read		Bits Written	Reference
		Average	Worst-Case	Worst-Case	
d	1	$2 - 2^{1-d}$	d	d	folklore
d	1	d	d	1	[2,3]
d	1	$O(\log d)$	d	1	Theorem 2
d	1	$O(\log^{(2c-1)} d)$	d	c	Theorem 3
d	1	17	d	$O(\log^* d)$	Corollary 1
$n+1$	1/2	$O(1)$	$\log n + 4$	4	[5]
$n + \log n$	$O(n^{-1})$	3	$\log n + 1$	$\log n + 1$	[6]
$n + \log n + 1$	$1/2 + O(n^{-1})$	4	$\log n + 1$	$\log n + 1$	[1]
$n + O(t \log n)$	$1 - O(n^{-t})$	$O(1)$	$O(t \log n)$	$O(t \log n)$	Theorem 4
$n + O(t \log n)$	$1 - O(n^{-t})$	$O(t \log n)$	$O(t \log n)$	3	Theorem 5
$n + O(t \log n)$	$1 - O(n^{-t})$	$O(\log^{(2c)} n)$	$O(t \log n)$	$2c + 1$	Theorem 6

Section 3.3 to create a d-dimensional DAT that reads on average $O(\log^{(2c-1)} d)$ bits, and writes at most c bits, for any constant $c \geq 1$. In section 3.4 we create a d-dimensional space-optimal DAT that reads at most 17 bits on average, and writes at most $O(\log^* d)$ bits. This reduces the average number of bits read to $O(1)$ for a space-optimal code, but increases the number of bits written to be slightly more than a constant.

Next, we consider quasi-Gray codes that are not space-optimal, but achieve space efficiency arbitrarily close to 1, and that read $O(\log d)$ bits in the worst case. In Section 3.5 we construct a DAT of dimension $d = n + O(t \log n)$ that reads and writes $O(t \log n)$ bits in the worst case, $O(1)$ on average, and has space efficiency $1 - O(n^{-t})$, for a constant $t > 0$. This improves the space efficiency dramatically of previous results where the worst-case number of bits written is $O(\log n)$. By combining a Gray code with this result, we produce a DAT of dimension $d = n + O(t \log n)$ that reads $O(t \log n)$ bits on average and in the worst case, but writes at most 3 bits. This reduces the worst-case number of bits written from $O(\log n)$ to $O(1)$. We then combine results from Section 3.3 to produce a DAT of dimension $d = n + O(t \log n)$ that reads on average $O(\log^{(2c)} n)$ bits, and writes at most $2c + 1$ bits, for any constant $c \geq 1$. This reduces the average number of bits read from $O(t \log d)$ to $O(\log \log d)$ when writing the same number of bits, and for each extra bits written, the average is further reduced by a logarithmic factor.

2   Decision Assignment Trees

In the Decision Assignment Tree (DAT) model, an algorithm is described as a binary tree. We use the DAT model to analyze algorithms for generating bit strings in a quasi-Gray code. We say that a DAT which reads and generates bit strings of length d has *dimension d*. Further, we refer to the bit string that the DAT reads and modifies as the *state* of the DAT between executions. Generally the initial state for a DAT will be the bit string 000...0.

Let T be a DAT of dimension d. Each internal node of T is labeled with a value $0 \leq i \leq d-1$ that represents reading bit i of the input bit string. The algorithm starts at the root of T, and reads the bit with which that node is labeled. Then it moves to a left or right child of that node, depending on whether the bit read was a 0 or a 1, respectively. This repeats recursively until a leaf node in the tree is reached. Each leaf node of T represents a subset of states where the bits read along the path to the leaf are in a fixed state. Each leaf node contains rules that describe which bits to update to generate the next bit string in the code. And each rule must set a single fixed bit directly to 0 or 1.

Under this model, we can measure the number of bits read to generate a bit string by the depth of the tree's leaves. We may use the average depth, weighted by the number of states in the generated code that reach each leaf, to describe the average number of bits read, and the tree's height to measure the worst-case number of bits read. The number of bits written can be measured by counting the rules in each leaf of the tree. Average and worst-case values for these can be found similarly. A trivial DAT, such as iterating through the standard binary representations of 0 to $2^d - 1$, in the worst case, will require reading and writing all d bits to generate the next bit string, but it may also read and write as few as one bit when the least-significant bit changes. On average, it reads and writes $2 - 2^{1-d}$ bits. Meanwhile, it is possible to create a DAT [2] that generates the Binary Reflected Gray Code [3]. This DAT would always write exactly one bit, but requires reading all d bits to generate the next bit string. This is because the least-siginificant bit is flipped if and only if the parity is even, which can only be determined by reading all d bits.

To generate a Gray code of dimension d with length $L = 2^d$, Fredman [2] shows that any DAT will require reading $\Omega(\log d)$ bits for some bit string. Fredman conjectures that for a Gray code of dimension d with $L = 2^d$, any DAT must read all d bits to generate at least one bit string in the code. That is, any DAT generating the code must have height d. This remains an open problem.[1]

3   Efficient Generation of Quasi-gray Codes

In this section we will address how to efficiently generate cyclic quasi-Gray codes of dimension d. First we present DATs that read up to d bits in the worst case, but read fewer bits on average. Then we present our lazy counters that read at

---

[1] In [5] the authors claim to have proven this conjecture true for "small" d by exhaustive search.

most $O(\log d)$ bits in the worst case and also read fewer bits on average, but with slightly reduced space-efficiency.

3.1 Recursive Partition Gray Code (RPGC)

We show a method for generating a cyclic Gray code of dimension d that requires reading an average of $6\log d$ bits to generate each successive bit string. First, assume that d is a power of two for simplicity. We use both an increment and decrement operation to generate the gray code, where these operations generate the next and previous bit strings in the code, respectively. Both increment and decrement operations are defined recursively, and they make use of each other. Pseudocode for these operations is provided in Algorithm 1 and 2.

Algorithm 1. RecurIncrement	**Algorithm 2.** RecurDecrement
Input: $b[]$, an array of n bits	**Input:** $b[]$, an array of n bits
1 **if** $n = 1$ **then**	1 **if** $n = 1$ **then**
2 **if** $b[0] = 1$ **then** $b[0] \leftarrow 0$;	2 **if** $b[0] = 1$ **then** $b[0] \leftarrow 0$;
3 **else** $b[0] \leftarrow 1$;	3 **else** $b[0] \leftarrow 1$;
4 **else**	4 **else**
5 Let $A = b[0...n/2 - 1]$;	5 Let $A = b[0...n/2 - 1]$;
6 Let $B = b[n/2...n - 1]$;	6 Let $B = b[n/2...n - 1]$;
7 **if** $A = B$ **then**	7 **if** $A = B + 1$ **then**
8 $RecurDecrement(B)$;	8 $RecurIncrement(B)$;
9 **else**	9 **else**
10 $RecurIncrement(A)$;	10 $RecurDecrement(A)$;
11 **end**	11 **end**
12 **end**	12 **end**

To perform an increment, we partition the bit string of dimension d into two substrings, A and B, each of dimension $d/2$. We then recursively increment A unless $A = B$, that is, unless the bits in A are in the same state as the bits in B, at which point we recursively decrement B.

To perform a decrement, we again partition a bit string of dimension d into two substrings, A and B, each of dimension $d/2$. We then recursively decrement A unless $A = B + 1$, that is, the bits of A are in the same state as the bits of B would be after they were incremented, at which time we recursively increment B instead.

Theorem 1. *Let $d \geq 2$ be a power of two. There exists a space-optimal DAT that generates a Gray code of dimension d, where generating the next bit string requires reading on average no more than $4\log d$ bits. In the worst case, d bits are read, and only 1 bit is written.*

Proof. The length L of the code is equal to the number of steps it takes to reach the initial state again. Because B is decremented if and only if $A = B$, then for each decrement of B, A will always be incremented $2^{d/2} - 1$ times in order to reach the state $A = B$ again. Thus the total number of states is the number of times B is decremented, plus the number of times A is incremented and $L = 2^{d/2} + (2^{d/2})(2^{d/2} - 1) = 2^d$. After L steps, the algorithm will output the bit string that was its initial input, creating a cyclic Gray code.

Since the RPGC has length $L = 2^d$, it is space-optimal, and the algorithm will be executed once for each possible bit string of dimension d. As such, we bound the average number of bits read by studying the expected number of bits read given a random bit string of dimension d. The proof is by induction on d. For the base case $d = 2$, in the worst case we read at most 2 bits, so the average bits read is at most $2 \le 4 \log d$. Then we assume it is true for all random bit strings $X \in \{0, 1\}^{d/2}$.

We define $|X|$ to denote the dimension of the bit string X. Let $C(A, B)$ be the number of bits read to determine whether or not $A = B$, where A and B are bit strings and $|A| = |B|$. Let $I(X)$ be the number of bits read to increment the bit string X. Let $D(X)$ be the number of bits read to decrement the bit string X. Note that since we are working in the DAT model, we read any bit at most one time, and $D(X) \le |X|$.

To finish the proof, we need to show that $\mathrm{E}[I(X)] \le 4 \log d$, when $X \in \{0, 1\}^d$ is a random bit strings. We can determine the expected value of $C(A, B)$ as follows. $C(A, B)$ must read two bits at a time, one from each of A and B, and compare them, only until it finds a pair that differs. Given two random bit strings, the probability that bit i is the first bit that differs between the two strings is $1/2^i$. If the two strings differ in bit i, then the function will read exactly i bits in each string. If $|A| = |B| = n$, then the expected value of $C(A, B)$ is $\mathrm{E}[C(A, B)] = 2 \sum_{i=1}^{n} i/2^i = (2^{n+1} - n - 2)/2^{n-1} = 4 - (n + 2)/2^{n-1}$.

Let $X = AB$, and $|A| = |B| = d/2$. Then $|X| = d$. For a predicate P, we define $\mathbf{1}_P$ to be the indicator random variable whose value is 1 when P is true, and 0 otherwise. Note that $I(A)$ is independent of $\mathbf{1}_{A=B}$ and $\mathbf{1}_{A \ne B}$. This is because the relation between A and B has no effect on the distribution of A (which remains uniform over $\{0, 1\}^{d/2}$).

The `RecurIncrement` operation only performs one increment or decrement action, depending on the condition $A = B$, thus the expected number of bits read by $I(X)$ is $\mathrm{E}[I(X)] = \mathrm{E}[C(A, B)] + \mathrm{E}[\mathbf{1}_{A=B} D(B)] + \mathrm{E}[\mathbf{1}_{A \ne B} I(A)] \le 4 - (n + 2)/2^{n-1} + (1/2^{d/2})(d/2) + (1 - 1/2^{d/2})\mathrm{E}[I(A)] \le 4 - (d/2 + 4)/2^{d/2} + 4 \log(d/2) \le 4 \log d$, as required.

The RPGC algorithm must be modified slightly to handle cases where d is not a power of two. We prove the following result in the full version of this paper.

Theorem 2. *Let $d \ge 2$. There exists a space-optimal DAT that generates a Gray code of dimension d, where generating the next bit string requires reading on average no more than $6 \log d$ bits. In the worst case, d bits are read, and only 1 bit is written.*

3.2 Composite Quasi-gray Code Construction

Lemma 1. *Let $d \ge 1$, $r \ge 3$, $w \ge 1$ be integers. Assume we have a space-optimal DAT for a quasi-Gray code of dimension d such that the following holds: Given* *a bit string of length d, generating the next bit string in the quasi-Gray code requires reading no more than r bits on average, and writing at most w bits in the worst case.*

Then there is a space-optimal DAT for a quasi-Gray code of dimension $d + \lceil \log r \rceil$, where generating each bit string requires reading at most $6 \log \lceil \log r \rceil + 3$ bits on average, and writing at most $w + 1$ bits. That is, the average number of bits read decreases from r to $O(\log \log r)$, while the worst-case number of bits written increases by 1.

Proof. We are given a DAT A that generates a quasi-Gray code of dimension d. We construct a DAT B for the RPGC of dimension d', as described in Section 3.1. B will read $O(\log d')$ bits on average, and writes only 1 bit in the worst case.

We construct a new DAT using A and B. The combined DAT generates bit strings of dimension $d + d'$. The last d' bits of the combined code, when updated, will cycle through the quasi-Gray code generated by B. The first d bits, when updated, will cycle through the code generated by A.

The DAT initially moves the last d' bits through $2^{d'}$ states according to the rules of B. When it leaves this final state, to generate the initial bit string of B again, the DAT also moves the first d bits to their next state according to the rules of A. During each generating step, the last d' bits are read and moved to their next state in the code generated by the rules of B, and checked to see if they have reached their initial position, which requires $6 \log d' + 2$ bits to be read on average and 1 bit to be written. However, the first d bits are only read and written when the last d' bits cycle back to their initial state - once for every $2^{d'}$ bit strings generated by the combined DAT.

If we let $d' = \lceil \log r \rceil$, then the RPGC B has dimension at least 2. Let r' be the average number of bits read by the d'-code. Then the average number of bits read by the combined quasi-Gray code is no more than $r' + 2 + r/2^{d'} \leq 6 \log d' + 2 + r/2^{\lceil \log r \rceil} \leq 6 \log \lceil \log r \rceil + 3$.

The number of bits written, in the worst case, is the number of bits written in DAT A and in DAT B together, which is at most $w + 1$.

3.3   RPGC-Composite Quasi-gray Code

We are able to use the RPGC from Theorem 2 with our Composite quasi-Gray code from Lemma 1 to construct a space-optimal DAT that generates a quasi-Gray code. By applying Lemma 1 to the RPGC, and then repeatedly applying it $c - 1$ more times to the resulting DAT, we create a DAT that generates a quasi-Gray code while reading on average no more than $6 \log^{(2c-1)} d + 14$ bits, and writing at most c bits to generate each bit string, for any constant $c \geq 1$.

Theorem 3. *Given integers d and $c \geq 1$, such that $\log^{(2c-1)} d \geq 14$. There exists a DAT of dimension d that generates a quasi-Gray code of length $L = 2^d$, where generating the next bit string requires reading on average no more than $6 \log^{(2c-1)} d + 14$ bits and writing in the worst case at most c bits.*

3.4 Reading a Constant Average Number of Bits

From Theorem 3, by taking c to be $O(\log^* d)$, it immediately follows that we can create a space-optimal DAT that reads a constant number of bits on average. This is a trade off, as the DAT requires writing at most $O(\log^* d)$ in the worst case, meaning the code generated by this DAT is no longer a quasi-Gray code.

Corollary 1. *For any integer $d > 24 + 2^{16}$, there exists a space-optimal DAT of dimension d that reads at most 17 bits on average, and writes no more than $\lfloor (\log^* d + 3)/2 \rfloor$ bits in the worst case.*

Proof. Let $d > 2^{16}$, and $c = \lfloor (\log^* d - 3)/2 \rfloor$. Then $c \geq 1$ and $\log^{(2c-1)} d \geq 14$ and it follows from Theorem 3 that there exists a space-optimal DAT of dimension d that reads on average at most $6 \log^{(\log^* d - 4)} d + 14 \leq 6 \cdot 2^{16} + 14$ bits, and writes at most $\lfloor (\log^* d - 1)/2 \rfloor$ bits.

Let $d > 19 + 2^{16}$, and $m = d - 19 > 2^{16}$ Use the DAT from our previous statement with Lemma 1, setting $r = 6 \cdot 2^{16} + 14$ and $w = \lfloor (\log^* d - 1)/2 \rfloor$. Then there exists a DAT of dimension $m + \lceil \log r \rceil = m + 19 = d$, that reads on average at most $6 \log \lceil \log r \rceil + 3 \leq 29$ bits and writes no more than $\lfloor (\log^* d + 1)/2 \rfloor$ bits.

If we apply this same technique again, we create a DAT of dimension $d > 24 + 2^{16}$ that reads on average at most 17 bits and writes no more than $\lfloor (\log^* d + 3)/2 \rfloor$ bits.

3.5 Lazy Counters

A lazy counter is a structure for generating a sequence of bit strings. In the first n bits, it counts through the standard binary representations of 0 to $2^n - 1$. However, this can require updating up to n bits, so an additional data structure is added to slow down these updates, making it so that each successive state requires fewer bit changes to be reached. We present a few known lazy counters, and then improve upon them, using our results to generate quasi-Gray codes.

Frandsen *et al.* [6] describe a lazy counter of dimension d that reads and writes at most $O(\log n) \leq O(\log d)$ bits for an increment operation. The algorithm uses $d = n + \log n$ bits, where the first n are referred to as b, and the last $\log n$ are referred to as i. A state in this counter is the concatenation of b and i, thus each state generates a bit string of dimension d. In the initial state, all bits in b and i are set to 0. The counter then moves through $2^{n+1} - 2$ states before cycling back to the initial state, generating a cyclic code.

The bits of b move through the standard binary numbers. However, moving from one such number to the next may require writing as many as n bits. The value in i is a pointer into b. For a standard binary encoding, the algorithm to move from one number to the next is as follows: starting at the right-most (least significant) bit, for each 1 bit, flip it to a 0 and move left. When a 0 bit is found, flip it to a 1 and stop. Thus the number of bit flips required to reach the next standard binary number is equal to one plus the position of the right-most 0. This counter simply uses i as a pointer into b such that it can flip a single 1 to a 0 each increment step until i points to a 0, at which point it flips the 0 to a

1, resets i to 0, and b has then reached the next standard binary number. The pseudocode is given in Algorithm 3.

Algorithm 3. LazyIncrement [6]
_____

    Input: $b[]$: an array of n bits; i: an integer of $\log n$ bits
1 **if** $b[i] = 1$ **then**
2      $b[i] \leftarrow 0$;
3      $i \leftarrow i + 1$;
4 **else**
5      $b[i] \leftarrow 1$;
6      $i \leftarrow 0$;
7 **end**

Lemma 2. *[6] There exists a DAT of dimension $d = n + \log n$, using the* `LazyIncrement` *algorithm, that generates $2^n - 1$ of a possible $n2^{n-1}$ bit strings, where in the limit $n \to \infty$ the space efficiency drops to 0. The DAT reads and writes in the worst case $\log n + 1$ bits to generate each successive bit string, and on average reads and writes 3 bits.*

An observation by Brodal [1] leads to a dramatic improvement in space efficiency over the previous algorithm by adding a single bit to the counter. This extra bit allows for the $\log n$ bits in i to spin through all their possible values, thus making better use of the bits and generating more bit strings with them. The variables b and i are unchanged from the counter in Lemma 2, and k is a single bit, making the counter have dimension $d = n + \log n + 1$. The pseudocode is given in Algorithm 4.

Algorithm 4. SpinIncrement [1]
_____

    Input: $b[]$: an array of n bits; i: an integer of $\log n$ bits; k: a single bit
1 **if** $k = 0$ **then**
2      $i \leftarrow i + 1$; // `spin i`
3   **if** $i = 0$ **then** $k \leftarrow 1$; // `the value of i has rolled over`
4 **else**
5   LazyIncrement($b[]$, i) ; // `really increment the counter`
6   **if** $i = 0$ **then** $k \leftarrow 0$;
7 **end**

Lemma 3. *[1] There exists a DAT of dimension $d = n + \log n + 1$, using the* `SpinIncrement` *algorithm, that generates $(n + 1)(2^n - 1)$ of a possible $2n2^n$ bit strings, where in the limit $n \to \infty$ the space efficiency converges to $1/2$. The DAT reads and writes in the worst case $\log n + 2$ bits to generate each successive bit string, and on average reads at most 4 bits.*

By generalizing the dimension of k, we are able to make the counter even more space efficient while keeping its $O(\log n) = O(\log d)$ worst-case bound for bits written and read. Let k be a bit array of dimension g, where $1 \leq g \leq t \log n$ and

$t > 0$. Then for a counter of dimension $d = n + \log n + g$, the new algorithm is given by Algorithm 5.

Algorithm 5. DoubleSpinIncrement

Input: $b[]$: an array of n bits; i: an integer of $\log n$ bits; k: an integer of g bits

1   **if** $k < 2^g - 1$ **then**
2      $i \leftarrow i + 1$; // happens in $(2^g - 1)/2^g$ of the cases
3   **if** $i = 0$ **then** $k \leftarrow k + 1$;
4   **else**
5   LazyIncrement($b[]$, i) ; // do a real increment
6   **if** $i = 0$ **then** $k \leftarrow 0$;
7   **end**

Theorem 4. *There exists a DAT of dimension $d = n + \log n + g$, where $1 \leq g \leq t \log n$ and $t > 0$, with space efficiency $1 - O(2^{-g})$. The DAT, using the* DoubleSpinIncrement *algorithm, reads and writes in the worst case $g + \log n + 1$ bits to generate each successive bit string, and on average reads and writes $O(1)$ bits.*

Proof. This counter generates an additional $2^g - 1$ states for each time it spins through the possible values of i. Thus the number of bit strings generated is $(2^g - 1)n(2^n - 1) + 2^n - 1$. Given the dimension of the counter, the possible number of bit strings generated is $n2^n 2^g$. When $g = 1$, we have exactly the same counter as given by Lemma 3. If $g > O(\log n)$, the worst-case number of bits read would increase. When $g = t \log n$, the space efficiency of this counter is
$$\frac{(n^{t+1} - n)(2^n - 1) + 2^n - 1}{n^{t+1}2^n} = 1 - O(n^{-t}).$$ Thus as n gets large, this counter becomes more space efficient, and is space-optimal as $n \to \infty$.

In the worst case, this counter reads and writes every bit in i and k, and a single bit in b, thus $g + \log n + 1 \leq (t + 1) \log n + 1$ bits. On average, the counter now reads and writes $O(1)$ bits. This follows from a similar argument to that made for Lemma 3, where each line modified in the algorithm still reads on average $O(1)$ bits.

Rahman and Munro [5] present a counter that reads at most $\log n + 4$ bits and writes at most 4 bits to perform an increment or decrement operation. The counter uses $n + 1$ bits to count through 2^n states, and has space efficiency $2^n/2^{n+1} = 1/2$. Compared to DoubleSpinIncrement, their counter writes fewer bits per generating step, but is less space efficient. By modifying our lazy counter to use Gray codes internally, the worst-case number of bits read remains asymptotically equivalent to the counter by Rahman and Munro, and the average number of bits we read increases. We are able to write a smaller constant number of bits per increment and retain a space efficiency arbitrarily close to 1.

We modify our counter in Theorem 4 to make i and k hold a cyclic Gray code instead of a standard binary number. The BRGC is a suitable Gray code for this purpose. Let $rank(j)$ be a function that returns the rank of the bit string j in the BRGC, and $next(j)$ be a function that returns a bit string k where $rank(k) = rank(j) + 1$. Algorithm 6 provides a lazy counter of dimension $d = n + \log n + g$, where $1 \leq g \leq t \log n$ and $t > 0$, that writes at most 3 bits, reads at most $g + \log n + 1$ bits to generate the next state, and has space efficiency arbitrarily close to 1.

Algorithm 6. WineIncrement[2]

Input: $b[]$: an array of n bits; i: a Gray code of $\log n$ bits; k: a Gray code of g bits

```
1  if k ≠ 100...00 then
2      i ← next(i) ; // happens in (2^g − 1)/2^g of the cases
3      if i = 0 then  k ← next(k);
4  else
5      if b[rank(i)] = 1 then
6          b[rank(i)] ← 0;
7          i ← next(i);
8          if i = 0 then  k ← 0 ; // wraps around to the initial state
9      else
10         b[rank(i)] ← 1;
11         k ← 0 ; // resets k to 0
12     end
13 end
```

Theorem 5. *There exists a DAT of dimension $d = n + \log n + g$, where $1 \leq g \leq t \log n$ and $t > 0$, with space efficiency $1 - O(2^{-g})$. The DAT, using the* WineIncrement *algorithm, reads in the worst case $g + \log n + 1$ bits and writes in the worst case 3 bits to generate each successive bit string, and on average reads at most $O(\log n + g)$ bits.*

While the previous counter reads at most $g + \log n + 1$ bits in the worst case, its average number of bits read is also $O(\log n)$. Using the quasi-Gray code counter from Theorem 3, we are able to bring the average number of bits read down as well. The worst case number of bits read remains $g + \log n + 1$, but on average, we only read at most $12 \log^{(2c)} n + O(1)$ bits, for any $c \geq 1$.

The algorithm does not need to change from its fourth iteration for these modifications. We simply make i a quasi-Gray code from Theorem 3 of dimension $\log n$ and k a similar quasi-Gray code of dimension $g \leq t \log n$, for a $t > 0$.

Theorem 6. *Let n be such that $\log^{(2c)} n \geq 14$ and g be such that $\log^{(2c-1)} g \geq 14$, for $1 \leq g \leq t \log n$ and $t > 0$. Then for any $c \geq 1$, there exists a DAT of*

---

[2] The name WineIncrement comes from the Caribbean dance known as Wineing, a dance that is centered on rotating your hips with the music. The dance is pervasive in Caribbean culture, and has been popularized elsewhere through songs and music videos such as Alison Hinds' "Roll It Gal", Destra Garcia's "I Dare You", and Fay-Ann Lyons' "Start Wineing".

dimension $d = n + \log n + g$ bits, using the `WineIncrement` algorithm, with space efficiency $1 - O(2^{-g})$, that reads in the worst case $g + \log n + 1$ bits, writes in the worst case $2c + 1$ bits, and reads on average no more than $12 \log^{(2c)} n + O(1)$ bits.

4   Conclusion

We have shown in this paper how to generate a Gray code, while reading significantly fewer bits on average than previously known algorithms, and how to efficiently generate a quasi-Gray code with the same worst-case performance and improved space efficiency. Our results give a tradeoff between space-optimality, and the worst-case number of bits written. This trade-off highlights the initial problem which motivated this work: a lower bound on the number of bits read in the worst case, for a space-optimal Gray code. After many hours and months spent on this problem, we are yet to find a tighter result than the $\Omega(\log d)$ bound shown by Fredman [2] (when d is more than a small constant). Our Recursive Partition Gray Code does provide a counter-example to any efforts to show a best-case bound of more than $\Omega(\log d)$, and our hope is that this work will contribute to a better understanding of the problem, and eventually, a tighter lower bound in the case of generating a space-optimal Gray code.

Acknowledgment. The fifth author would like to thank Peter Bro Miltersen for introducing us to this class of problems.

References

1. Brodal, G.S.: Personal Communication (2009)
2. Fredman, M.L.: Observations on the complexity of generating quasi-gray codes. Siam Journal of Computing 7(2), 134–146 (1978)
3. Gray, F.: Pulse Code Communications. U.S. Patent 2632058 (1953)
4. Minsky, M., Papert, S.: Perceptrons. MIT Press, Cambridge (1969)
5. Ziaur Rahman, M., Ian Munro, J.: Integer representation and counting in the bit probe model. Algorithmica (December 2008)
6. Frandsen, G.S., Miltersen, P.B., Skyum, S.: Dynamic word problems. J. ACM 44(2), 257–271 (1997)

The MST of Symmetric Disk Graphs Is Light

A. Karim Abu-Affash*, Rom Aschner*,
Paz Carmi*, and Matthew J. Katz**

Department of Computer Science, Ben-Gurion University, Beer-Sheva 84105, Israel
{abuaffas,romas,carmip,matya}@cs.bgu.ac.il

Abstract. Symmetric disk graphs are often used to model wireless communication networks. Given a set S of n points in $\mathbb{R}^d$ (representing n transceivers) and a transmission range assignment $r : S \rightarrow \mathbb{R}$, the *symmetric disk graph* of S (denoted $SDG(S)$) is the undirected graph over S whose set of edges is $E = \{(u,v) \,|\, r(u) \geq |uv|$ and $r(v) \geq |uv|\}$, where $|uv|$ denotes the Euclidean distance between points u and v. We prove that the weight of the MST of *any* connected symmetric disk graph over a set S of n points in the plane, is only $O(\log n)$ times the weight of the MST of the complete Euclidean graph over S. We then show that this bound is tight, even for points on a line.

Next, we prove that if the number of different ranges assigned to the points of S is only k, $k << n$, then the weight of the MST of $SDG(S)$ is at most $2k$ times the weight of the MST of the complete Euclidean graph. Moreover, in this case, the MST of $SDG(S)$ can be computed efficiently in time $O(kn \log n)$.

We also present two applications of our main theorem, including an alternative and simpler proof of the Gap Theorem, and a result concerning range assignment in wireless networks.

Finally, we show that in the non-symmetric model (where $E = \{(u,v) \,|\, r(u) \geq |uv|\}$), the weight of a minimum spanning subgraph might be as big as $\Omega(n)$ times the weight of the MST of the complete Euclidean graph.

1 Introduction

Symmetric disk graphs are often used to model wireless communication networks. Given a set S of n points in $\mathbb{R}^d$ (representing n transceivers) and a transmission range assignment $r : S \rightarrow \mathbb{R}$, the *symmetric disk graph* of S (denoted $SDG(S)$) is the undirected graph over S whose set of edges is $E = \{(u,v) \,|\, r(u) \geq |uv|$ and $r(v) \geq |uv|\}$, where $|uv|$ denotes the Euclidean distance between points u and v. If $r(u) \geq diam(S)$, for each $u \in S$, then $SDG(S)$ is simply the complete Euclidean graph over S. However, usually, the transmission ranges are much shorter than $diam(S)$.

---

* Partially supported by the Lynn and William Frankel Center for Computer Sciences.
** Partially supported by the MAGNET program of the Israel Ministry of Industry, Trade & Labor (CORNET consortium), and by the Lynn and William Frankel Center for Computer Sciences.

H. Kaplan (Ed.): SWAT 2010, LNCS 6139, pp. 236–247, 2010.

The Minimum Spanning Tree (MST) of a connected Euclidean graph G is an extremely important substructure of G. In the context of wireless networks, the MST is especially important. Besides its role in various routing protocols, it is also used to obtain good approximations when the problem being considered is NP-hard; see, e.g., the power assignment problem.

It is usually impossible to use the Euclidean MST of S (denoted $MST(S)$), under the symmetric disk graph model, simply because some of the edges of $MST(S)$ are not present in $SDG(S)$. Instead, it is natural to use the MST of $SDG(S)$ (denoted $MST_{SDG(S)}$). However, it is still desirable to (tightly) bound the approximation ratio also with respect to the weight of $MST(S)$ (and not only with respect to the weight of $MST_{SDG(S)}$). The main result of this paper makes this possible. We prove the following, somewhat surprising, theorem. For *any* set S of points in the plane and for *any* assignment of ranges to the points of S, such that $SDG(S)$ is connected, the weight (i.e., the sum of the edge lengths) of $MST_{SDG(S)}$ is $O(\log n)$ times the weight of $MST(S)$.

Disk graphs and especially unit disk graphs have received much attention, especially in the context of wireless networks. Notice that the unit disk graph of S is the symmetric disk graph of S that is obtained when $r(u) = 1$, for each point $u \in S$. The disk graph of S, on the other hand, is a directed graph, where there is an arc from u to v if $r(u) \geq |uv|$. Despite their importance, symmetric disk graphs have not received as much attention as (unit) disk graphs. Before describing our results concerning the MST of symmetric disk graphs, we mention two applications of our main result (stated above).

The Gap Theorem. The proof of our main result is based on a cute property of $MST_{SDG(S)}$ (see Lemma 1). This property also allows us to obtain an alternative and simpler proof of the, so called, Gap Theorem, stated and proved by Chandra et al. [4]. The gap theorem is used to show that the weight of the greedy spanner is $O(\log n)$ times the weight of $MST(S)$ [1,8].

Range assignment. A *range assignment* is an assignment of transmission ranges to each of the nodes of a network, so that the induced communication graph is connected and the total power consumption is minimized. The power consumed by a node v is $r(v)^{\alpha}$, where $r(v)$ is the range assigned to v and $\alpha \geq 1$ is some constant. The range assignment problem was first studied by Kirousis et al. [7], who did not impose any restriction on the potential transmission range of a node. They proved that the problem is NP-hard in three-dimensional space, assuming $\alpha = 2$. Subsequently, Clementi et al. [5] proved that the problem remains NP-hard in two-dimensional space. Kirousis et al. [7] also presented a simple 2-approximation algorithm, based on $MST(S)$.

It is more realistic to study the range assignment problem under the symmetric disk graph model. That is, the potential transmission range of a node u is bounded by some maximum range $r(u)$, and two nodes u, v can directly communicate with each other if and only if v lies within the range assigned to u and vise versa. The range assignment problem under this model was studied in [3,2]. Blough et al. [2] show that this version of the problem is also NP-hard in 2-dimensional and in 3-dimensional space. Our main theorem enables us, assuming $\alpha = 1$, to bound the weight of an optimal range assignment with limits on the ranges with respect to an optimal range assignment without such limits.

1.1   Our Results

In this paper, we prove several results concerning the minimum spanning tree of symmetric disk graphs. In Section 2, we prove that the weight of the MST of *any* connected symmetric disk graph $SDG(S)$ is bounded by $O(\log n)$ times the weight of $MST(S)$. Or, in our notation, $wt(MST_{SDG(S)}) = O(\log n) \cdot wt(MST(S))$. We also show that this bound is tight, in the sense that there exists a symmetric disk graph, such that $wt(MST_{SDG(S)}) = \Omega(\log n) \cdot wt(MST(S))$. If the ratio between the maximum range and minimum range is bounded by some constant, then we show that $wt(MST_{SDG(S)}) = O(wt(MST(S)))$. In Section 3, we consider the common case where the number of different ranges is only k, for $k << n$. We prove that in this case $wt(MST_{SDG(S)}) \leq 2k \cdot wt(MST(S))$. Moreover, we present an algorithm for computing $MST_{SDG(S)}$ in this case in time $O(kn \log n)$. In Section 4, we discuss the two applications mentioned above. In particular, we provide an alternative and simpler proof of the Gap Theorem. In Section 5, we consider disk graphs. We prove that the weight of a minimum spanning subgraph of a disk graph is bounded by $O(n)$ times the weight of $MST(S)$, and give an example where this bound is tight.

2   Symmetric Disk Graphs

Given a set S of n points in the plane and a function $r : S \to \mathbb{R}$, the *symmetric disk graph* of S, denoted $SDG(S)$, is the undirected graph over S whose set of edges is $E = \{(u, v) \mid r(u) \geq |uv| \text{ and } r(v) \geq |uv|\}$, where $|uv|$ denotes the Euclidean distance between points u and v. The *weight*, $wt(e)$, of an edge $e = (u, v) \in E$ is $|uv|$, and the weight, $wt(E')$, of $E' \subseteq E$ is $\sum_{e \in E'} wt(e)$.

We denote by $MST_{SDG(S)}$ the minimum spanning tree of $SDG(S)$. In this section, we show that $wt(MST_{SDG(S)}) = \Theta(\log n) \cdot wt(MST(S))$, where $MST(S)$ is the Euclidean minimum spanning tree of S (i.e., the minimum spanning tree of the complete Euclidean graph over S). More precisely, we show that if $SDG(S)$ is connected, then $wt(MST_{SDG(S)}) = O(\log n) \cdot wt(MST(S))$, and there exists a connected symmetric disk graph (over some set of points S) whose weight is $\Omega(\log n) \cdot wt(MST(S))$.

Lemma 1. *Let $SDG(S) = (S, E)$ be a symmetric disk graph over S. Let (a, b), $(c, d) \in E(MST_{SDG(S)})$ be two edges of $MST_{SDG(S)}$ that do not share an endpoint, such that $0 < |ab| \leq |cd|$. Then at most one edge from the set $A = \{(a, c), (b, c), (a, d), (b, d)\}$ is shorter than (a, b).*

Proof. Assume that there are two edges $e', e'' \in A$ that are shorter than (a, b). Since e' is shorter than (a, b) (and therefore also from (c, d)), it belongs to $SDG(S)$. Similarly, e'' belongs to $SDG(S)$. Therefore the edges e', e'' together

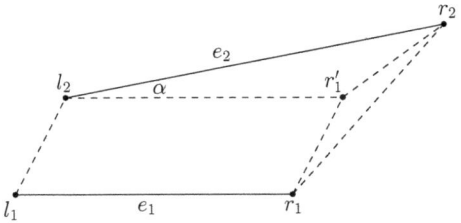

Fig. 1. Proof of Lemma 2

with $(a, b), (c, d)$ form a cycle in $SDG(S)$, implying that (a, b) or (c, d) is not in $E(MST_{SDG(S)})$ — a contradiction.

Lemma 2. *Let $e_1 = (l_1, r_1), e_2 = (l_2, r_2)$ be two edges of $MST_{SDG(S)}$, where r_i is to the right of l_i, $i = 1, 2$, such that (i) $1 \le \frac{|e_2|}{|e_1|} \le \frac{5}{4}$, and (ii) the difference α between the orientations of e_1 and e_2 is in the range $[0, \frac{\pi}{9}]$. Then $|l_1 l_2| \ge \frac{1}{2}|e_1|$.*

Proof. Assume that e_1 and e_2 do not share an endpoint. If they do, then the proof becomes much easier. Assume that $|l_1 l_2| < \frac{1}{2}|e_1|$. Let r_1' be the point to the right of l_2, such that (l_1, r_1) and (l_2, r_1') are parallel to each other and $|l_1 r_1| = |l_2 r_1'|$; see Figure 1. By the triangle inequality,

$$|l_1 l_2| + |r_1' r_2| = |r_1 r_1'| + |r_1' r_2| \ge |r_1 r_2| \,.$$

Since $|e_1| \le |e_2|$ and $|l_1 l_2| < |e_1|$, we know, by Lemma 1, that $|r_1 r_2| \ge |e_1|$. Thus, we get that

$$|r_1' r_2| \ge |e_1| - |l_1 l_2| > |e_1| - \frac{1}{2}|e_1| = \frac{1}{2}|e_1| \,.$$

By the law of cosines,

$$|r_1' r_2|^2 = |e_1|^2 + |e_2|^2 - 2|e_1||e_2| \cos(\alpha)$$

and therefore

$$|e_1|^2 + |e_2|^2 - 2|e_1||e_2| \cos(\alpha) > \frac{1}{4}|e_1|^2$$

or

$$\frac{3}{4} + \frac{|e_2|^2}{|e_1|^2} - 2\frac{|e_2|}{|e_1|} \cos(\alpha) > 0 \,. \tag{1}$$

We now show that this is impossible. Substituting $\frac{|e_2|}{|e_1|}$ with x in (1), we get $x^2 - 2\cos(\alpha)x + 3/4 > 0$. The solutions of the equation $x^2 - 2\cos(\alpha)x + 3/4 = 0$ are $x_{1,2} = \cos(\alpha) \pm \sqrt{\cos^2(\alpha) - \frac{3}{4}}$. Notice that since $0 < \alpha \le \frac{\pi}{9}$, we have $\cos(\alpha) > 37/40$, and therefore $x_1 > 5/4$ and $x_2 < 1$. Thus, for any x in the interval $[1, \frac{5}{4}]$, the left side of inequality (1) is non-positive. But this contradicts the assumption that $1 \le \frac{|e_2|}{|e_1|} \le 5/4$. We conclude that $|l_1 l_2| \ge \frac{1}{2}|e_1|$.

We are ready to prove our main theorem.

Theorem 1. *(SDG theorem)*

1. *The weight of the minimum spanning tree of a connected symmetric disk graph over a set S of n points in the plane is $O(\log n) \cdot wt(MST(S))$, where $MST(S)$ is the Euclidean minimum spanning tree of S.*
2. *There exists a set S of n points on a line, such that $wt(MST_{SDG(S)}) = \Omega(\log n) \cdot wt(MST(S))$.*

We prove the first part (i.e., the upper bound) in Section 2.1, and the second part (i.e., the lower bound) in Section 2.2.

2.1   Upper Bound

Let $SDG(S) = (S, E)$ be a connected symmetric disk graph over a set S of n points in the plane. We prove that $wt(MST_{SDG(S)}) = O(\log n) \cdot wt(MST(S))$, where $MST(S)$ is the Euclidean minimum spanning tree of S.

We partition the edge set of $MST_{SDG(S)}$ into two subsets. Let $E' = \{e \in MST_{SDG(S)} \mid |e| > wt(MST(S))/n\}$ and let $E'' = \{e \in MST_{SDG(S)} \mid |e| \leq wt(MST(S))/n\}$. Since $MST_{SDG(S)}$ has $n - 1$ edges,

$$wt(E'') = \sum_{e \in E''} |e| \leq (n - 1) \cdot \frac{wt(MST(S))}{n} < wt(MST(S)) \, .$$

In order to bound $wt(E')$, we divide the edges of E' into $k \geq 9$ classes $\{C_1, \ldots, C_k\}$, according to their orientation (which is an angle in the range $(-\pi/2, \pi/2]$). Within each class we divide the edges into $O(\log n)$ buckets, according to their length. For $1 \leq i \leq k$ and $1 \leq j \leq \log_p n$, where $p = 5/4$, let

$$B_{i,j} = \left\{ e \in E' \cap C_i \,\middle|\, |e| \in \left(\frac{wt(MST(S))}{n} \cdot p^{j-1}, \frac{wt(MST(S))}{n} \cdot p^j \right] \right\} \, .$$

(Notice that for each $e \in E', |e| \leq diam(S) \leq wt(MST(S))$.) Finally, let $S_{i,j} = \{s \in S \mid (s, t) \in B_{i,j}\}$.

Let $s, s' \in S_{i,j}$, and let $t, t' \in S$ such that $e = (s, t)$ and $e' = (s', t')$ are edges in $B_{i,j}$. Since e, e' belong to the same class, the difference between their orientations is less than $\frac{\pi}{9}$, and since they also belong to the same bucket, we may apply Lemma 2 and obtain that $|s, s'| \geq \frac{1}{2} \cdot \min\{|e|, |e'|\}$.

We now show that $wt(B_{i,j}) = O(wt(MST(S)))$. First notice that

$$wt(MST(S_{i,j})) \geq (|S_{i,j}| - 1) \cdot \min_{e \in MST(S_{i,j})} \{|e|\} \geq \frac{(|S_{i,j}| - 1)}{2} \cdot \min_{e \in B_{i,j}} \{|e|\} \, ,$$

and since for any $e_1, e_2 \in B_{i,j}$, $\min\{|e_1|, |e_2|\} \geq \frac{1}{p} \cdot \max\{|e_1|, |e_2|\}$ we get that

$$wt(MST(S_{i,j})) \geq \frac{(|S_{i,j}| - 1)}{2p} \cdot \max_{e \in B_{i,j}} \{|e|\} \geq \frac{1}{2p} \cdot \left(wt(B_{i,j}) - \max_{e \in B_{i,j}} \{|e|\} \right) \, .$$

Rearranging,

$$wt(B_{i,j}) \leq 2p \cdot wt(MST(S_{i,j})) + \max_{e \in B_{i,j}}\{|e|\} \leq 2p \cdot wt(MST(S_{i,j})) + p \cdot \min_{e \in B_{i,j}}\{|e|\}$$

$$\leq 2p \cdot wt(MST(S_{i,j})) + 2p \cdot \max_{e \in MST(S_{i,j})}\{|e|\} \leq 2p \cdot wt(MST(S_{i,j}))$$

$$+ 2p \cdot wt(MST(S_{i,j}))$$

$$\leq 4p \cdot wt(MST(S_{i,j})) \, .$$

Referring to $S \setminus S_{i,j}$ as Steiner points we get $wt(MST(S_{i,j})) \leq 2 \cdot wt(MST(S))$, and therefore $wt(B_{i,j}) \leq 8p \cdot wt(MST(S))$.

It follows that

$$wt(E') = \sum_{i=1}^{k} \sum_{j=1}^{\log_p n} wt(B_{i,j}) \leq 8pk \log_p n \cdot wt(MST(S)) \, , \text{ and}$$

$$wt(E) = wt(E') + wt(E'') \leq 8pk \log_p n \cdot wt(MST(S)) + wt(MST(S)$$
$$= O(\log n) \cdot wt(MST(S)).$$

A more delicate upper bound. We showed that the weight of the minimum spanning tree of a connected symmetric disk graph is bounded by $O(\log n) \cdot wt(MST(S))$, whereas the weight of the minimum spanning tree of a connected *unit* disk graph (UDG) is known to be $O(wt(MST(S)))$. A more delicate bound that depends also on r_{max} and r_{min}, the maximum and minimum ranges, bridges between the two upper bounds.

This bound is obtained by changing the proof of Theorem 1 in the following manner. Let $l_1 = \max\{r_{min}, wt(MST(S))/n\}$ and $l_2 = \min\{r_{max}, wt(MST(S))\}$. Define $E'' = \{e \in MST_{SDG(S)} \,|\, |e| \leq l_1\}$, and $E' = \{e \in MST_{SDG(S)} \,|\, |e| > l_1\}$. Now, if $l_1 = r_{min}$, then we get that $E'' \subseteq E(MST(S))$ and therefore $wt(E'') \leq wt(MST(S))$, and if $l_1 = wt(MST(S))/n$, then we get that $wt(E'') \leq (n-1) \cdot wt(MST(S))/n < wt(MST(S))$. Thus, in both cases we get that $wt(E'') \leq wt(MST(S))$.

Concerning E', we slightly modify the division into buckets, so that $B_{i,j} = \{e \in E' \cap C_i \,|\, |e| \in (l_1 p^{j-1}, l_1 p^j]\}$. Since the weight of any edge in $MST_{SDG(S)}$ is at most l_2, we get that the number of buckets is $\log_p(l_2/l_1)$. The asymptotic weight of each bucket remains $O(wt(MST(S)))$ (as before).

Therefore, the new bound on the weight of $MST_{SDG(S)}$ is $O(\log(l_2/l_1) + 1) \cdot wt(MST(S))$. The following theorem summarize our result.

Theorem 2

$$wt(MST_{SDG(S)}) = O(\log\left(\frac{\min\{r_{max}, wt(MST(S))\}}{\max\{r_{min}, wt(MST(S))/n\}}\right) + 1) \cdot wt(MST(S)) \, ,$$

where r_{max} and r_{min} are the maximum and minimum ranges, respectively. In particular, if (r_{max}/r_{min}) is bounded by some constant, then $wt(MST_{SDG(S)}) = O(wt(MST(S)))$.

2.2 Lower Bound

Consider the following set of $n + 1$ points $S = (v_0, v_1, \ldots, v_n)$ on a line, where $n = 2^k$ for some positive integer k. The distance between two adjacent points v_i and v_{i+1} is $1 + i\varepsilon$, where $\varepsilon = O(1/n)$, for $i = 0, \ldots, n - 1$. We assign a range $r(v_i)$ to each of the points $v_i \in S$; see Figure 2.

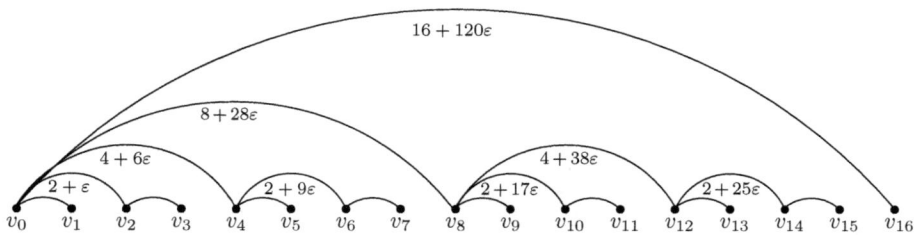

Fig. 2. Theorem 1 — the lower bound

Set
$$r(v_0) = n + \frac{(n-1)n}{2}\varepsilon .$$

That is, v_0's range is the distance between the two extreme points v_0 and v_n. For $i \neq 0$, let $m = 2^l$ be the largest power of two that divides i. Set

$$r(v_i) = m + \frac{m(2i - m - 1)}{2}\varepsilon .$$

Consider the induced symmetric disk graph, $SDG(S)$, depicted in Figure 2. $SDG(S)$ is a tree and therefore $MST_{SDG(S)}$ is simply $SDG(S)$.

$$wt(MST_{SDG(S)}) > \frac{n}{2} \cdot 1 + \frac{n}{4} \cdot 2 + \frac{n}{8} \cdot 4 + \cdots + \frac{n}{n} \cdot \frac{n}{2} = \frac{n}{2} \cdot \log n = \Omega(n \log n) .$$

On the other hand, $MST(S)$ is simply the path $v_0, v_1, \ldots, v_n$, and therefore

$$wt(MST(S)) = n + \sum_{i=1}^{n-1} i\varepsilon = n + \frac{(n-1)n}{2}\varepsilon = O(n) .$$

Therefore, in this example, $wt(MST_{SDG(S)}) = \Omega(\log n) \cdot wt(MST(S))$.

3 k-Range Symmetric Disk Graphs

In this section we consider the common case where the number of different ranges assigned to the points of S is only k, for $k \ll n$. That is, the function $r : S \to \mathbb{R}$ assumes only k different values, denoted $r_1 < r_2 < \cdots < r_k$. We first prove that in this case the weight of the minimum spanning tree of $SDG(S)$ is at most $2k \cdot wt(MST(S))$. Next, we present an efficient $O(kn \log n)$ algorithm for computing this minimum spanning tree. Thus, assuming k is some constant, we get that $wt(MST_{SDG(S)}) = O(wt(MST(S)))$ and $MST_{SDG(S)}$ can be constructed in time $O(n \log n)$.

3.1   The Weight of the Minimum Spanning Tree

Let $SDG(S)$ be a k-range symmetric disk graph. Let $MST_{SDG(S)}$ be the minimum spanning tree of $SDG(S)$, and let E be the set of edges of $MST_{SDG(S)}$. We divide the edges of E into k subsets according to their length. Notice that, by definition, the length of the longest edge in E is at most r_k. Put $r_0 = 0$, and let $E_i = \{e \in E \mid r_{i-1} < |e| \le r_i\}$, for $i = 1, \ldots, k$. Also, let $S_i = \{v \in S \mid r_i \le r(v)\}$, for $i = 1, \ldots, k$. Then, $S = S_1 \supseteq S_2 \supseteq \cdots \supseteq S_k$.

Claim. $E_i \subseteq E(MST(S_i))$, for $i = 1, \ldots, k$.

Proof. Let $e = (u, v) \in E_i$. We first observe that $u, v \in S_i$. Indeed, since $e \in E(SDG(S))$, we know that $r(u) \ge |e|$ and $r(v) \ge |e|$. Now, since $|e| > r_{i-1}$, it follows that $r(u) \ge r_i$ and $r(v) \ge r_i$ and therefore $u, v \in S_i$.

Let u', v' be any two vertices in S_i, such that $|u'v'| < |e|$. Then, $e' = (u', v') \in E(SDG(S))$ (since $r(u') \ge r_i \ge |e| > |e'|$ and $r(v') \ge r_i \ge |e| > |e'|$). Thus, in the construction of $MST_{SDG(S)}$ by Kruskal's algorithm (see [6]), the edge e', as well as all other edges of $SDG(S)$ with both endpoints in S_i and shorter than e, were considered before e. Nevertheless, e was selected, since there was still no path between u and v. Therefore, in the construction of $MST(S_i)$, when e is considered, there is still no path between its endpoints, and it is selected, i.e., $e \in E(MST(S_i))$.

Theorem 3. $wt(MST_{SDG(S)}) \le 2k \cdot wt(MST(S))$.

Proof. For each $1 \le i \le k$, $E_i \subseteq MST(S_i)$ (by the claim above), and therefore $wt(E_i) \le wt(MST(S_i))$. Referring to $S \setminus S_i$ as Steiner points, we get that $wt(E_i) \le 2 \cdot wt(MST(S))$. Thus, $wt(MST_{SDG(S)}) = \sum_{i=1}^{k} wt(E_i) \le \sum_{i=1}^{k} 2 \cdot wt(MST(S)) = 2k \cdot wt(MST(S))$.

3.2   Constructing the Minimum Spanning Tree

We describe below an $O(kn \log n)$ algorithm for computing the minimum spanning tree of a k-range symmetric disk graph. The algorithm applies Kruskal's minimum spanning tree algorithm to a subset of the edges of $SDG(S)$ (see [6]). We then prove that the subset E of edges that were selected by Kruskal's algorithm is $E(MST_{SDG(S)})$.

Lemma 3. $E = E(MST_{SDG(S)})$.

Proof. We prove that $E(MST_{SDG(S)}) \subseteq E$, and since the algorithm assures that E does not contain any cycle (line 9), we conclude that $E(MST_{SDG(S)}) = E$.

Let $e = (u, v) \in E(MST_{SDG(S)})$ and let i, $1 \le i \le k$, such that $r_{i-1} < |e| \le r_i$. (Recall that $r_0 = 0$.) Since e is an edge of $SDG(S)$, we have that $r(u) \ge |e|$ and $r(v) \ge |e|$, implying that $r(u), r(v) \ge r_i$ and therefore $u, v \in S_i$.

We show that $e \in E(MST(S_i))$. Assume that $e \notin E(MST(S_i))$, then $E(MST(S_i)) \cup \{e\}$ contains a cycle C. For each $e' \in C$, $e' \ne e$, we have that

---

Algorithm 1. Computing the MST of a k-range symmetric disk graph

---

Input: S; $r_1 < r_2 < \cdots < r_k$; $r : S \rightarrow \{r_1, \ldots, r_k\}$
Output: $MST_{SDG(S)}$
  1: $E \leftarrow \emptyset$
  2: **for** $i = 1$ to k **do**
  3:     $S_i \leftarrow \emptyset; E_i \leftarrow \emptyset$
  4:     **for each** $s \in S$ such that $r(s) \geq r_i$ **do**
  5:         $S_i \leftarrow S_i \cup \{s\}$
  6:     $DT(S_i) \leftarrow$ Delaunay triangulation of S_i
  7:     **for each** $e \in DT(S_i)$ such that $r_{i-1} < |e| \leq r_i$ **do**
  8:         $E_i \leftarrow E_i \cup \{e\}$
  9: $E \leftarrow Kruskal(\bigcup_{i=1}^{k} E_i)$
 10: **return** E

$|e'| < |e| \leq r_i$. That is, e' is an edge of $SDG(S_i)$, and therefore also an edge of $SDG(S)$. Now, assume we apply Kruskal's minimum spanning tree algorithm to $SDG(S)$. Then, since each of the edges of $C - \{e\}$ is considered before e, e is not selected as an edge of $MST_{SDG(S)}$ — a contradiction.

Since $E(MST(S_i)) \subseteq E(DT(S_i))$, e also belongs to $E(DT(S_i))$, and therefore (since $r_{i-1} < |e| \leq r_i$) $e \in E_i$. To complete the proof, notice that $E_i \subseteq E(SDG(S))$, for $i = 1, \ldots, k$, and since, by assumption, $e \in E(MST_{SDG(S)})$, we conclude that $e \in Kruskal(\bigcup_{i=1}^{k} E_i) = E$.

Theorem 4. *The minimum spanning tree of a k-range symmetric disk graph of n points can be computed in time $O(kn \log n)$.*

Proof. In each of the k iterations of the main loop, we compute the Delaunay triangulation of a subset of S. This can be done in $O(n \log n)$ time. Finally, we apply Kruskal's algorithm to a set of size $O(kn)$. Thus, the total running time is $O(kn \log n)$.

4   Applications

4.1   An Alternative Proof of the Gap Theorem

Let $w \geq 0$ be a real number, and let E be a set of *directed* edges in $\mathbb{R}^d$. We say that E has the *w-gap property* if for any two distinct edges (p, q) and $(r, s) \in E$, we have $|pr| > w \cdot \min\{|pq|, |rs|\}$. The gap property was introduced by Chandra et al.[4], who also proved the Gap Theorem; see below. The gap theorem bounds the weight of any set of edges that satisfies the gap property.

Theorem 5. *(Gap Theorem) Let $w > 0$, let S be a set of n points in $\mathbb{R}^d$, and let $E \subseteq S \times S$ be a set of directed edges that satisfies the w-gap property. Then $wt(E) = O(\log n) \cdot wt(MST(S))$.*

Chandra et al. use in their proof a shortest traveling salesperson tour $TSP(S)$ of S. They charge the lengths of the edges in E to portions of $TSP(S)$, and prove that $wt(E) < (1+2/w) \log n \cdot wt(MST(S))$. We give an alternative, simpler, proof of the gap theorem, which is similar to the proof of the first part of Theorem 1.

We now present our proof. Let $E' = \{e \in E \mid wt(e) > wt(MST(S))/n\}$ and let $E'' = \{e \in E \mid wt(e) \leq wt(MST(S))/n\}$. Since $w > 0$, each point of S is the source of at most one edge of E, which implies that there are at most n edges in E. Therefore,

$$wt(E'') = \sum_{e \in E''} w(e) \leq n \cdot \frac{wt(MST(S))}{n} = wt(MST(S)) \,.$$

As for $wt(E')$, notice that for each $e \in E'$, $wt(e) \leq diam(S) \leq wt(MST(S))$. We divide the edges of E' into $\log n$ buckets according to their size. For $1 \leq i \leq \log n$, let $B_i = \{e \in E' \mid wt(e) \in (\frac{wt(MST(S))}{n} \cdot 2^{i-1}, \frac{wt(MST(S))}{n} \cdot 2^i]\}$ and let $S_i = \{s \in S \mid (s,t) \in B_i\}$.

Let $s, s' \in S_i$, and let $t, t' \in S$ such that $e = (s,t)$ and $e' = (s',t')$ are edges in B_i. By the w-gap property, $wt(s,s') > w \cdot \min\{wt(s,t), wt(s',t')\} \geq \frac{w}{2} \cdot \max\{wt(s,t), wt(s',t')\}$. We now show that $wt(B_i) \leq \frac{8}{w} \cdot wt(MST(S))$. We omit the details; however, this claim is very similar to the analogous claim in the proof of the first part of the SDG Theorem.

It follows that

$$wt(E') = \sum_{i=1}^{\log n} wt(B_i) \leq \frac{8}{w} \log n \cdot wt(MST(S)) \,, \text{ and therefore}$$

$$wt(E) \leq \frac{8}{w} \log n \cdot wt(MST(S)) + wt(MST(S)) = O(\log n) \cdot wt(MST(S)) \,.$$

4.2   Range Assignment

Let S be a set of n points in the plane (representing transceivers). For each $v_i \in S$, let r_i be the maximum transmission range of v_i, and put $r = (r_1, \ldots, r_n)$. The following problem is known as *The Range Assignment Problem*. Assign a transmission range d_i, $d_i \leq r_i$, to each of the points v_i of S, such that (i) the induced symmetric disk graph (using the ranges $d_1, \ldots, d_n$) is connected, and (ii) $\sum_{i=1}^{n} d_i$ is minimized. Below, we compute a range assignment, such that the sum of ranges of the assignment is bounded by $O(\log n)$ times the sum of ranges of an optimal assignment, computed under the assumption that $r_1 = \cdots = r_n = diam(S)$.

Let $SDG(S)$ be the symmetric disk graph of S. We first compute $MST_{SDG(S)}$. Next, for each $v_i \in S$, let d_i be the weight of the heaviest edge incident to v_i in $MST_{SDG(S)}$. Notice that the induced symmetric disk graph (using the ranges $d_1, \ldots, d_n$) is connected, since it contains $E(MST_{SDG(S)})$. It remains to bound the sum of ranges of the assignment with respect to $wt(MST(S))$, where $MST(S)$ is the Euclidean minimum spanning tree of S. Let $OPT(S)$ denote an optimal range assignment with respect to the complete Euclidean graph of S. It is easy to see that $wt(MST(S)) < wt(OPT(S)) < 2 \cdot wt(MST(S))$ (see Kirousis et al. [7]). Thus, $\sum_{i=1}^{n} d_i < 2 \cdot wt(MST_{SDG(S)}) = O(\log n) \cdot wt(MST(S)) = O(\log n) \cdot wt(OPT(S))$. (Of course, we also know that $\sum_{i=1}^{n} d_i < 2 \cdot wt(OPT_r(S))$, where $OPT_r(S)$ is an optimal range assignment with respect to $SDG(S)$.)

5   Disk Graphs

Given a set S of n points in $\mathbb{R}^d$ and a function $r : S \rightarrow \mathbb{R}$, the *Disk Graph* of S, denoted $DG(S)$, is a directed graph over S whose set of edges is $E = \{(u, v) \mid r(u) \geq |uv|\}$. In this section we show that, unlike symmetric disk graphs, the weight of a minimum spanning subgraph of a disk graph might be much bigger than that of $MST(S)$.

Notice that if the corresponding symmetric disk graph $SDG(S)$ is connected, then the weight of a minimum spanning subgraph of $DG(S)$ (denoted $MST_{DG(S)}$) is bounded by $wt(MST_{DG(S)}) \leq 2 \cdot wt(MST_{SDG(S)}) = O(\log n) \cdot wt(MST(S))$.

We now state the main theorem of this section.

Theorem 6. *Let $DG(S)$ be a strongly connected disk graph over a set S of n points in $\mathbb{R}^d$. Then, (i) $wt(MST_{DG(S)}) = O(n) \cdot wt(MST(S))$, and (ii) there exists a set of n points in the plane, such that $wt(MST_{DG(S)}) = \Omega(n) \cdot wt(MST(S))$.*

Proof. We first prove part (i), the upper bound, and then part (ii), the lower bound.

Upper bound. Since $\max_{e \in E(DG(S))}\{wt(e)\} \leq wt(MST(S))$ and since the number of edges in $MST_{DG(S)}$ is less than $2n$,

$$wt(MST_{DG(S)}) < 2n \cdot \max_{e \in E(DG(S))}\{wt(e)\} \leq 2n \cdot wt(MST(S)) = O(n) \cdot wt(MST(S)).$$

Lower bound. Consider the following set S of $n + 1$ points in the plane, where $n = 3k$ for some positive integer k. We place $\frac{2}{3}n + 1$ points on the line $y = 0$, such that the distance between any two adjacent points is $1 - \varepsilon$, where $0 < \varepsilon < 1/2$, and we place $\frac{1}{3}n$ points on the line $y = 1$, such that the distance between any two adjacent points is $2 - 2\varepsilon$; see Figure 3. For each point u on the top line, set $r(u) = 1$, and for each point v on the bottom line, except for the rightmost point s, set $r(v) = 1 - \varepsilon$. Set $r(s)$ so that it reaches all points on the top line.

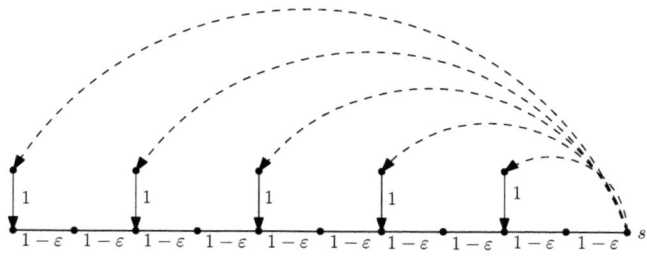

Fig. 3. Theorem 6 — the lower bound

We show that $wt(MST_{DG(S)}) = \Omega(n) \cdot wt(MST(S))$. First notice that the Euclidean minimum spanning tree of S has the shape of a comb, and therefore

$$wt(MST(S)) = \frac{2}{3}n \cdot (1 - \varepsilon) + \frac{1}{3}n \cdot 1 < n = O(n).$$

Next notice that for each point u on the top line, the minimum spanning subgraph of $DG(S)$ must include the edge (s, u), since this is the only edge that enters u. The total weight of these $n/3$ edges is at least $2+4+\cdots+2n/3 = \Omega(n^2)$. Therefore, $wt(MST_{DG(S)}) = \Omega(n^2) = \Omega(n) \cdot wt(MST(S))$.

References

1. Althöfer, I., Das, G., Dobkin, D., Joseph, D., Soares, J.: On sparse spanners of weighted graphs. Discrete and Computational Geometry 9(1), 81–100 (1993)
2. Blough, D.M., Leoncini, M., Resta, G., Santi, P.: On the symmetric range assignment problem in wireless ad hoc networks. In: TCS'02: Proc. of the IFIP 17th World Computer Congress – TC1 Stream / 2nd IFIP International Conference on Theoretical Computer Science, pp. 71–82 (2002)
3. Călinescu, G., Măndoiu, I.I., Zelikovsky, A.: Symmetric connectivity with minimum power consumption in radio networks. In: TCS'02: Proc. of the IFIP 17th World Computer Congress – TC1 Stream / 2nd IFIP International Conference on Theoretical Computer Science, pp. 119–130 (2002)
4. Chandra, B., Das, G., Narasimhan, G., Soares, J.: New sparseness results on graph spanners. Internat. J. of Computational Geometry and Applications 5, 125–144 (1995)
5. Clementi, A.E.F., Penna, P., Silvestri, R.: Hardness results for the power range assignment problem in packet radio networks. In: Hochbaum, D.S., Jansen, K., Rolim, J.D.P., Sinclair, A. (eds.) RANDOM 1999 and APPROX 1999. LNCS, vol. 1671, pp. 197–208. Springer, Heidelberg (1999)
6. Cormen, T.H., Leiserson, C.E., Rivest, R.L., Stein, C.: Introduction to Algorithms, 3rd edn. The MIT Press, Cambridge (2009)
7. Kirousis, L., Kranakis, E., Krizanc, D., Pelc, A.: Power consumption in packet radio networks. Theoretical Computer Science 243(1-2), 289–305 (2000)
8. Narasimhan, G., Smid, M.: Geometric Spanner Networks. Cambridge University Press, Cambridge (2007)

Vector Bin Packing with Multiple-Choice*
(Extended Abstract)

Boaz Patt-Shamir** and Dror Rawitz

School of Electrical Engineering,
Tel-Aviv University, Tel-Aviv 69978, Israel
{boaz,rawitz}@eng.tau.ac.il

Abstract. We consider a variant of *bin packing* called *multiple-choice vector bin packing*. In this problem we are given a set of n items, where each item can be selected in one of several D-dimensional *incarnations*. We are also given T bin types, each with its own cost and D-dimensional size. Our goal is to pack the items in a set of bins of minimum overall cost. The problem is motivated by scheduling in networks with guaranteed quality of service (QoS), but due to its general formulation it has many other applications as well. We present an approximation algorithm that is guaranteed to produce a solution whose cost is about $\ln D$ times the optimum. For the running time to be polynomial we require $D = O(1)$ and $T = O(\log n)$. This extends previous results for *vector bin packing*, in which each item has a single incarnation and there is only one bin type. To obtain our result we also present a PTAS for the multiple-choice version of *multidimensional knapsack*, where we are given only one bin and the goal is to pack a maximum weight set of (incarnations of) items in that bin.

1   Introduction

Bin packing, where one needs to pack a given set of items using the least number of limited-space containers (called bins), is one of the fundamental problems of combinatorial optimization (see, e.g., [1]). In the *multidimensional* flavor of bin packing, each item has sizes in several dimensions, and the bins have limited size in each dimension [2]. In this paper we consider a natural generalization of multidimensional bin packing that occurs frequently in practice, namely *multiple-choice* multidimensional bin packing. In this variant, items and space are multidimensional, and in addition, each item may be selected in one of a few *incarnations*, each with possibly different sizes in the different dimensions. Similarly, bins can be selected from a set of types, each bin type with its own size cap in each dimension, and possibly different cost. The problem is to select incarnations of the items and to assign them to bins so that the overall cost of bins is minimized.

---

* Research supported in part by the Next Generation Video Consortium, Israel.
** Supported in part by the Israel Science Foundation, Grant 664/05.

H. Kaplan (Ed.): SWAT 2010, LNCS 6139, pp. 248–259, 2010.

Multidimensionality models the case where the objects to pack have costs in several incomparable budgets. For example, consider a distribution network (e.g., a cable-TV operator), which needs to decide which data streams to provide. Streams typically have prescribed bandwidth requirements, monetary costs, processing requirements etc., while the system typically has limited available bandwidth, a bound on the amount of funds dedicated to buying content, bounded processing power etc. The multiple-choice variant models, for example, the case where digital objects (such as video streams) may be taken in one of a variety of formats with different characteristics (e.g., bandwidth and processing requirements), and similarly, digital bins (e.g., server racks) may be configured in more than one way. The multiple-choice multidimensional variant is useful in many scheduling applications such as communication under Quality of Service (QoS) constraints, and including work-plans for nursing personnel in hospitals [3].

Specifically, in this paper we consider the problem of *multiple-choice vector bin packing* (abbreviated MVBP, see Section 2 for a formal definition). The input to the problem is a set of n *items* and a set of T *bin types*. Each item is represented by at most m *incarnations*, where each incarnation is characterized by a D-dimensional vector representing the size of the incarnation in each dimension. Each bin type is also characterized by a D-dimensional vector representing the capacity of that bin type in each dimension. We are required to pack all items in the minimal possible number of bins, i.e., we need to select an incarnation for each item, select a number of required bins from each type, and give an assignment of item incarnations to bins so that no bin exceeds its capacity in any dimension. In the weighted version of this problem each bin type has an associated cost, and the goal is to pack item incarnations into a set of bins of minimum cost.

Before stating our results, we note that naïve reductions to the single-choice model do not work. For example, consider the case where $n/2$ items can be packed together in a single type-1 bin but require $n/2$ type-2 bins, while the other $n/2$ items fit together in a single type-2 bin but require $n/2$ type-1 bins. If one uses only one bin type, the cost is dramatically larger than the optimum—even with one incarnation per item. Regarding the choice of item incarnation, one may try to use only a cost-effective incarnation for each item (using some natural definition). However, it is not difficult to see that this approach results in approximation ratio $\Omega(D)$ even when there is only one bin type.

Our results. We present a polynomial-time approximation algorithm for the multiple-choice vector bin packing problem in the case where D (the number of dimensions) is a constant. The approximation ratio for the general weighted version is $\ln 2D + 3$, assuming that T (the number of bin types) satisfies $T = O(\log n)$. For the unweighted case, the approximation ratio can be improved to $\ln 2D + 1 + \varepsilon$, for any constant $\varepsilon > 0$, if $T = O(1)$ as well. Without any assumption on T, we can guarantee, in the unweighted case, cost of $(\ln(2D) + 1)\text{OPT} + T + 1$, where OPT denotes the optimal cost. To the best of our knowledge, this is the first approximation algorithm for the problem with multiple choice, and it is as good as the best solution for single-choice vector bin packing (see below).

As an aside, to facilitate our algorithm we consider the multiple-choice multi-dimensional *knapsack* problem (abbreviated MMK), where we are given a single bin and the goal is to load it with the maximum weight set of (incarnations of) items. We present a polynomial-time approximation scheme (PTAS) for MMK for the case where the dimension D is constant. The PTAS for MMK is used as a subroutine in our algorithm for MVBP.

Related work. Classical bin packing (BP) (single dimension, single choice) admits an asymptotic PTAS [4] and an asymptotic fully polynomial-time approximation scheme (asymptotic FPTAS) [5]. Friesen and Langston [6] presented constant factor approximation algorithms for a more general version of BP in which a fixed collection of bin sizes is allowed, and the cost of a solution is the sum of sizes of used bins. For more details about this version of BP see [7] and references therein. Correa and Epstein [8] considered BP with controllable item sizes. In this version of BP each item has a list of pairs associated with it. Each pair consists of an allowed size for this item, and a nonnegative penalty. The goal is to select a pair for each item so that the number of bins needed to pack the sizes plus the sum of penalties is minimized. Correa and Epstein [8] presented an asymptotic PTAS that uses bins of size slightly larger than 1.

Regarding multidimensionality, it has been long known that vector bin packing (VBP, for short) can be approximated to within a factor of $O(D)$ [9,4]. More recently, Chekuri and Khanna [10] presented an $O(\log D)$-approximation algorithm for VBP, for the case where D is constant. They also showed that approximating VBP for arbitrary dimension is as hard as graph coloring, implying that it is unlikely that VBP admits approximation factor smaller than $\sqrt{D}$. The best known approximation ratio for VBP is due to Bansal, Caprara and Sviridenko [11], who gave a polynomial-time approximation algorithm for constant dimension D with approximation ratio arbitrarily close to $\ln D + 1$. Our algorithm for MVBP is based on their ideas.

Knapsack admits an FPTAS [12,13]. Frieze and Clarke [14] presented a PTAS for the (single-choice) multidimensional variant of knapsack, but obtaining an FPTAS for multidimensional knapsack is NP-hard [15]. Shachnai and Tamir [16] use the approach of [14] to obtain a PTAS for a special case of 2-dimensional multiple-choice knapsack. Our algorithm for MMK extends their technique to the general case. MMK was studied extensively by practitioners. There are many specialized heuristics for MMK, see, e.g., [17,18,19,20]. Branch and bound techniques for MMK are studied in [21,22]. From the algorithmic viewpoint, the first relevant result is by Chandra et al. [23], who present a PTAS for single-dimension, multiple-choice knapsack. The reader is referred to [24] for a comprehensive treatment of knapsack problems.

Paper organization. The remainder of this paper is organized as follows. In Section 2 we formalize the problems. Our approximation algorithm for the MVBP problem is given in Section 3. This algorithm uses the PTAS for the MMK problem that is presented in Section 4.

2   Problem Statements

We now formalize the optimization problems we deal with. For a natural number n, let $[n] \stackrel{\text{def}}{=} \{1, 2, \ldots, n\}$ (we use this notation throughout the paper).

Multiple-Choice Multidimensional Knapsack problem (MMK).

Instance: A set of n *items*, where each item is a set of m or fewer D-dimensional *incarnations*. Incarnation j of item i has size $a_{ij} \in (\mathbb{R}^+)^D$, in which the dth dimension is a real number $a_{ijd} \geq 0$.
In the *weighted* version, each incarnation j of item i has weight $w_{ij} \geq 0$.
Solution: A set of incarnations, at most one of each item, such that the total size of the incarnations in each dimension d is at most 1.
Goal: Maximize the number (total weight) of incarnations in a solution.

When $D = m = 1$, this is the classical Knapsack problem (KNAPSACK).

Multiple-Choice Vector Bin Packing (MVBP).

Instance: Same as for unweighted MMK, with the addition of T *bin types*, where each bin type t is characterized by a vector $b_t \in (\mathbb{R}^+)^D$. The dth coordinate of b_t is called the *capacity* of type t in dimension d, and denoted by b_{td}.
In the *weighted* version, each bin type t has a weight $w_t \geq 0$.
Solution: A set of bins, each assigned a bin type and a set of item incarnations, such that exactly one incarnation of each item is assigned to any bin, and such that the total size of incarnations assigned to a bin does not exceed its capacity in any dimension
Goal: Minimize number (total weight) of assigned bins.

When $m = 1$ we get VBP, and the special case where $D = m = 1$ is the classical bin packing problem (BP).

3   Multiple-Choice Vector Bin Packing

It is possible to obtain approximation ratio of $O(\ln D)$ for MVBP by extending the technique of [10], under the assumption that bin types are of unit weight and that both D and T are constants. In this section we present a stronger result, namely an $O(\log D)$-approximation algorithm for MVBP, assuming that $D = O(1)$ and $T = O(\log n)$. Our algorithm is based on and extends the work of [11].

The general idea is as follows. We first encode MVBP using a covering linear program with exponentially many variables, but polynomially many constraints. We find a near optimal fractional solution to this (implicit) program using a separation oracle of the dual program. (The oracle is implemented by the MMK algorithm from Section 4.) We assign some incarnations to bins using a greedy rule based on some "well behaved" dual solution (the number of greedy assignments depends on the value of the solution to the primal program). Then we are left with a set of unassigned items, but due to our greedy rule we can assign these remaining items to a relatively small number of bins.

3.1   Covering Formulation

We start with the transformation of MVBP to weighted Set Cover (SC). An instance of SC is a family of sets $\mathcal{C} = \{C_1, C_2, \ldots\}$ and a cost $w_C \geq 0$ for each $C \in \mathcal{C}$. We call $\bigcup_{C \in \mathcal{C}} C$ the *ground set* of the instance, and usually denote it by I. The goal in SC is to choose sets from $\mathcal{C}$ whose union is I and whose overall cost is minimal. Clearly, SC is equivalent to the following integer program:

$$\min \sum_{C \in \mathcal{C}} w_C \cdot x_C$$
$$\text{s.t. } \sum_{C \ni i} x_C \geq 1 \quad \forall i \in I \tag{P}$$
$$x_C \in \{0, 1\} \quad \forall C \in \mathcal{C}$$

where x_C indicates whether the set C is in the cover. A linear program relaxation is obtained by replacing the integrality constraints of (P) by positivity constraints $x_C \geq 0$ for every $C \in \mathcal{C}$. The above formulation is very general. We shall henceforth call problems whose instances can be formulated as in (P) for some $\mathcal{C}$ and w_C values, *(P)-problems*.

In particular, MVBP is a (P)-problem, as the following reduction shows. Let $\mathcal{I}$ be an instance of MVBP. Construct an instance $\mathcal{C}$ of SC as follows. The ground set of $\mathcal{C}$ is the set of items in $\mathcal{I}$, and sets in $\mathcal{C}$ are the subsets of items that can be assigned to some bin. Formally, a set C of items is called *compatible* if and only if there exists a bin type t and an incarnation mapping $f : C \to [m]$ such that $\sum_{i \in C} a_{if(i)d} \leq b_{td}$ for every dimension d, i.e., if there is a way to accommodate all members of C is the same bin. In the instance of SC, we let $\mathcal{C}$ be the collection of all compatible item sets. Note that a solution to set cover does not immediately solve MVBP, because selecting incarnations and bin-types is an NP-hard problem in its own right. To deal with this issue we have one variable for each possible *assignment* of incarnations and bin type. Namely, we may have more than one variable for a compatible item subset.

3.2   Dual Oblivious Algorithms

We shall be concerned with approximation algorithms for (P)-problems which have a special property with respect to the dual program. First, we define the dual to the LP-relaxation of (P):

$$\max \sum_{i \in I} y_i$$
$$\text{s.t. } \sum_{i \in C} y_i \leq w_C \quad \forall C \in \mathcal{C} \tag{D}$$
$$y_i \geq 0 \quad \forall i \in I$$

Next, for an instance $\mathcal{C}$ of set cover and a set S, we define the *restriction of $\mathcal{C}$ to S* by $\mathcal{C}|_S \stackrel{\text{def}}{=} \{C \cap S \mid C \in \mathcal{C}\}$, namely we project out all elements not in S. Note that for any S, a solution to $\mathcal{C}$ is also a solution to $\mathcal{C}|_S$: we may only discard some of the constraints in (P). We now arrive at our central concept.

Definition 1 (Dual Obliviousness). *Let Π be a* (P)-*problem. An algorithm A for Π is called ρ-dual oblivious if there exists a constant δ such that for every instance $\mathcal{C} \in \Pi$ there exists a dual solution $y \in \mathbb{R}^n$ to* (D) *satisfying, for all $S \subseteq I$, that*

$$A(\mathcal{C}|_S) \leq \rho \cdot \sum_{i \in S} y_i + \delta \ .$$

Let us show that the First-Fit (FF) heuristic for BP is dual oblivious (we use this property later). In FF, the algorithm scans the items in arbitrary order and places each item in the left most bin which has enough space to accommodate it, possibly opening a new bin if necessary. A newly open bin is placed to the right of rightmost open bin.

Observation 1. *First-Fit is a 2-dual oblivious algorithm for bin packing.*

Proof. In any solution produced by FF, all non-empty bins except perhaps one are more than half-full. Furthermore, this property holds throughout the execution of FF, and regardless of the order in which items are scanned. It follows that if we let $y_i = a_i$, where a_i is the size of the ith item, then for every $S \subseteq I$ we have $\text{FF}(S) \leq \max\{2 \sum_{i \in S} y_i, 1\} \leq 2 \sum_{i \in S} y_i + 1$, and hence FF is dual oblivious for BP with $\rho = 2$ and $\delta = 1$. $\square$

The usefulness of dual obliviousness is expressed in the following result. Let Π be a (P)-problem, and suppose that APPR is a ρ-dual oblivious algorithm for Π. Suppose further that we can efficiently find the dual solution y promised by dual obliviousness. Under these assumptions, Algorithm 1 below solves any instance $\mathcal{C}$ of Π.

Algorithm 1

1: Find an optimal solution x^* to (P). Let OPT* denote its value.
2: Let $\mathcal{C}^+ = \{C \ : \ x_C^* > 0\}$. Let $\mathcal{G} \leftarrow \emptyset, S \leftarrow I$.
3: **while** $\sum_{C \in \mathcal{G}} w_c < \ln \rho \cdot \text{OPT}^*$ **do**
4:     Find $C \in \mathcal{C}^+$ for which $\dfrac{1}{w_C} \sum_{i \in S \setminus C} y_i$ is maximized;
5:     $\mathcal{G} \leftarrow \mathcal{G} \cup \{C\}, S \leftarrow S \setminus C$.
6: **end while**
7: Apply APPR to the residual instance S, obtaining solution $\mathcal{A}$.
8: **return** $\mathcal{G} \cup \mathcal{A}$.

We bound the weight of the solution $\mathcal{G} \cup \mathcal{A}$ that is computed by Algorithm 1.

Theorem 1. *Let Π be a* (P)-*problem. Then for any instance of Π with optimal fractional solution* OPT*, *Algorithm 1 outputs $\mathcal{G} \cup \mathcal{A}$ satisfying*

$$w(\mathcal{G} \cup \mathcal{A}) \leq (\ln \rho + 1)\text{OPT}^* + \delta + w_{\max} \ ,$$

where $w_{\max} = \max_t w_t$.

Proof. Clearly, $w(\mathcal{G}) < \ln \rho \cdot \text{OPT}^* + w_{\max}$. It remains to bound the weight of $\mathcal{A}$. Let S' be the set of items not covered by $\mathcal{G}$. We prove that $\sum_{i \in S'} y_i \leq \frac{1}{\rho} \sum_{i \in I} y_i$, which implies

$$w(\mathcal{A}) \; \leq \; \rho \sum_{i \in S'} y_i + \delta \; \leq \; \rho e^{-\ln \rho} \sum_{i=1}^{n} y_i + \delta \; \leq \; \text{OPT}^* + \delta \;,$$

proving the theorem.

Let $C_k \in \mathcal{C}^+$ denote the kth subset added to $\mathcal{G}$ during the greedy phase, and let $S_k \subseteq I$ be the set of items not covered after the kth subset was chosen. Define $S_0 = I$. We prove, by induction on $|\mathcal{G}|$, that for every k,

$$\sum_{i \in S_k} y_i \leq \prod_{q=1}^{k} \left(1 - \frac{w_{C_q}}{\text{OPT}^*}\right) \cdot \sum_{i \in I} y_i \tag{1}$$

For the base case we have trivially $\sum_{i \in S_0} y_i \leq \sum_{i \in I} y_i$. For the inductive step, assume that

$$\sum_{i \in S_{k-1}} y_i \leq \prod_{q=1}^{k-1} \left(1 - \frac{w_{C_q}}{\text{OPT}^*}\right) \cdot \sum_{i \in I} y_i \;.$$

By the greedy rule and the pigeonhole principle, we have that

$$\frac{1}{w_{C_k}} \sum_{i \in S_{k-1} \cap C_k} y_i \geq \frac{1}{\text{OPT}^*} \sum_{i \in S_{k-1}} y_i \;.$$

It follows that

$$\sum_{i \in S_k} y_i = \sum_{i \in S_{k-1}} y_i - \sum_{i \in S_{k-1} \cap C_k} y_i \leq \left(1 - \frac{w_{C_k}}{\text{OPT}^*}\right) \sum_{i \in S_{k-1}} y_i \leq \prod_{q=1}^{k} \left(1 - \frac{w_{C_q}}{\text{OPT}^*}\right) \cdot \sum_{i \in I} y_i \;,$$

completing the inductive argument. The theorem follows, since by (1) we have

$$\sum_{i \in S'} y_i \; \leq \; (1 - \ln \rho / k)^k \cdot \sum_{i \in I} y_i \; \leq \; e^{-\ln \rho} \sum_{i \in I} y_i \;,$$

as required. $\qquad\square$

Note that if x^* can be found in polynomial time, and if APPR is a polynomial-time algorithm, then Algorithm 1 runs in polynomial time. Also observe that Theorem 1 holds even if x^* is not an optimal solution of (P), but rather a $(1 + \varepsilon)$-approximation. We use this fact later.

In this section we defined the notion of *dual obliviousness* of an algorithm. We note that Bansal et al. [11] defined a more general property of algorithms called *subset obliviousness*. (For example, a subset oblivious algorithm is associated with several dual solutions.) Furthermore, Bansal et al. showed that the asymptotic PTAS for BP from [4] with minor modifications is subset oblivious and used it to obtain a subset oblivious $(D + \varepsilon)$-approximation algorithm for MVBP. This paved the way to an algorithm for VBP, whose approximation guarantee is arbitrarily close to $\ln D + 1$. However, in the case of MVBP, using the above APTAS for BP (at least in a straightforward manner) would lead to a subset oblivious algorithm whose approximation guarantee is $(DT + \varepsilon)$. In the next section we present a $2D$-dual oblivious algorithm for weighted MVBP that is based on First-Fit.

3.3   Algorithm for Multiple-Choice Vector Bin Packing

We now apply the framework of Theorem 1 to derive an approximation algorithm for MVBP. There are several gaps we need to fill.

First, we need to solve (P) for MVBP, which consists of a polynomial number of constraints (one for each item), but an exponential number of variables. We circumvent this difficulty as follows. Consider the dual of (P). The *separation problem* of the dual program in our case is to find (if it exists) a subset C with $\sum_{i \in C} y_i > w_C$ for given item profits $y_1, \ldots, y_n$. The separation problem can therefore be solved by testing, for each bin type, whether the optimum is greater than w_t, which in turn is simply an MMK instance, for which we present a PTAS in Section 4. In other words, the separation problem of the dual (D) has a PTAS, and hence there exists a PTAS for the LP-relaxation of (P) [25,26].

Second, we need to construct a dual oblivious algorithm for MVBP. To do that, we introduce the following notation. For every item $i \in I$, incarnation j, dimension d, and bin type t we define the *load* of incarnation j of i on the dth dimension of bins of type by $\ell_{ijtd} = a_{ijd}/b_{td}$. For every item $i \in I$ we define the *effective load* of i as

$$\bar{\ell}_i = \min_{1 \le j \le m, 1 \le t \le T} \left\{ w_t \cdot \max_d \ell_{ijtd} \right\} .$$

Also, let $t(i)$ denote the bin type that can contain the most (fractional) copies of some incarnation of item i, where $j(i)$ and $d(i)$ are the incarnation and dimension that determine this bound. Formally:

$$j(i) = \operatorname*{argmin}_j \min_t \{ w_t \cdot \max_d \ell_{ijtd} \}$$

$$t(i) = \operatorname*{argmin}_t \{ w_t \cdot \max_d \ell_{ij(i)td} \}$$

$$d(i) = \operatorname*{argmax}_d \ell_{ij(i)t(i)d} .$$

Assume that $j(i)$, $t(i)$ and $d(i)$ are the choices of j, t and d that are taken in the definition of $\bar{\ell}_i$.

Our dual oblivious algorithm APPR for MVBP is as follows:

1. Divide the item set I into T subsets by letting $I_t \overset{\text{def}}{=} \{i \ : \ t(i) = t\}$.
2. Pack each subset I_t in bins of type t using FF, where the size of each item i is $a_{ij(i)d(i)}$.

Observe that the size of incarnation $j(i)$ of item i in dimension $d(i)$ is the largest among all other sizes of this incarnation. Hence, the solution computed by FF is feasible for I_t.

We now show that this algorithm is $2D$-dual oblivious.

Lemma 2. *Algorithm APPR above is a polynomial time $2D$-dual oblivious algorithm for MVBP.*

Proof. Consider an instance of MVBP with item set I, and let the corresponding set cover problem instance be $\mathcal{C}$. We show that there exists a dual solution $y \in \mathbb{R}^n$ such that for any $S \subseteq I$, $\text{APPR}(\mathcal{C}|_S) \leq 2D \cdot \sum_{i \in S} y_i + \sum_{t=1}^{T} w_t$. Define $y_i = \bar{\ell}_i/D$ for every i. We claim that y is a feasible solution to (D). Let $C \in \mathcal{C}$ be a compatible item set. C induces some bin type t, and an incarnation $j'(i)$ for each $i \in C$. Let $d'(i) = \text{argmax}_d \{a_{ij'(i)d}/b_{td}\}$, i.e., $d'(i)$ is a dimension of bin type t that receives maximum load from (incarnation $j'(i)$ of) item i. Then

$$
\sum_{i \in C} y_i = \sum_{d=1}^{D} \sum_{\substack{i \in C \\ i:d'(i)=d}} \frac{\bar{\ell}_i}{D} \leq \frac{1}{D} \sum_{d=1}^{D} \sum_{\substack{i \in C \\ i:d'(i)=d}} w_t \cdot \ell_{ij'(i)td}
$$

$$
= \frac{w_t}{D} \sum_{d=1}^{D} \sum_{\substack{i \in C \\ i:d'(i)=d}} \frac{a_{ij'(i)d}}{b_{td}} \leq \frac{w_t}{D} \sum_{d=1}^{D} \frac{1}{b_{td}} \cdot b_{td} = w_t \ ,
$$

where the last inequality follows from the compatibility of C.

Now, since FF computes bin assignments that occupy at most twice the sum of bin sizes, we have that $\text{FF}(I_t) \leq w_t \cdot \max\{2\sum_{i \in I_t} \bar{\ell}_i/w_t, 1\} \leq 2\sum_{i \in I_t} \bar{\ell}_i + w_t$. Hence, for every instance $\mathcal{I}$ of MVBP we have

$$
\text{APPR}(\mathcal{I}) = \sum_{t=1}^{T} \text{FF}(I_t) \leq \sum_{t=1}^{T} \left(2\sum_{i \in I_t} \bar{\ell}_i + w_t \right) = 2\sum_{i \in I} \bar{\ell}_i + \sum_{t=1}^{T} w_t
$$

$$
= 2D \sum_{i \in I} y_i + \sum_{t=1}^{T} w_t \leq 2D \cdot \text{OPT}^* + \sum_{t=1}^{T} w_t \ .
$$

Furthermore, for every $S \subseteq I$ we have

$$
\text{APPR}(\mathcal{C}|_S) = \sum_{t=1}^{T} \text{FF}(S \cap I_t) \leq 2\sum_{i \in S} \bar{\ell}_i + \sum_{t=1}^{T} w_t = 2D \sum_{i \in S} y_i + \sum_{t=1}^{T} w_t \ ,
$$

and we are done.   □

Based on Theorem 1 and Lemma 2 we obtain our main result.

Theorem 2. *If $D = O(1)$, then there exists a polynomial time algorithm for MVBP with T bin types that computes a solution whose size is at most*

$$
(\ln 2D + 1)\text{OPT}^* + \sum_{t=1}^{T} w_t + w_{\max} \ .
$$

Note that while the approximation ratio is logarithmic in D, the running time of the algorithm is exponential in D.

Theorem 2 implies the following result for unweighted MVBP:

Corollary 1. *If $D = O(1)$, then there exists a polynomial time algorithm for unweighted* MVBP *that computes a solution whose size is at most* $(\ln 2D + 1)$OPT$^* + T + 1$. *Furthermore, if $T = O(1)$, then there exists a polynomial time* $(\ln 2D + 1 + \varepsilon)$-*approximation algorithm for unweighted* MVBP, *for every $\varepsilon > 0$.*

We also have the following for weighted MVBP.

Corollary 2. *If $D = O(1)$ and $T = O(\log n)$, then there exists a polynomial time* $(\ln 2D + 3)$-*approximation algorithm for* MVBP.

Proof. The result follows from the fact that as we show, we may assume that $\sum_t w_t \leq$ OPT. In this case, due to Theorem 2 we have that the cost of the computed solution is at most

$$(\ln 2D + 1)\text{OPT}^* + \sum_{t=1}^{T} w_t + w_{\max} \leq (\ln 2D + 3)\text{OPT} .$$

The above assumption is fulfilled by the following wrapper for our algorithm: Guess which bin types are used in some optimal solution. Iterate through all $2^T - 1$ guesses, and for each guess, compute a solution for the instance that contains only the bin types in the guess. Output the best solution. Since our algorithm computes a $(\ln 2D + 3)$-approximate solution for the right guess, the best solution is also a $(\ln 2D + 3)$-approximation. □

4   Multiple-Choice Multidimensional Knapsack

In this section we present a PTAS for weighted MMK for the case where D is a constant. Our construction extends the algorithms of Frieze and Clarke [14] and of Shachnai and Tamir [16].

We first present a linear program of MMK, where the variables x_{ij} indicate whether the jth incarnation of the ith item is selected.

$$
\begin{aligned}
\max \ & \sum_{i=1}^{n} \sum_{j=1}^{m} w_{ij} x_{ij} \\
\text{s.t.} \ & \sum_{i=1}^{n} \sum_{j=1}^{m} a_{ijd} x_{ij} \leq 1 \quad && \forall d \in [D] \\
& \sum_{j=1}^{m} x_{ij} \leq 1 \quad && \forall i \in [n] \\
& x_{ij} \geq 0 \quad && \forall i \in [n], j \in [m]
\end{aligned}
\tag{MMK}
$$

The first type of constraints make sure that the load on the knapsack in each dimension is bounded by 1; the second type of constraints ensures that at most one copy of each element is taken into the solution. Constraints of the third type indicate the relaxation: the integer program for MMK requires that $x_{ij} \in \{0, 1\}$.

Our PTAS for MMK is based on the linear program (MMK). Let $\varepsilon > 0$. Suppose we somehow guess the heaviest q incarnations that are packed in the knapsack by some optimal solution, for $q = \min\{n, \lceil D/\varepsilon \rceil\}$. Formally, assume we are given a set $G \subseteq [n]$ of at most q items and a function $g : G \to [m]$ that selects incarnations of items in G. In this case we can assign values to some variables of (MMK) as follows:

$$x_{ij} = \begin{cases} 1 \ , & \text{if } i \in G \text{ and } j = g(i) \\ 0 \ , & \text{if } i \in G \text{ and } j \neq g(i) \\ 0 \ , & \text{if } i \notin G \text{ and } w_{ij} > \min\{w_{\ell g(\ell)} \mid \ell \in G\} \end{cases}$$

That is, if we guess that incarnation j of item i is in the optimal solution, then $x_{ij} = 1$ and $x_{ij'} = 0$ for $j' \neq j$; also, if the jth incarnation of item i weighs more than some incarnation in our guess, then $x_{ij} = 0$. Denote the resulting linear program MMK(G, g).

Let $x^*(G, g)$ be an optimal (fractional) solution of MMK(G, g). The idea of Algorithm 2 below is to simply round down the values of $x^*(G, g)$. We show that if G and g are indeed the heaviest incarnations in the knapsack, then the rounded-down solution is very close to the optimum. Therefore, in the algorithm we loop over all possible assignments of G and g and output the best solution.

Algorithm 2

1: **for all** $G \subseteq [n]$ such that $|G| \leq q$ and $g : G \to [m]$ **do**
2:     $b_d(G, g) \leftarrow 1 - \sum_{i \in G} a_{ig(i)d}$ for every $d \in [D]$
3:     **if** $b_d(G, g) \geq 0$ for every d **then**
4:         Compute an optimal basic solution $x^*(G, g)$ of MMK(G, g)
5:         $x_{ij}(G, g) \leftarrow \lfloor x_{ij}^*(G, g) \rfloor$ for every i and j
6:     **end if**
7:     $x \leftarrow \text{argmax}_{x(G,g)} \ w \cdot x(G, g)$
8: **end for**
9: **return** x

The proof of the next theorem is omitted for lack of space.

Theorem 3. *If $D = O(1)$, then Algorithm 2 is a PTAS for* MMK.

References

1. Papadimitriou, C.H., Steiglitz, K.: Combinatorial optimization: algorithms and complexity. Prentice-Hall, Englewood Cliffs (1981)
2. Kellerer, H., Pferschy, U., Pisinger, D.: Knapsack Problems. Springer, Berlin (2004)
3. Warner, D., Prawda, J.: A mathematical programming model for scheduling nursing personnel in a hospital. Manage. Sci. (Application Series Part 1) 19, 411–422 (1972)
4. Fernandez de la Vega, W., Lueker, G.S.: Bin packing can be solved within 1+epsilon in linear time. Combinatorica 1(4), 349–355 (1981)
5. Karmarkar, N., Karp, R.M.: An efficient approximation scheme for the one-dimensional bin-packing problem. In: 23rd IEEE Annual Symposium on Foundations of Computer Science, pp. 312–320 (1982)

6. Friesen, D.K., Langston, M.A.: Variable sized bin packing. SIAM J. Comput. 15(1), 222–230 (1986)
7. Seiden, S.S., van Stee, R., Epstein, L.: New bounds for variable-sized online bin packing. SIAM J. Comput. 32(2), 455–469 (2003)
8. Correa, J.R., Epstein, L.: Bin packing with controllable item sizes. Information and Computation 206(18), 1003–1016 (2008)
9. Garey, M.R., Graham, R.L., Johnson, D.S., Yao, A.C.: Resource constrained scheduling as generalized bin packing. J. Comb. Theory, Ser. A 21(3), 257–298 (1976)
10. Chekuri, C., Khanna, S.: On multidimensional packing problems. SIAM J. Comput. 33(4), 837–851 (2004)
11. Bansal, N., Caprara, A., Sviridenko, M.: Improved approximation algorithms for multidimensional bin packing problems. In: 47th IEEE Annual Symposium on Foundations of Computer Science, pp. 697–708 (2006)
12. Sahni, S.: Approximate algorithms for the 0/1 knapsack problem. J. ACM 22(1), 115–124 (1975)
13. Ibarra, O.H., Kim, C.E.: Fast approximation algorithms for the knapsack and sum of subset problems. J. ACM 22(4), 463–468 (1975)
14. Frieze, A.M., Clarke, M.R.B.: Approximation algorithms for the m-dimensional $0-1$ knapsack problem: worst-case and probabilistic analyses. Eur. J. Oper. Res. 15, 100–109 (1984)
15. Magazine, M.J., Chern, M.S.: A note on approximation schemes for multidimensional knapsack problems. Math. Oper. Res. 9(12), 244–247 (1984)
16. Shachnai, H., Tamir, T.: Approximation schemes for generalized 2-dimensional vector packing with application to data placement. In: Arora, S., Jansen, K., Rolim, J.D.P., Sahai, A. (eds.) RANDOM 2003 and APPROX 2003. LNCS, vol. 2764, pp. 165–177. Springer, Heidelberg (2003)
17. Akbar, M. M., Manning, E.G., Shoja, G.C., Khan, S.: Heuristic solutions for the multiple-choice multi-dimension knapsack problem. In: International Conference on Computational Science-Part II, pp. 659–668 (2001)
18. Hifi, M., Michrafy, M., Sbihi, A.: Heuristic algorithms for the multiple-choice multidimensional knapsack problem. J. Oper. Res. Soc. 55, 1323–1332 (2004)
19. Parra-Hernández, R., Dimopoulos, N.J.: A new heuristic for solving the multi-choice multidimensional knapsack problem. IEEE Trans. on Systems, Man, and Cybernetics—Part A: Systems and Humans 35(5), 708–717 (2005)
20. Akbara, M.M., Rahman, M.S., Kaykobad, M., Manning, E.G., Shoja, G.C.: Solving the multidimensional multiple-choice knapsack problem by constructing convex hulls. Computers & Operations Research 33, 1259–1273 (2006)
21. Khan, M.S.: Quality Adaptation in a Multisession Multimedia System: Model, Algorithms and Architecture. PhD thesis, Dept. of Electrical and Computer Engineering (1998)
22. Sbihi, A.: A best first search exact algorithm for the multiple-choice multidimensional knapsack problem. J. Comb. Optim. 13(4), 337–351 (2007)
23. Chandra, A.K., Hirschberg, D.S., Wong, C.K.: Approximate algorithms for some generalized knapsack problems. Theoretical Computer Science 3(3), 293–304 (1976)
24. Kellerer, H., Pferschy, U., Pisinger, D.: Knapsack Problems. Springer, Heidelberg (2004)
25. Plotkin, S.A., Shmoys, D.B., Tardos, É.: Fast approximation algorithms for fractional packing and covering problems. Math. Oper. Res. 20, 257–301 (1995)
26. Grötschel, M., Lovasz, L., Schrijver, A.: Geometric Algorithms and Combinatorial Optimization. Springer, Heidelberg (1988)

Bin Packing with Fixed Number of Bins Revisited

Klaus Jansen[1,*], Stefan Kratsch[2], Dániel Marx[3,**], and Ildikó Schlotter[4,***]

[1] Institut für Informatik, Christian-Albrechts-Universität Kiel, 24098 Kiel, Germany
[2] Max-Planck-Institut für Informatik, 66123 Saarbrücken, Germany
[3] Tel Aviv University, Israel
[4] Budapest University of Technology and Economics, H-1521 Budapest, Hungary

Abstract. As BIN PACKING is NP-hard already for $k = 2$ bins, it is unlikely to be solvable in polynomial time even if the number of bins is a fixed constant. However, if the sizes of the items are polynomially bounded integers, then the problem can be solved in time $n^{O(k)}$ for an input of length n by dynamic programming. We show, by proving the W[1]-hardness of UNARY BIN PACKING (where the sizes are given in unary encoding), that this running time cannot be improved to $f(k) \cdot n^{O(1)}$ for any function $f(k)$ (under standard complexity assumptions). On the other hand, we provide an algorithm for BIN PACKING that obtains in time $2^{O(k \log^2 k)} + O(n)$ a solution with additive error at most 1, i.e., either finds a packing into $k + 1$ bins or decides that k bins do not suffice.

1 Introduction

The aim of this paper is to clarify the exact complexity of BIN PACKING for a small fixed number of bins. An instance of BIN PACKING consists of a set of rational item sizes, and the task is to partition the items into a minimum number of bins with capacity 1. Equivalently, we can define the problem such that the sizes are integers and the input contains an integer B, the capacity of the bins.

Complexity investigations usually distinguish two versions of BIN PACKING. In the general version, the item sizes are arbitrary integers encoded in binary, thus they can be exponentially large in the size n of the input. In the *unary* version of the problem, the sizes are bounded by a polynomial of the input size; formally, this version requires that the sizes are given in unary encoding.

In the general (not unary) case, a reduction from PARTITION shows that BIN PACKING is NP-hard [7]. Thus it is hard to decide whether a given set of items can be packed into exactly two bins. Apart from NP-hardness, this has a number of other known implications. First of all, unless P = NP, it is impossible to achieve a better polynomial-time approximation ratio than 3/2, matching the best known algorithm [16].

---

* Research supported by the EU project AEOLUS, Contract Number 015964.
** Supported by ERC Advanced Grant DMMCA.
*** Supported by the Hungarian National Research Fund (OTKA 67651).

H. Kaplan (Ed.): SWAT 2010, LNCS 6139, pp. 260–272, 2010.
© Springer-Verlag Berlin Heidelberg 2010

In contrast, however, there are much better approximation results when the optimum number of bins is larger [4,12]. De la Vega and Lueker [4] found an asymptotic polynomial-time approximation scheme (APTAS) for BIN PACKING with ratio $(1 + \epsilon)OPT(I) + 1$ and running time $O(n) + f(1/\epsilon)$ (if the items are sorted). To bound the function f, one has to consider the integer linear program (ILP) used implicitly in [4]. This ILP has $2^{O(1/\epsilon \log(1/\epsilon))}$ variables and length $2^{O(1/\epsilon \log(1/\epsilon))} \log n$. Using the algorithm by Lenstra [13] or Kannan [11], this ILP can be solved within time $2^{2^{O(1/\epsilon \log(1/\epsilon))}} O(\log n) \le 2^{2^{O(1/\epsilon \log(1/\epsilon))}} + O(\log^2 n)$. Thus, the algorithm of [4] can be implemented such that the additive term $f(1/\epsilon)$ in the running time is double exponential in $1/\epsilon$. Setting $\epsilon = \frac{1}{OPT(I)+1}$, this algorithm computes a packing into at most $OPT(I) + 1$ bins in time $O(n) + 2^{2^{OPT(I) \log(OPT(I))}}$.

Using ideas in [6,10], the algorithm of de la Vega and Lueker can be improved to run in time $O(n) + 2^{O(1/\epsilon^3 \log(1/\epsilon))}$. Setting again $\epsilon = \frac{1}{OPT(I)+1}$, we obtain an additive 1-approximation that runs in $O(n) + 2^{O(OPT(I)^3 \log(OPT(I)))}$ time.

Karmarkar and Karp [12] gave an asymptotic fully polynomial-time approximation scheme (AFPTAS) that packs the items into $(1 + \epsilon)OPT(I) + O(1/\epsilon^2)$ bins. The AFPTAS runs in time polynomial in n and $1/\epsilon$, but has a larger additive term $O(1/\epsilon^2)$. Plotkin, Shmoys and Tardos [14] achieved a running time of $O(n \log(1/\epsilon) + \epsilon^{-6}) \log^6(1/\epsilon))$ and a smaller additive term $O(1/\epsilon \log(1/\epsilon))$.

BIN PACKING remains NP-hard in the unary case as well [7]. However, for every fixed k, UNARY BIN PACKING can be solved in polynomial time: a standard dynamic programming approach gives an $n^{O(k)}$ time algorithm. Although the running time of this algorithm is polynomial for every fixed value of k, it is practically useless even for, say, $k = 10$, as an n^{10} time algorithm is usually not considered efficient. Our first result is an algorithm with significantly better running time that approximates the optimum within an additive constant of 1:

Theorem 1. *There is an algorithm for* BIN PACKING *which computes for each instance I of length n a packing into at most $OPT(I) + 1$ bins in time*

$$2^{O(OPT(I) \log^2 OPT(I))} + O(n).$$

Note that the algorithm works not only for the unary version, but also for the general BIN PACKING as well, where the item sizes can be exponentially large.

It is an obvious question whether the algorithm of Theorem 1 can be improved to an exact algorithm with a similar running time. As the general version of BIN PACKING is NP-hard for $k = 2$, the question makes sense only for the unary version of the problem. By proving that UNARY BIN PACKING is W[1]-hard parameterized by the number k of bins, we show that there is no exact algorithm with running time $f(k) \cdot n^{O(1)}$ for any function $f(k)$ (assuming the standard complexity hypothesis FPT $\neq$ W[1]).

Theorem 2. UNARY BIN PACKING *is* W[1]-*hard, parameterized by the number of bins.*

Thus no significant improvement over the $n^{O(k)}$ dynamic programming algorithm is possible for UNARY BIN PACKING. From the perspective of parameterized complexity, the general (not unary) BIN PACKING is not even known to be contained in the class XP, when parameterized by the number of bins.

Finally, let us mention that the existence of a polynomial-time algorithm with additive error of 1 (or any other constant) is a fundamental open problem, discussed for example in [5, Section 2.2.9]. Such a polynomial-time algorithm would be a significant breakthrough even in the unary case. Our algorithm shows that obtaining such an approximation is easy in the unary case for fixed number of bins. Thus an important consequence of our result is that any hardness proof ruling out the possibility of constant additive error approximation for the unary version has to consider instances with an unbounded number of bins.

For proofs omitted due to lack of space, marked by a $\star$, see the full paper.

2 Additive 1-Approximation for Bin Packing in FPT Time

This section deals with the following version of BIN PACKING: given an integer K and a set I of items with rational item sizes (encoded in binary), and the task is to pack the items into K bins of capacity 1. We prove Theorem 1 by describing an algorithm for this problem which uses at most $K+1$ bins for each I, provided that $OPT(I) = K$, where $OPT(I)$ is the minimum number of bins needed for I.

Our algorithm computes a packing into K or $K+1$ bins. We suppose $K \geq 2$; otherwise we pack all items into a single bin. We divide the instance I into three groups:

$$
\begin{aligned}
I_{large} &= \{a \in I \mid size(a) > \tfrac{1}{2x}\tfrac{1}{\log(K)}\}, \\
I_{medium} &= \{a \in I \mid \tfrac{1}{y}\tfrac{1}{K} \leq size(a) \leq \tfrac{1}{2x}\tfrac{1}{\log(K)}\}, \\
I_{small} &= \{a \in I \mid size(a) < \tfrac{1}{y}\tfrac{1}{K}\},
\end{aligned}
$$

where x, y are constants specified later. In the first phase of our algorithm we consider the large items. Since each bin has at most $\lfloor 2x \log(K) \rfloor$ large items and $OPT(I) = K$, the total number of large items in I is at most $K\lfloor 2x \log(K) \rfloor$. Suppose that $I_{large} = \{a_1, \ldots, a_\ell\}$ where $\ell \leq K\lfloor 2x \log(K) \rfloor$. We can assign large items to bins via a mapping $f : \{1, \ldots, \ell\} \to \{1, \ldots, K\}$. A mapping f is feasible, if and only if $\sum_{i|f(i)=j} size(a_i) \leq 1$ for all $j = 1, \ldots, K$. The total number of feasible mappings or assignments of large items to bins is at most $K^{K\lfloor 2x \log(K) \rfloor} = 2^{O(K \log^2(K))}$. Each feasible mapping f generates a pre-assignment $preass(b_j) \in [0,1]$ for the bins $b_j \in \{b_1, \ldots, b_K\}$; i.e. $preass(b_j) = \sum_{i|f(i)=j} size(a_i) \leq 1$. Notice that at least one of the $2^{O(K \log^2(K))}$ mappings corresponds to a packing of the large items in an optimum solution.

In the second phase we use a geometric rounding for the medium items. This method was introduced by Karmarkar and Karp [12] for the BIN PACKING problem. Let I_r be the set of all items from I_{medium} whose sizes lie in $(2^{-(r+1)}, 2^{-r}]$ where $2^r > x \log(K)$ or equivalently $\frac{1}{x}\frac{1}{\log(K)} > \frac{1}{2^r}$ (see Fig. 1(a) for an example I_r where we have divided the set I_r into groups of size $\lceil \frac{2^r}{x \log(K)} \rceil$). Let $r(0)$ be the smallest integer r such that $2^r > x \log(K)$. Then, $2^{r(0)-1} \leq x \log(K)$ and $2^{r(0)} >$

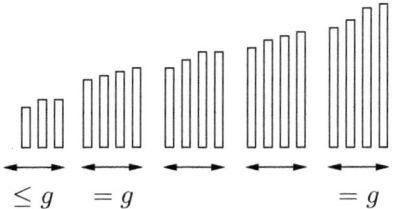

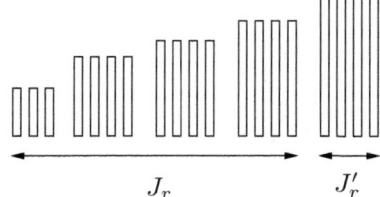

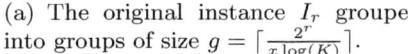

(a) The original instance I_r grouped into groups of size $g = \lceil \frac{2^r}{x \log(K)} \rceil$.

(b) The rounded instance J_r and J'_r.

Fig. 1. The original and the rounded instances for the interval $(2^{-(r+1)}, 2^{-r}]$

$x \log(K)$. This implies that the interval with the smallest index $r(0)$ contains items of size in $(1/2^{r(0)+1}, 1/2^{r(0)}]$ and that $\frac{1}{2x} \frac{1}{\log(K)} \in (1/2^{r(0)+1}, 1/2^{r(0)}]$.

For each $r \geq r(0)$ let J_r and J'_r be the instances obtained by applying linear grouping with group size $g = \lceil \frac{2^r}{x \log(K)} \rceil$ to I_r. To do this we divide each instance I_r into groups $G_{r,1}, G_{r,2}, \ldots, G_{r,q_r}$ such that $G_{r,1}$ contains the g largest items in I_r, $G_{r,2}$ contains the next g largest items and so on (see Fig. 1(a)). Each group of items is rounded up to the largest size within the group (see also Fig. 1). Let $G'_{r,i}$ be the multi-set of items obtained by rounding the size of each item in $G_{r,i}$. Then, $J_r = \bigcup_{i \geq 2} G'_{r,i}$ and $J'_r = G'_{r,1}$.

Furthermore, let $J = \bigcup J_r$ and $J' = \bigcup J'_r$. Then, $J_r \leq I_r \leq J_r \cup J'_r$ where $\leq$ is the partial order on BIN PACKING instances with the interpretation that $I_A \leq I_B$ if there exists a one-to-one function $h : I_A \to I_B$ such that $size(x) \leq size(h(x))$ for all items $x \in I_A$. Furthermore, J'_r consists of one group of items with the largest medium items in $(2^{-(r+1)}, 2^{-r}]$. The cardinality of each group (with exception of maybe the smallest group in I_r) is equal to $\lceil \frac{2^r}{x \log(K)} \rceil$.

Lemma 1. *For $K \geq 2$ and $x \geq 4$, we have $size(J') \leq \frac{\log(yK)}{x \log(K)}$.*

Proof. Each non-empty set J'_r contains at most $\lceil \frac{2^r}{x \log(K)} \rceil$ items each of size at most $1/2^r$. Hence $size(J'_r) \leq (\frac{2^r}{x \log(K)} + 1)\frac{1}{2^r} = \frac{1}{x \log(K)} + \frac{1}{2^r}$. This implies that the total size $size(J') = \sum_{r \geq r(0)} size(J'_r) \leq \sum_{r \geq r(0)}(\frac{1}{x \log(K)} + \frac{1}{2^r})$. Let $r(1)$ be the index with $\frac{1}{yK} \in (2^{-(r(1)+1)}, 2^{-r(1)}]$. This implies that $r(1) \leq \lfloor \log(yK) \rfloor$. Then, the number of indices $r \in \{r(0), \ldots, r(1)\}$ is equal to the number of intervals $(2^{-(r+1)}, 2^{-r}]$ which may contain a medium item. Since $\frac{1}{x \log(K)} > 1/2^{r(0)}$ and $K \geq 2$, we have $2^{r(0)} > x \log(K) \geq x$ or equivalently $r(0) > \log(x \log(K)) \geq \log(x)$. For $x \geq 4$ we obtain $r(0) \geq 3$.

Thus, the number of such intervals is $r(1) - r(0) + 1 \leq r(1) - 2 \leq \lfloor \log(yK) \rfloor - 2$. Using $\frac{1}{2^{r(0)}} < \frac{1}{x \log(K)}$ and $\sum_{r \geq r(0)} 1/2^r \leq 1/2^{r(0)-1}$, we get

$$size(J') \leq (\lfloor \log(yK) \rfloor - 2)\frac{1}{x \log(K)} + \sum_{r \geq r(0)} \frac{1}{2^r}$$
$$\leq \frac{\log(yK) - 2}{x \log(K)} + \frac{1}{2^{r(0)-1}} \leq \frac{\log(yK) - 2}{x \log(K)} + \frac{2}{x \log(K)} \leq \frac{\log(yK)}{x \log(K)}. \qquad \square$$

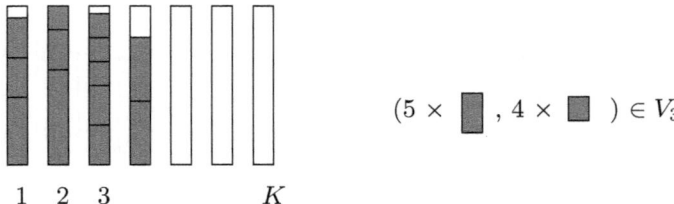

$$(5 \times \blacksquare \, , 4 \times \blacksquare \,) \in V_3$$

1 2 3   ...   K

Fig. 2. The dynamic program for rounded medium items $J = \cup_j J_r$

The lemma above implies $OPT(J') = 1$ for $x \geq 4$, $K \geq 2$ and $\log(y) \leq (x - 1)$, since these items have total size at most 1. A possible choice is $x = 4$ and $y \leq 8$.

By $\bigcup_{r \geq r(0)} J_r \leq I_{medium} \leq \bigcup_{r \geq r(0)} (J_r \cup J'_r)$ and $J' = \bigcup_{r \geq r(0)} J'_r$, we obtain:

Lemma 2

$$OPT(I_{large} \cup \bigcup_{r \geq r(0)} J_r \cup I_{small}) \leq OPT(I_{large} \cup I_{medium} \cup I_{small})$$

$$OPT(I_{large} \cup I_{medium} \cup I_{small}) \leq OPT(I_{large} \cup \bigcup_{r \geq r(0)} J_r \cup I_{small}) + 1.$$

Lemma 3. *There are at most $O(K \log(K))$ different rounded sizes for medium items for $x \geq 1$ and $K \geq 2$.*

Proof. Let $n(I_r)$ be the number of medium items in I_r, and let $m(I_r)$ be the number of groups (or rounded sizes) generated by the linear grouping for I_r. Then, $size(I_r) \geq \frac{1}{2^{r+1}} n(I_r) \geq \frac{1}{2^{r+1}} [(m(I_r) - 1) \lceil \frac{2^r}{x \log(K)} \rceil]$. Notice that one group may have less than $\lceil \frac{2^r}{x \log(K)} \rceil$ items. This implies that

$$m(I_r) - 1 \leq \frac{2^{r+1} size(I_r)}{\lceil \frac{2^r}{x \log(K)} \rceil}.$$

Using $\lceil a \rceil \geq a$ for $a \geq 0$, $m(I_r) \leq 2x \log(K) size(I_r) + 1$. For $x \geq 1$ and $K \geq 2$, we have $r(0) > \log(x) \geq 0$ and, therefore, $r(0) \geq 1$. Since the number of intervals for the medium items is at most $r(1) - r(0) + 1 \leq r(1) \leq \lfloor \log(yK) \rfloor$, the total number of rounded medium sizes $\sum_{r \geq r(0)} m(I_r) \leq \sum_{r \geq r(0)} (2x \log(K) size(I_r) + 1) \leq 2x \log(K) \sum_{r \geq r(0)} size(I_r) + \log(yK)$. Since all medium items fit into K bins and x, y are constants, $size(I_{medium}) = \sum_{r \geq r(0)} size(I_r) \leq K$ and

$$\sum_{r \geq r(0)} m(I_r) \leq 2xK \log(K) + \log(yK) \leq O(K \log(K)).$$

$\square$

Now we describe the third phase of our algorithm. The rounded medium item sizes lie in the interval $[\frac{1}{yK}, \frac{1}{2x \log(K)}]$ and there are at most $R \leq O(K \log(K))$ many different rounded item sizes. For each $j = 1, \ldots, R$ let k_j be the number of items for each rounded item size x_j. Since $x_j \geq \frac{1}{yK}$ and $OPT(I) = K$, the number $k_j \leq K/x_j \leq K^2y$ for each item size x_j. To describe a packing for one bin b we use a mapping $p : \{1, \ldots, R\} \to \{0, \ldots, yK\}$ where $p(j)$ gives the number of items of size x_j in b. A mapping p is feasible, if and only if $\sum_j p(j)x_j + preass(b) \leq 1$ where $preass(b)$ is the total size of large items assigned to b in the first phase of the algorithm. The total number of feasible mappings for one bin is at most $(yK + 1)^{O(K \log(K))} = 2^{O(K \log^2(K))}$. Using a dynamic program we go over the bins from b_1 up to b_K. For each $A = 1, \ldots, K$, we compute a set V_A of vectors $(a_1, \ldots, a_R)$ where a_j gives the number of items of size x_j used for the bins $b_1, \ldots, b_A$ (see also Fig. 2). The cardinality of each set V_A is at most $(K^2y + 1)^{O(K \log(K))} = 2^{O(K \log^2(K))}$. The update step from one bin to the next (computing the next set V_{A+1}) can be implemented in time

$$2^{O(K \log^2(K))} \cdot 2^{O(K \log^2(K))} \cdot poly(K) \leq 2^{O(K \log^2(K))}.$$

If there is a solution for our Bin Packing instance I into K bins, then the set V_K contains the vector $(n_1, \ldots, n_R)$ that corresponds to the number of rounded medium item sizes in $\bigcup_{r \geq r(0)} J_r$. Notice that the other set $\bigcup_{r \geq r(0)} J'_r$ will be placed into the additional bin b_{K+1}. We can also compute a packing of the medium items into the bins as follows. First, we compute all vector sets V_A for $A = 1, \ldots, K$. If for two vectors $a = (a_1, \ldots, a_K) \in V_A$ and $a' = (a'_1, \ldots, a'_K) \in V_{A+1}$ the medium items given by the difference $a' - a$ and the preassigned large items fit into bin b_{A+1}, we store the corresponding pair (a, a') in a set S_{A+1}. By using a directed acyclic graph $D = (V, E)$ with vertex set $V = \{[a, A] | a \in V_A, A = 1, \ldots, K\}$ and $E = \{([a, A], [a', A + 1]) | (a, a') \in S_{A+1}, A = 1, \ldots, K-1\}$, we may compute a feasible packing of large and medium rounded items into the bins $b_1, \ldots, b_K$. This can be done via depth first search starting with the vector $(n_1, \ldots, n_R) \in V_K$ that corresponds to the number of rounded medium item sizes. The algorithm to compute the directed acyclic graph and the backtracking algorithm can be implemented in time

$$2^{O(K \log^2(K))}.$$

In the last phase of our algorithm we add the small items via a greedy algorithm to the bins. Consider a process which starts with a given packing of the original large and medium items into the bins $b_1, \ldots, b_{K+1}$. We insert the small items of size at most $\frac{1}{yK}$ with the greedy algorithm Next Fit (NF) into the bins $b_1, \ldots, b_{K+1}$. The correctness of this step is proved in the next lemma. Notice that NF can be implemented in linear time with at most $O(n)$ operations.

Lemma 4. *If $OPT(I) = K$, $K \geq 2$, and $y \geq 2$, then our algorithm packs all items into at most $K + 1$ bins.*

Proof. Assume by contradiction that we use more than $K + 1$ bins for the small items. In this case, the total size of the items packed into the bins is more than $(K+1)(1 - \frac{1}{yK}) = K + 1 - \frac{K+1}{yK}$. Note that $\frac{K+1}{yK} \leq \frac{K+1}{2K} < 1$ by $y \geq 2$ and $K \geq 2$. Thus, the total size of the items is larger than K, yielding $OPT(I) > K$. $\qquad \square$

The algorithm for BIN PACKING for $OPT(I) = K$ works as follows:

(1) Set $x = 4$, $y = 2$ and divide I into three groups I_{large}, I_{medium}, and I_{small}.
(2) Assign the large items to K bins considering all feasible pre-assignments.
(3) Use geometric rounding on the sizes of the medium items; for each interval $(2^{-(r+1)}, 2^{-r}]$ apply linear grouping with group size $\lceil \frac{2^r}{x \log(K)} \rceil$ to the item set I_r and compute rounded item sets J_r and J'_r.
(4) For each pre-assignment apply the dynamic program to assign the medium items in $\bigcup J_r$ to the bins $b_1, \ldots, b_K$, and place the set $\bigcup_j J'_r$ into b_{K+1}.
(5) Take a feasible packing into $K + 1$ bins for one pre-assignment (there is one by $OPT(I) = K$), replace the rounded items by their original sizes and afterwards assign the small items via a greedy algorithm to the bins $b_1, \ldots, b_{K+1}$.

3 Parameterized Hardness of Bin Packing

The aim of this section is to prove that UNARY BIN PACKING is W[1]-hard, parameterized by the number k of bins. In this version of BIN PACKING, we are given a set of integer item sizes encoded in unary, and two integers b and k. The task is to decide whether the items can be packed into k bins of capacity b.

To prove the W[1]-hardness of this problem when parameterized by the number of bins, we use the hardness of an intermediate problem, a variant of UNARY BIN PACKING involving vectors of constant length c and bins of varying sizes.

Let $[c] = \{1, \ldots, c\}$ for any $c \in \mathbb{N}$, and let $\mathbb{N}^c$ be the set of vectors with c coordinates, each in $\mathbb{N}$. We use boldface letters for vectors. Given vectors $\mathbf{v}, \mathbf{w} \in \mathbb{N}^c$, we write $\mathbf{v} \leq \mathbf{w}$, if $v^j \leq w^j$ for each $j \in [c]$, where v^j is the j-th coordinate of $\mathbf{v}$.

For a fixed c, we will call the following problem c-UNARY VECTOR BIN PACKING. We are given a set of n items having sizes $\mathbf{s}_1, \mathbf{s}_2, \ldots, \mathbf{s}_n$ with each $\mathbf{s}_i \in \mathbb{N}^c$ encoded in unary, and k vectors $\mathbf{B}_1, \mathbf{B}_2, \ldots, \mathbf{B}_k$ from $\mathbb{N}^c$ representing bin sizes. The task is to decide whether $[n]$ can be partitioned into k sets $J_1, J_2, \ldots, J_k$ such that $\sum_{h \in J_i} \mathbf{s}_h \leq \mathbf{B}_i$ for each $i \in [k]$.

Lemma 5 shows that Theorem 2 follows from the W[1]-hardness of c-UNARY VECTOR BIN PACKING, for any fixed c. In Section 3.1, we introduce two concepts, k-non-averaging and k-sumfree sets, that will be useful tools in the hardness proof. The reduction itself appears in Section 3.2.

Lemma 5 ($\star$). *For every fixed integer $c \geq 1$, there is a parameterized reduction from c-UNARY VECTOR BIN PACKING to UNARY BIN PACKING, where both problems are parameterized by the number of bins.*

3.1 Non-averaging and Sumfree Sets

Given an integer k, we are going to construct a set $\mathcal{A}$ containing n non-negative integers with the following property: for any k elements $a_1, a_2, \ldots, a_k$ in $\mathcal{A}$ it holds that their arithmetical mean $\frac{1}{k} \sum_{i=1}^{k} a_i$ can only be contained in $\mathcal{A}$ if all of them are equal, i.e. $a_1 = a_2 \cdots = a_k$. Sets having this property will be called k-*non-averaging*. Such sets have already been studied by several researchers [1,3]. Although, up to our knowledge, the construction presented here does not appear in the literature in this form, it applies only standard techniques[1].

First, let us fix an arbitrary integer d. (In fact, it will suffice to assume $d = 2$, but for completeness, we present the case for an arbitrary d.) Depending on d, let us choose m to be the smallest integer for which $m^d \geq n$, i.e. let $m = \lceil n^{1/d} \rceil$. We construct a set $\mathcal{X}$ containing each vector $(x_1, x_2, \ldots, x_d, y)$ where $0 \leq x_i \leq m-1$ for all $i \in [d]$ and $\sum_{i=1}^{d} x_i^2 = y$. Clearly, $|\mathcal{X}| = m^d$, so in particular, $|\mathcal{X}| \geq n$.

Lemmas 6 and 7 show that we can easily construct a non-averaging set $\mathcal{A}$ from $\mathcal{X}$, having n elements. Setting $d = 2$, we get that the maximal element in $\mathcal{A}$ is smaller than $2^5 k^2 n^2 = O(k^2 n^2)$. Also, $\mathcal{A}$ can be constructed in $O(k^2 n^3)$ time.

Lemma 6 ($\star$). *If $\mathbf{u}_1, \mathbf{u}_2, \ldots, \mathbf{u}_k$ and $\mathbf{v}$ are elements of $\mathcal{X}$ and $\mathbf{v} = \frac{1}{k} \sum_{i=1}^{k} \mathbf{u}_i$, then $\mathbf{u}_1 = \mathbf{u}_2 = \cdots = \mathbf{u}_k = \mathbf{v}$.*

Lemma 7 ($\star$). *If $b = k(m-1)+1$, then the set $\mathcal{A} = \{v^1 + v^2 b + \cdots + v^c b^{c-1} \mid \mathbf{v} \in \mathcal{X}\}$ is k-non-averaging. Moreover, the largest element N in $\mathcal{A}$ is smaller than $4d(2k)^d n^{1+2/d}$, and $\mathcal{A}$ can be constructed in time linear in $O(2^d n N)$.*

A set $\mathcal{F}$ is k-*sumfree*, if for any two sets $S_1, S_2 \subseteq \mathcal{F}$ of the same size $k' \leq k$, $\sum_{x \in S_1} x = \sum_{x \in S_2} x$ holds if and only if $S_1 = S_2$. Such sets have been studied extensively in the literature, also under the name B_k-sequences [8,9,15].

It is easy to verify that the set $S = \{(k+1)^i \mid 0 \leq i < n\}$ is a k-sumfree set of size n. The maximal element in such a set is of course $(k+1)^{n-1}$. Intuitively, the elements of S are the $(k+1)$-base representations of those 0-1 vectors V_S in $\mathbb{N}^n$ that have exactly one coordinate of value 1. Since no vector in $\mathbb{N}^n$ can be obtained in more than one different ways as the sum of at most k vectors from V_S, an easy argument shows that S is k-sumfree.

Although this will be sufficient for our purposes, we mention that a construction due to Bose and Chowla [2] shows that a k-sumfree set of size n with maximum element at most $(2n)^k$ can also be found (see also [9], Chapter II). If k is relatively small compared to n, the bound $(2n)^k$ is a considerable reduction on the bound $(k+1)^{n-1}$ of the construction above.

Lemma 8 ([2]). *For any integers n and k, there exists a k-sumfree set having n elements, with the maximum element being at most $(2n)^k$.*

3.2   Hardness of the Vector Problem

The following lemma contains the main part of the hardness proof. By Lemma 5, it immediately implies Theorem 2.

Lemma 9. 10-UNARY VECTOR BIN PACKING *is* W[1]-*hard.*

---

[1] We thank Imre Ruzsa for explaining us these techniques.

Proof. We present an FPT reduction from the W[1]-hard CLIQUE parameterized by the size of the desired clique. Let $G = (V, E)$ and k be the input graph and the parameter given for CLIQUE. We assume $V = [n]$ and $|E| = m$, and we write $x_h y_h$ for the h-th edge of G according to some arbitrary ordering. We construct an instance $\mathcal{I}$ of 10-UNARY VECTOR BIN PACKING with $\binom{k}{2} + k + 1$ bins.

Item sizes. The sizes of the items in $\mathcal{I}$ will be contained in $S \cup T$, where $S = \bigcup_{(i,j) \in \binom{[k]}{2}} S_{i,j}$, $T = T^= \cup T^< \cup T^>$, $T^= = \bigcup_{i \in [k]} T_{i,i}$, $T^< = \bigcup_{(i,j) \in \binom{[k]}{2}} T_{i,j}$, and $T^> = \bigcup_{(j,i) \in \binom{[k]}{2}} T_{i,j}$; here we use $\binom{[k]}{2} = \{(i,j) \mid 1 \leq i < j \leq k\}$. For each possible pair of i and j, $|S_{i,j}| = m$ and $|T_{i,j}| = n$ will hold.

To determine the item sizes, we first construct a k-non-averaging set $\mathcal{A}$ of size n, using Lemma 7. Let $\mathcal{A}$ contain the elements $a_1 \leq a_2 \leq \cdots \leq a_n$. By Lemma 7, we know $a_n = O(k^2 n^2)$. Let $A = \sum_{h \in [n]} a_h$, clearly $A = O(k^2 n^3)$.

We also construct a k-sumfree set $\mathcal{F}$ containing k^2 elements, using Lemma 8. Let us index the elements of $\mathcal{F}$ by pairs of the form $(i,j) \in [k]^2$, so let $\mathcal{F} = \{f_{i,j} \mid (i,j) \in [k]^2\}$. We also assume that $f_{i_1,j_1} < f_{i_2,j_2}$ holds if and only if (i_1, j_1) is lexicographically smaller than (i_2, j_2). By Lemma 8, we know $f_{k,k} = O(k^{2k})$. Again, we let $F = \sum_{f \in \mathcal{F}} f$, so $F = O(k^{2k+2})$.

For some $1 \leq i < j \leq k$, let $S_{i,j} = \bigcup_{h=1}^{m} \mathbf{s}_{i,j}(h)$, and for some $1 \leq i, j \leq k$ let $T_{i,j} = \bigcup_{h=1}^{n} \mathbf{t}_{i,j}(h)$. The exact values of $\mathbf{s}_{i,j}(h)$ and $\mathbf{t}_{i,j}(h)$ are as follows:

$$\mathbf{s}_{i,j}(h) = (ik + j, 1, 0, 0, 0, 0, 0, 0, a_{x_h}, a_{y_h}) \quad \text{if } (i,j) \in \binom{[k]}{2}, h \in [m],$$
$$\mathbf{t}_{i,i}(h) = (0, 0, f_{i,i}, 1, 0, 0, 0, 0, (k-i)a_h, (i-1)a_h) \quad \text{if } i \in [k], h \in [n],$$
$$\mathbf{t}_{i,j}(h) = (0, 0, 0, 0, f_{i,j}, 1, 0, 0, a_h, 0) \quad \text{if } (i,j) \in \binom{[k]}{2}, h \in [n],$$
$$\mathbf{t}_{i,j}(h) = (0, 0, 0, 0, 0, 0, f_{i,j}, 1, 0, a_h) \quad \text{if } (j,i) \in \binom{[k]}{2}, h \in [n].$$

Bin capacities. We define $\binom{k}{2} + k + 1$ bins as follows: we introduce bins $\mathbf{p}_{i,j}$ for each $(i,j) \in \binom{[k]}{2}$, bins $\mathbf{q}_i$ for each $i \in [k]$, and one additional bin $\mathbf{r}$. The capacities of the bins $\mathbf{p}_{i,j}$ and $\mathbf{q}_i$ are given below (depending on i and j). To define $\mathbf{q}_i$, we let $F_i^< = \sum_{i < j \leq k} f_{i,j}$ and $F_i^> = \sum_{1 \leq j < i} f_{i,j}$ for each $i \in [k]$. Finally, we set the capacity of $\mathbf{r}$ in a way that the total size of the bins equals the total size of the items. Hence, any solution must completely fill all bins.

$$\mathbf{p}_{i,j} = (ik + j, 1, 0, 0, (n-1)f_{i,j}, n-1, (n-1)f_{j,i}, n-1, A, A)$$
$$\mathbf{q}_i = (0, 0, (n-1)f_{i,i}, n-1, F_i^<, k-i, F_i^>, i-1, (k-i)A, (i-1)A)$$

It is easy to see that $\mathbf{r} \in \mathbb{N}^{10}$. Observe that $|S \cup T| = m\binom{k}{2} + nk^2$, the unary encoding of the item sizes in S needs a total of at most $O(mk^4 + nk^2 A)$ bits, and the unary encoding of the item sizes in T needs a total of at most $O(nF + k^3 A)$ bits. By the bounds on A and F, the reduction given is indeed an FPT reduction.

Main idea. At a high-level abstraction, we think of the constructed instance as follows. First, a bin $\mathbf{q}_i$ requires $n-1$ items from $T_{i,i}$, which means that we need all items from $T_{i,i}$, except for some item $\mathbf{t}_{i,i}(h)$. Choosing an index $h \in [n]$ for each i will correspond to choosing k vertices from G. Next, we have to fill up the bin $\mathbf{q}_i$, by taking altogether $k-1$ items from $T^<$ and $T^>$ in a way such that the sum of their last two coordinates equals the last two coordinates of $\mathbf{t}_{i,i}(h)$. The sumfreeness of $\mathcal{F}$ and the non-averaging property of $\mathcal{A}$ will imply that the chosen items must be of the form $\mathbf{t}_{i,j}(h)$ and $\mathbf{t}_{j,i}(h)$ for some j.

This can be thought of as "copying" the information about the chosen vertices, since as a result, each bin $\mathbf{p}_{i,j}$ will miss only those items from $T_{i,j}$ and from $T_{j,i}$ that correspond to the i-th and j-th chosen vertex in G. Suppose $\mathbf{p}_{i,j}$ contains all items from $T_{i,j}$ and $T_{j,i}$ except for the items, say, $\mathbf{t}_{i,j}(h_a)$ and $\mathbf{t}_{j,i}(h_b)$. Then, we must fill up the last two coordinates of $\mathbf{p}_{i,j}$ exactly by choosing one item from $S_{i,j}$. But choosing the item $\mathbf{s}_{j,i}(e)$ will only do if the edge corresponding to $e \in [m]$ connects the vertices corresponding to h_a and h_b, ensuring that the chosen vertices form a clique.

Correctness. Now, let us show formally that $\mathcal{I}$ is solvable if and only if G has a clique of size k. Clearly, $\mathcal{I}$ is solvable if and only if each of the bins can be filled exactly. Thus, a solution for $\mathcal{I}$ means that the items in $S \cup T$ can be partitioned into sets $\{P_{i,j} \mid (i,j) \in \binom{[k]}{2}\}$, $\{Q_i \mid i \in [k]\}$, and R such that

$$\mathbf{p}_{i,j} = \sum_{v \in P_{i,j}} \mathbf{v} \quad \text{for each } (i,j) \in \binom{[k]}{2}, \tag{1}$$

$$\mathbf{q}_i = \sum_{v \in Q_i} \mathbf{v} \quad \text{for each } i \in [k], \text{ and} \tag{2}$$

$$\mathbf{r} = \sum_{v \in R} \mathbf{v}. \tag{3}$$

Direction $\Rightarrow$. First, we argue that if G has a clique of size k, then $\mathcal{I}$ is solvable. Suppose that $c_1, c_2, \ldots, c_k$ form a clique in G. Let $d_{i,j}$ be the number for which $c_i c_j$ is the $d_{i,j}$-th edge of G. Using this, we set $P_{i,j}$ for each $(i,j) \in \binom{[k]}{2}$ and Q_i for each $i \in [k]$ as follows, letting R include all the remaining items.

$$P_{i,j} = \{\mathbf{t}_{i,j}(h) \mid h \neq c_i\} \cup \{\mathbf{t}_{j,i}(h) \mid h \neq c_j\} \cup \{\mathbf{s}_{i,h}(d_{i,j})\}. \tag{4}$$

$$Q_i = \{\mathbf{t}_{i,j}(c_i) \mid j \neq i\} \cup \{\mathbf{t}_{i,i}(h) \mid h \neq c_i\}. \tag{5}$$

It is easy to see that the sets $P_{i,j}$ for some $(i,j) \in \binom{[k]}{2}$ and the sets Q_i for some $i \in [k]$ are all pairwise disjoint. Thus, in order to verify that this indeed yields a solution, it suffices to check that (1) and (2) hold, since in that case, (3) follows from the way $\mathbf{r}$ is defined. For any $(i,j) \in \binom{[k]}{2}$, using

$$\sum_{h \neq c_i} \mathbf{t}_{i,j}(h) = \sum_{h \neq c_i} (0,0,0,0,f_{i,j},1,0,0,a_h,0)$$
$$= (0,0,0,0,(n-1)f_{i,j}, n-1, 0, 0, A - a_{c_i}, 0),$$

$$\sum_{h \neq c_j} \mathbf{t}_{j,i}(h) = \sum_{h \neq c_j} (0,0,0,0,0,0,f_{j,i},1,0,a_h)$$
$$= (0,0,0,0,0,0,(n-1)f_{i,j}, n-1, 0, A - a_{c_j}),$$

$$\mathbf{s}_{i,h}(d_{i,j}) = (ik+j, 1, 0, 0, 0, 0, 0, 0, a_{c_i}, a_{c_j}),$$

we get (1) by the definition of $P_{i,j}$. To see (2), we only have to use the definition of Q_i, and sum up the equations below:

$$\sum_{i<j\leq k} \mathbf{t}_{i,j}(c_i) = \sum_{i<j\leq k} (0,0,0,0,f_{i,j},1,0,0,a_{c_i},0)$$
$$= (0,0,0,0,F_i^<,k-i,0,0,(k-i)a_{c_i},0),$$

$$\sum_{1\leq j<i} \mathbf{t}_{i,j}(c_i) = \sum_{1\leq j<i} (0,0,0,0,0,0,f_{i,j},1,0,a_{c_i})$$
$$= (0,0,0,0,0,0,F_i^>,i-1,0,(i-1)a_{c_i}),$$

$$\sum_{h\neq c_i} \mathbf{t}_{i,i}(h) = \sum_{h\neq c_i} (0,0,f_{i,i},1,0,0,0,0,(k-i)a_h,(i-1)a_h)$$
$$= (0,0,(n-1)f_{i,i},n-1,0,0,0,0,(k-i)(A-a_{c_i}),(i-1)(A-a_{c_i})).$$

Direction $\Leftarrow$. To prove the other direction, suppose that a solution exists, meaning that some sets $\{P_{i,j} \mid (i,j) \in \binom{[k]}{2}\}$, $\{Q_i \mid i \in [k]\}$ and R fulfill the conditions of (1), (2), and (3). We show that this implies a clique of size k in G.

Let X denote the set of items that are contained in some particular bin $\mathbf{x}$. Observing the second, fourth, sixth, and eighth coordinates of the items in $S \cup T$ and the bin $\mathbf{x}$, we can immediately count the number of items from S, $T^=$, $T^<$, and $T^>$ that are contained in X. The following table shows the information obtained by this argument for each possible X.

	$|X \cap S|$	$|X \cap T^=|$	$|X \cap T^<|$	$|X \cap T^>|$
$X = P_{i,j}$ for some (i,j)	1	0	$n-1$	$n-1$
$X = Q_i$ for some i	0	$n-1$	$k-i$	$i-1$
$X = R$	$(m-1)\binom{k}{2}$	k	0	0

Next, observe that $r^3 = \sum_{i\in[k]} f_{i,i}$. This means that R contains exactly k vectors from $\bigcup_{i\in[k]} T_{i,i}$ such that the third coordinate of their sum is $\sum_{i\in[k]} f_{i,i}$. But since $\mathcal{F}$ is k-sumfree, this can only happen if R contains exactly one vector from each of $T_{1,1}, T_{2,2}, \ldots, T_{k,k}$. Let these vectors be $\{\mathbf{t}_{i,i}(c_i) \mid i \in [k]\}$. We claim that the vertices $\{c_i \mid i \in [k]\}$ form a clique in G.

Using $q_i^3 = (n-1)f_{i,i}$, the table above, and $f_{1,1} < f_{2,2} < \cdots < f_{k,k}$ we obtain that Q_i must contain every item in $T_{i,i} \setminus \{\mathbf{t}_{i,i}(c_i)\}$, for each $i \in [k]$. Also, we know that Q_i must contain $k-i$ items from $T^<$ and $i-1$ from $T^>$, so from the values of q_i^5 and q_i^7 and the fact that $\mathcal{F}$ is k-sumfree, we also obtain that Q_i must contain exactly one item from each of the sets $T_{i,j}$ where $j \neq i$. Note that apart from these $(n-1)+(k-1)$ vectors, Q_i cannot contain any other items.

Now, note that the last two coordinates of the sum $\sum_{h\neq c_i} \mathbf{t}_{i,i}(h)$ are $(k-i)(A-a_{c_i})$ and $(i-1)(A-a_{c_i})$. Since the last two coordinates of $\mathbf{q}_i$ are $(k-i)A$ and $(i-1)A$, we get that $\sum_{\mathbf{v}\in Q_i\setminus T_{i,i}} \mathbf{v}$ must have $(k-i)a_{c_i}$ and $(i-1)a_{c_i}$ at the last two coordinates. As argued above, $Q_i \setminus T_{i,i}$ contains exactly one item from each of the sets $T_{i,j}$ where $j \neq i$. Fixing some i and letting $T_{i,j} \cap Q_i = \{t_{i,i}(h_j)\}$, this implies $\sum_{i<j\leq k} a_{h_j} = (k-i)a_{c_i}$ and $\sum_{1\leq j<i} a_{h_j} = (i-1)a_{c_i}$. But as $\mathcal{A}$ is k-non-averaging, this yields $h_j = c_i$ for each $j \neq i$. This means that (5) holds.

Next, let us consider the set $P_{i,j}$ for some $(i,j) \in \binom{[k]}{2}$. First, the first two coodinates of $\mathbf{p}_{i,j}$ imply that $P_{i,j}$ must contain exactly one element of $S_{i,j}$. Let us define $d_{i,j}$ such that $P_{i,j} \cap S_{i,j} = \mathbf{s}_{i,j}(d_{i,j})$. Furthermore, the table above and the result (5) shows that $P_{i,j}$ must contain $(n-1)$ items from both of the sets $T^<$ and $T^>$. Recall that $\{\mathbf{t}_{i,j}(c_i) \mid (i,j) \in [k]^2\} \subseteq \bigcup_{i\in[k]} Q_i$. Using $p_{i,j}^5 = (n-1)f_{i,j}$ and $p_{i,j}^7 = (n-1)f_{j,i}$, and taking into account the ordering of the elements of $\mathcal{F}$, it follows that (4) holds.

Finally, let us focus on the last two coordinates of the sum $\sum_{\mathbf{v}\in P_{i,j}} \mathbf{v}$. Clearly, if $i < j$ then the sum of the vectors in $T_{i,j} \setminus \{\mathbf{t}_{i,j}(c_i)\}$ has $A - a_{c_i}$ and 0 as the last two coordinates, and similarly, the sum of the vectors in $T_{j,i} \setminus \{\mathbf{t}_{j,i}(c_j)\}$ has 0 and $A - a_{c_j}$ in the last two coordinates. From this, (4) and the definition of $\mathbf{p}_{i,j}$ yield that $\mathbf{s}_{i,j}(d_{i,j})$ must contain a_{c_i} and a_{c_j} in the last two coordinates. But by the definition of $S_{i,j}$, this can only hold if (c_i, c_j) is an edge in G. This proves the second direction of the correctness of the reduction. $\square$

References

1. Alon, N., Ruzsa, I.Z.: Non-averaging subsets and non-vanishing transversals. J. Comb. Theory, Ser. A 86(1), 1–13 (1999)
2. Bose, R.C., Chowla, S.: Theorems in the additive theory of numbers. Comment. Math. Helv. 37(1), 141–147 (1962-1963)
3. Bosznay, A.P.: On the lower estimation of non-averaging sets. Acta Math. Hung. 53, 155–157 (1989)
4. de la Vega, W.F., Lueker, G.: Bin packing can be solved in within $1 + \epsilon$ in linear time. Combinatorica 1, 349–355 (1981)
5. Coffman, J.E.G., Garey, M.R., Johnson, D.S.: Approximation algorithms for bin packing: A survey. In: Hochbaum, D. (ed.) Approximation Algorithms for NP-Hard Problems, pp. 46–93. PWS Publishing, Boston (1997)
6. Eisenbrand, F., Shmonin, G.: Caratheodory bounds for integer cones. OR Letters 34, 564–568 (2006)
7. Garey, M.R., Johnson, D.S.: Computers and Intractability: A Guide to the Theory of NP-Completeness. W. H. Freeman, New York (1979)
8. Graham, S.W.: B_h sequences. In: Analytic number theory. Progr. Math., vol. 1 (1995), vol. 138, pp. 431–449. Birkhäuser, Boston (1996)
9. Halberstam, H., Roth, K.F.: Sequences. Springer, New York (1983)
10. Jansen, K.: An EPTAS for scheduling jobs on uniform processors: using an MILP relaxation with a constant number of integral variables. In: ICALP '09: 36th International Colloquium on Automata, Languages and Programming, pp. 562–573 (2009)
11. Kannan, R.: Minkowski's convex body theorem and integer programming. Math. of OR 12, 415–440 (1987)

12. Karmarkar, N., Karp, R.: An efficient approximation scheme for the one-dimensional bin-packing problem. In: FOCS 1982: 23rd IEEE Symposium on Foundations of Computer Science, pp. 312–320 (1982)
13. Lenstra, H.: Integer programming with a fixed number of variables. Math. of OR 8, 538–548 (1983)
14. Plotkin, S., Tardos, D., Tardos, E.: Fast approximation algorithms for fractional packing and covering problems. Math. of OR 20, 257–301 (1995)
15. Ruzsa, I.Z.: Solving a linear equation in a set of integers. I. Acta Arith. 65(3), 259–282 (1993)
16. Simchi-Levi, D.: New worst-case results for the bin-packing problem. Naval Res. Logist. 41(4), 579–585 (1994)

Cops and Robber Game without Recharging

Fedor V. Fomin[1], Petr A. Golovach[2,*], and Daniel Lokshtanov[1]

[1] Department of Informatics, University of Bergen, PB 7803, 5020 Bergen, Norway
{Fedor.Fomin|daniello}@ii.uib.no
[2] School of Engineering and Computing Sciences, Durham University, South Road,
DH1 3LE Durham, UK
petr.golovach@durham.ac.uk

Abstract. Cops & Robber is a classical pursuit-evasion game on undirected graphs, where the task is to identify the minimum number of cops sufficient to catch the robber. In this work, we consider a natural variant of this game, where every cop can make at most f steps, and prove that for each $f \geq 2$, it is PSPACE-complete to decide whether k cops can capture the robber.

1 Introduction

The study of pursuit-evasion games is driven by many real-world applications where a team of agents/robots must reach a moving target. The mathematical study of such games has a long history, tracing back to the work of Pierre Bouguer, who in 1732 studied the problem of a pirate ship pursuing a fleeing merchant vessel. In 1960s the study of pursuit-evasion games, mostly motivated by military applications like missile interception, gave a rise to the theory of Differential Games [10]. Besides the original military motivations, pursuit-evasion games have found many applications reaching from law enforcement to video games and thus were studied within different disciplines and from different perspectives. The necessity of algorithms for pursuit tasks occur in many real-world domains. In the Artificial Intelligence literature many heuristic algorithms for variations of the problem like Moving Target Search have been studied extensively [7,11,12,16,17]. In computer games, for instance, computer-controlled agents often pursue human-controlled players and making a good strategy for pursuers is definitely a challenge [14]. The algorithmic study of pursuit-evasion games is also an active area in Robotics [9,21] and Graph Algorithms [15,5].

One of the classical pursuit-evasion problems is the Man and Lion problem attributed to Rado by Littlewood in [13]: A lion (pursuer) and a man (evader) in a closed arena have equal maximum speeds. What tactics should the lion employ to be sure of his meal? See also for more recent results on this problem [3,20]. The discrete version of the Man and Lion problem on graphs was introduced by Winkler and Nowakowski [18] and Quilliot [19]. Aigner and Fromme [1] initiated the study of the problem with several pursuers. This game, named

---

* Supported by EPSRC under project EP/G043434/1.

Cops & Robber, is played by two players: cop and robber on an undirected graph. The cop-player has a team of cops who attempt to capture the robber. At the beginning of the game cop-player selects vertices and puts cops on these vertices. Then the robber player puts the robber on a vertex. The players take turns starting with the cop-player. At every move each of the cops can be either moved to an adjacent vertex or kept on the same vertex. Similarly, the robber player responds by moving the robber to an adjacent vertex or keeping him on the same vertex. The cop-player wins if at some step of the game he succeeds to catch the robber, i.e. to put one of his cops on a vertex occupied by the robber. The game was studied intensively and there is an extensive literature on this problem. We refer to surveys [2,5] for references on different pursuit-evasion and search games on graphs.

In the game of Cops & Robber there are no restrictions on the number of moves the players can make. Such model is not realistic for most of the applications: No lion can pursuit a man without taking a nap and no robot can move permanently without recharging batteries. In this work, we introduce a more realistic scenario of Cops & Robber, the model capturing the fact that each of the cops has a limited amount of power or fuel.

We also find the Cops & Robber problem with restricted power interesting from combinatorial point of view because it generalizes the *Minimum Dominating Set* problem, one of the fundamental problems in Graph Theory and Graph Algorithms. Indeed, with fuel limit 1 every cop can make at most one move, then k cops can win on a graph G if and only if G has a dominating set of size k. Thus two classical problems— *Minimum Dominating Set* (fuel limit is 1) and Cops & Robber (unlimited fuel) are the extreme cases of our problem. It would be natural to guess that if the amount of fuel the cops possess is some fixed integer f, then the problem is related to distance f domination. Indeed, for some graph classes (e.g. for trees), the problems coincide. Surprisingly, the intuition that Cops & Robber and *Minimum f-Dominating Set* (the classical NP-complete problem) should be similar from the computational complexity point of view is wrong. The main result of this paper is that the problem deciding if k cops can win on an undirected graph is PSPACE-complete even for $f = 2$. Another motivation for our work is the long time open question on the computational complexity of the Cops & Robber problem (without power constrains) on undirected graphs. In 1995, Goldstein and Reingold [8] have shown that the classical Cops & Robber game is EXPTIME-hard on *directed* graphs and conjectured that similar holds for undirected graphs. However, even NP-hardness of the problem was not known until very recently [4]. By our results, in the game on an n-vertex undirected graph if the number of steps each cop is allowed to make is at most some polynomial of k, then deciding if k cops can win is PSPACE-complete.

2   Basic Definitions and Preliminaries

We consider finite undirected graphs without loops or multiple edges. The vertex set of a graph G is denoted by $V(G)$ and its edge set by $E(G)$, or simply by

V and E if this does not create confusion. If $U \subseteq V(G)$ then the subgraph of G induced by U is denoted by $G[U]$. For a vertex v, the set of vertices which are adjacent to v is called the *(open) neighborhood* of v and denoted by $N_G(v)$. The *closed neighborhood* of v is the set $N_G[v] = N_G(v) \cup \{v\}$. If $U \subseteq V(G)$ then $N_G[U] = \bigcup_{v \in U} N_G[v]$. The *distance* $\operatorname{dist}_G(u, v)$ between a pair of vertices u and v in a connected graph G is the number of edges in a shortest u, v-path in G. For a positive integer r, $N_G^r[v] = \{u \in V(G) \colon \operatorname{dist}_G(u, v) \le r\}$. Whenever there is no ambiguity we omit the subscripts.

The Cops & Robber game can be defined as follows. Let G be a graph, and let $f > 0$ be an integer. The game is played by two players: the cop-player $\mathcal{C}$ and the robber player $\mathcal{R}$, which make moves alternately. The cop-player $\mathcal{C}$ has a team of k cops who attempt to capture the robber. At the beginning of the game this player selects vertices and put cops on these vertices. Then $\mathcal{R}$ puts the robber on a vertex. The players take turns starting with $\mathcal{C}$. At every turn each of the cops can be either moved to an adjacent vertex or kept on the same vertex, and during the whole game each of the cops can be moved from a vertex to another vertex at most f times in total. In other words, each of the cops has an amount of fuel which allows him to make at most f steps. Let us note that several cops can occupy the same vertex at some move. Similarly, $\mathcal{R}$ responds by moving the robber to an adjacent vertex or keeping him on same vertex. It is said that a cop *catches* (or captures) the robber at some move if at that move they occupy the same vertex. Notice that even if a cop cannot move to adjacent vertex (run out of fuel), he is still active and the robber cannot move to the vertex occupied by the cop without being caught. The cop-player wins if one of his cops catches the robber. Player $\mathcal{R}$ wins if he can avoid such a situation, or equivalently, to survive for $kf + 1$ moves, since it can be assumed that at least one cop is moved at each step (otherwise the robber can either keep his position or improve it). For an integer f and a graph G, we denote by $c_f(G)$ the minimum number k of cops sufficient for $\mathcal{C}$ to win on graph G.

We define the *position* of a cop as a pair (v, s) where $v \in V(G)$ and s is an integer, $0 \le s \le f$. Here v is the vertex occupied by the cop, and s is the number of moves along edges (amount of fuel) which the cop can do. The *position of a team* of k cops (or *position of cops*) is a multiset $((v_1, s_1), \ldots, (v_k, s_k))$, where (v_i, s_i) is the position of the i-th cop. For the *initial* position, all $s_i = f$. The *position of the robber* is a vertex of the graph occupied by him.

We consider the following COPS AND ROBBER decision problem:

Input: A connected graph G and two positive integers k, f.

Question: Is $c_f(G) \le k$?

Let us finish the section on preliminaries with the proof of relations between Cops & Robber and r-domination announced in Introduction. The Cops & Robber problem with restricted power is closely related to domination problems. Let r be a positive integer. A set of vertices $S \subset V(G)$ of a graph G is called an r-*dominating set* if for any $v \in V(G)$, there is $u \in S$ such that $\operatorname{dist}(u, v) \le r$. The *r-domination number* $\gamma_r(G)$ is the minimum k such that there is an r-dominating set with at most k vertices. Then $\gamma_1(G)$ is the domination number of G.

The proof of the following observation is straightforward.

Observation 1. *For any connected graph G, $c_1(G) = \gamma_1(G)$.*

For $f > 1$, the values $c_f(G)$ and $\gamma_f(G)$ can differ arbitrarily. Consider, for example, the graph G which is the union of k complete graphs K_k with one additional vertex joined with all vertices of these copies of complete graphs by paths of length f. It can be easily seen that $\gamma_f(G) = 1$ but $c_f(g) = k$. Still, for some graph classes (e.g. for trees) these numbers are equal. Recall, that the *girth* of a graph G, denoted by $g(G)$, is the length of a shortest cycle in G (if G is acyclic then $g(G) = \infty$).

Theorem 1. *Let $f > 0$ be an integer and let G be a connected graph of girth at least $4f - 1$. Then $c_f(G) = \gamma_f(G)$.*

Proof. The proof of $\gamma_f(G) \leq c_f(G)$ is trivial. To prove that $c_f(G) \leq \gamma_f(G)$, we give a winning strategy of $\gamma_f(G)$ cops. Suppose that S is an f-dominating set in G of size $\gamma_f(G)$. The cops are placed on the vertices of S. Suppose that the robber occupies a vertex u. Then the cops from vertices of $S \cap N_G^{2f-1}$ move towards the vertex occupied by the robber at the current moment along the shortest paths. We claim that the robber is captured after at most f moves of the cops. Notice that the robber can move at distance at most $f-1$ from u before the cops make f moves. Because $g(G) \geq 4f - 1$, the paths along which the cops move are unique. Suppose that the robber is not captured after $f - 1$ moves of the cop-player, and the robber occupies a vertex w after his $f-1$ moves. Since S is an f-dominating set, there is a vertex $z \in S$ such that $\mathrm{dist}_G(w, z) \leq f$. Using the fact that $g(G) \geq 4f - 1$, and since the robber was not captured before, we observe that the cop from z moved to w along the shortest path between z and w and by his f-th move he has to enter w and capture the robber.

3   PSPACE-Completeness

It immediately follows from Observation 1 that it is NP-complete to decide whether $c_1(G) \leq k$. Here we prove that for any $f \geq 2$, the problem is much more difficult.

Theorem 2. *For any $f \geq 2$, the* Cops and Robber *problem is* PSPACE-*complete.*

Proof. We reduce the PSPACE-complete Quantified Boolean Formula in Conjunctive Normal Form (QBF) problem [6]. For a set of Boolean variables $x_1, x_2, \ldots, x_n$ and a Boolean formula $F = C_1 \wedge C_2 \wedge \cdots \wedge C_m$, where C_j is a clause, the QBF problem asks whether the expression $\phi = Q_1 x_1 Q_2 x_2 \cdots Q_n x_n F$ is *true*, where for every i, Q_i is either $\forall$ or $\exists$. For simplicity, we describe the reduction for the case $f = 2$. For $f > 2$, the proof uses the same ideas, but the construction is slightly more involved. We provide more details for the case $f > 2$ right after we finish the proof of Lemma 5.

Given a quantified Boolean formula ϕ, we construct an instance G, k of our problem in several steps. We first construct a graph $G^{(1)}$ and show that if the considered strategies are restricted to some specific conditions, then ϕ is true if and only if the cop-player can win on $G^{(1)}$ with a a specific number of cops.

Constructing $G^{(1)}$. For every $Q_i x_i$ we introduce a gadget graph G_i. For $Q_i = \forall$, we define the graph $G_i(\forall)$ with vertex set

$$\{u_{i-1}, u_i, x_i, \overline{x}_i, y_i, \overline{y}_i, z_i\}$$

and edge set

$$\{u_{i-1}y_i, y_iu_i, u_{i-1}\overline{y}_i, \overline{y}_iu_i, z_ix_i, x_iy_i, z_i\overline{x}_i, \overline{x}_i\overline{y}_i\}.$$

For $Q_i = \exists$, we define $G_i(\exists)$ as the graph with vertex set

$$\{u_{i-1}, u_i, x_i, \overline{x}_i, y_i, z_i\}$$

and edge set

$$\{u_{i-1}y_i, y_iu_i, z_ix_i, x_iy_i, z_i\overline{x}_i, \overline{x}_iy_i, x_i\overline{x}_i\}.$$

The graphs $G_i(\forall)$ and $G_i(\exists)$ are shown in Fig 1. Observe that the vertex u_i appears both in the gadget graph G_i and in the gadget G_{i+1} for $i \in \{1, 2, \ldots, n-1\}$. Let $U_i = \{u_0, \ldots, u_i\}$, $Y_i = \{y_1, \ldots, y_i\}$ and $\overline{Y}_i = \{\overline{y}_j | 1 \le j \le i\}$ for $1 \le i \le n$. The graph $G^{(1)}$ also has vertices $C_1, C_2, \ldots, C_m$ corresponding to clauses. The vertex x_i is joined with C_j by an edge if C_j contains the literal x_i, and $\overline{x}_i$ is joined with C_j if C_j contains the literal $\overline{x}_i$. The vertex u_n is connected with all vertices $C_1, C_2, \ldots, C_m$ by edges. An example of $G^{(1)}$ for $\phi = \exists x_1 \forall x_2 \, (x_1 \vee \overline{x}_2) \wedge (\overline{x}_1 \vee x_2)$ is shown in Fig 1.

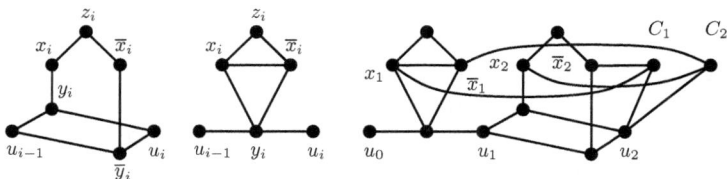

Fig. 1. Graphs $G_i(\forall)$, $G_i(\exists)$ and $G^{(1)}$

We proceed to prove several properties of $G^{(1)}$.

Lemma 1. *Suppose that the robber can use only strategies with the following properties:*

- *he starts from u_0,*
- *he cannot remain in vertices $u_0, \ldots, u_n$,*

- *he moves along edges $u_{i-1}y_i, y_iu_i, u_{i-1}\overline{y}_i, \overline{y}_iu_i$ only in the direction induced by this ordering, i.e. these edges are "directed" for him.*

Assume also that n cops on $G^{(1)}$ use strategy with the following restrictions:

- *they start from vertices $z_1, \ldots, z_n$,*
- *the cop on z_i cannot move until the robber reaches vertices y_i or $\overline{y}_i$.*

Then $\phi = true$ if and only if n cops have a winning strategy on $G^{(1)}$.

Proof. Assume that $\phi = true$. We describe a winning strategy for the cop-player. The cops start by occupying vertices $z_1, \ldots, z_n$. Suppose that at some point during the game the robber moves to vertex u_{i-1}. Since he cannot stay in this vertex, we have that he has to move to y_i or $\overline{y}_i$. If the robber moves to y_i from u_{i-1} of $G_i(\forall)$, then the cop occupying z_i moves to x_i and the corresponding variable x_i is set to *true*. If the robber moves to $\overline{y}_i$, then the cop moves to $\overline{x}_i$ and we set $x_i = false$. It means that for a quantified variable $\forall x_i$, the robber chooses the value of x_i. Notice that the robber cannot stay on y_i or $\overline{y}_i$ because a cop which still has fuel occupies an adjacent vertex. Therefore he has to move to u_i. If the robber moves to y_i of $G_i(\exists)$ from u_{i-1}, then the cop player replies by moving a cop from z_i to x_i or $\overline{x}_i$, and this represents the value of the variable x_i. Hence for a quantified variable $\exists x_i$, the cops choose the value of x_i. Then the robber is forced to move to u_i—it is senseless for him to move to x_i or $\overline{x}_i$ or stay in y_i. Since $\phi = true$, we have that the cops in $G_i(\exists)$ gadgets can move in such a way that when the robber occupies the vertex u_n, every vertex C_j has at least one neighbor occupied by a cop. If the robber moves to some vertex C_j then a cop moves to C_j and the robber is captured. Thus the cops win in this case.

Suppose that $\phi = false$. We describe a winning strategy for the robber-player against cops occupying vertices $z_1 \ldots, z_n$. The robber starts moving from u_0 toward the vertex u_n along some path in $G^{(1)}$. Every time the robber steps on a vertex y_i of $G_i(\forall)$, there should be a cop responding to this move by moving to x_i from s_i. Otherwise the robber can stay in this vertex, and since cops from $z_1, \ldots, z_{i-1}$ do not have enough fuel to reach y_i and the cops from $z_{i+1}, \ldots, z_n$ cannot move because of our restrictions, the robber-player wins in this case. By the same arguments, if the robber occupies $\overline{y}_i$, then the cop from z_i has to move to $\overline{x}_i$. It means that in the same way as above the robber chooses the value of the variable x_i. Similarly, if the robber occupies the vertex y_i in $G_i(\exists)$, then a cop is forced to move from z_i to x_i or $\overline{x}_i$, and this cop can choose which vertex from x_i and $\overline{x}_i$ to occupy, and now the cop-player chooses the value of the variable x_i. Since $\phi = false$, we have that the robber can choose between y_i and $\overline{y}_i$ in gadgets $G_i(\forall)$ such that no matter how the cop-player chooses to place the cops on x_i or $\overline{x}_i$ in gadgets $G_i(\exists)$, when the robber arrives at u_n at least one vertex C_j is within distance two from vertices x_i and $\overline{x}_i$ which were occupied by cops when the robbers visited y_i or $\overline{y}_i$. Therefore, cops cannot reach this vertex. Then the robber moves to C_j and remains there.

Now we are going to introduce gadgets that force the players to follow the constraints on moves described in Lemma 1. Then we gradually eliminate all constraints. At first, we construct a gadget F which forces a cop to occupy a given vertex and forbids him to leave it until some moment.

Constructing $F(W,t)$. Let W be a set of vertices, and let $t \notin W$ be a vertex (we use these vertices to attach $F(W,t)$ to other parts of our constructions). We introduce four vertices a, b, c, d, join a and b with t and the vertices of W by edges, and join c and d with t by paths of length two (see Fig 2).

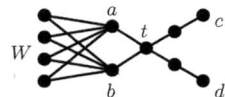

Fig. 2. Graph $F(W,t)$

Properties of $F(W,t)$ are summarized in the following lemma.

Lemma 2. *Let H be a graph such that $V(H) \cap V(F(W,t)) = W \cup \{t\}$. For any winning strategy for the cops on the graph $H' = H \cup F(W,t)$, at least one cop have to be placed on vertices $V(F(W,t)) \setminus (W \cup \{a,b\})$ in the initial position, and if exactly one cop is placed there then he has to occupy t. Moreover, if one cop is placed on vertices $V(F(W,t)) \setminus (W \cup \{a,b\})$ and there are no other cops at distance two from a, b, then the cop cannot leave t while the robber is on one of the vertices of W.*

Proof. The first claim follows from the observation that at least one cop should be placed at distance at most two from c and d. Otherwise the robber can occupy one of these vertices, and he cannot be captured. To prove the second claim, note that only the cop from t can visit vertices a and b. If the cop leaves t then at least one of these vertices is not occupied by cops, and the robber can move there from vertices of W. After that he wins since no cop can reach this vertex.

Our next step is to force restrictions on strategies of the cop-player.

Constructing $G^{(2)}$. We consider the graph $G^{(1)}$. For each $1 \leq i \leq n$, the gadget $F(U_{i-1} \cup Y_{i-1} \cup \overline{Y}_{i-1}, z_i)$ is added. Denote the obtained graph by $G^{(2)}$ (see Fig 3).

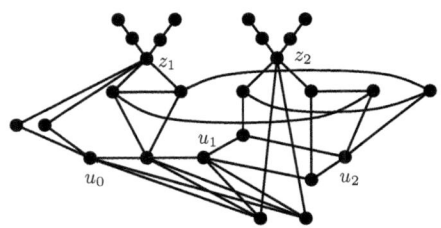

Fig. 3. Graph $G^{(2)}$ for $\phi = \exists x_1 \forall x_2 \ (x_1 \vee \overline{x}_2) \wedge (\overline{x}_1 \vee x_2)$

Lemma 3. *Suppose that the robber can use only strategies with the following properties:*

- *he starts from u_0 if cops are placed on $z_1, \ldots, z_n$,*
- *he cannot remain in vertices $u_0, \ldots, u_n$,*
- *he moves along edges $u_{i-1}y_i, y_iu_i, u_{i-1}\overline{y}_i, \overline{y}_iu_i$ only in the direction induced by this ordering, i.e. these edges are "directed" for him.*

Then $\phi = true$ if and only if n cops have a winning strategy on $G^{(2)}$.

Proof. If $\phi = true$ then n cops can use exactly the same winning strategy as in the proof of Lemma 3. We should only note that it makes no sense for the robber to move to vertices a and b of gadgets F since he would be immediately captured. Suppose that $\phi = false$. If the cops are not placed on $z_1, \ldots, z_n$, then by Lemma 2 the robber wins by staying in one of the pendent vertices of gadgets F. If cops occupy vertices $z_1, \ldots, z_n$, then we can use the same winning strategy for the robber as the proof of Lemma 3. Indeed, by Lemma 2, the cop on z_i cannot move until the robber reaches vertices y_i or $\overline{y}_i$.

In the next stage we add a gadget that forces the robber to occupy u_0 in the beginning of the game.

Constructing $G^{(3)}$. We construct $G^{(2)}$. Then we add vertices p, q, w_1, w_2 and edges w_1w_2, w_2u_0, and then join p and q with w_1 by paths of length two. Finally, we make w_1 to be adjacent to vertices $u_1, \ldots, u_n$ and $C_1, \ldots, C_m$. Denote the obtained graph by $G^{(3)}$ (see Fig 4).

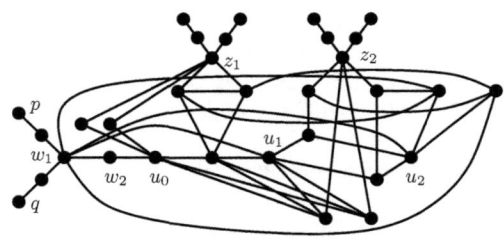

Fig. 4. Graph $G^{(3)}$ for $\phi = \exists x_1 \forall x_2 \ (x_1 \vee \overline{x}_2) \wedge (\overline{x}_1 \vee x_2)$

Lemma 4. *Suppose that the robber can use only strategies with the following properties:*

- *he cannot remain in vertices $u_1, \ldots, u_n$,*
- *he moves along edges $u_{i-1}y_i, y_iu_i, u_{i-1}\overline{y}_i, \overline{y}_iu_i$ only in the direction induced by this ordering, i.e. these edges are "directed" for him.*

Then $\phi = true$ if and only if $n+1$ cops have a winning strategy on $G^{(3)}$.

Proof. Suppose that $\phi = true$. We place $n + 1$ cops on vertices $w_1, z_1, \ldots, z_n$. If the robber chooses vertices of $N_{G^{(3)}}[\{w_1, z_1, \ldots, z_n\}]$, then he can be captured right the next step. If he occupies vertices p, q or pendent vertices of gadgets F, then he can be clearly captured in at most two steps. Suppose that the robber is placed on some vertex y_i or $\overline{y}_i$. If he tries to move to u_{i-1} or u_i, then he is captured by the cop from the vertex w_1. If he moves to some vertex a or b of gadget F attached to a vertex z_j, $j > i$, then he is captured by the cop from z_j. Otherwise he is captured by the cop from the vertex z_i in at most two steps. Thus the only remaining possibility for the robber to avoid the capture is to occupy u_0. In this case the robber from w_1 moves to w_2. Then the robber should leave the vertex u_0, and the cop-player can use the same strategy as before (see Lemma 3). Finally, the robber cannot move to w_1 from vertices $u_1, \ldots, u_n$ and $C_1, \ldots, C_m$, since he would be captured by the cop standing in w_2.

Suppose that $\phi = false$. By Lemma 2 and by construction of $G^{(3)}$, we can assume that the cops are placed on $w_1, z_1, \ldots, z_n$ (otherwise the robber wins by choosing a pendent vertex within distance at least three from the cops). We describe a winning strategy for the robber-player against the cops occupying these vertices. The robber is placed on u_0. Then he waits until some cop is moved to an adjacent vertex. If the cop from w_1 is moved to some vertex different from w_2, then the robber responds by moving to w_2, and he wins by staying in this vertex. Suppose that a cop stays in the vertex w_1 and another cop, say the cop from z_i, moves to an adjacent vertex. The robber responds by moving to one of the vertices a or b of the gadget F attached to z_i and not occupied by cops. Then only the cop from w_1 can try to capture him by moving to some vertex u_j, $j < i$, but in this case the robber can return to u_0 and stay there. It remains to consider the case when a cop moves from w_1 to w_2, but now the robber can use the same winning strategy described before in Lemma 3.

Finally, we attach gadgets to force the remaining restrictions on actions of the robber.

Constructing $G^{(4)}$. We consider $G^{(3)}$. For each $1 \leq i \leq n$, we add vertices f_i, g_i and the edge $f_i g_i$. Then f_i is joined by edges with u_i, y_i and $\overline{y}_i$ (if it exists). Finally, we add the gadget $F(U_{i-1} \cup Y_i \cup \overline{Y}_i, g_i)$. The construction is shown in Fig 5.

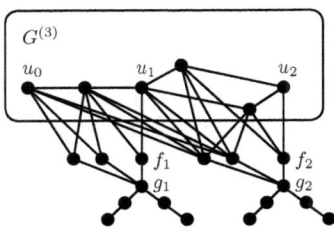

Fig. 5. Graph $G^{(4)}$ for $\phi = \exists x_1 \forall x_2 \ (x_1 \vee \overline{x}_2) \wedge (\overline{x}_1 \vee x_2)$

Now we are in the position to prove the SPACE-hardness result.

Lemma 5. *For the constructed graph $G^{(4)}$, we have $\phi = true$ if and only if $2n + 1$ cops have a winning strategy on $G^{(4)}$.*

Proof. Suppose that $\phi = true$. The cops are placed on the vertices $w_1, z_1, \ldots, z_n$, $g_1, \ldots, g_n$. The winning strategy for the cop-player is constructed as in Lemma 4 with one addition: if the robber reaches the vertex u_i, then a cop is moved from g_i to f_i. Then the robber cannot stay in u_i or move to y_i or to $\overline{y}_i$. Notice also that if the robber is on y_i or $\overline{y}_i$ then he cannot move to u_{i-1} because he would be captured in one step by the cop from f_{i-1} or w_2 if $i = 1$.

Let $\phi = false$. We can assume that the cops are occupying $w_1, z_1, \ldots, z_n$, $g_1, \ldots, g_n$ because otherwise the robber wins by selecting one of the pendent vertices at distance 2 from one of the cop-free vertices. By Lemma 2, no cop can leave g_i for $1 \leq i \leq n$, before the robber reaches the vertex u_i. But then the robber wins by using exactly same strategy we described in Lemma 4. $\qquad \square$

This concludes the proof of the PSPACE-hardness for $f = 2$. For $f > 2$, the proof is very similar and here we sketch only the most important differences. In particular, the graph $G^{(1)}$ should be modified in the following way: for each $1 \leq i \leq n$, we add a vertex z'_i and join it with the vertex z_i by a path of length $f - 2$. For this graph, it is possible to prove the claim similar to Lemma 1 with the difference that the cops should start from vertices $z'_1 \ldots, z'_n$ and with additional condition that the robber cannot leave u_0 until some cop enters one of the vertices $z_0, \ldots, z_n$, and then to "enforce" special strategies for the players.

To complete the proof of the theorem, it remains to show that our problem is in PSPACE.

Lemma 6. *For every integers $f, k \geq 1$ and an n-vertex graph G, it is possible to decide whether $c_f(G) \leq k$ by making use of space $O(kfn^{O(1)})$.*

Proof. The proof is constructive. We describe a recursive algorithm which solves the problem. Note that we can consider only strategies of the cop-player such that at least one cop is moved to an adjacent vertex. Otherwise, if all cops are staying in old positions, the robber can only improve his position.

Our algorithm uses a recursive procedure $W(P, u, l)$, which for a non negative integer l, position of the cops $P = ((v_1, s_1), \ldots, (v_k, s_k))$ such that $l = s_1 + \ldots + s_k$, and a vertex $u \in V(G)$, returns $true$ if k cops can win starting from the position P against the robber which starts from the vertex u, and the procedure returns $false$ otherwise. Clearly, k cops can capture the robber on G if and only if there is an initial position P_0 such that for any $u \in V(G)$, $W(P_0, u, l) = true$ for $l = kf$.

If $l = 0$ then $W(P, u, l) = true$ if and only if $u = v_i$ for some $1 \leq i \leq k$. Suppose that $l > 0$. Then $W(P, u, l) = true$ in the following cases:

- $u = v_i$ for some $1 \leq i \leq k$,
- $u \in N_G(v_i)$ and $s_i > 0$ for some $1 \leq i \leq k$,

- there is a position $P' = ((v'_1, s'_1), \ldots, (v'_k, s'_k))$ such that the cops can go from P to P' in one step, and for any $u' \in N_G[u]$, $W(P', u', l') = true$ where $l' = s'_1 + \ldots + s'_k < l$.

Since all positions can be listed (without storing them) by using polynomial space, the number of possible moves of the robber is at most n, and the depth of the recursion is at most kf, the algorithm uses space $O(kfn^{O(1)})$.

Now the proof of the theorem follows from Lemmata 5 and 6.

Notice that the PSPACE-hardness proof also holds for the case when f is a part of the input. However, our proof only shows that the problem is in PSPACE only for $f = n^{O(1)}$.

4 Conclusion

In this paper we introduced the variant of the Cops & Robber game with restricted resources and have shown that the problem is PSPACE-complete for every $f > 1$. In fact, our proof also shows that the problem is PSPACE-complete even when f is at most some polynomial of the numer of cops. One of the long standing open questions in Cops & Robber games, is the computational complexity of the classical variant of the game on undirected graphs without restrictions on the power of cops. In 1995, Goldstein and Reingold [8] conjectured that this problem is EXPTIME-hard. On the other hand, we do not know any example, where to win cops are required to make exponential number of steps (or fuel). This lead to a very natural question: How many steps along edges each cop needs in the Cops & Robber game without fuel restrictions?

References

1. Aigner, M., Fromme, M.: A game of cops and robbers. Discrete Appl. Math. 8, 1–11 (1984)
2. Alspach, B.: Searching and sweeping graphs: a brief survey. Matematiche (Catania) 59, 5–37 (2006)
3. Dumitrescu, A., Suzuki, I., Zylinski, P.: Offline variants of the "lion and man" problem. In: Proceedings of the 23d annual symposium on Computational Geometry (SCG 07), pp. 102–111. ACM, New York (2007)
4. Fomin, F.V., Golovach, P.A., Kratochvíl, J.: On tractability of cops and robbers game. In: Ausiello, G., Karhumäki, J., Mauri, G., Ong, C.-H.L. (eds.) IFIP TCS. IFIP, vol. 273, pp. 171–185. Springer, Heidelberg (2008)
5. Fomin, F.V., Thilikos, D.M.: An annotated bibliography on guaranteed graph searching. Theor. Comput. Sci. 399, 236–245 (2008)
6. Garey, M.R., Johnson, D.S.: Computers and intractability. W. H. Freeman and Co, San Francisco (1979); A guide to the theory of NP-completeness, A Series of Books in the Mathematical Sciences
7. Goldenberg, M., Kovarsky, A., Wu, X., Schaeffer, J.: Multiple agents moving target search. In: Proceedings of the 18th international joint conference on Artificial Intelligence (IJCAI '03), pp. 1536–1538. Morgan Kaufmann Publishers Inc., San Francisco (2003)

8. Goldstein, A.S., Reingold, E.M.: The complexity of pursuit on a graph. Theoret. Comput. Sci. 143, 93–112 (1995)
9. Guibas, L.J., Claude Latombe, J., Lavalle, S.M., Lin, D., Motwani, R.: A visibility-based pursuit-evasion problem. International Journal of Computational Geometry and Applications 9, 471–494 (1996)
10. Isaacs, R.: Differential games. A mathematical theory with applications to warfare and pursuit, control and optimization. John Wiley & Sons Inc., New York (1965)
11. Ishida, T., Korf, R.E.: Moving target search. In: Proceedings of the International joint conference on Artificial Intelligence (IJCAI'91), pp. 204–211 (1991)
12. Ishida, T., Korf, R.E.: Moving-target search: A real-time search for changing goals. IEEE Trans. Pattern Anal. Mach. Intell. 17, 609–619 (1995)
13. Littlewood, J.E.: Littlewood's miscellany. Cambridge University Press, Cambridge (1986); Edited and with a foreword by Béla Bollobás
14. Loh, P.K.K., Prakash, E.C.: Novel moving target search algorithms for computer gaming. Comput. Entertain. 7, 1–16 (2009)
15. Megiddo, N., Hakimi, S.L., Garey, M.R., Johnson, D.S., Papadimitriou, C.H.: The complexity of searching a graph. J. Assoc. Comput. Mach. 35, 18–44 (1988)
16. Moldenhauer, C., Sturtevant, N.R.: Evaluating strategies for running from the cops. In: Proceedings of the 21st International Joint Conference on Artificial Intelligence (IJCAI 2009), pp. 584–589 (2009)
17. Moldenhauer, C., Sturtevant, N.R.: Optimal solutions for moving target search. In: Proceedings of the 8th International Joint Conference on Autonomous Agents and Multiagent Systems (AAMAS 2009), IFAAMAS, pp. 1249–1250 (2009)
18. Nowakowski, R., Winkler, P.: Vertex-to-vertex pursuit in a graph. Discrete Math. 43, 235–239 (1983)
19. Quilliot, A.: Some results about pursuit games on metric spaces obtained through graph theory techniques. European J. Combin. 7, 55–66 (1986)
20. Sgall, J.: Solution of David Gale's lion and man problem. Theor. Comput. Sci. 259, 663–670 (2001)
21. Sugihara, K., Suzuki, I., Yamashita, M.: The searchlight scheduling problem. SIAM J. Comput. 19, 1024–1040 (1990)

Path Schematization for Route Sketches

Daniel Delling[1], Andreas Gemsa[2], Martin Nöllenburg[2,3,*], and Thomas Pajor[2]

[1] Microsoft Research Silicon Valley, 1065 La Avenida, Mountain View, CA 94043
dadellin@microsoft.com
[2] Karlsruhe Institute of Technology, P.O. Box 6980, 76128 Karlsruhe, Germany
{gemsa, noellenburg, pajor}@kit.edu
[3] Department of Computer Science, University of California, Irvine, CA 92697-3435

Abstract. Motivated from drawing route sketches, we consider the following path schematization problem. We are given a simple embedded polygonal path $P = (v_1, \ldots, v_n)$ and a set $\mathcal{C}$ of admissible edge orientations including the coordinate axes. The problem is to redraw P schematically such that all edges are drawn as line segments that are parallel to one of the specified orientations. We also require that the path preserves the orthogonal order and that it remains intersection-free. Finally, we want the drawing to maximize the number of edges having their preferred edge direction and to minimize the path length.

In this paper we first present an efficient two-step approach for schematizing *monotone* paths. It consists of an $O(n^2)$-time algorithm to assign edge directions optimally and a subsequent linear program to minimize the path length. In order to schematize non-monotone paths we propose a heuristic that first splits the input into k monotone subpaths and then combines the optimal embeddings of the monotone subpaths into a single, intersection-free embedding of the initial path in $O(k^2 + n)$ time.

1 Introduction

Simplification and schematization of map objects are well-known operators in cartographic *generalization*, i.e., the process to adapt map content to its scale and use. Simplification usually reduces unnecessary complexity, e.g., by removing extraneous vertices of a polygonal line while still maintaining its overall appearance. Schematization, however, may abstract more drastically from geographic reality as long as the intended map use allows for it. Public transport maps are good examples of schematization, where edge orientations are limited to a small number of slopes and edge lengths are no longer drawn to scale [10]. In spite of all distortions, such maps usually work well.

In this paper we consider a path schematization problem that is motivated from visualizing routes in road networks. Routes typically begin and end in residential or commercial areas, where roads are mostly used only for short distances of a few meters up to a few hundred meters. As soon as the route leaves the city limits, however, country roads and highways tend to be used for distances

---

* Supported by grant NO 899/1-1 of the German Research Foundation (DFG).

H. Kaplan (Ed.): SWAT 2010, LNCS 6139, pp. 285–296, 2010.

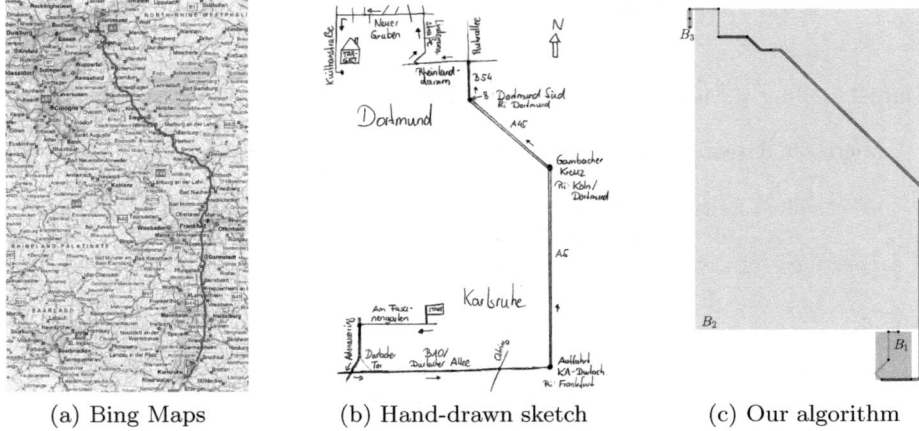

(a) Bing Maps   (b) Hand-drawn sketch   (c) Our algorithm

Fig. 1. Comparison of different methods for drawing a route

ranging from a few up to hundreds of kilometers. Moreover, optimal routes tend to follow a general driving direction and deviations from this direction are rare.

Commercial route planners typically present driving directions for such routes as a graphical overview of the route highlighted in a traditional road map (see Fig. 1a) in combination with a textual step-by-step description. The overview map is good for giving a general idea of the route, but due to its small scale it often does not succeed in showing details of the route, in particular for short roads in the vicinity of start and destination and off the main highways. Textual descriptions are accurate when used at the right moment but there is a high risk of loss of context. On the other hand, a manually drawn route sketch often shows the whole route in a single picture, where each part of the route has its own appropriate scale: important turning points along the route and short residential roads are enlarged while long stretches of highways and country roads are shortened. Edges are often aligned with a small set of orientations rather than being geographically accurate [12]. Figure 1b gives an example. In spite of the cartographic error, such route sketches are often easier to read than textual descriptions and traditional road maps—at least if the user's *mental* or *cognitive map*, i. e., a rough idea of the geographic reality, is preserved [8,11].

We formalize the application problem of drawing route sketches as a geometric path schematization problem. Given a plane embedding of a path P, the goal is to find a short *schematic* embedding of P that is as similar to the input embedding as possible but uses only a restricted set $\mathcal{C}$ of edge orientations. We call such an embedding $\mathcal{C}$-*oriented*. For our application of route sketches, the path P is given by n important points along the route. These important points can be turns, important junctions, highway ramps, etc.

Related Work. Similar path schematization problems have been studied before. Neyer [9] proposed an algorithm to solve the $\mathcal{C}$-oriented line simplification problem, where a $\mathcal{C}$-oriented simplification Q of a polygonal path P is to be computed that uses a minimum number of edges. Furthermore, Q must have Fréchet distance at most ε from P. For a constant-size set $\mathcal{C}$ the algorithm has a running time of $O(kn^2 \log n)$, where n is the number of vertices of P and k is the number of vertices of Q. Merrick and Gudmundsson [7] studied a slightly relaxed version of the same problem and gave an $O(n^2|\mathcal{C}|^3)$-time algorithm to compute a $\mathcal{C}$-oriented simplification of P that is within Hausdorff distance at most ε of P. Agrawala and Stolte [1] designed a system called *LineDrive* that uses heuristic methods based on simulated annealing in order to render route maps similar to hand-drawn sketches. While their system allows distortion of edge lengths and angles, the resulting paths are neither $\mathcal{C}$-oriented nor can hard quality guarantees be given. They did, however, implement and evaluate the system in a study that showed that users generally preferred LineDrive route maps over traditional route maps. Brandes and Pampel [2] studied the path schematization problem in the presence of orthogonal order constraints [8] in order to preserve the mental map. They showed that deciding whether a rectilinear schematization of a path P exists that preserves the orthogonal order of P is NP-hard. They also showed that schematizing a path using arbitrarily oriented unit-length edges is NP-hard.

Our Contribution. Due to the NP-hardness of rectilinear path schematization [2], we cannot hope for an algorithm that solves the general $\mathcal{C}$-oriented path schematization problem efficiently. Rather, we present an efficient algorithm to solve the corresponding *monotone* path schematization problem, in which the input is restricted to x- or y-monotone paths (Section 3). The algorithm consists of two steps: First, we compute in quadratic time a $\mathcal{C}$-oriented schematization of the input path that preserves the orthogonal order of the input and has minimum schematization cost (to be defined). Next, we use a linear program to minimize the total path length such that the schematization cost remains minimum.

In order to use this algorithm to generate route sketches for non-monotone input paths, we present a three-step heuristic approach (Section 4): We first split the path in linear time into a minimum number k of monotone subpaths, then we use the previous algorithm to optimally schematize each subpath, and finally we combine the k schematized subpaths into a single intersection-free route sketch for the non-monotone input path in $O(k^2 + n)$ time. Note that routes in practice tend to follow a general direction given by the straight line connecting start and destination. Thus if a path is not monotone itself, then it usually consists of a very small number of monotone subpaths (see the example in Fig. 1c, which decomposes into three monotone subpaths).

For omitted proofs and details we refer to the full version of this paper [4].

2   Preliminaries

Let $P = (v_1, \ldots, v_n)$ be a path with edges $v_i v_{i+1}$ for $1 \leq i \leq n-1$. For a vertex v and an edge e of P we say $v \in P$ and $e \in P$. A *plane embedding* $\pi : P \to \mathbb{R}^2$ maps each vertex $v_i \in P$ to a point $\pi(v_i) = (x_\pi(v_i), y_\pi(v_i))$ and each edge $uv \in P$ to the line segment $\pi(uv) = \overline{\pi(u)\pi(v)}$ such that π is a simple polygonal path with vertex set $\{\pi(v_1), \ldots, \pi(v_n)\}$. We denote the length of an edge e in π as $|\pi(e)|$. An *embedded path* is a pair (P, π) of a path P and a plane embedding π of P. Let $\mathcal{C} = \{\gamma_1, \ldots, \gamma_k\}$ be a set of angles w.r.t. the x-axis that represents the admissible edge orientations. We require that $\{0°, 90°, 180°, 270°\} \subseteq \mathcal{C}$. Reasonable sets of edge directions for route sketches are, e.g., multiples of 30 or 45 degrees. A plane embedding π of a path is called $\mathcal{C}$-*oriented* if the direction of each edge in π corresponds to an angle in $\mathcal{C}$. For an embedding π of P and an edge $e \in P$ we denote by $\alpha_\pi(e)$ the angle of $\pi(e)$ w.r.t. the x-axis. For the input embedding π, we similarly denote by $\omega_\mathcal{C}(e)$ the *preferred angle* $\gamma \in \mathcal{C}$, i.e., the angle in $\mathcal{C}$ that is closest to $\alpha_\pi(e)$. For a $\mathcal{C}$-oriented embedding ρ of P and an edge $e \in P$ the *direction cost* $c_\rho(e)$ captures by how much the angle $\alpha_\rho(e)$ deviates from $\omega_\mathcal{C}(e)$. Then, we define the *schematization cost* $c(\rho)$ as $c(\rho) = \sum_{e \in P} c_\rho(e)$.

Following Misue et al. [8], we say that an embedding ρ of a path P preserves the *orthogonal order* of another embedding π of P if for any two vertices v_i and $v_j \in P$ we have $x_\pi(v_i) \leq x_\pi(v_j)$ if and only if $x_\rho(v_i) \leq x_\rho(v_j)$ and $y_\pi(v_i) \leq y_\pi(v_j)$ if and only if $y_\rho(v_i) \leq y_\rho(v_j)$. In other words, any two vertices keep their above-below and left-right relationship.

3   Monotone Path Schematization

In this section, we solve the monotone $\mathcal{C}$-oriented path schematization problem:

Problem 1. Given an embedded x- or y-monotone path (P, π), a set $\mathcal{C}$ of edge orientations and a minimum length $\ell_{\min}(e)$ for each edge $e \in P$, find a plane $\mathcal{C}$-oriented embedding ρ of P that

   (i)   preserves the orthogonal order of the input embedding π,
  (ii)   minimizes the schematization cost $c(\rho)$,
 (iii)   respects the individual minimum edge lengths $|\rho(e)| \geq \ell_{\min}(e)$, and
 (iv)   minimizes the total path length $\sum_{e \in P} |\rho(e)|$.

Note that schematization cost and total path length are two potentially conflicting optimization criteria. Primarily, we want to find an embedding that minimizes the schematization cost (see Section 3.1). In a second step, we minimize the total path length of that embedding without changing the previously assigned edge directions (see Section 3.2). The rationale for preserving the orthogonal order of the input is to maintain the user's mental map [5,2,8].

3.1   Minimizing the Schematization Cost

The goal in the first step of our algorithm is to find an embedding with minimum schematization cost. Here we assume that the input path (P, π) is x-monotone; y-monotone paths are schematized analogously. We assign the preferred angle $\omega_\mathcal{C}(e) = \gamma$ to each edge $e \in P$, where $\gamma \in \mathcal{C}$ is the angle closest to $\alpha_\pi(e)$. This takes constant time per edge. It could, however, result in the following conflict.

Consider two subsequent edges e_1, e_2 with $\{\omega_{\mathcal{C}}(e_1), (\omega_{\mathcal{C}}(e_2)\} = \{90°, 270°\}$. Assigning such preferred angles would result in an overlap of e_1 and e_2. In this case, we either set $\omega_{\mathcal{C}}(e_1)$ or $\omega_{\mathcal{C}}(e_2)$ to its next best value, depending on which edge is closer to it. This neither changes the solution nor creates new conflicts since in a plane embedding not both edges can have their preferred direction.

The output embedding ρ must be x-monotone, too, as it preserves the orthogonal order of π. So we can assume that $P = (v_1, \ldots, v_n)$ is ordered from left to right in both embeddings. Let ρ' be any orthogonal-order preserving embedding of P. We start with the observation that in ρ' every edge $e = v_i v_{i+1}$ with $\omega_{\mathcal{C}}(e) \neq 0°$ and $y_{\rho'}(v_i) \neq y_{\rho'}(v_{i+1})$ can be embedded with its preferred direction $\alpha_{\rho'}(e) = \omega_{\mathcal{C}}(e)$. This is achieved by horizontally shifting the whole embedding ρ' right of $x_{\rho'}(v_{i+1})$ (including v_{i+1}) to the left or to the right until the slope of e satisfies $\alpha_{\rho'}(e) = \omega_{\mathcal{C}}(e)$. Due to the x-monotonicity of P no other edges are affected by this shift. We now group all edges $e = uv$ of P into four categories:

1. if $\omega_{\mathcal{C}}(e) = 0°$ and $y_\pi(u) \neq y_\pi(v)$ then e is called horizontal edge (or h-edge);
2. if $y_\pi(u) = y_\pi(v)$ then e is called strictly horizontal edge (or sh-edge);
3. if $\omega_{\mathcal{C}}(e) \neq 0°$ and $x_\pi(u) \neq x_\pi(v)$ then e is called vertical edge (or v-edge);
4. if $x_\pi(u) = x_\pi(v)$ then e is called strictly vertical edge (or sv-edge).

Using these categories, we define the direction cost as follows. All edges e with $\alpha_\rho(e) = \omega_{\mathcal{C}}(e)$ are drawn according to their preferred angle and we assign the cost $c_\rho(e) = 0$. For all edges e with $\alpha_\rho(e) \neq \omega_{\mathcal{C}}(e)$ we assign the cost $c_\rho(e) = 1$. An exception are the sh- and sv-edges, which must be assigned their preferred angle due to the orthogonal ordering constraints. Consequently, we set $c_\rho(e) = \infty$ for any sh- or sv-edge e with $\alpha_\rho(e) \neq \omega_{\mathcal{C}}(e)$. Using the above horizontal shifting argument, the cost $c_\rho(e)$ of any edge e depends only on the vertical distance between its endpoints. So, the schematization cost of an x-monotone embedding ρ is already fully determined by assigning y-coordinates $y_\rho(v)$ to all v of P.

In order to determine an embedding with minimum schematization cost we define $m \leq n - 1$ closed and vertically bounded horizontal strips $s_1, \ldots, s_m$ induced by the set $\{y = y_\pi(v_i) \mid 1 \leq i \leq n\}$ of horizontal lines through the vertices of (P, π). Let these strips be ordered from top to bottom as shown in Fig. 2a. Furthermore we define a dummy strip s_0 above s_1 that is unbounded on its upper side. We say that an edge $e = uv$ crosses a strip s_i and conversely that s_i affects e if $\pi(u)$ and $\pi(v)$ lie on opposite sides of s_i. In fact, to determine the cost of an embedding ρ it is enough to know for each strip whether it has a positive height or not. Our algorithm will assign a symbolic height $h(s_i) \in \{0, 1\}$ to each strip s_i such that the schematization cost is minimum. Note that sh-edges do not cross any strip but rather coincide with some strip boundary. Hence all sh-edges are automatically drawn horizontally and have no direction costs. We can therefore assume that there are no sh-edges in (P, π).

Let $S[i, j] = \bigcup_{k=i}^{j} s_k$ be the union of the strips $s_i, \ldots, s_j$ and let $\mathcal{I}(i, j)$ be the subinstance of the path schematization problem containing all edges that lie completely within $S[i, j]$. Note that $\mathcal{I}(1, m)$ corresponds to the original instance (P, π), whereas in general $\mathcal{I}(i, j)$ is no longer a connected path but a collection of edges. The following lemma is a key to our algorithm.

Lemma 1. *Let $\mathcal{I}(i, j)$ be a subinstance of the path schematization problem and let $s_k \subseteq S[i, j]$ be a strip for some $i \leq k \leq j$. If we assign $h(s_k) = 1$ then $\mathcal{I}(i, j)$ decomposes into the two independent subinstances $\mathcal{I}(i, k-1)$ and $\mathcal{I}(k+1, j)$. The direction costs of all edges affected by s_k are determined by setting $h(s_k) = 1$.*

Proof. We first show that the cost of any edge $e = uv$ that crosses s_k is determined by setting $h(s_k) = 1$. Since u and v lie on opposite sides of s_k we know that $y_\rho(u) \neq y_\rho(v)$. So if e is a v- or sv-edge it can be drawn with its preferred angle and $c_\rho(e) = 0$ regardless of the height of any other strip crossed by e. Conversely, if e is an h-edge it is impossible to draw e horizontally regardless of the height of any other strip crossed by e and $c_\rho(e) = 1$. Recall that sh-edges do not cross any strips. Assume that $k = 2$ in Fig. 2a and we set $h(s_2) = 1$; then edges $v_3 v_4$ and $v_5 v_6$ cross strip s_2 and none of them can be drawn horizontally.

The remaining edges of $\mathcal{I}(i, j)$ do not cross s_k and are either completely contained in $S[i, k-1]$ or in $S[k+1, j]$. Since the costs of all edges affected by s_k are independent of the heights of the remaining strips in $S[i, j] \setminus \{s_k\}$, we solve the two subinstances $\mathcal{I}(i, k-1)$ and $\mathcal{I}(k+1, j)$ independently, see Fig. 2a. $\square$

Our Algorithm. We can now describe our algorithm for assigning symbolic heights to all strips $s_1, \ldots, s_m$ such that the induced embedding ρ has minimum schematization cost. The main idea is to recursively compute an optimal solution for each instance $\mathcal{I}(1, i)$ by finding the best $k \leq i$ such that $h(s_k) = 1$ and $h(s_j) = 0$ for $j = k + 1, \ldots, i$. By using dynamic programming we can compute an optimal solution for $\mathcal{I}(1, m) = (P, \pi)$ in $O(n^2)$ time.

Let $C(k, i)$ for $1 \leq k \leq i$ denote the schematization cost of all edges in the instance $\mathcal{I}(1, i)$ that either cross s_k or have both endpoints in $S[k+1, i]$ if we set $h(s_k) = 1$ and $h(s_j) = 0$ for $j = k+1, \ldots, i$. Let $C(0, i)$ denote the schematization cost of all edges in the instance $\mathcal{I}(1, i)$ if $h(s_j) = 0$ for all $j = 1, \ldots, i$. We use an array T of size $m + 2$ to store the minimum schematization cost $T[i]$ of the instance $\mathcal{I}(1, i)$. Then $T[i]$ is recursively defined as follows

$$T[i] = \begin{cases} \min_{0 \leq k \leq i} (T[k-1] + C(k, i)) & \text{if } 1 \leq i \leq m \\ 0 & \text{if } i = 0 \text{ or } i = -1. \end{cases} \tag{1}$$

Together with $T[i]$ we store the index k that achieves the minimum value in the recursive definition of $T[i]$ as $k[i] = k$. This allows us to compute the actual strip

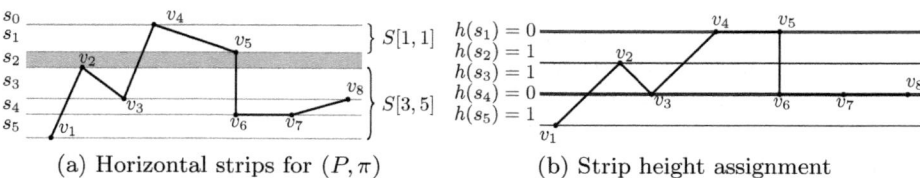

    (a) Horizontal strips for (P, π) (b) Strip height assignment

Fig. 2. Example of (a) an x-monotone embedded input path (P, π) and (b) a $\mathcal{C}$-oriented (multiples of $45°$) orthogonal-order preserving output path (P, ρ)

heights using backtracking. Note that $T[m] < \infty$ since, e.g., the solution that assigns height 1 to every strip induces cost 0 for all sv-edges. Obviously, we need $O(m)$ time to compute each entry in T assuming that the schematization cost $C(k,i)$ is available in $O(1)$ time. This yields a total running time of $O(m^2)$.

The next step is to precompute the schematization cost $C(k,i)$ for any $0 \le k \le i \le m$. This cost is composed of two parts. The first part is the schematization cost of all edges that are affected by s_k. As observed in Lemma 1, all v- and sv-edges crossing s_k have no cost. On the other hand, every h-edge that crosses s_k has cost 1. So we need to count all h-edges in $\mathcal{I}(1,i)$ that cross s_k. The second part is the cost of all edges that are completely contained in $S[k+1,i]$. Since $h(s_{k+1}) = \ldots = h(s_i) = 0$ we observe that any h-edge in $S[k+1,i]$ is drawn horizontally at no cost. In contrast, no v- or sv-edge e in $S[k+1,i]$ attains its preferred angle $\omega_C(e) \ne 0°$. Hence every v-edge in $S[k+1,i]$ has cost 1 and every sv-edge has cost ∞. So we need to check whether there is an sv-edge contained in $S[k+1,i]$ and if this is not the case count all v-edges contained in $S[k+1,i]$.

In order to efficiently compute the values $C(k,i)$ we assign to each strip s_i three sets of edges. Let $H(i)$ (resp. $V(i)$ or $SV(i)$) be the set of all h-edges (resp. v-edges or sv-edges) whose lower endpoint lies on the lower boundary of s_i. We can compute $H(i)$, $V(i)$, and $SV(i)$ in $O(n)$ time for all strips s_i. Then for $k \le i$ the number of h-edges in $H(i)$ that cross s_k is denoted by $\sigma_H(k,i)$ and the number of v-edges in $V(i)$ that do *not* cross s_k is denoted by $\sigma_V(k,i)$. Finally, let $\sigma_{SV}(k,i)$ be the number of sv-edges in $SV(i)$ that do *not* cross s_k. This allows us to recursively compute the values $C(k,i)$, $0 \le k \le i \le m$:

$$C(k,i) = \begin{cases} \infty & \text{if } \sigma_{SV}(k,i) \ge 1 \\ C(k,i-1) + \sigma_H(k,i) + \sigma_V(k,i) & \text{if } k \le i-1 \\ \sigma_H(k,k) & \text{if } k = i. \end{cases} \tag{2}$$

Since each edge appears in exactly one of the sets $H(i)$, $V(i)$, or $SV(i)$ for some i it is counted towards at most m values $\sigma_H(\cdot,i)$, $\sigma_V(\cdot,i)$, or $\sigma_{SV}(\cdot,i)$, respectively. Thus for computing all these values we need $O(nm)$ time. The values $C(k,i)$ can be precomputed in $O(m^2)$ time and require a table of size $O(m^2)$. This can be reduced, however, to $O(m)$ space as follows. We compute and store the values $T[i]$ in the order $i = 1, \ldots, m$. For computing the entry $T[i]$ we use only the values $C(\cdot,i)$. To compute the next entry $T[i+1]$ we first compute the values $C(\cdot,i+1)$ from $C(\cdot,i)$ and then discard all $C(\cdot,i)$. This reduces the required space to $O(m)$. Since $m \le n$ we obtain

Theorem 1. *Our algorithm to compute the array T of path schematization costs requires $O(n^2)$ time and $O(n)$ space.*

It remains to determine the strip height assignments corresponding to the schematization cost in $T[m]$ and show the optimality of that solution. We initialize all heights $h(s_i) = 0$ for $i = 1, \ldots, m$. Recall that $k[i]$ equals the index k that minimized the value $T[i]$ in (1). To find all strips with height 1 we initially set $j = m$. If $k[j] = 0$ we stop; otherwise we assign $h(s_{k[j]}) = 1$, update $j = k[j] - 1$, and continue with the next index $k[j]$ until we hit $k[j] = 0$ for some j encountered in this process. Let ρ be the $\mathcal{C}$-oriented embedding of P induced by this strip height assignment, see Fig. 2b. We now show the optimality of ρ in terms of the schematization cost.

Theorem 2. *Given an x-monotone embedded path (P, π) and a set $\mathcal{C}$ of edge orientations, our algorithm computes a plane $\mathcal{C}$-oriented embedding ρ of P that preserves the orthogonal order of π and has minimum schematization cost $c(\rho)$.*

Proof. Since the path is x-monotone and by construction there are no two adjacent edges with preferred angles $90°$ and $270°$ the embedding ρ is plane. By construction, ρ is $\mathcal{C}$-oriented and it also preserves the orthogonal order of π since the x- and y-ordering of the vertices of P is not altered.

We show that ρ has minimum schematization cost by structural induction. For an instance with a single strip s there are only two possible solutions of which our algorithm chooses the better one. The induction hypothesis is that our algorithm finds an optimal solution for any instance with at most m strips. So let's consider an instance with $m + 1$ strips and let ρ' be any optimal plane $\mathcal{C}$-oriented and orthogonal-order preserving solution for this instance. If all strips s in ρ' have height $h(s) = 0$ then by (1) it holds that $c(\rho) = T[m+1] \leq C(0, m+1) = c(\rho')$. Otherwise, let k be the largest index for which $h(s_k) = 1$ in ρ'. When computing $T[m+1]$ our algorithm also considers the case where s_k is the bottommost strip of height 1, which has a cost of $T[k-1] + C(k, m+1)$. If $h(s_k) = 1$ we can split the instance into two independent subinstances to both sides of s_k by Lemma 1. The schematization cost $C(k, m+1)$ contains the cost for all edges that cross s_k and this cost is obviously the same as in ρ' since $h(s_k) = 1$ in both embeddings. Furthermore, $C(k, m+1)$ contains the cost of all edges in the subinstance below s_k, for which we have by definition $h(s_{k+1}) = \ldots = h(s_{m+1}) = 0$. Since k is the largest index with $h(s_k) = 1$ in ρ' this is also exactly the same cost that this subinstance has in ρ'. Finally, the independent subinstance above s_k has at most m strips and hence $T[k-1]$ is the minimum cost for this subinstance by induction. It follows that $c(\rho) = T[m+1] \leq T[k-1] + C(k, m+1) \leq c(\rho')$. This concludes the proof. □

3.2   Minimizing the Path Length

In the first step of our algorithm we obtained a $\mathcal{C}$-oriented and orthogonal-order preserving embedding ρ with minimum schematization cost for an embedded input path (P, π). The strip heights assigned in that step are either 0 or 1, but this does not yet take into account the actual edge lengths induced by ρ. So in the second step, we adjust ρ such that the total path length is minimized and $|\rho(e)| \geq \ell_{\min}(e)$ for all $e \in P$. We make sure, however, that the orthogonal order and all angles $\alpha_\rho(e)$ remain unchanged.

Note that we can immediately assign the minimum length $\ell_{\min}(e)$ to every horizontal edge e in the input (P, ρ) by horizontally shifting the subpaths on both sides of e. For any non-horizontal edge $e = uv$ the length $|\rho(e)|$ depends only on the vertical distance $\Delta_y(e) = |y_\rho(u) - y_\rho(v)|$ of its endpoints and the angle

$\alpha_\rho(e)$. In fact, $|\rho(e)| = \Delta_y(e)/\sin\alpha_\rho(e)$. So in order to minimize the path length we need to find y-coordinates for all strip boundaries such that $\sum_{e\in P}|\rho(e)|$ is minimized. These y-coordinates together with the given angles for all edges $e \in P$ induce the corresponding x-coordinates of all vertices of P.

So for each strip s_i ($i = 0, \ldots, m$) let y_i denote the y-coordinate of its lower boundary. For every edge $e \in P$ let $t(e)$ and $b(e)$ denote the index of the top- and bottommost strip, respectively, that is crossed by e. Then $\Delta_y(e) = y_{t(e)-1} - y_{b(e)}$. We propose the following linear program (LP) to minimize the path length of a given $\mathcal{C}$-oriented embedded path (P, ρ).

$$\text{Minimize} \qquad \sum_{e\in P,\, \alpha_\rho(e)\neq 0^\circ} \left[\frac{1}{\sin\alpha_\rho(e)} \cdot (y_{t(e)-1} - y_{b(e)})\right]$$

$$\begin{array}{llll}
\text{subject to} & y_{t(e)-1} - y_{b(e)} & \geq & \sin\alpha_\rho(e) \cdot \ell_{\min}(e) \qquad \forall e \in P,\, \alpha_\rho(e) \neq 0^\circ \\
& y_{i-1} - y_i & \geq & 0 \qquad\qquad\qquad\quad \forall s_i \text{ with } h(s_i) = 1 \\
& y_{i-1} - y_i & = & 0 \qquad\qquad\qquad\quad \forall s_i \text{ with } h(s_i) = 0
\end{array}$$

We assign to all vertices their corresponding y-coordinates from the solution of the LP. In a left-to-right pass over P we compute the correct x-coordinates of each vertex v_i from the vertical distance to its predecessor vertex v_{i-1} and the angle $\alpha_\rho(v_{i-1}v_i)$. This yields a modified embedding ρ' that satisfies all our requirements: the path length is minimized; the orthogonal order is preserved due to the x-monotonicity of P and the constraints in the LP to maintain the y-order; by construction the directions of all edges are the same in ρ and ρ'; no edge e is shorter than its minimum length $\ell_{\min}(e)$. Hence, ρ' solves Problem 1 and together with Theorems 1 and 2 we obtain

Theorem 3. *The monotone $\mathcal{C}$-oriented path schematization problem (Problem 1) for a monotone input path of length n can be solved by an $O(n^2)$-time algorithm to compute an embedding ρ of minimum schematization cost followed by solving a linear program to minimize the path length of ρ.*

Note that linear programs can be solved efficiently in $O(n^{3.5}L)$ time [6], where n is the number of variables and L is the number of input bits.

4   Extension to General Simple Paths

In the last section, we showed how to schematize a monotone path. Unfortunately, some routes in road networks are neither x- nor y-monotone, however, they can be decomposed into a (very) limited number of x- and y- monotone subpaths. So, we propose the following three-step heuristic to schematize general simple paths: We first split the input path (P, π) into a minimum number of x- or y-monotone subpaths (P_i, π_i), where π_i equals π restricted to the subpath P_i. We embed each (P_i, π_i) separately according to Section 3. Then, we concatenate the subpaths such that the resulting path (P', ξ) is a simple $\mathcal{C}$-oriented path. Note that this heuristic does not guarantee to preserve the orthogonal order between node pairs of different subpaths.

Splitting an embedded simple path $P = (v_1, \ldots, v_n)$ into the minimal number k of subpaths $P_i, 1 \leq i \leq k$ with the property that each P_i is an x- or y-monotone path can be done in straightforward greedy fashion, starting from v_1. We traverse P until we find the last vertices v' and v'' which are not violating the x- and y-monotonicity, respectively. If v' appears later than v'' on P, we set $P_1 = (v_1, \ldots, v')$, otherwise $P_1 = (v_1, \ldots, v'')$. We continue this procedure until we reach the end of P. This algorithm runs in $O(n)$ time and returns the minimal number k of x- or y-monotone subpaths.

After splitting the input path, we schematize each subpath (P_i, π_i) according to Section 3. We obtain a $\mathcal{C}$-oriented and orthogonal-order preserving embedding ρ_i with minimum schematization cost and minimum path length for each (P_i, π_i). For concatenating these subpaths, we must solve the following problem.

Problem 2. Given a sequence of k embedded x- or y-monotone paths (P_i, ρ_i) with $1 \leq i \leq k$, find an embedding ξ of $P' = P_1 \oplus \cdots \oplus P_k$, where $\oplus$ denotes the concatenation of paths, such that

(i) for each subpath (P_i, ξ_i), the embedding ξ_i is a translation of ρ_i and
(ii) (P', ξ) is a simple $\mathcal{C}$-oriented path.

Our approach is based on iteratively embedding the subpaths $P_1, \ldots, P_k$. We ensure that in each iteration i the embedding of $P_1 \oplus \ldots \oplus P_i$ remains conflict-free, i.e., it has no self-intersections. We achieve this by adding up to three new *path-link edges* between any two adjacent subpaths P_i and P_{i+1}. For each $1 \leq i \leq k$ let B_i denote the bounding box of (P_i, ρ_i). A key operation of the algorithm is *shifting* a subpath P_i (or equivalently a bounding box B_i) by an offset $\Delta = (\Delta_x, \Delta_y) \in \mathbb{R}^2$. This is done by defining the lower left corner of each bounding box B_i as its origin o_i and storing the coordinates of P_i relative to o_i, i.e., $\xi(v) = o_i + \rho_i(v)$. Note that shifting preserves all local properties of (P_i, ρ_i), i.e., the orthogonal order as well as edge lengths and orientations.

Each iteration of our algorithm consists of two steps. First, we *attach* the subpath P_i to its predecessor P_{i-1}. To that end, we initially place (P_i, ξ_i) such that the last vertex u of P_{i-1} and the first vertex v of P_i coincide. Then we add either two path-link edges (if the monotonicity directions of P_{i-1} and P_i are orthogonal) or three path-link edges (if P_{i-1} runs in the opposite direction of P_i) between u and v and shift B_i by finding appropriate lengths for the new edges such that $B_{i-1} \cap B_i = \emptyset$. Paths P_{i-1} and P_i are now conflict-free, but there may still exist conflicts between P_i and paths $P_j (j < i-1)$. These are resolved in a second step that "pushes" any conflicting bounding boxes away from B_i by stretching some of the path-link edges.

Attaching a Subpath. Without loss of generality, we restrict ourselves to the case that P_{i-1} is an x-monotone path from left to right. Let u be the last vertex of P_{i-1} and v be the first vertex of P_i. If P_i is y-monotone we add a horizontal edge $e_1 = uu'$ with $\alpha_\xi(e_1) = 0°$ connecting u to a new vertex u'. Then we also add a vertical edge $e_2 = u'v$ with $\alpha_\xi(e_1) = 90°$ if P_i is upward directed and $\alpha_\xi(e_1) = 270°$ if it is a downward path. Otherwise, if P_i is x-monotone from right to left,

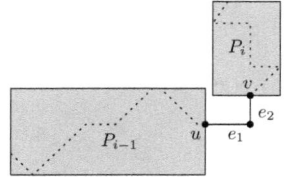

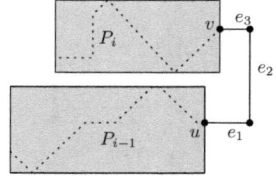

(a) Paths with orthogonal directions   (b) Oppositely directed paths

Fig. 3. Two examples for attaching P_i to P_{i-1} by inserting path-link edges

we add two vertices u' and u'' and three path-link edges $e_1 = uu'$, $e_2 = u'u''$, and $e_3 = u''v$ with $\alpha_\xi(e_1) = 0°$, $\alpha_\xi(e_2) = 90°$ if P_i is above P_{i-1} in π or $\alpha_\xi(e_2) = 270°$ otherwise, and $\alpha_\xi(e_3) = 180°$. Note that we treat each path-link edge as having its own bounding box with zero width or height. It remains to set the lengths of the path-link edges such that $B_i \cap B_j = \emptyset$ by computing the vertical and horizontal overlap of B_{i-1} and B_i. Figure 3 illustrates both situations.

Resolving Conflicts. After adding P_i we have $B_{i-1} \cap B_i = \emptyset$. However, there may still exist conflicts with any B_j, $1 \leq j < i-1$. In order to free up the space required to actually place B_i without overlapping any other bounding box, we push away all conflicting boxes in three steps. For illustration, let P_i be x-monotone from left to right, and let v be the first vertex of P_i. Each bounding box B is defined by its lower left corner $ll(B) = (ll_x(B), ll_y(B))$ and its upper right corner $ur(B) = (ur_x(B), ur_y(B))$. In the first step we identify the leftmost box B' (if any) that is intersected by a line segment that extends from $\xi(v)$ to the right with length equal to the width of B_i. For this box B' we have $ll_y(B') \leq y_\xi(v) \leq ur_y(B')$ and $ll_x(B_i) \leq ll_x(B') \leq ur_x(B_i)$. If there is such a B' let the offset be $\Delta_x = ur_x(B_i) - ll_x(B')$. Now we shift *all* bounding boxes B that lie completely to the right of $ll_x(B')$ to the right by Δ_x. All horizontal path-link edges (which are also considered bounding boxes by themselves) that connect a shifted with a non-shifted path are stretched by Δ_x to keep the two paths connected. Note that there is always a horizontal path-link edge between any two subsequent paths. Next, we inflate B_i, which is currently a horizontal line segment, downwards: we first determine the topmost conflicting box B'' (if any) below a horizontal line through $\xi(v)$, i.e., a box B'' whose x-range intersects the x-range of B_i and for which $ll_y(B_i) \leq ur_y(B'') \leq y_\xi(v)$. If we find such a B'' we define the vertical offset $\Delta_{y1} = ur_y(B'') - ll_y(B_i)$. We shift all bounding boxes B that lie completely below $ur_y(B'')$ downwards by Δ_{y1}. All vertical path-link edges that connect a shifted with a non-shifted box are stretched by Δ_{y1} in order to keep the two boxes connected. Again, there is always a vertical path-link edge between any two subsequent paths. Finally, we inflate B_i upwards, which is analogous to the downward inflation. Figure 4 shows an example and Theorem 4 sums up the insights gained in this section. The proof and more details can be found in the full paper [4].

Theorem 4. *Our algorithm computes a solution (P', ξ) to Problem 2 by adding at most $3(k-1)$ path-link edges to P in $O(k^2 + n)$ time.*

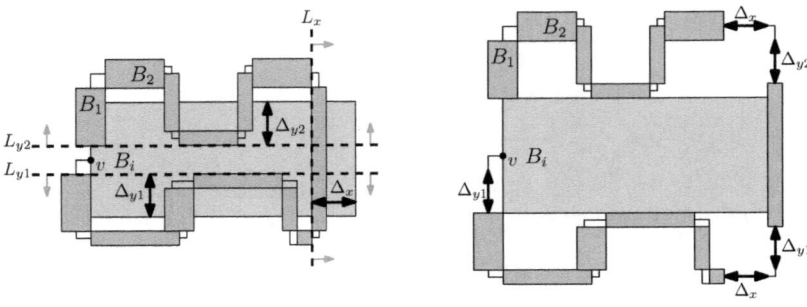

(a) Before resolving the conflict   (b) After resolving the conflict

Fig. 4. Example for iteratively resolving conflicts induced by attaching P_i

Acknowledgments. David Eppstein, Bastian Katz, Maarten Löffler, Ignaz Rutter, and Markus Völker for discussions on the edge-length minimization problem.

References

1. Agrawala, M., Stolte, C.: Rendering effective route maps: Improving usability through generalization. In: Fiume, E. (ed.) SIGGRAPH, pp. 241–249. ACM Press, New York (2001)
2. Brandes, U., Pampel, B.: On the hardness of orthogonal-order preserving graph drawing. In: Tollis, I.G., Patrignani, M. (eds.) GD 2008. LNCS, vol. 5417, pp. 266–277. Springer, Heidelberg (2009)
3. de Berg, M., Cheong, O., van Kreveld, M., Overmars, M.: Computational Geometry: Algorithms and Applications, 3rd edn. Springer, Heidelberg (2008)
4. Delling, D., Gemsa, A., Nöllenburg, M., Pajor, T.: Path Schematization for Route Sketches. Technical Report 2010-02, Karlsruhe Institute of Technology (2010)
5. Dwyer, T., Koren, Y., Marriott, K.: Stress majorization with orthogonal ordering constraints. In: Healy, P., Nikolov, N.S. (eds.) GD 2005. LNCS, vol. 3843, pp. 141–152. Springer, Heidelberg (2006)
6. Karmarkar, N.: A new polynomial-time algorithm for linear programming. In: STOC'84, pp. 302–311. ACM, New York (1984)
7. Merrick, D., Gudmundsson, J.: Path simplification for metro map layout. In: Kaufmann, M., Wagner, D. (eds.) GD 2006. LNCS, vol. 4372, pp. 258–269. Springer, Heidelberg (2007)
8. Misue, K., Eades, P., Lai, W., Sugiyama, K.: Layout adjustment and the mental map. J. Visual Languages and Computing 6(2), 183–210 (1995)
9. Neyer, G.: Line simplification with restricted orientations. In: Dehne, F., Gupta, A., Sack, J.-R., Tamassia, R. (eds.) WADS 1999. LNCS, vol. 1663, pp. 13–24. Springer, Heidelberg (1999)
10. Ovenden, M.: Metro Maps of the World. Capital Transport Publishing (2003)
11. Tversky, B.: Cognitive maps, cognitive collages, and spatial mental models. In: Campari, I., Frank, A.U. (eds.) COSIT 1993. LNCS, vol. 716, pp. 14–24. Springer, Heidelberg (1993)
12. Tversky, B., Lee, P.U.: Pictorial and verbal tools for conveying routes. In: Freksa, C., Mark, D.M. (eds.) COSIT 1999. LNCS, vol. 1661, pp. 51–64. Springer, Heidelberg (1999)

Approximation Algorithms for Free-Label Maximization

Mark de Berg and Dirk H.P. Gerrits

Department of Mathematics and Computer Science
Technische Universiteit Eindhoven, The Netherlands
mdberg@win.tue.nl, dirk@dirkgerrits.com

Abstract. Inspired by air traffic control and other applications where moving objects have to be labeled, we consider the following (static) point labeling problem: given a set P of n points in the plane and labels that are unit squares, place a label with each point in P in such a way that the number of free labels (labels not intersecting any other label) is maximized. We develop efficient constant-factor approximation algorithms for this problem, as well as PTASs, for various label-placement models.

1 Introduction

Air traffic controllers have the important job of monitoring airplanes and warning pilots to change course on any potential collision. They do this using computer screens that show each airplane as a moving point with an associated textual label. The labels hold important information (such as altitude and velocity) that needs to remain readable. As the airplanes move, however, labels may start to intersect. Currently this means air traffic controllers spend a lot of their time moving labels around by hand. We are interested in developing algorithms to automate this process.

Label models. A good labeling for a point set has legible labels, and an unambiguous association between the labels and the points. The latter puts restrictions on the shape of labels and the way they can be placed in relation to points. Various such *label models* have been proposed, most often with labels assumed to be axis-aligned rectangles slightly larger than the text they contain.

In the *fixed-position models*, every point has a finite number of *label candidates* (often 4 or 8), each being a rectangle having the point on its boundary. In particular, in the 1-position (1P) model one designated corner of the label must coincide with the point, in the 2-position (2PH, 2PV) models there is a choice between two adjacent corners, and the 4-position (4P) model allows any corner of the label to coincide with the point (see the upper-left 2x2 block in Figure 1). The *slider models*, introduced by Van Kreveld et al. [11] generalize this. In the 1-slider (1SH, 1SV) models one side of the label is designated, but the label may contain the point anywhere on this side. In the 2-slider (2SH, 2SV) models there

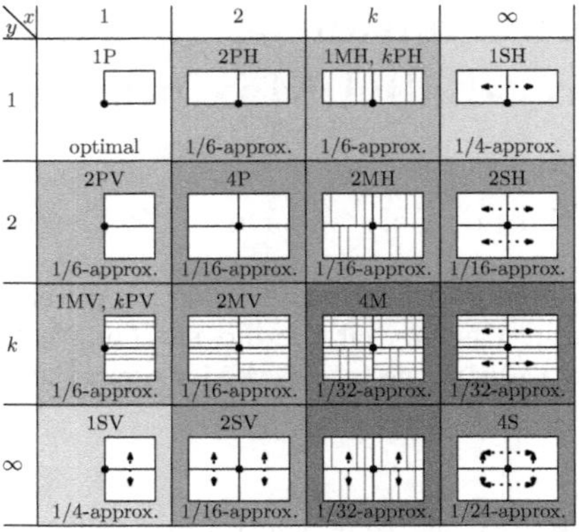

$\underset{y}{\overset{x}{\diagdown}}$	1	2	k	∞
1	1P — optimal	2PH — 1/6-approx.	1MH, kPH — 1/6-approx.	1SH — 1/4-approx.
2	2PV — 1/6-approx.	4P — 1/16-approx.	2MH — 1/16-approx.	2SH — 1/16-approx.
k	1MV, kPV — 1/6-approx.	2MV — 1/16-approx.	4M — 1/32-approx.	— 1/32-approx.
∞	1SV — 1/4-approx.	2SV — 1/16-approx.	— 1/32-approx.	4S — 1/24-approx.

Fig. 1. The fixed-position and slider models, and our constant-factor approximation results for them for the free-label-maximization problem (assuming unit-square labels). The x-axis (y-axis) indicates the number of allowed horizontal (vertical) positions for a label.

is a choice between two opposite sides of the label, and in the 4-slider (4S) model the label can have the point anywhere on its boundary (see the fourth row and column in Figure 1). Erlebach et al. [6] introduced terminology analogous to the slider models for fixed-position models with a non-constant number of positions (1MH, 1MV, 2MH, 2MV, 4M; see the third row and column in Figure 1).

Previous work. A lot of research has gone into labeling static points (as well as polylines and polygons) on cartographic maps. See for instance the on-line Map Labeling Bibliography [15], which currently contains 371 references. This research has focused mostly on two optimization problems. The *size-maximization problem* asks for a labeling of all points with pairwise non-intersecting labels by scaling down all labels uniformly by the least possible amount. This problem is APX-hard (except in the 1P model), even for unit-square labels [7]. Constant-factor approximation algorithms exist for various label models [7,10]. The more widely studied *number-maximization problem* asks for a maximum-cardinality *subset* of the n points to be labeled with pairwise non-intersecting labels of *given* dimensions. Even if all labels are unit squares, this problem is known to be strongly NP-hard for the 1P [8], 4P [7,12], and 4S models [11]. A generalization of this problem concerns weighted points [13] and asks for a maximum-weight subset of the points to be labeled so that, for example, a big city will more likely get a label than a small town. For unit-height rectangular labels this problem admits a polynomial-time approximation scheme (PTAS) for static points in all fixed-position and slider models, both in the unweighted [3,11] and the weighted case [6,13]. For arbitrary rectangles in the unweighted case an $O(1/\log\log n)$-approximation algorithm is known for the fixed-position models [2], but the slider models, the weighted case, and the (non-)existence of a PTAS remain open problems.

Despite the large body of work on labeling static points, virtually no results have been published on labeling moving points. Been et al. [1] studied the unweighted number-maximization problem for static points under continuous zooming in and out by the viewer, which can be seen as points moving on a very specific kind of trajectories. Rostamabadi and Ghodsi [14] studied how to quickly flip and scale the labels of static points to avoid one moving point.

Free-label maximization. As just discussed, previous work has focused on the size-maximization and number-maximization versions of the label-placement problem. By either shrinking the labels, or removing some of them, a labeling is produced without any intersections. However, European air traffic safety regulations require all airplanes to be labeled at all times, with labels of fixed sizes [4]. Thus we must allow label intersections, and naturally want as few of them as possible.

The decision problem of determining whether a labeling without intersections exists for a static point set is strongly NP-complete [7, 12], even if all labels are unit squares. This immediately implies that finding a labeling with the least number of intersecting labels admits no polynomial-time approximation algorithm unless P = NP. Thus we instead seek a labeling with the greatest number of labels that are not intersected, and we call such labels *free*. As this *free-label-maximization problem* had not been previously studied, we have first investigated it for static points, leaving the case of moving points to future research.

Our results. As a first step towards the automatic labeling of moving points in air traffic control we have studied the free-label-maximization problem for static points. For unit-square labels we have developed a simple $O(n \log n)$-time, $O(n)$-space constant-factor approximation algorithm, as well as a PTAS. (In fact, our algorithms work if all labels are translates of a fixed rectangle, since a suitable scaling can transform this case to the case of unit-square labels.) This makes free-label maximization easier than size maximization, as the latter is APX-hard even for unit-square labels. In contrast, techniques used for (approximate) number maximization for unit-square labels easily extend to unit-height labels of differing widths, which seems not to be the case for free-label maximization. Thus the complexity of free-label maximization seems to fall in between that of the size- and number-maximization problems.

We present our constant-factor approximation algorithm in Section 2, and our PTAS in Section 3. The former's approximation guarantees for the various label models are listed in Figure 1. We will only discuss the 2PH, 4P, 1SH, 2SH, and 4S label models; the algorithms and proofs for the other models are analogous. Throughout the paper we assume that no two points have the same x- or y-coordinate, and that labels are open sets (their boundaries may intersect). Neither assumption is essential, but they make our exposition simpler.

2   Constant-Factor Approximations for Unit Squares

Consider the algorithm GREEDYSWEEP, which works as follows. Going through the points from left to right, we label them one-by-one. We call a label candidate

ℓ for a point being processed *freeable* if none of the previously placed labels intersect ℓ, and every point still to be labeled has at least one label candidate that does not intersect ℓ or any previously placed freeable label. We always choose a freeable label candidate if possible, and then also call the resulting label freeable. If a point has no freeable label candidate we pick a non-freeable label candidate that does not intersect any previously placed freeable label (which is always possible by the definition of freeable). In case of ties, we pick the label candidate farthest to the left. (Further ties between equally leftmost label candidates can be broken arbitrarily.)

Lemma 1. *For the free-label-maximization problem with unit-square labels, algorithm* GREEDYSWEEP *gives a 1/4-approximation for the 1SH model and a 1/6-approximation for the 2PH model, and both ratios are tight.*

Proof. Let OPT be some optimal solution, and let ALG be the solution computed by GREEDYSWEEP. Now suppose a point p is labeled with a free label ℓ_p^{OPT} in OPT, but that the label candidate ℓ_p^{OPT} was not freeable when p was being processed by GREEDYSWEEP. Call a label candidate for a point *rightmost* if it is farthest to the right of all label candidates for that point, and define *leftmost* analogously. Since p and all points that already have a label lie to the left of every unprocessed point p', their labels cannot intersect the rightmost label candidate for p' without intersecting all other label candidates for p' as well. Thus all unprocessed points can be labeled with their rightmost label candidate without intersecting ℓ_p^{OPT}. Hence, ℓ_p^{OPT} not being freeable must be caused by a label $\ell_{p'}^{\text{ALG}}$ (either freeable or not) that was placed earlier. We note that $\ell_{p'}^{\text{ALG}}$ cannot be leftmost. (If the leftmost label candidate for a point p' left of p intersects ℓ_p^{OPT}, then all other label candidates for p' do as well, contradicting that ℓ_p^{OPT} is free in OPT.) That $\ell_{p'}^{\text{ALG}}$ is not leftmost can mean two things. Either $\ell_{p'}^{\text{ALG}}$ is freeable, in which case we charge ℓ_p^{OPT} to $\ell_{p'}^{\text{ALG}}$, or making $\ell_{p'}^{\text{ALG}}$ leftmost will cause it to intersect some freeable label $\ell_{p''}^{\text{ALG}}$, in which case we charge ℓ_p^{OPT} to $\ell_{p''}^{\text{ALG}}$.

Note that in both cases we charge ℓ_p^{OPT} to a free label in ALG that lies relatively close to p. A packing argument therefore shows that any free label in ALG is charged $O(1)$ times. With a more careful analysis, one can argue that at most four free labels of OPT get charged to a single freeable label of ALG by the above scheme for the 1SH model (see Figure 3(c)), and at most six for the 2PH model (see Figure 3(a)). Figure 2 shows that the resulting ratio is tight.

We still need to consider the case where a point p has a free label ℓ_p^{OPT} in OPT that is also a freeable label candidate when p is being processed by GREEDY-SWEEP. Then ℓ_p^{ALG} must also be a free label, and we charge ℓ_p^{OPT} to ℓ_p^{ALG}. The label ℓ_p^{ALG} can at most be as far to the right as ℓ_p^{OPT}, otherwise GREEDYSWEEP would have picked ℓ_p^{OPT} over ℓ_p^{ALG}. Thus if any label candidate ℓ for a point p' to the right of p intersects ℓ_p^{ALG}, then ℓ will also intersect ℓ_p^{OPT}. One can argue that this implies ℓ_p^{ALG} will only be charged once.   □

Already for the 4P model, GREEDYSWEEP can be as bad as an $O(1/\sqrt{n})$-approximation. We instead take the best solution over running GREEDYSWEEP

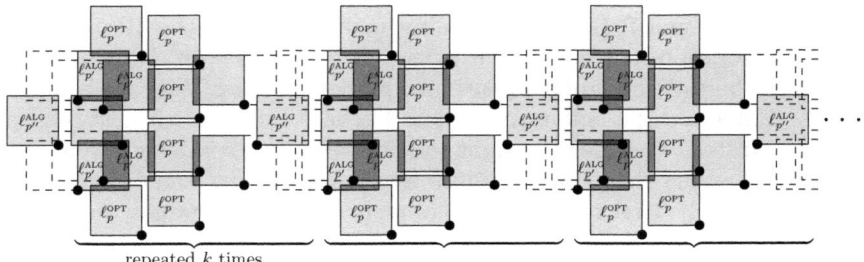

Fig. 2. A labeling computed by GREEDYSWEEP for the 2PH model, where the $k + 1$ labels marked $\ell_{p''}^{\mathrm{ALG}}$ are free. In the optimal solution the $6k$ labels marked ℓ_p^{OPT} are free. Thus the 1/6-approximation is tight for the 2PH model, and a similar example shows the 1/4-approximation for the 1SH model is also tight.

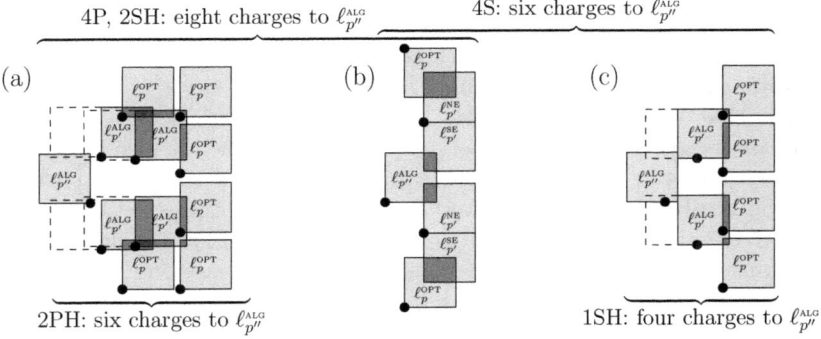

Fig. 3. (b) If every ℓ_p^{OPT} charged to $\ell_{p''}^{\mathrm{ALG}}$ is intersected by labels placed *later*, $\ell_{p''}^{\mathrm{ALG}}$ is charged at most twice. If every ℓ_p^{OPT} charged to $\ell_{p''}^{\mathrm{ALG}}$ is intersected by labels placed *earlier*, $\ell_{p''}^{\mathrm{ALG}}$ is charged (a) at most six times for the 2PH, 4P, and 2SH models, and (c) at most four times for the 1SH and 4S models.

several times with different sweep directions. For the 4P model we do one left-to-right sweep (as before) and one right-to-left sweep (preferring rightmost label candidates). For the 2SH model we do one top-to-bottom sweep (preferring topmost label candidates) and one bottom-to-top sweep (preferring bottommost label candidates). For the 4S model we sweep in all four of these directions. This always yields a constant-factor approximation:

Theorem 1. *There are $O(n \log n)$-time and $O(n)$-space algorithms for free-label maximization on n points with unit-square labels, having the following approximation ratios: 1/4 (tight) for the 1SH model, 1/6 (tight) for the 2PH model, 1/16 for the 4P and 2SH models, and 1/24 for the 4S model.*

Proof. We will prove the approximation ratio for the 4P model; the proofs for the 2SH and 4S models are similar, and the ratio for the 2PH and 1SH models was proved in Lemma 1. Let OPT be an optimal solution for the 4P model, and consider the solution ALG computed in the left-to-right sweep. We can assume that at least half of the labels in OPT are placed in one of the two rightmost positions. (If not, at least half must be placed in one of the two leftmost positions and we can instead consider the right-to-left sweep in a completely symmetric way.) We will argue that the rightmost free labels in OPT can be charged to free labels of ALG so that no label receives more than eight charges, yielding the stated 1/16-approximation.

Suppose p is a point with a rightmost free label ℓ_p^{OPT} in OPT, but with a non-free label ℓ_p^{ALG} in ALG. At the time p was being processed, the label candidate ℓ_p^{OPT} must not have been freeable, either because some unprocessed point would inevitably get a label intersecting ℓ_p^{OPT}, or because some processed point already had a label intersecting ℓ_p^{OPT}. We consider these two cases separately.

- Suppose every label candidate of some unprocessed point p' intersects either ℓ_p^{OPT} or some previously placed freeable label. (This cannot occur in the 2PH and 1SH models.) Of the rightmost label candidates for p' one must be topmost, say $\ell_{p'}^{\mathrm{NE}}$, and one must be bottommost, say $\ell_{p'}^{\mathrm{SE}}$. Since p and all points that already have a label lie to the left of p', if ℓ_p^{OPT} or a freeable label intersects a rightmost label candidate for p', then it also intersects the label candidate(s) for p' with the same y-coordinate but lying more to the left. So if all rightmost label candidates for p' are intersected by previously placed freeable labels, then all label candidates for p' are intersected by previously placed freeable labels, meaning that at least one of them was in fact not freeable. Thus ℓ_p^{OPT} must intersect some rightmost label candidate of p'. This implies that ℓ_p^{OPT} does not intersect the horizontal line through p', for otherwise ℓ_p^{OPT} would contain p'. Thus ℓ_p^{OPT} intersects either $\ell_{p'}^{\mathrm{NE}}$ or $\ell_{p'}^{\mathrm{SE}}$ but not both, so there must be a freeable label $\ell_{p''}^{\mathrm{ALG}}$ in ALG which intersects $\ell_{p'}^{\mathrm{SE}}$ if ℓ_p^{OPT} intersects $\ell_{p'}^{\mathrm{NE}}$, or vice versa. Charge ℓ_p^{OPT} to $\ell_{p''}^{\mathrm{ALG}}$. One can argue that any freeable label can be charged at most twice this way (see Figure 3(b)).
- Suppose some already processed point p' has a label $\ell_{p'}^{\mathrm{ALG}}$ (either freeable or not) that intersects ℓ_p^{OPT}. Because ℓ_p^{OPT} is rightmost, $\ell_{p'}^{\mathrm{ALG}}$ cannot be leftmost. So either $\ell_{p'}^{\mathrm{ALG}}$ is freeable, and we charge ℓ_p^{OPT} to $\ell_{p'}^{\mathrm{ALG}}$, or making $\ell_{p'}^{\mathrm{ALG}}$ leftmost will cause it to intersect some freeable label $\ell_{p''}^{\mathrm{ALG}}$, and we charge ℓ_p^{OPT} to $\ell_{p''}^{\mathrm{ALG}}$. One can argue that any freeable label can be charged at most six times this way for the 4P model (see Figure 3(a)).

Combining the charges of these two cases yields at most eight charges per free label for the 4P model, and we argued that at least one half the free labels in OPT could be charged, yielding the claimed 1/16-approximation. We have not yet charged free labels in OPT which label points that also have a free label in ALG. One can argue that charging such labels does not cost us extra charges, as one of the charges to $\ell_{p''}^{\mathrm{ALG}}$ in Figure 3(b) must disappear if $\ell_{p''}^{\mathrm{OPT}}$ is free.

The proofs for the 2SH and 4S models are similar, but each free label is only charged at most six times for the 4S model (see Figure 3(b)–(c)). In the 2SH model every free label in OPT is either topmost or bottommost so that we can again charge at least half of them, but in the 4S model a label can also be leftmost or rightmost so that we can charge only one fourth.

With some clever use of standard data structures, similar to the 1/2-approximation algorithm for number maximization by Van Kreveld et al. [11], GREEDY-SWEEP can be implemented to run in $O(n \log n)$ time and $O(n)$ space. We omit the details. $\qquad\square$

3 PTASs for Unit Squares

We can obtain a PTAS for the case of unit-square labels by applying the "shifting technique" of Hochbaum and Maass [9]. Imagine a grid of unit squares overlaying the plane such that no point is on a grid line and call this the 1-*grid*. If, for some integer $k > 4$ to be specified later, we leave out all but every k^{th} horizontal and vertical grid line this forms a coarser k-*grid*. By varying the offsets at which we start counting the lines, we can form k^2 different k-grids $G_1, \ldots, G_{k^2}$ out of the 1-grid. Consider one of them, say G_i. For any $k \times k$ square cell $\text{C} \in G_i$, let $\overline{\text{C}} \subset \text{C}$ be the smaller $(k-4) \times (k-4)$ square with the same midpoint as C (see Figure 4(a)). We call $\overline{\text{C}}$ the *inner cell* of C. For a given set P of n points, let $P_{\text{C}} := \text{C} \cap P$, $P_{\overline{\text{C}}} := \overline{\text{C}} \cap P$, and $P_{\text{in}}(G_i) := \bigcup_{\text{C} \in G_i} P_{\overline{\text{C}}}$. We call a labeling $\mathcal{L}$ for P *inner-optimal* with respect to G_i if $\mathcal{L}$ maximizes the number of points in $P_{\text{in}}(G_i)$ that get a free label. Note that if $\text{C}, \text{C}' \in G_i$ are distinct cells, then a point $p \in \overline{\text{C}}$ can never have a label intersecting the label for a point $p' \in \text{C}'$ (see Figure 4(a)). Hence an inner-optimal labeling for P can be obtained by computing an inner-optimal labeling on P_{C} independently for each cell $\text{C} \in G_i$. We will show below how to do this in time polynomial in n (but exponential in k). By itself this does not help us, as any particular k-grid G_i may have many points that lie outside of inner cells. We claim, however, that computing an inner-optimal labeling for all k-grids $G_1, \ldots, G_{k^2}$ and then taking the best one still yields a $(1 - \varepsilon)$-approximation for suitably chosen k:

Lemma 2. *For all fixed position and slider models, the best inner-optimal labeling for P with respect to all k^2 different k-grids $G_1, \ldots, G_{k^2}$ yields a $(1-\varepsilon)$-approximation to free-label maximization with unit-square labels if $k \geqslant 8/\varepsilon$.*

Proof. Let OPT be some optimal solution, and let $F \subseteq P$ be the set of points with a free label in OPT. In any k-grid the inner cells are separated from each other by horizontal and vertical strips with a width of four 1-grid cells (see Figure 4(a)). Thus any point in F lies in an inner cell for $(k-4)^2$ of the k^2 different k-grids. By the pigeon-hole principle, there must be a k-grid G_i for which $|F \cap P_{\text{in}}(G_i)| \geqslant (k-4)^2/k^2 \cdot |F| = (1 - 4/k)^2 \cdot |F|$. An inner-optimal labeling for P with respect to G_i will have at least $|F \cap P_{\text{in}}(G_i)|$ free labels. Hence we get a $(1 - \varepsilon)$-approximation if $(1 - 4/k)^2 = 1 - 8/k + 4/k^2 \geqslant 1 - \varepsilon$, which is satisfied if $k \geqslant 8/\varepsilon$. $\qquad\square$

To complete the PTAS we need to show how to compute an inner-optimal labeling for the set P_{C} of points inside a $k \times k$ cell C. We say that a subset $F \subset P_{\overline{\text{C}}}$ is

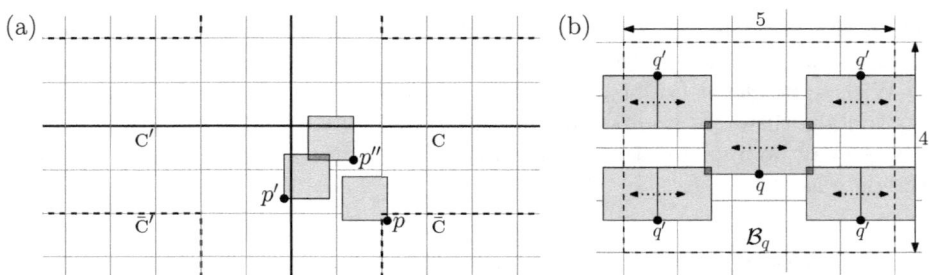

Fig. 4. (a) The cells (bold lines) and inner cells (dashed bold lines) of a k-grid. The underlying 1-grid is shown in thin gray lines. (b) Any two points q, q' with intersecting labels must lie in a region $\mathcal{B}_q$ of 20 cells of the 1-grid.

freeable if we can label the points in P_{C} such that all points in F get a free label. The key insight is that, by a packing argument, not too many of the points $P_{\overline{\mathrm{C}}}$ in the inner cell $\overline{\mathrm{C}}$ can get a free label. Thus there is a limited number of freeable subsets. We first bound the number of potentially freeable subsets that we need to consider, and then show how to test each one for feasibility.

In many applications, there will not be too many points that are very close together (with respect to the label sizes). To take this into account, we will not just use the total number of points (n) in our analysis, but also their "density" (Δ). More precisely, let $\Delta \leqslant n$ denote the maximum number of points in P contained in any unit square. If $\Delta = 1$ then labeling every point with its topleft label candidate, say, yields a solution where all labels are free. So assume $\Delta \geqslant 2$.

Lemma 3. *Let* C *be a cell in a k-grid, and let* P_{C} *be the set of points inside* C. *Then there is a collection* $\mathcal{F}$ *of subsets of* $P_{\overline{\mathrm{C}}}$ *such that for any freeable subset* $F \subseteq P_{\overline{\mathrm{C}}}$ *we have* $F \in \mathcal{F}$. *We can compute* $\mathcal{F}$ *in* $O(\sum_{F \in \mathcal{F}} |F|)$ *time, and have*

- $|\mathcal{F}| \leqslant \Delta^{2(k-4)^2}$ *for the 2PH and 1SH models, and*
- $|\mathcal{F}| \leqslant \Delta^{4(k-4)^2}$ *for the 4P, 2SH, and 4S models.*

Proof. A 1-grid cell contains at most Δ points, and $\overline{\mathrm{C}}$ consists of $(k-4)^2$ cells of the 1-grid. In the 2PH and 1SH models, no more than two points from the same 1-grid cell can be simultaneously labeled with non-intersecting labels. Thus any freeable subset $F \subseteq P_{\overline{\mathrm{C}}}$ can be constructed by taking at most two points from each 1-grid cell. Hence, there are at most

$$\left(\binom{\Delta}{0} + \binom{\Delta}{1} + \binom{\Delta}{2} \right)^{(k-4)^2} \leqslant \Delta^{2(k-4)^2}$$

potentially freeable subsets F, where the inequality follows from the assumption that $\Delta \geqslant 2$. Similarly, no more than four points from the same 1-grid cell can be simultaneously labeled with non-intersecting labels in the 4P, 2SH, and 4S models, leading to at most

$$\left(\tbinom{\Delta}{0} + \tbinom{\Delta}{1} + \tbinom{\Delta}{2} + \tbinom{\Delta}{3} + \tbinom{\Delta}{4} \right)^{(k-4)^2} \leqslant \Delta^{4(k-4)^2}$$

potentially freeable subsets F. $\qquad\square$

Lemma 4. *Given the set P_C of all $n_\mathrm{C} \leqslant \Delta k^2$ points contained in a k-grid cell C, and a subset $F \subseteq P_{\overline{\mathrm{C}}}$ of those points, we can decide in $O(n_\mathrm{C} \log n_\mathrm{C})$ time for the 2PH and 1SH models whether there exists a labeling $\mathcal{L}$ for P_C where all points in F have a free label, and if so produce $\mathcal{L}$.*

Proof. Go through the points from left to right and label them one-by-one. For every point $p \in F$ we pick the leftmost label candidate that does not intersect a previously placed label, and for every point $p \in P_\mathrm{C} \setminus F$ we pick the leftmost label candidate that does not intersect a previously placed label for a point in F. If we can process all points in P_C in this way then clearly we have found a suitable labeling $\mathcal{L}$. If we instead encounter a point p for which no label candidate can be chosen, then we report that no such labeling $\mathcal{L}$ exists. This is correct, because the partial labeling constructed by this algorithm has all labels at least as far to the left as $\mathcal{L}$ would have, so p cannot be correctly labeled in $\mathcal{L}$ either. The above is simply a somewhat simplified version of the GREEDYSWEEP algorithm from Section 2, and can be implemented to run in $O(n_\mathrm{C} \log n_\mathrm{C})$ time. $\qquad\square$

Lemma 5. *Given the set P_C of all $n_\mathrm{C} \leqslant \Delta k^2$ points contained in a k-grid cell C, and a subset $F \subseteq P_{\overline{\mathrm{C}}}$ of f of those points, we can decide in $O((n_\mathrm{C} - f)4^f)$ time for the 4P model whether there exists a labeling $\mathcal{L}$ for P_C where all points in F have a free label, and if so produce $\mathcal{L}$.*

Proof. Enumerate all 4^f labelings of the points in F, and check for each such labeling $\mathcal{L}'$ whether it can be extended into a labeling $\mathcal{L}$ for all points in P_C. This entails checking whether each point $p \in P_\mathrm{C} \setminus F$ has a label candidate that does not intersect any label of $\mathcal{L}'$. For this we only need to look at labels for points $p' \in F$ that lie in the 3×3 square of 1-grid cells centered at the 1-grid cell containing p. Since each 1-grid cell contains at most four points of F, this check can be done in $O(1)$ time for each of the $n_\mathrm{C} - f$ points in $P_\mathrm{C} \setminus F$. $\qquad\square$

For the 2SH models we can neither use the greedy labeling of F (as for the 2PH and 1SH models), nor try all labelings (as for the 4P model). Instead we proceed as follows. Try all 2^f ways of restricting the labels for the points in F to be either topmost or bottommost. The problem is then to decide whether F can be labeled with free labels in the 1SH model, while labeling $P_\mathrm{C} \setminus F$ with (free and/or non-free) labels in the 2SH model. The position of a label along a 1-slider can be modeled as a number between 0 and 1, making the configuration space $\mathcal{C}$ of possible labelings for F the f-dimensional unit hypercube. Let $\mathcal{C}_{\mathrm{nonint}} \subseteq \mathcal{C}$ be the subspace of labelings for F where the labels of the points in F are disjoint, and for any point $p \in P_\mathrm{C} \setminus F$ let $\mathcal{C}_p \subseteq \mathcal{C}$ be the subspace of labelings $\mathcal{L}' \in \mathcal{C}_p$ where p can still get a label without intersecting labels in $\mathcal{L}'$. We then need to decide whether $\mathcal{C}_{\mathrm{free}} := \mathcal{C}_{\mathrm{nonint}} \cap \bigcap_{p \in P_\mathrm{C} \setminus F} \mathcal{C}_p$ is non-empty, and if so construct a feasible labeling $\mathcal{L}' \in \mathcal{C}_{\mathrm{free}}$ for F and extend it into a labeling $\mathcal{L}$ for P_C. We will show how this can be done using an arrangement of $O(n_\mathrm{C})$ hyperplanes in $\mathcal{C}$.

Lemma 6. *Given the set P_C of all $n_C \leqslant \Delta k^2$ points contained in a k-grid cell C, and a subset $F \subseteq P_{\overline{C}}$ of f of those points, we can decide in $O(n_C)^f$ time for the 2SH and 4S models whether there exists a labeling $\mathcal{L}$ for P_C where all points in F have a free label, and if so produce $\mathcal{L}$.*

Proof. We discuss only the 2SH model; the proof for the 4S model is analogous. If two points $q, q' \in F$ have intersecting labels, then they must be fairly close. Specifically, there is a 4×5 rectangle $\mathcal{B}_q$ around q, consisting of 20 cells of the 1-grid, such that $\mathcal{B}_q$ contains q' (see Figure 4(b)). Preventing q and q' from having intersecting labels introduces a linear constraint on the slider coordinates of q and q'. Since F has at most four points in any 1-grid cell, $\mathcal{B}_q$ contains at most 80 points of F (including q itself). Hence $\mathcal{C}_{\mathrm{nonint}}$ is the intersection of at most $f \cdot (80 - 1) \cdot 1/2 = 79f/2$ half-spaces.

For any point $p \in P_C \setminus F$, let $\mathcal{C}_p^{\mathrm{top}} \subseteq \mathcal{C}$ be the subspace of labelings for F which still allow p to get a topmost label. Now consider a labeling for F that is not in $\mathcal{C}_p^{\mathrm{top}}$. Thus any topmost label for p will intersect at least one label for a point in F. We claim (and will argue later) that then there exists a subset $F' \subseteq F$ with $|F'| \leqslant 2$ such that any topmost label of p intersects a label of a point in F'. Hence, $\mathcal{C}_p^{\mathrm{top}}$ can be constructed as $\mathcal{C}_p^{\mathrm{top}} = \bigcap_{F' \subseteq F, |F'| \leqslant 2} \mathcal{C}_p^{\mathrm{top}}(F')$, where $\mathcal{C}_p^{\mathrm{top}}(F')$ is the subspace of labelings for F where p has at least one topmost label candidate not intersecting the labels for F'. Since we can assume that $F' \subseteq \mathcal{B}_p$, there are at most $\binom{80}{1} + \binom{80}{2} = 3240$ sets F' to be considered. For any $q \in F$, the subspace $\mathcal{C}_p^{\mathrm{top}}(\{q\})$ is defined by a linear constraint on q's slider coordinate, giving it a minimum or maximum value. For any $q, q' \in F$, with q left of p and q' right of p, the subspace $\mathcal{C}_p^{\mathrm{top}}(\{q, q'\})$ is defined by the linear constraint that the horizontal distance between the labels of q and q' should be at least 1 (and if q and q' are on the same side of p, then $\mathcal{C}_p^{\mathrm{top}}(\{q, q'\}) = \mathcal{C}_p^{\mathrm{top}}(\{q\}) \cup \mathcal{C}_p^{\mathrm{top}}(\{q'\})$). Hence, $\mathcal{C}_p^{\mathrm{top}}$ can be constructed as the intersection of at most 3240 half-spaces. The same is true for $\mathcal{C}_p^{\mathrm{bottom}}$, the subspace of labelings for F which allow p to get a bottommost label. Since $\mathcal{C}_p = \mathcal{C}_p^{\mathrm{top}} \cup \mathcal{C}_p^{\mathrm{bottom}}$, we can find $\mathcal{C}_{\mathrm{free}} = \mathcal{C}_{\mathrm{nonint}} \cap \bigcap_{p \in P_C \setminus F} \mathcal{C}_p$ as the union of some of the cells in an arrangement of (at most) $h := 79f/2 + 2 \cdot 3240 n_C = O(n_C)$ hyperplanes. We can construct this arrangement in $O(h^f)$ time [5], and in the same time test whether $\mathcal{C}_{\mathrm{free}}$ is non-empty and if so construct a labeling $\mathcal{L}' \in \mathcal{C}_{\mathrm{free}}$ for F. Greedily extending this into a labeling $\mathcal{L}$ for P_C does not increase the running time.

To substantiate the claim that we can ignore sets F' with three or more elements, consider a labeling $\mathcal{L}'$ for F which intersects all topmost label candidates for p. Let ℓ be the rightmost label in $\mathcal{L}'$ that intersects p's top-leftmost label candidate, and let ℓ' be the leftmost label in $\mathcal{L}'$ that intersects p's top-rightmost label candidate (possibly with $\ell' = \ell$). Then ℓ and ℓ' together must intersect all topmost label candidates for p, otherwise p would have a free topmost label candidates horizontally between ℓ and ℓ'. $\qquad \square$

Putting together the above lemmas yields the following result:

Theorem 2. *For any fixed $\varepsilon > 0$, and for each of the fixed-position and slider models, there exists a polynomial-time algorithm that computes a $(1 - \varepsilon)$-approximation to free-label maximization with unit-square labels.*

Proof. Compute a 1-grid in $O(n \log n)$ time [9]. Let $k = \lceil 8/\varepsilon \rceil$ and generate all k^2 possible k-grids $G_1, \ldots, G_{k^2}$ out of the 1-grid. For each k-grid G_i, we compute an inner-optimal labeling for the (at most n) cells containing points. This is done for a cell $\text{C} \in G_i$ by enumerating the potentially freeable subsets F of $P_{\overline{\text{C}}}$ (Lemma 3), and checking for each subset F whether P_C can be labeled so that all points in F have a free label (Lemmas 4, 5, and 6). The best out of these k^2 solutions is a $(1 - \varepsilon)$-approximation (Lemma 2). The resulting running time is

$$O(n \log n) + nk^2 \cdot \begin{cases} \Delta^{2(k-4)^2} \cdot O\big(\Delta k^2 \log(\Delta k)\big) & \text{for the 2PH and 1SH models,} \\ \Delta^{4(k-4)^2} \cdot O\big(\Delta k^2 4^{(k-3)^2}\big) & \text{for the 4P model,} \\ \Delta^{4(k-4)^2} \cdot O\big(\Delta k^2\big)^{(k-3)^2} & \text{for the 2SH and 4S models.} \end{cases} \qquad \square$$

4   Conclusion

Air traffic controllers monitor airplanes on computer screens as moving points with associated textual labels, and warn pilots to change course on potential collisions. Currently they spend a lot of their time moving labels around by hand to prevent labels from intersecting one another and becoming unreadable. Algorithms from the cartographic map labeling literature do not apply, as these solve a different problem. To this end we have introduced the *free-label-maximization problem* as a new variant of the labeling problem, and have studied it for static points as a first step. In free-label maximization we must label all points with labels of fixed dimensions and seek to maximize the number of *free* labels (labels that do not intersect other labels). We have presented a simple and efficient constant-factor approximation algorithm, as well as a PTAS, for free-label maximization under the commonly assumed model that labels are directly attached to their points. In air traffic control, however, labels are usually connected to their point by means of a short line segment (a *leader*). Our constant-factor approximation can be extended to this case, and we believe the same may be true for our PTAS.

Our algorithms work if all labels are unit squares (or, equivalently, all labels are translates of a fixed rectangle). The cases of labels being unit-height rectangles or arbitrary rectangles are still open. For the number-maximization problem these cases allow, respectively, a PTAS [3] and an $O(1/\log \log n)$-approximation [2]. The former achieves a $(1-1/k)$-approximation to number maximization in only $O(n \log n + n\Delta^{k-1})$ time, while the running time of our PTAS for free-label maximization is completely impractical. It would be interesting to see if these results for number maximization can be matched for free-label maximization. If not, then free-label maximization is strictly harder than number maximization, while easier than size maximization. The weighted version of the free-label-maximization problem is another interesting direction for future research.

The most important area for future research, however, is the labeling of moving points. Even outside of air traffic control applications, we believe that free-label maximization is a better model for this than the size- and number-maximization problems. Continuously scaling labels under size maximization would be hard to read, and the (dis)appearance of a label under number maximization is an inherently discrete event which can be disturbing for the viewer. It is fairly simple to kinetically maintain the labeling of our constant-factor approximation algorithm as the points move. This is not enough to obtain a good result, however, as labels will sometimes "jump" from place to place. We would prefer to "smooth out" the label trajectories so that labels move continuously at finite speeds, but it is not yet clear how to do this.

References

1. Been, K., Nöllenburg, M., Poon, S.-H., Wolff, A.: Optimizing active ranges for consistent dynamic map labeling. Comput. Geom. Theory Appl. 43(3), 312–328 (2010)
2. Chalermsook, P., Chuzhoy, J.: Maximum independent set of rectangles. In: Mathieu, C. (ed.) Proc. 20th ACM-SIAM Sympos. on Discrete Algorithms (SODA'09), New York, pp. 892–901 (2009)
3. Chan, T.: A note on maximum independent set in rectangle intersection graphs. Information Processing Letters 89, 19–23 (2004)
4. Dorbes, A.: Requirements for the implementation of automatic and manual label anti-overlap functions. EEC Note No. 21/00, EUROCONTROL Experimental Centre (2000)
5. Edelsbrunner, H., O'Rourke, J., Seidel, R.: Constructing arrangements of lines and hyperplanes with applications. SIAM Journal on Computing 15(2), 341–363 (1986)
6. Erlebach, T., Hagerup, T., Jansen, K., Minzlaff, M., Wolff, A.: Trimming of graphs, with an application to point labeling. In: Albers, S., Weil, P. (eds.) Proc. 25th Internat. Sympos. Theoretical Aspects Comput. Sci (STACS'08), Bordeaux, pp. 265–276 (2008)
7. Formann, M., Wagner, F.: A packing problem with applications to lettering of maps. In: Proc. 7th Annu. ACM Sympos. Comput. Geom (SoCG'91), North Conway, pp. 281–288 (1991)
8. Fowler, R., Paterson, M., Tanimoto, S.: Optimal packing and covering in the plane are NP-complete. Information Processing Letters 12, 133–137 (1981)
9. Hochbaum, D., Maass, W.: Approximation schemes for covering and packing problems in image processing and VLSI. Journal of the ACM 32(1), 130–136 (1985)
10. Jiang, M., Bereg, S., Qin, Z., Zhu, B.: New bounds on map labeling with circular labels. In: Fleischer, R., Trippen, G. (eds.) ISAAC 2004. LNCS, vol. 3341, pp. 606–617. Springer, Heidelberg (2004)
11. van Kreveld, M., Strijk, T., Wolff, A.: Point labeling with sliding labels. Comput. Geom. Theory Appl. 13, 21–47 (1999)
12. Marks, J., Shieber, S.: The computational complexity of cartographic label placement. Technical Report TR-05-91, Harvard CS (1991)
13. Poon, S.-H., Shin, C.-S., Strijk, T., Uno, T., Wolff, A.: Labeling points with weights. Algorithmica 38(2), 341–362 (2003)
14. Rostamabadi, F., Ghodsi, M.: A fast algorithm for updating a labeling to avoid a moving point. In: Proc. 16th Canadian Conf. Comput. Geom. (CCCG'04), pp. 204–208 (2004)
15. Wolff, A., Strijk, T.: The Map Labeling Bibliography (2009),
    http://liinwww.ira.uka.de/bibliography/Theory/map.labeling.html

Phase Transitions in Sampling Algorithms
and the Underlying Random Structures

Dana Randall

College of Computing, Georgia Institute of Technology
Atlanta, GA 30332-0765
randall@cc.gatech.edu

Sampling algorithms based on Markov chains arise in many areas of computing, engineering and science. The idea is to perform a random walk among the
elements of a large state space so that samples chosen from the stationary distribution are useful for the application. In order to get reliable results, we require
the chain to be rapidly mixing, or quickly converging to equilibrium. For example, to sample independent sets in a given graph G, the so-called hard-core
lattice gas model, we can start at any independent set and repeatedly add or
remove a single vertex (if allowed). By defining the transition probabilities of
these moves appropriately, we can ensure that the chain will converge to a useful distribution over the state space Ω. For instance, the Gibbs (or Boltzmann)
distribution, parameterized by $\Lambda > 0$, is defined so that $p(\Lambda) = \pi(I) = \Lambda^{|I|}/Z$,
where $Z = \sum_{J \in \Omega} \Lambda^{|J|}$ is the normalizing constant known as the partition func-
tion. An interesting phenomenon occurs as Λ is varied. For small values of Λ,
local Markov chains converge quickly to stationarity, while for large values, they
are prohibitively slow. To see why, imagine the underlying graph G is a region of
the Cartesian lattice. Large independent sets will dominate the stationary distribution π when Λ is sufficiently large, and yet it will take a very long time to move
from an independent set lying mostly on the odd sublattice to one that is mostly
even. This phenomenon is well known in the statistical physics community, and
characterizes by a phase transition in the underlying model.

In general, phase transitions occur in models where a small microscopic change
to some parameter suddenly causes a macroscopic change to the system. This
phenomenon is pervasive in physical systems, but often lacks rigorous analysis. Colloids, which are mixtures of two disparate substances in suspension, are
an interesting example. At low enough density the two types of particles will
be uniformly interspersed, but at sufficiently high density the particles of each
substance will cluster together, effectively separating. Unlike seemingly related
models where the clustering occurs because like particles are drawn together by
enthalpic forces, such as the ferromagnetic Ising model, the behavior of colloids
is purely entropic – the particles separate because because the overwhelming
majority of configurations in the stationary distribution exhibit such a separation. In this talk I will give an overview of some techniques used to prove the
existence of two distinct phases in various sampling algorithms and the underlying physical models. Finally, I will suggest some potential approaches for using
our understanding of these two phases to inform the design of more efficient
algorithms.

H. Kaplan (Ed.): SWAT 2010, LNCS 6139, p. 309, 2010.
© Springer-Verlag Berlin Heidelberg 2010

Polynomial Kernels for Hard Problems on Disk Graphs

Bart Jansen*

Utrecht University, PO Box 80.089, 3508 TB Utrecht, The Netherlands
bart@cs.uu.nl

Abstract. Kernelization is a powerful tool to obtain fixed-parameter tractable algorithms. Recent breakthroughs show that many graph problems admit small polynomial kernels when restricted to sparse graph classes such as planar graphs, bounded-genus graphs or H-minor-free graphs. We consider the intersection graphs of (unit) disks in the plane, which can be arbitrarily dense but do exhibit some geometric structure. We give the first kernelization results on these dense graph classes. CONNECTED VERTEX COVER has a kernel with $12k$ vertices on unit-disk graphs and with $3k^2 + 7k$ vertices on disk graphs with arbitrary radii. RED-BLUE DOMINATING SET parameterized by the size of the smallest color class has a linear-vertex kernel on planar graphs, a quadratic-vertex kernel on unit-disk graphs and a quartic-vertex kernel on disk graphs. Finally we prove that H-MATCHING on unit-disk graphs has a linear-vertex kernel for every fixed graph H.

1   Introduction

Motivation. The theory of parameterized complexity [1] has made it possible to analyze the strength of preprocessing schemes through the concept of kernelization [2]. An instance of a parameterized problem is a tuple (x, k) where x is an encoding of the classical input (for example a graph) and k is a non-negative integer that measures some structural property of x (such as the desired solution size). A kernelization algorithm (or kernel) for a parameterized problem is a mapping that transforms an instance (x, k) into an equivalent instance (x', k') in time $p(|x| + k)$ for some polynomial p, such that $|x'|, k' \leq f(k)$ for a computable function f and the answer to the decision problem is preserved. The function f is the *size* of the kernel, and much effort has been invested in finding kernels of polynomial size. A rich theoretical framework is in development, yielding both positive and negative results regarding the limits of kernelization. A celebrated result by Bodlaender et al. [3] states that there are problems in FPT that cannot have kernels of polynomial size, unless some unlikely complexity-theoretic collapse occurs. Recent results also show that all problems that satisfy certain compactness and distance properties admit polynomial kernels on restricted graph classes such as planar graphs [4], graphs of bounded genus [5]

---

* This work was supported by the Netherlands Organisation for Scientific Research (NWO), project "KERNELS: Combinatorial Analysis of Data Reduction".

H. Kaplan (Ed.): SWAT 2010, LNCS 6139, pp. 310–321, 2010.

and H-minor-free graphs [6]. The common theme in these frameworks is that they yield kernels for problems on *sparse graphs*, in which the number of edges is bounded linearly in the number of vertices.

Our work focuses on graph classes that exhibit some geometric structure but can be arbitrarily dense: the intersection graphs of disks in the plane ("disk graphs"). Each vertex is represented by a disk, and two vertices are adjacent if and only if their disks intersect. Disks that touch eachother but do not overlap are also said to be intersecting. It is a well-known consequence of the Koebe-Andreev-Thurston theorem that planar graphs are a strict subclass of disk graphs [7, Section 3.2]. If all the disks have the same radius then by a scaling argument we may assume that this radius is 1: the intersection graphs of such systems are therefore called *unit-disk* graphs. It is easy to see that (unit)disk graphs may contain arbitrarily large cliques. Many classical graph problems remain NP-complete when restricted to unit-disk graphs [8], and the natural parameterizations of several important problems such as INDEPENDENT SET and DOMINATING SET remain $W[1]$-hard [9,10] for unit-disk graphs. In this paper we will show how the structure of disk graphs graphs may be exploited to obtain polynomial kernels.

Previous Work. Hardly any work has been done on kernelization for disk graphs: only a single result in this direction is known to us. Alber and Fiala [11] use the concept of a *geometric problem kernel* as in their work on INDEPENDENT SET: they obtain a reduced instance in which the *area of the union of all disks* is bounded in k. This geometric kernel leads to subexponential exact algorithms, and to an FPT algorithm by applying restrictions that bound the maximum degree of the graph. In the context of kernels for dense graph classes it is interesting to point out the work of Philip et al. [12] who recently obtained polynomial kernels for DOMINATING SET on graph classes that exclude $K_{i,j}$ as a subgraph for some $i, j \geq 1$; these graphs can have a superlinear number of edges.

Our Results. We believe we are the first to present polynomial problem kernels on dense disk graph classes. We show that the CONNECTED VERTEX COVER problem has a trivial $12k$-vertex kernel on unit-disk graphs, and obtain a more complex kernel with $3k^2 + 7k$ vertices for disk graphs with arbitrary radii. We prove that RED-BLUE DOMINATING SET parameterized by the size of the smallest color class admits a kernel of linear size on planar graphs, of quadratic size on unit-disk graphs and of quartic size on disk graphs with arbitrary radii. Note that neither of these two problems admit polynomial kernels on general graphs unless the polynomial hierarchy collapses [13]. We also present a linear kernel for the H-MATCHING problem on unit-disk graphs, which asks whether a unit-disk graph G contains at least k vertex-disjoint copies of a fixed connected graph H. In the conclusion we identify a property of graph classes that implies the existence of polynomial kernels for CONNECTED VERTEX COVER and RED-BLUE DOMINATING SET, and show that not only (unit)disk graphs but also $K_{i,j}$-subgraph-free graphs have this property - thereby showing how some of our results carry over to $K_{i,j}$-subgraph-free graphs.

2   Preliminaries

All graphs are undirected, finite and simple unless explicitly stated otherwise. Let $G = (V, E)$ be a graph. For $v \in V$ we denote the *open* (resp. closed) neighborhoods of v by $N_G(v)$ and $N_G[v]$. The degree of a vertex v in graph G is denoted by $\deg_G(v)$. We write $G' \subseteq G$ if G' is a subgraph of G. For $X \subseteq V$ we denote by $G[X]$ the subgraph of G that is induced by the vertices in X. We identify the set of connected components of G by $\mathrm{COMP}(G)$. We treat geometric objects as closed sets of points in the Euclidean plane. If v is a vertex of a geometric intersection graph, then we use $\mathcal{D}(v)$ to denote the geometric representation of v; usually this is just a disk. We will write $\mathcal{D}(V')$ for $V' \subseteq V$ to denote the union of the geometric objects representing the vertices V'. Throughout this paper we parameterize problems by the desired solution size k unless explicitly stated otherwise. Some proofs had to be omitted due to space restrictions.

Lemma 1. *Let $G = (X \cup Y, E)$ be a planar bipartite graph. If for all distinct $v, v' \in Y$ it holds that $N_G(v) \not\subseteq N_G(v')$ then $|Y| \leq 5|X|$.*   □

Lemma 2 (Compare to Lemma 3.4 of [14]). *If v is a vertex in a unit-disk graph G then there must be a clique of size at least $\lceil \deg_G(v)/6 \rceil$ in $G[N_G(v)]$.*   □

3   Connected Vertex Cover

The CONNECTED VERTEX COVER problem asks for a given connected graph $G = (V, E)$ and integer k whether there is a subset $S \subseteq V$ with $|S| \leq k$ such that every edge in G has at least one endpoint in S, and such that $G[S]$ is connected. Guo and Niedermeier gave a kernel with $14k$ vertices for this problem [4] restricted to planar graphs. More recently it was shown that CONNECTED VERTEX COVER on general graphs does not admit a polynomial kernel unless the polynomial hierarchy collapses [13]; this situation is in sharp contrast with the regular VERTEX COVER problem, which has a kernel with $2k$ vertices [2]. It is not hard to prove that a vertex cover for a connected unit-disk graph on n vertices must have size at least $n/12$ [15, Theorem 2] which yields a trivial linear kernel.

Observation 1. CONNECTED VERTEX COVER has a kernel with $12k$ vertices on unit-disk graphs.

The situation becomes more interesting when we consider disk graphs with arbitrary radii, where we show the existence of a quadratic kernel. To simplify the exposition we start by giving a kernel for an annotated version of the problem. Afterwards we argue that the annotation can be undone to yield a kernel for the original problem. An instance of the problem ANNOTATED CONNECTED VERTEX COVER is a tuple $\langle G, k, C \rangle$ where G is the intersection graph of a set of disks in the plane, k is a positive integer and $C \subseteq V(G)$ is a subset of vertices. The question is whether there is a connected vertex cover $S \subseteq V(G)$ of cardinality at most k such that $C \subseteq S$. We introduce some notation to facilitate the description of the reduction rules.

Definition 1. *Let* $\langle G, k, C \rangle$ *be an instance of* ANNOTATED CONNECTED VER-TEX COVER. *A vertex* $v \in C$ *is* marked; *vertices in* $V(G) \setminus C$ *are* unmarked. *For unmarked vertices we distinguish between two types: an unmarked vertex is* dead *if all its neighbors are marked, and it is* live *if it has an unmarked neighbor. We call an edge* covered *if it is incident on a marked vertex, and* uncovered *otherwise.*

Observe that all edges incident on dead unmarked vertices will be covered by the marked vertices in any solution. Therefore the dead vertices can only be useful in a solution to ensure connectivity. The live unmarked vertices may be needed to cover additional edges.

Reduction Rule 1. If there is an unmarked vertex v with more than k neighbors, then add v to C (i.e. mark v).

This is an annotated analogue of Buss' rule for the VERTEX COVER problem. It is easy to see that v must be part of any solution of size at most k; for if v is not taken in a vertex cover then all its $> k$ neighbors must be taken - note that this also holds if some neighbors of v are already marked. We give a new definition that will simplify the exposition of the next reduction rule.

Definition 2. *We define the* component graph *that corresponds to the instance* $\langle G, k, C \rangle$ *with dead vertices* D *to be the bipartite graph* $G_C := (\text{COMP}(G[C]) \cup D, E)$ *where there is an edge between a connected component* $X \in \text{COMP}(G[C])$ *and a dead vertex* $d \in D$ *if and only if* $N_G(d) \cap V(X) \neq \emptyset$, *i.e. if* d *is adjacent in* G *to a vertex in the connected component* X.

Reduction Rule 2. If there are two distinct dead vertices u and v such that $N_{G_C}(u) \subseteq N_{G_C}(v)$ then delete u.

For the correctness of this rule observe that if S is a solution containing u, then $(S \setminus \{u\}) \cup \{v\}$ is also a solution: since all neighbors of u are marked, the removal of u does not cause edges to become uncovered. Removal of u cannot cause the vertex cover to become disconnected because all components of $G[C]$ that were connected through u remain connected through v.

Lemma 3. *Let* $\langle G, k, C \rangle$ *be a reduced instance of* ANNOTATED CONNECTED VERTEX COVER *for a disk graph* G, *and let* D *denote the set of dead vertices. There is a vertex set* R *and a bipartite graph* $G^* = (R \cup D, E^*)$ *with the following properties:*

(i) *The graph* G^* *is planar.*
(ii) *If* $N_{G^*}(d) \subseteq N_{G^*}(d')$ *for* $d, d' \in D$ *then* $N_{G_C}(d) \subseteq N_{G_C}(d')$.
(iii) *The number of vertices in* R *is at most* $|C|$.

Proof. Assume the conditions stated in the lemma hold. Fix some realization of G by intersecting disks in the plane. The vertex set R corresponds to disjoint regions that are subsets of the maximally connected regions in $\mathcal{D}(C)$. We define the regions R in two parts by setting $R := R_1 \cup R_2$.

<table>
<tr><td>(a) Subgraph induced by D and C. Disks of dead vertices are drawn with broken lines.</td><td>(b) Regions R_1 colored grey, R_2 colored black.</td><td>(c) Embedding of bipartite graph G^*.</td></tr>
</table>

Fig. 1. Construction of the planar bipartite graph G^* from an instance $\langle G, k, C \rangle$ with dead vertices D

- For every pair consisting of a component $X \in \mathrm{COMP}(G[C])$ and dead vertex $d \in D$ such that $\mathcal{D}(V(X)) \subseteq \mathcal{D}(d)$ we let $\mathcal{D}(V(X))$ be a region in R_1.
- The set R_2 consists of the maximally connected regions of the plane that are obtained by taking the union of all disks of vertices in C, and subtracting the interior of all disks of vertices in D. Observe that a region $\mathcal{D}(V(X))$ for $X \in \mathrm{COMP}(G[C])$ may be split into multiple regions by subtracting the interiors of $\mathcal{D}(D)$; see Figure 1(b).

Since the constructed regions R are subsets of the maximally connected regions induced by $\mathcal{D}(C)$, we know that for each $r \in R$ there is a unique $X \in \mathrm{COMP}(G[C])$ such that $r \subseteq \mathcal{D}(V(X))$. Define the parent $\pi(r)$ of r to be the component X for which this holds, and define the parent of a set of regions $R' \subseteq R$ to be the union $\bigcup_{r \in R'} \{\pi(r)\}$. Let $d \in D$ be a dead vertex and consider the regions of R intersected by $\mathcal{D}(d)$. It is not difficult to verify that by construction of R, the disk $\mathcal{D}(d)$ intersects at least one region in R for every component $X \in \mathrm{COMP}(G[C])$ that d is adjacent to in G. Since the regions of R are subsets of $\mathcal{D}(C)$, the disk $\mathcal{D}(d)$ can *only* intersect a region $r \in R$ if d is adjacent in G_C to $\pi(r)$. Hence we establish that if R' is the set of regions in R intersected by $\mathcal{D}(d)$, then $\pi(R') = N_{G_C}(d)$. We will refer to this as the *parent property*.

We define the bipartite graph G^* with partite sets D and R as the intersection graph of D and R: there is an edge between a vertex $d \in D$ and a vertex corresponding to a region $r \in R$ if and only if $\mathcal{D}(d)$ intersects region r. We will show that G^* is planar by embedding it in the geometric model of the disk graph G. The planarity of G^* can be derived from the following observations:

(a) If some region intersects only one region in the other partite set, then it will become a degree-1 vertex in G^* and it will never violate planarity. So we only need to consider regions that intersect at least two regions in the other partite set.

(b) Every region $r \in R_1$ is completely contained within some $\mathcal{D}(d)$ for $d \in D$ and hence r only intersects one region in D since disks of dead vertices are disjoint; therefore we can ignore the regions in R_1 by the previous observation.

(c) For every disk $\mathcal{D}(d)$ for $d \in D$ the interior of the disk is not intersected by any other region in $D \cup R_2$, by construction of R_2. Similarly, for every region $r \in R_2$ the interior of that region is not intersected by any other region in $D \cup R_2$.

We show how to create an embedding in the plane of the intersection graph of $D \cup R_2$, which suffices to prove that G^* is planar by observation (b). For every region in $D \cup R_2$ we select a location for the corresponding vertex in the interior of that region. Since the interior of every region is connected, we can draw edges from these points to all intersected neighboring regions. Since no three regions have a common intersection we can draw these edges in the plane without crossings - see Figure 1(c). This proves that G^* is planar and establishes (i). The parent property of the regions R establishes (ii), since the neighborhood of a dead vertex in G^* corresponds to the set of regions of R it intersects. Hence if two vertices $d, d' \in D$ satisfy $N_{G^*}(d) \subseteq N_{G^*}(d')$ then $\pi(N_{G^*}(d)) \subseteq \pi(N_{G^*}(d'))$ and therefore $N_{G_C}(d) \subseteq N_{G_C}(d')$.

As the last part of the proof we establish that $|R| \leq |C|$ by showing that we can charge every region in r to a vertex of C such that no vertex is charged twice. A region $r \in R_1$ corresponds to a component $X \in \text{COMP}(G[C])$ such that $\mathcal{D}(V(X))$ is contained in $\mathcal{D}(d)$ for some $d \in D$; for every such connected component X there is exactly one region in R_1, and we can charge the cost of vertex r to one vertex in the connected component X. For the regions in R_2 the situation is slightly more involved. Consider some vertex $v \in C$. It is not hard to see that if we start with the region $\mathcal{D}(v)$ and subtract the interiors of mutually *disjoint* disks from that region, then the result must be either an empty region or a connected region. This implies that for every vertex $v \in C$, there is at most one region $r \in R_2$ that has a non-empty intersection with $\mathcal{D}(v)$. Since every region of R_2 is a subset of $\mathcal{D}(C)$, every region $r \in R_2$ intersects $\mathcal{D}(v)$ for at least one $v \in C$, and by the previous argument r is the only region of R_2 for which this holds. Therefore we can charge the region r to such a vertex v and we can be sure that we never charge to a vertex of C twice; this proves that $|R| = |R_1| + |R_2| \leq |C|$, which establishes (iii) and completes the proof. $\square$

Theorem 1. ANNOTATED CONNECTED VERTEX COVER *has a kernel with* $2k^2 + 6k$ *vertices on disk graphs with arbitrary radii.*

Proof. Given an instance $\langle G, k, C \rangle$ we first exhaustively apply the reduction rules; it is not hard to verify that this can be done in polynomial time. Let $\langle G', k', C' \rangle$ be the resulting reduced instance. Since the reduction rules do not change the value of k we have $k' = k$. If the reduced instance contains more than k^2 uncovered edges then the answer to the decision problem must be NO and we can output a trivial NO-instance: all uncovered edges need to be covered by an unmarked vertex, and any unmarked vertex may cover at most k edges since it has degree at most k by Rule [1] - therefore any vertex cover must consist of more than k unmarked vertices if there are more than k^2 uncovered edges. If the number of uncovered edges is at most k^2 then the number of live vertices is at most $2k^2$ since every live vertex is incident on at least one uncovered edge.

If the number of marked vertices C exceeds k then clearly there is no solution set containing C of size at most k, and we output a NO instance. Otherwise the number of marked vertices is at most k. Consider the dead vertices D in the reduced instance, and the bipartite planar graph $G^* = (R \cup D, E^*)$ whose existence is guaranteed by Lemma 3. By Rule [2] we know that $N_{G_C}(d) \not\subseteq N_{G_C}(d')$ for distinct vertices $d, d' \in D$, and by (ii) of the lemma this implies that $N_{G^*}(d) \not\subseteq N_{G^*}(d')$. Therefore we may apply Lemma 1 to graph G^* where R plays the role of X, and D plays the role of Y, to conclude that $|D| \leq 5|R|$. Using condition (iii) of Lemma 3 gives $|D| \leq 5|C| \leq 5k$, which is the final ingredient of the size bound.

In summary there are at most k marked vertices, at most $2k^2$ live vertices and at most $5k$ dead vertices which shows that $|V(G')| \leq 2k^2 + 6k$. Therefore we can output the reduced instance $\langle G', k', C' \rangle$ as the result of the kernelization; by the safety of the reduction rules we know that $\langle G, k, C \rangle$ is a YES-instance if and only if $\langle G', k', C' \rangle$ is.   □

It is not hard to undo the annotation to obtain a kernel for the original problem: for every marked vertex we arbitrarily re-add some of its deleted dead-vertex neighbors, until every marked vertex has degree more than k.

Theorem 2. CONNECTED VERTEX COVER *has a kernel with $3k^2 + 7k$ vertices on disk graphs with arbitrary radii.*   □

The resulting kernel can be lifted to more general geometric graph classes. For ease of presentation we have presented the kernel for the intersection graphs of disks, but the stated results should easily generalize to intersection graphs of connected geometric objects in pseudo-disk relation [16].

4 H-Matching

The H-MATCHING problem asks whether a given graph G contains at least k vertex-disjoint subgraphs that are isomorphic to some fixed connected graph H. For ease of notation we define $|H|$ as short-hand for $|V(H)|$. Subgraph matching problems have received considerable attention from the FPT community, resulting in $O(k)$ vertex kernels for H-MATCHING on sparse graphs [5,6] and a kernel with $O(k^{|H|-1})$ vertices on general graphs [17]. The restriction to unit-disk graphs has been considered in the context of approximation algorithms [18]. For every fixed H we give a linear-vertex kernel for H-MATCHING in the case that G is required to be a unit-disk graph.

Reduction Rule 3. Delete all vertices that are not contained in a H-subgraph of G.

This rule is clearly correct. We can obtain a reduced graph by trying all $|V(G)|^{|H|}$ possible ordered vertex sets of size $|H|$ and marking the vertices for which the guessed set forms a H-subgraph. Afterwards we delete all vertices that were not marked. Since we assume H to be fixed, this can be done in polynomial time.

Theorem 3. *Let G be a reduced unit-disk instance of H-MATCHING. If there is a maximal H-matching in G that consists of k^* copies of H, then $|V(G)| \in O(k^*)$.*

Proof. Let G be a reduced graph and consider a maximal H-matching in G consisting of k^* copies. Let S be the vertices that occur in a matched copy of H. Since the selected copies are vertex-disjoint we have $|S| = k^*|H|$. Let $W := V(G) \setminus S$ be the vertices not used in a copy of H in the matching. We will prove that the size of W is bounded in $|S|$.

Define $\delta_G(u, v)$ as the number of edges on a shortest path between u and v in G, or $+\infty$ if u and v are not connected in G. Let d_H be the diameter of the graph H, i.e. $d_H := \max_{u,v \in V(H)} \delta_H(u, v)$. Since we assume H to be connected we have $d_H \leq |H|$. We measure the distance of a vertex $v \in V(G)$ to the closest vertex in S by $\delta_G(v, S) := \min_{u \in S} \delta_G(u, v)$. Now suppose that there is some $w \in W$ with $\delta_G(w, S) > d_H$. By Rule [3] the vertex w must be contained in some subgraph $G' \subseteq G$ that is isomorphic to H. All vertices $v \in V(G')$ involved in this subgraph have a distance to w in G' (and therefore in G) of at most d_H by the definition of diameter. Since $\delta_G(w, S) > d_H$ none of the vertices in $V(G')$ are in S. But that means that we can add G' as an extra copy of H to the matching, contradicting the assumption that we started from a *maximal* H-matching. Therefore we may conclude that $\delta_G(w, S) \leq d_H$ for all $w \in W$.

For the next step in the analysis we show that any vertex has a bounded number of neighbors in W. If $v \in V(G)$ has more than $6(|H| - 1)$ neighbors in W, then by Lemma 2 there is a clique of size $|H|$ in $G[N_G(v) \cap W]$. This clique must contain a subgraph isomorphic to H and hence it can be added to the H-matching, again contradicting the assumption that we started with a maximal matching. Therefore every vertex in G has at most $6(|H|-1)$ neighbors in W. Since there are exactly $k^*|H|$ vertices in S this shows that there are at most $k^*|H| \cdot 6(|H|-1)$ vertices $w \in W$ for which $\delta_G(w, S) = 1$. Now observe that a vertex v with $\delta_G(v, S) = 2$ must be adjacent to some vertex u with $\delta_G(u, S) = 1$. Since all such vertices u have a bounded number of neighbors in W, we find that there are at most $k^*|H| \cdot (6(|H|-1))^2$ vertices $w \in W$ for which $\delta_G(w, S) = 2$. By generalizing this step we obtain a bound of $k^*|H| \cdot (6(|H| - 1))^r$ on the number of vertices v that have $\delta_G(v, S) = r$. Since we established $\delta_G(w, S) \leq d_H$ for all $w \in W$ we can conclude that $|W| \leq \sum_{i=1}^{d_H} k^*|H| \cdot (6(|H| - 1))^i$ which implies that $|W| \in O(k^*(6|H|)^{d_H})$. Now we can bound the size of the instance G by noting that $|V(G)| = |S| + |W|$ and therefore $|V(G)| \in O(k^*(6|H|)^{d_H})$. Since the diameter d_H of H is at most $|H|$ this completes the proof: for every fixed H the term $(6|H|)^{d_H} \leq (6|H|)^{|H|}$ is constant and hence $|V(G)| \in O(k^*)$. $\qquad \square$

Theorem 3 leads to a linear-vertex kernel for H-MATCHING since we can find a maximal H-matching in polynomial time for fixed H. If the maximal matching has size $k^* \geq k$ we have solved the problem; if not then there is a maximal matching of size $k^* < k$ and the size of the instance is bounded.

Corollary 1. *H-MATCHING in unit-disk graphs has a kernel with $O(k)$ vertices for every fixed H.* $\qquad \square$

5   Red-Blue Dominating Set

In the RED-BLUE DOMINATING SET problem we are given a graph $G = (V, E)$, a positive integer k and a partition of the vertices into red and blue color classes R, B such that $V = R \cup B$ and $R \cap B = \emptyset$; the goal is to determine whether there is a set $S \subseteq R$ consisting of at most k red vertices such that every blue vertex in B has at least one neighbor in S. In the literature it is often assumed that G is bipartite with the red and blue classes as the partite sets; we explicitly do not require this here since bipartite disk graphs are planar [19, Lemma 3.4]. Under these assumptions the RED-BLUE DOMINATING SET problem on disk graphs graphs is not easier than DOMINATING SET on those graphs when parameterized by the size of the solution set, since we can reduce from the regular DOMINATING SET problem by making two copies of every vertex, marking one of them as red and the other as blue: hence RED-BLUE DOMINATING SET is $W[1]$-hard on unit-disk graphs when parameterized by k [10]. The situation changes when we parameterize by $|R|$ or by $|B|$, which causes the problem to become fixed parameter tractable on general graphs. Dom et al. [13] have shown that the RED-BLUE DOMINATING SET problem parameterized by $|R| + k$ or $|B| + k$ does not have a polynomial kernel on general graphs, unless the polynomial hierarchy collapses. We show that the situation is different for disk graphs by proving that RED-BLUE DOMINATING SET when parameterized by the size of the smallest color class has polynomial kernels on planar graphs and (unit)disk graphs. We use the same reduction rules for all three graph classes.

Reduction Rule 4. If there are distinct red vertices $u, v \in R$ such that $N_G(u) \cap B \subseteq N_G(v) \cap B$ then delete u.

It is easy to see that this rule is correct: the red vertex v can dominate all blue vertices that can be dominated by u, and hence there is always a smallest dominating set that does not contain u. The following rule is similar in spirit, but works on the other vertex set.

Reduction Rule 5. If there are distinct blue vertices $u, v \in B$ such that $N_G(u) \cap R \subseteq N_G(v) \cap R$ then delete v.

In this case we may delete v because whenever u is dominated by some red $x \in N_G(u)$, the vertex v must be dominated as well since $x \in N_G(v)$. These reduction rules immediately lead to a kernel with $O(\min(|R|, |B|))$ vertices on planar graphs by invoking Lemma 1. For unit-disk graphs and disk graphs with arbitrary radii we need the following structural results.

Theorem 4. Let $G = (V, E)$ be a unit-disk graph whose vertex set is partitioned into red and blue color classes by $V = R \cup B$ with $R \cap B = \emptyset$. If for all distinct $u, v \in R$ it holds that $N_G(u) \cap B \nsubseteq N_G(v) \cap B$ then $|R| \in O(|B|^2)$.   □

The proof of Theorem 4 relates the maximum number of red disks to the complexity of the arrangement of the blue disk in the plane. A well-known construction from the area of computational geometry (see [20, Section 5.2]) shows that the bound is asymptotically tight.

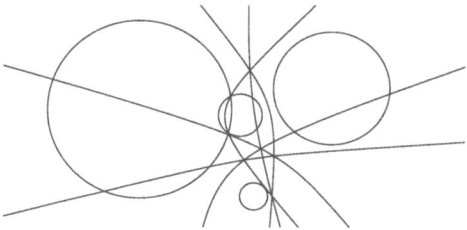

Fig. 2. Four circles and the bisector curves for all pairs

Theorem 5. *Let $G = (V, E)$ be a disk graph with arbitrary radii whose vertex set is partitioned into red and blue color classes by $V = R \cup B$ with $R \cap B = \emptyset$. If for all distinct $u, v \in R$ it holds that $N_G(u) \cap B \nsubseteq N_G(v) \cap B$ then $|R| \in O(|B|^4)$.*

Proof. Assume the conditions stated in the theorem hold, and fix a realization of the disk graph G that specifies a disk for every vertex. For a pair of blue disks $\mathcal{D}(b_1), \mathcal{D}(b_2)$ we consider all points for which the distances to the boundaries of $\mathcal{D}(b_1)$ and $\mathcal{D}(b_2)$ are the same, i.e. the points that are equidistant to the boundaries of $\mathcal{D}(b_1)$ and $\mathcal{D}(b_2)$. If the disks do not completely coincide then this set of points forms a curve in the plane. The type of curve depends on the relative orientation of the two disks. If the two disks have equal radius then the curve is a line. If the radii differ then the curve is an ellipse if one disk completely contains the other, and otherwise the curve is a branch of a hyperbola. Consider the arrangement in the plane that is induced by the set of all possible $\binom{|B|}{2}$ bisector curves for pairs of blue disks, and let F be a face of this arrangement (see Figure 2).

Suppose we choose a point p in the face F and grow a disk around p by gradually increasing its radius. If the disk grows large enough then it will intersect some of the blue disks. Once it intersects a blue disk $\mathcal{D}(b)$ at a certain radius, then naturally it will keep intersecting $\mathcal{D}(b)$ as its radius increases. Observe that the *order* in which the blue disks are encountered as the radius increases does not depend on where we place p within face F: the relative order in which two blue disks $\mathcal{D}(b_1)$ and $\mathcal{D}(b_2)$ are encountered is determined by the relative position of point p to the bisector curve of $\mathcal{D}(b_1)$ and $\mathcal{D}(b_2)$. Since F is a face in the arrangement of all bisector curves, this relative position will be the same for all points $p \in F$, and therefore all choices of a point p in F yield the same order in which blue vertices are encountered. Now observe that for every possible order of encountering the blue disks there can be at most one red vertex: if two red vertices are placed on positions that yield the same order, then the neighborhood of one must be a subset of the neighborhood of the other - which is not possible by the assumption in the statement of the theorem. Therefore F can contain at most one center of a red disk. The same argument shows that every edge or vertex in the arrangement induced by the bisector curves can contain at most one red vertex. Hence we can bound the number of red vertices by bounding the total number of vertices, edges and faces of the arrangement (i.e. the complexity of the arrangement).

The arrangement of curves consists of lines, branches of hyperbolas and ellipses. It is not hard to verify that each pair of these objects can intersect at most a constant number of times. In the area of computational geometry it is a well-known fact [21] that the complexity of an arrangement of n curves is $O(n^2)$ if the number of intersections between each pair of curves is bounded by a constant. In our case we find that there are $|B|^2$ objects and hence the complexity is $O(|B|^4)$. Since we have at most one red vertex per element in the arrangement, the claim follows.   □

By combining the structure of the graph after the reduction rules with the results of Theorem 4 and Theorem 5 we obtain the following corollary.

Corollary 2. *The* RED-BLUE DOMINATING SET *problem admits a kernel with* $O(\min(|R|, |B|))$ *vertices on planar graphs, with* $O(\min(|R|, |B|)^2)$ *vertices on unit-disk graphs and with* $O(\min(|R|, |B|)^4)$ *vertices on disk graphs with arbitrary radii.*

6   Conclusion

We have shown that the geometric structure of dense (unit)disk graphs can be exploited to obtain polynomial size problem kernels for several hard problems. It would be interesting to see whether other hard graph problems allow polynomial kernels on disk graphs; potential candidates are EDGE CLIQUE COVER and PARTITION INTO CLIQUES. We also leave it as an open problem to determine whether the bound from Theorem 5 can be asymptotically improved.

Some of the ideas used in this paper to obtain polynomial kernels on disk graphs can also be used to obtain polynomial kernels on other classes of graphs. Let $G = (V, E)$ be an undirected graph and let the sets R, B partition V. If $N := \{N_G(v) \cap R \mid v \in B\}$ is an antichain (i.e. no set in N is a subset of another set in N) then R, B form an *antichain partition* of the graph G. If $|B| \leq |R|^c$ then we say that the antichain partition is c-balanced. If $\mathcal{G}$ is a class of undirected graphs such that all antichain partitions of all graphs in $\mathcal{G}$ are c-balanced for a fixed value c, then we say that $\mathcal{G}$ is *c-antichain-balanced*. If such a c-antichain-balanced class of graphs $\mathcal{G}$ is closed under vertex deletions then CONNECTED VERTEX COVER parameterized by k has a kernel with $O(k^{\max(2,c)})$ vertices and RED-BLUE DOMINATING SET parameterized by the smallest color class has a kernel with $O(\min(|R|, |B|)^c)$ vertices, when these problems are restricted to graphs of class $\mathcal{G}$. Theorem 4 and Theorem 5 show that unit-disk graphs and disk-graphs are 2- and 4-antichain-balanced respectively, which explains the existence of some of the polynomial kernels we obtained. Now consider the class of $K_{i,j}$-subgraph-free graphs for some fixed $i \leq j$; an elementary combinatorial argument shows this class is i-antichain-balanced and hence there are polynomial kernels for the mentioned parameterizations of CONNECTED VERTEX COVER and RED-BLUE DOMINATING SET on $K_{i,j}$-subgraph-free graphs. It will be interesting to see if the property of antichain-balance can be shown to be a sufficient condition for the existence of polynomial kernels for other problems.

Acknowledgements. I would like to thank Manu Basavaraju, Hans Bodlaender, Marc van Kreveld, Erik Jan van Leeuwen and Jan van Leeuwen for insightful discussions.

References

1. Downey, R., Fellows, M.R.: Parameterized complexity. Monographs in Computer Science. Springer, New York (1999)
2. Guo, J., Niedermeier, R.: Invitation to data reduction and problem kernelization. SIGACT News 38, 31–45 (2007)
3. Bodlaender, H.L., Downey, R.G., Fellows, M.R., Hermelin, D.: On problems without polynomial kernels. J. Comput. Syst. Sci. 75, 423–434 (2009)
4. Guo, J., Niedermeier, R.: Linear problem kernels for NP-hard problems on planar graphs. In: Proc. 34th ICALP, pp. 375–386 (2007)
5. Bodlaender, H.L., Fomin, F.V., Lokshtanov, D., Penninkx, E., Saurabh, S., Thilikos, D.M.: (Meta) Kernelization. In: Proc. 50th FOCS, pp. 629–638 (2009)
6. Fomin, F., Lokshtanov, D., Saurabh, S., Thilikos, D.M.: Bidimensionality and kernels. In: Proc. 21st SODA, pp. 503–510 (2010)
7. van Leeuwen, E.J.: Optimization and Approximation on Systems of Geometric Objects. PhD thesis, CWI Amsterdam (2009)
8. Clark, B.N., Colbourn, C.J., Johnson, D.S.: Unit disk graphs. Discrete Mathematics 86, 165–177 (1990)
9. Marx, D.: Efficient approximation schemes for geometric problems? In: Proc. 13th ESA, pp. 448–459 (2005)
10. Marx, D.: Parameterized complexity of independence and domination on geometric graphs. In: Proc. 2nd IWPEC, pp. 154–165 (2006)
11. Alber, J., Fiala, J.: Geometric separation and exact solutions for the parameterized independent set problem on disk graphs. J. Algorithms 52, 134–151 (2004)
12. Philip, G., Raman, V., Sikdar, S.: Solving dominating set in larger classes of graphs: Fpt algorithms and polynomial kernels. In: Proc. 17th ESA, pp. 694–705 (2009)
13. Dom, M., Lokshtanov, D., Saurabh, S.: Incompressibility through colors and ids. In: Proc. 36th ICALP, pp. 378–389 (2009)
14. Marathe, M., Breu, H., Iii, H.B.H., Ravi, S.S., Rosenkrantz, D.J.: Simple heuristics for unit disk graphs. Networks 25, 59–68 (1995)
15. Wiese, A., Kranakis, E.: Local PTAS for independent set and vertex cover in location aware unit disk graphs. In: Proc. 4th DCOSS, pp. 415–431 (2008)
16. Kratochvíl, J., Pergel, M.: Intersection graphs of homothetic polygons. Electronic Notes in Discrete Mathematics 31, 277–280 (2008)
17. Moser, H.: A problem kernelization for graph packing. In: Proc. 35th SOFSEM, pp. 401–412 (2009)
18. Hunt, H.B., Marathe, M.V., Radhakrishnan, V., Ravi, S.S., Rosenkrantz, D.J., Stearns, R.E.: NC-approximation schemes for NP- and PSPACE-hard problems for geometric graphs. Journal of Algorithms 26, 238–274 (1998)
19. Chan, T.M., Har-Peled, S.: Approximation algorithms for maximum independent set of pseudo-disks. In: Proceedings 25th SCG, pp. 333–340 (2009)
20. Efrat, A., Sharir, M., Ziv, A.: Computing the smallest k-enclosing circle and related problems. Comput. Geom. 4, 119–136 (1994)
21. de Berg, M., Speckmann, B.: Computational geometry: Fundamental structures. In: Handbook of Data Structures and Applications, pp. 62.1–62.20 (2004)

Faster Parameterized Algorithms
for Minor Containment

Isolde Adler[1,*], Frederic Dorn[2,**], Fedor V. Fomin[2,**],
Ignasi Sau[3,***], and Dimitrios M. Thilikos[4,†]

[1] Institut für Informatik, Goethe-Universität, Frankfurt, Germany
`iadler@informatik.uni-frankfurt.de`
[2] Department of Informatics, University of Bergen, Norway
`{frederic.dorn,fedor.fomin}@ii.uib.no`
[3] Department of Computer Science, Technion, Haifa, Israel
`ignasi@cs.technion.ac.il`
[4] Department of Mathematics, National and
Kapodistrian University of Athens, Greece
`sedthilk@math.uoa.gr`

Abstract. The theory of Graph Minors by Robertson and Seymour is one of the deepest and significant theories in modern Combinatorics. This theory has also a strong impact on the recent development of Algorithms, and several areas, like Parameterized Complexity, have roots in Graph Minors. Until very recently it was a common belief that Graph Minors Theory is mainly of theoretical importance. However, it appears that many deep results from Robertson and Seymour's theory can be also used in the design of practical algorithms. Minor containment testing is one of algorithmically most important and technical parts of the theory, and minor containment in graphs of bounded branchwidth is a basic ingredient of this algorithm. In order to implement minor containment testing on graphs of bounded branchwidth, Hicks [NETWORKS 04] described an algorithm, that in time $\mathcal{O}(3^{k^2} \cdot (h + k - 1)! \cdot m)$ decides if a graph G with m edges and branchwidth k, contains a fixed graph H on h vertices as a minor. That algorithm follows the ideas introduced by Robertson and Seymour in [J'CTSB 95]. In this work we improve the dependence on k of Hicks' result by showing that checking if H is a minor of G can be done in time $\mathcal{O}(2^{(2k+1) \cdot \log k} \cdot h^{2k} \cdot 2^{2h^2} \cdot m)$. Our approach is based on a combinatorial object called *rooted packing*, which captures the properties of the potential models of subgraphs of H that we seek in our dynamic programming algorithm. This formulation with rooted packings allows us to speed up the algorithm when G is embedded in a fixed surface, obtaining the first single-exponential algorithm for minor containment testing. Namely, it runs in time $2^{\mathcal{O}(k)} \cdot h^{2k} \cdot 2^{\mathcal{O}(h)} \cdot n$, with $n = |V(G)|$. Finally, we show that slight modifications of our algorithm permit to solve some related problems within the same time bounds, like induced minor or contraction minor containment.

* Supported by the Norwegian Research Council.
** Supported by the Norwegian Research Council.
*** Supported by the Israel Science Foundation, grant No. 1249/08.
† Supported by the project "Kapodistrias" (AΠ 02839/28.07.2008) of the National and Kapodistrian University of Athens (project code: 70/4/8757).

H. Kaplan (Ed.): SWAT 2010, LNCS 6139, pp. 322–333, 2010.

Keywords: Graph minors, branchwidth, parameterized complexity, dynamic programming, graphs on surfaces.

1 Introduction

Robertson and Seymour asserted Wagner's conjecture by showing that each minor-closed graph property can be characterized by a finite set of forbidden minors [14, 15]. Suppose that **P** is a property on graphs that is *minor-closed*, that is, if a graph has this property then all its minors have it too. Graph Minors Theory implies that there is a *finite* set $\mathcal{F}$ of *forbidden minors* such that a graph G has property **P** if and only if G does not have any of the graphs in $\mathcal{F}$ as a minor. This result also has a strong impact on algorithms, since it implies that testing for minor closed properties can be done in polynomial time, namely by finitely many calls to an $\mathcal{O}(n^3)$-time algorithm (introduced in [14]) checking whether the input graph G contains some fixed pattern H as a minor. As a consequence, several graph problems have been shown to have polynomial-time algorithms, some of which were previously not even known to be decidable [8]. However, these algorithmic results are *non-constructive*. This triggered an ongoing quest in the Theory of Algorithms since then –next to the simplification of the the 23-papers proof of the Graph Minors Theorem– for extracting constructive algorithmic results out of Graph Minors (e.g., [4,5,2,12]) and for making its algorithmic proofs practical. Minor containment is one of the important steps in the technique of minor-closed property testing. Unfortunately the hidden constants in the polynomial-time algorithm of [14] are immense even for very simple patterns, which makes the algorithm absolutely impractical.

A basic algorithmic tool introduced in the Graph Minors series is *branchwidth* that servers (together with its twin parameter of *treewidth*) as a measure of the topological resemblance of a graph to the structure of a tree. The algorithmic importance of branchwidth resides in Courcelle's theorem [3] stating that all graph problems expressible by some Monadic Second Order Logic (MSOL) formula ϕ can be solved in $f(\mathbf{bw}(k), \phi) \cdot n$ steps (we denote by $\mathbf{bw}(G)$ the branchwidth of the graph G). As minor checking (for fixed patterns) can be expressed in MSOL, we obtain the existence of a $f(k, h) \cdot |V(G)|$ step algorithm for the following (parameterized) problem (throughout the paper, we let $n = |V(G)|$, $m = |E(G)|$, and $h = |V(H)|$):

H-MINOR CONTAINMENT
Input: A graph G (the host graph).
Parameter: $k = \mathbf{bw}(G)$.
Question: Does G contain a minor isomorphic to H (the pattern graph)?

This fact is one of the basic subroutines required by the algorithm in [14], and every attempt to improve its efficiency requires the improvement of the parameter dependance $f(k, h)$. A significant step in this direction was done by Hicks [11], who provided an $\mathcal{O}(3^{k^2} \cdot (h + k - 1)! \cdot m)$ step algorithm for H-MINOR CONTAINMENT, exploiting the ideas sketched by Robertson and Seymour in [14].

Note that when H is not fixed, determining whether G contains H as a minor is NP-complete even if G has bounded branchwidth [13].

The objective of this paper is to provide parameterized algorithms for the H-MINOR CONTAINMENT problem with better parameter dependance.

Our results. We present an algorithm for H-MINOR CONTAINMENT with running time $\mathcal{O}(2^{(2k+1)\cdot\log k} \cdot h^{2k} \cdot 2^{2h^2} \cdot m)$, where k is the branchwidth of G, which improves the bound that follows from [14] (explicitly described in [11]). When we restrict the host graph to be embeddable in a fixed surface, we provide an algorithm with running time $2^{\mathcal{O}(k)} \cdot h^{2k} \cdot 2^{\mathcal{O}(h)} \cdot n$. This is the first algorithm for H-MINOR CONTAINMENT with single-exponential dependence on branchwidth. Finally, we show how to modify our algorithm to explicitly find –within the same time bounds– a model of H in G, as well as for solving some related problems, like finding a model of smallest size, or solving the induced and contraction minor containment problems.

Our techniques. We introduce a dynamic programming technique based on a combinatorial object called *rooted packing* (defined in Subection 3.1). The idea is that rooted packings capture how potential models of H (defined in Section 2) are intersecting the separators that the algorithm is processing. We present the algorithm for general host graphs in Subsection 3.2. When the host graph G is embedded in a surface, this formulation with rooted packings allows us to apply the framework introduced in [16] to obtain single-exponential algorithms for H-MINOR CONTAINMENT. In this framework we use a new type of branch decomposition, called *surface cut decomposition*, which generalizes sphere cut decompositions for planar graphs introduced by Seymour and Thomas [17]. Our algorithms are robust, in the sense that slight variations permit us to solve several related problems within the same time bounds. Due to space constraints, the details of the algorithm for graphs on surfaces and for the variations of the minor containment problem are not included in this extended abstract, and can be found in [1]. Finally, we present some lines for further research in Section 4.

2   Definitions

Graphs and minors. We use standard graph terminology, see for instance [6]. All the graphs considered in this article are simple and undirected. Given a graph G, we denote the vertex set of G by $V(G)$ and the edge set of G by $E(G)$. A graph F is a subgraph of a graph G, $F \subseteq G$, if $V(F) \subseteq V(G)$ and $E(F) \subseteq E(G)$. For a subset $X \subseteq V(G)$, we use $G[X]$ to denote the subgraph of G *induced* by X, i.e. $V(G[X]) := X$ and $E(G[X]) := \{\{u,v\} \subseteq X \mid \{u,v\} \in E(G)\}$. For a subset $Y \subseteq E(G)$ we let $G[Y]$ be the graph with $V(G[Y]) := \{v \in V(G) \mid v \in e \text{ for some } e \in Y\}$ and $E(G[Y]) := Y$. *Hypergraphs* generalize graphs by allowing edges to be arbitrary subsets of the vertex set. Let H be a hypergraph. A *path* in H is a sequence $v_1, \ldots, v_n$ of vertices of H, such that for every two consecutive vertices there exists a distinct hyperedge of H containing both. In this way, the notions of connectivity, connected component, etc. are transferred from graphs to hypergraphs. Given a subset $S \subseteq V(G)$, we define $N_G[S]$ to be the set of vertices of $V(G)$ at distance at most 1 from at least one vertex of S. If $S = \{v\}$, we simply use the notation $N_G[v]$. We also define $N_G(v) = N_G[v] \setminus \{v\}$ and $E_G(v) = \{\{v, u\} \mid u \in N_G(v)\}$. Let $e = \{x, y\} \in E(G)$. Given G, let $G \backslash e := (V(G), E(G) \setminus \{e\})$, and let

$G/e =$

$$((V(G) \setminus \{x, y\}) \,\dot\cup\, \{v_{x,y}\}, (E(G) \setminus (E_G(x) \cup E_G(y))) \cup \{\{v_{xy}, z\} \mid z \in N_G[\{x, y\}]\}),$$

where $v_{xy} \notin V(G)$ is a new vertex, not contained in $V(G)$. If H can be obtained from a subgraph of G by a (possibly empty) sequence of edge contractions, we say that H is a *minor* of G. If H can be obtained from an induced subgraph of G (resp. the whole graph G) by a (possibly empty) sequence of edge contractions, we say that H is an *induced minor* (resp. a *contraction minor*) of G.

Branch decompositions. A *branch decomposition* (T, μ) of a graph G consists of a ternary tree T (i.e., all internal vertices are of degree three) and a bijection $\mu : L \rightarrow E(G)$ from the set L of leaves of T to the edge set of G. We define for every edge e of T the *middle set* $\mathbf{mid}(e) \subseteq V(G)$ as follows: Let T_1 and T_2 be the two connected components of $T \setminus \{e\}$. Then let G_i be the graph induced by the edge set $\{\mu(f) : f \in L \cap V(T_i)\}$ for $i \in \{1, 2\}$. The *middle set* is the intersection of the vertex sets of G_1 and G_2, i.e., $\mathbf{mid}(e) = V(G_1) \cap V(G_2)$. Note that for each $e \in E(T)$, $\mathbf{mid}(e)$ is a separator of G (unless $\mathbf{mid}(e) = \emptyset$). The *width* of (T, μ) is the maximum order of the middle sets over all edges of T, i.e., $width(T, \mu) := \max\{|\mathbf{mid}(e)| : e \in E(T)\}$. The *branchwidth* of G is defined as $\mathbf{bw}(G) := \min\{width(T, \mu) \mid (T, \mu) \text{ branch decomposition of } G\}$. Intuitively, a graph G has small branchwidth, if G is close to being a tree.

In our algorithms, we need to root a branch decomposition (T, μ) of G. For this, we pick an arbitrary edge $e^* \in E(T)$, we subdivide it by adding a new vertex v_{new} and then add a new vertex r and make it adjacent to v_{new}. We extend μ by setting $\mu(r) = \emptyset$ (thereby slightly extending the definition of a branch decomposition). Now vertex r is the root. For each $e \in E(T)$ let T_e be the tree of the forest $T \backslash e$ that does not contain r as a leaf (i.e., the tree that is "below" e in the rooted tree T) and let E_e be the edges that are images, via μ, of the leaves of T that are also leaves of T_e. Let $G_e := G[E_e]$ Observe that, if $e_r = \{v_{\text{new}}, r\}$, then $G_{e_r} = G$ unless G has isolated vertices.

Models and potential models. A *model* of H in G [14] is a mapping ϕ, that assigns to every edge $e \in E(H)$ an edge $\phi(e) \in E(G)$, and to every vertex $v \in V(H)$ a non-empty connected subgraph $\phi(v) \subseteq G$, such that

(i) the graphs $\{\phi(v) \mid v \in V(H)\}$ are mutually vertex-disjoint and the edges $\{\phi(e) \mid e \in E(H)\}$ are pairwise distinct,
(ii) for $e = \{u, v\} \in E(H)$, $\phi(e)$ has one end-vertex in $V(\phi(u))$ and the other in $V(\phi(v))$.

Thus, H is isomorphic to a minor of G if and only if there exists a model of H in G.

Remark 1. We can assume that for each vertex $v \in V(H)$, the subgraph $\phi(v) \subseteq G$ is a tree. Indeed, if for some $v \in V(H)$, $\phi(v)$ is not a tree, then by replacing $\phi(v)$ with a spanning tree of $\phi(v)$ we obtain another model with the desired property.

For each $v \in V(H)$, we call the graph $\phi(v)$ a *vertex-model* of v. With slight abuse of notation, the subgraph $M \subseteq G$ defined by the union of $\{\phi(v) \mid v \in V(H)\}$ and $\{\phi(e) \mid e \in E(H)\}$ is also called a *model* of H in G. For each edge $e \in E(H)$, the edge $\phi(e) \in E(G)$ is called a *realization* of e.

In the course of dynamic programming along a branch decomposition, we will need to search for potential models of subgraphs of H in G, which we proceed to define. For graphs $\bar{H}$ and G, a set $R \subseteq V(\bar{H})$, and a (possibly empty) set $X \subseteq E(\bar{H}[R])$, an (R, X)-*potential model* of $\bar{H}$ in G is a mapping ϕ, that assigns to every edge $e \in (E(\bar{H}) \setminus E(\bar{H}[R])) \cup X$ an edge $\phi(e) \in E(G)$, and to every vertex $v \in V(\bar{H})$ a non-empty subgraph $\phi(v) \subseteq G$, such that

(i) the graphs $\{\phi(v) \mid v \in V(\bar{H})\}$ are mutually vertex-disjoint and the edges $\{\phi(e) \mid e \in E(\bar{H})\}$ are pairwise distinct;
(ii) for every $e = \{u, v\} \in (E(\bar{H}) \setminus E(\bar{H}[R])) \cup X$, the edge $\phi(e)$ has one end-vertex in $V(\phi(u))$ and the other in $V(\phi(v))$;
(iii) for every $v \in V(\bar{H}) \setminus R$ the graph $\phi(v)$ is connected in G.

For the sake of intuition, we can think of an (R, X)-potential model of $\bar{H}$ as a candidate of becoming a model of $\bar{H}$ in further steps of the dynamic programming, if the missing edges (that is, those in $E(\bar{H}[R]) \setminus X$) can be realized, and if the graphs $\{\phi(v) \mid v \in R\}$ get eventually connected.

We say that ϕ is an R-*potential model* of $\bar{H}$ in G, if ϕ is an (R, X)-potential model of $\bar{H}$ in G for some $X \subseteq E(\bar{H}[R])$, and we say that ϕ is a *potential model* of $\bar{H}$ in G, if ϕ is an R-potential model of $\bar{H}$ in G for some $R \subseteq V(\bar{H})$. Note that a $\emptyset$-potential model of $\bar{H}$ in G is a model of $\bar{H}$ in G. Again slightly abusing notation, we also say that the subgraph $M \subseteq G$ defined by the union of $\{\phi(v) \mid v \in V(\bar{H})\}$ and $\{\phi(e) \mid e \in (E(\bar{H}) \setminus E(\bar{H}[R])) \cup X\}$ is an (R, X)-*potential model* of $\bar{H}$ in G.

3   Dynamic Programming for General Graphs

Roughly speaking, in each edge of the branch decomposition, the tables of our dynamic programming algorithm store all the potential models of H in the graph processed so far. While the vertex-models of H are required to be connected in G, in potential models, they may have several connected components, and we need to keep track of them. In order to do so, we introduce *rooted packings* of the middle sets (defined in Subsection 3.1). A rooted packing encodes the trace of the components of a potential model in the middle set, together with a mapping of the components to vertices of H. We denote the *empty set* by $\emptyset$ and the *empty function* by $\varnothing$.

3.1   Rooted Packings

Let $S \subseteq V(H)$ be a subset of the vertices of the pattern H, and let $R \subseteq S$. Given a middle set $\mathbf{mid}(e)$ corresponding to an edge e of a branch decomposition (T, μ) of G, we define a *rooted packing* of $\mathbf{mid}(e)$ as a quintuple $\mathbf{rp} = (\mathcal{A}, S, R, \psi, \chi)$, where $\mathcal{A}$ is a (possible empty) collection of mutually disjoint non-empty subsets of $\mathbf{mid}(e)$ (that is, a *packing* of $\mathbf{mid}(e)$), $\psi : \mathcal{A} \to R$ is a surjective mapping (the *rooting*) assigning vertices of R to the sets in $\mathcal{A}$, and $\chi : R \times R \to \{0, 1\}$ is a binary symmetric function between pairs of vertices in R.

The intended meaning of a rooting packing $(\mathcal{A}, S, R, \psi, \chi)$ is as follows. In a given middle set $\mathbf{mid}(e)$, a packing $\mathcal{A}$ represents the intersection of the connected components of the potential model with $\mathbf{mid}(e)$. The subsets $R, S \subseteq V(H)$ and the function χ indicate that we are looking for an $(R, \{\{u, v\} \mid u, v \in R, \chi(u, v) = 1\})$-potential model M of $H[S]$ in G_e, Intuitively, the function χ captures which edges of $H[S]$ have been realized so far. Since we allow the vertex-models intersecting $\mathbf{mid}(e)$ to be disconnected, we need to keep track of their connected components. The subset $R \subseteq S$ tells us which vertex-models intersect $\mathbf{mid}(e)$, and the function ψ associates the sets in $\mathcal{A}$ to the vertices in R. We can think of ψ as a coloring that colors the subsets in $\mathcal{A}$ with colors given by the vertices in R. Note that several subsets in $\mathcal{A}$ can have the same color $u \in R$, which means that the vertex-model of u in G_e is not connected yet, but it may get connected in further steps of the dynamic programming, if the necessary edges appear from other branches of the branch decomposition of G. Note that we distinguish between two types of edges of $H[S]$, namely those with both end-vertices in R, and the rest. The key observation is that if the desired R-potential model of $H[S]$ exists, then all the edges in $E(H[S]) \setminus E(H[R])$ must have already been realized in G_e. Indeed, as $\mathbf{mid}(e)$ is a separator of G and no vertex-model of a vertex in $S \setminus R$ intersects $\mathbf{mid}(e)$, the edges in $E(H[S]) \setminus E(H[R])$ cannot appear in $G \setminus G_e$. Therefore, we make sure that the edges in $E(H[S]) \setminus E(H[R])$ have already been realized, and we only need to keep track of the edges in $E(H[R])$. I.e., for two distinct vertices $u, v \in R$, we let $\chi(u, v) = 1$ if and only if $\{u, v\} \in E(H)$ and there exist two subsets $A, B \in \mathcal{A}$, with $\psi(A) = u$ and $\psi(B) = v$, such that there is an edge in the potential model M between a vertex in A and a vertex in B. In that case, it means that we have a realization of the edge $\{u, v\} \in E(H)$ in $M \subseteq G_e$. A rooted packing $\mathbf{rp} = (\mathcal{A}, S, R, \psi, \chi)$ defines a unique subgraph $H_{\mathbf{rp}}$ of H, with $V(H_{\mathbf{rp}}) = S$ and $E(H_{\mathbf{rp}}) = E(H[S]) \setminus E(H[R]) \cup \{\{u, v\} \mid u, v \in R, \chi(u, v) = 1\}$. An example of the intended meaning of a rooted packing is illustrated in Fig. 1.

In the sequel, it will be convenient to think of a packing $\mathcal{A}$ of $\mathbf{mid}(e)$ as a hypergraph $\mathcal{G} = (\mathbf{mid}(e), \mathcal{A})$. Note that, by definition, $\mathcal{A}$ is a matching in $\mathcal{G}$. We use the notation $\bigcup \mathcal{A} := \bigcup_{X \in \mathcal{A}} X$.

Operations with rooted packings. Let $\mathbf{rp}_1 = (\mathcal{A}_1, S_1, R_1, \psi_1, \chi_1)$ and $\mathbf{rp}_2 = (\mathcal{A}_2, S_2, R_2, \psi_2, \chi_2)$ be rooted packings of two middle sets $\mathbf{mid}(e_1)$ and $\mathbf{mid}(e_2)$, such that e_1 and e_2 are the children edges of an edge $e \in E(T)$. We say that $\mathbf{rp}_1$ and $\mathbf{rp}_2$ are *compatible* if

(i) $E(H_{\mathbf{rp}_1}) \cap E(H_{\mathbf{rp}_2}) = \emptyset$;
(ii) $S_1 \cap S_2 = R_1 \cap R_2$;

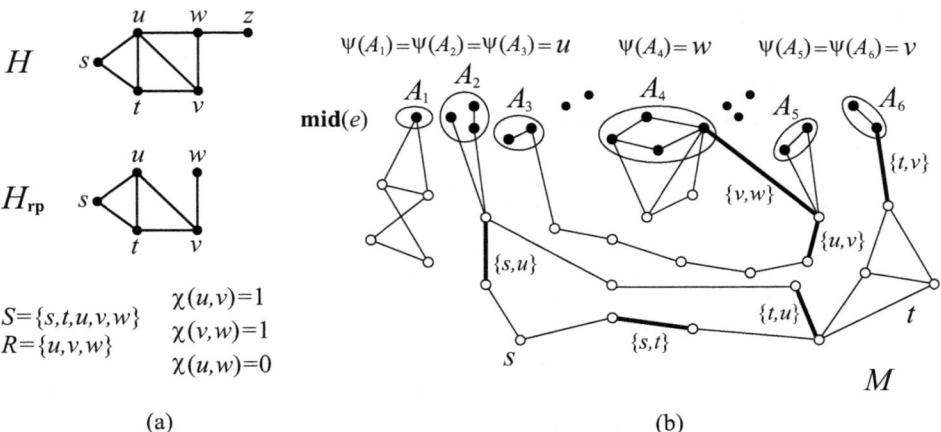

H

$H_{\mathbf{rp}}$

$S=\{s,t,u,v,w\}$
$R=\{u,v,w\}$

$\chi(u,v)=1$
$\chi(v,w)=1$
$\chi(u,w)=0$

(a)   (b)

Fig. 1. (a) A pattern H and a subgraph $H_{\mathbf{rp}} \subseteq H$ associated to a rooted packing $\mathbf{rp} = (\mathcal{A}, S, R, \psi, \chi)$. We have $V(H) = \{s,t,u,v,w,z\}$, $S = \{s,t,u,v,w\}$, and $R = \{u,v,w\}$. The function χ is given by $\chi(u,v) = \chi(v,w) = 1$ and $\chi(u,w) = 0$, which defines the edges in $H_{\mathbf{rp}}$. (b) An R-potential model $M \subseteq G_e$ corresponding to the rooted packing $\mathbf{rp}$ of $\mathbf{mid}(e)$. Full dots represent vertices in $\mathbf{mid}(e)$, and the ovals indicate the subsets of the packing $\mathcal{A} = \{A_1, A_2, A_3, A_4, A_5, A_6\}$. Above the ovals, the coloring ψ is shown. The thick edges in M correspond to realizations of edges in $E(H_{\mathbf{rp}})$, which are explicitly labeled in the figure. Note that the vertex-models in M corresponding to vertices $s, t \in S \setminus R$ are connected, as required.

(iii) for any $A_1 \in \mathcal{A}_1$ and $A_2 \in \mathcal{A}_2$ such that $A_1 \cap A_2 \neq \emptyset$, we have $\psi_1(A_1) = \psi_2(A_2)$.

In other words, two rooted packings $\mathbf{rp}_1$ and $\mathbf{rp}_2$ are compatible if the edge-sets of the corresponding subgraphs $H_{\mathbf{rp}_1}$ and $H_{\mathbf{rp}_2}$ are disjoint, if their intersection is given by the intersection of R_1 and R_2, and if their colorings coincide in the common part. Note that whether two rooted packings are compatible can be easily checked in time linear on the sizes of the middle sets.

Given two hypergraphs H_1 and H_2 of H, we define $H_1 \cup H_2$ as the graph with vertex set $V(H_1) \cup V(H_2)$ and edge set $E(H_1) \cup E(H_2)$. Given two compatible rooted packings $\mathbf{rp}_1 = (\mathcal{A}_1, S_1, R_1, \psi_1, \chi_1)$ and $\mathbf{rp}_2 = (\mathcal{A}_2, S_2, R_2, \psi_2, \chi_2)$, we define $\mathbf{rp}_1 \oplus \mathbf{rp}_2$ as the rooted packing $(\mathcal{A}, S, R, \psi, \chi)$, where

- $\mathcal{A}$ is the packing of $\mathbf{mid}(e)$ defined by the connected components of the hypergraph $(\mathbf{mid}(e_1) \cup \mathbf{mid}(e_2), \mathcal{A}_1 \cup \mathcal{A}_2)$. I.e., the sets of the packing $\mathcal{A}$ are the vertex sets corresponding to the connected components of the hypergraph $(\mathbf{mid}(e_1) \cup \mathbf{mid}(e_2), \mathcal{A}_1 \cup \mathcal{A}_2)$;
- $S = S_1 \cup S_2$;
- $R = R_1 \cup R_2$;
- for any subset $A \in \mathcal{A}$, $\psi(A)$ is defined as

$$\psi(A) = \begin{cases} \psi_1(A_1), & \text{if there exists } A_1 \in \mathcal{A}_1 \text{ such that } A \cap A_1 \neq \emptyset. \\ \psi_2(A_2), & \text{if there exists } A_2 \in \mathcal{A}_2 \text{ such that } A \cap A_2 \neq \emptyset. \end{cases}$$

Note that the mapping ψ is well-defined. Indeed, if there exist both $A_1 \in \mathcal{A}_1$ and $A_2 \in \mathcal{A}_2$ intersecting a subset A, then by definition of $\mathcal{A}$ it holds $A_1 \cap A_2 \neq \emptyset$, and therefore $\psi_1(A_1) = \psi_2(A_2)$ because by assumption the rooted packings $\mathbf{rp}_1$ and $\mathbf{rp}_2$ are compatible;

- for any two vertices $u, v \in R$, $\chi(u, v)$ is defined as

$$\chi(u, v) = \begin{cases} 1, & \text{if either } u, v \in R_1 \text{ and } \chi_1(u, v) = 1, \\ & \quad \text{or } u, v \in R_2 \text{ and } \chi_2(u, v) = 1. \\ 0, & \text{otherwise.} \end{cases}$$

Note that if $\mathbf{rp}_1$ and $\mathbf{rp}_2$ are two compatible rooted packings, then $H_{\mathbf{rp}_1 \oplus \mathbf{rp}_2} = H_{\mathbf{rp}_1} \cup H_{\mathbf{rp}_2}$.

If $(\mathcal{A}, S, R, \psi, \chi)$ is a rooted packing of a middle set $\mathbf{mid}(e)$ and $B \subseteq \mathbf{mid}(e)$, we define $(\mathcal{A}, S, R, \psi, \chi)|_B$ as the rooted packing $(\mathcal{A}', S', R', \psi', \chi')$ of B, where

- $\mathcal{A}' = \{X \cap B \mid X \in \mathcal{A}\} \setminus \{\emptyset\}$;
- $S' = S$;
- for a set $X \cap B \in \mathcal{A}'$ with $X \in \mathcal{A}$ we let $\psi'(X \cap B) = \psi(X)$;
- R' is defined as the image of ψ', that is, $R' = \{\psi'(A) \mid A \in \mathcal{A}'\}$;
- χ' is defined at the restriction of χ to $R' \times R'$, that is, for two vertices $u, v \in R'$, $\chi'(u, v) = \chi(u, v)$.

Note that the property of being a rooted packing is closed under the two operations defined above.

How to encode a potential model. Let $\mathscr{P}_e$ be the collection of all rooted packings $(\mathcal{A}, S, R, \psi, \chi)$ of $\mathbf{mid}(e)$. We use the notation $\mathcal{C}(F)$ for the set of connected components of a graph (or hypergraph) F. Given a rooted packing $(\mathcal{A}, S, R, \psi, \chi) \in \mathscr{P}_e$ we define the boolean variable $\mathbf{mod}_e(\mathcal{A}, S, R, \psi, \chi)$, saying whether G_e contains a potential model with the conditions given by the rooted packing. Namely, the variable is set to $\mathtt{true}$ if the required potential model exists, and to $\mathtt{false}$ otherwise: $\mathbf{mod}_e(\mathcal{A}, S, R, \psi, \chi) =$

$$= \begin{cases} \mathtt{true}, & \text{if there exist a subgraph } M \subseteq G_e \text{ and a partition of} \\ & \quad V(M) \text{ into } |S| \text{ sets } \{V_u \mid u \in S\} \text{ such that} \\ & \\ & \text{(i) for every } u \in S \setminus R, \ |\mathcal{C}(M[V_u])| = 1 \text{ and } V_u \cap \mathbf{mid}(e) = \emptyset; \\ & \text{(ii) for every } u \in R, \\ & \quad \{V(M') \cap \mathbf{mid}(e) \mid M' \in \mathcal{C}(M[V_u])\} = \psi^{-1}(u); \\ & \text{(iii) for every two vertices } u, v \in S \text{ with } \{u, v\} \in E(H) \\ & \quad \text{and such that } \{u, v\} \not\subseteq R, \text{ there exist } u^* \in V_u \text{ and} \\ & \quad v^* \in V_v \text{ such that } \{u^*, v^*\} \in E(M); \\ & \text{(iv) for every two vertices } u, v \in R, \ \chi(u, v) = 1 \text{ if} \\ & \quad \text{and only if } \{u, v\} \in E(H) \text{ and there exist} \\ & \quad u^* \in V_u \text{ and } v^* \in V_v \text{ such that } \{u^*, v^*\} \in E(M). \\ & \\ \mathtt{false}, & \text{otherwise.} \end{cases}$$

Note that since $\mathcal{A}$ does not contain the empty set, in (ii) we implicitly require every connected component of V_u to have a non-empty intersection with $\mathbf{mid}(e)$. The following lemma follows immediately from the definitions.

Lemma 1. *Let G and H be graphs, let e be an edge in a rooted branch decomposition of G.*

1. *If $\mathbf{mod}_e(\mathcal{A}, S, R, \psi, \chi) = \mathtt{true}$, then G_e contains an $(R, \{\{u, v\} \mid u, v \in R, \chi(u, v) = 1\})$-potential model of $H[S]$.*
2. *If G_e contains an R-potential model of $H[S]$, then there exist $\mathcal{A}$, ψ, and χ such that $\mathbf{mod}_e(\mathcal{A}, S, R, \psi, \chi) = \mathtt{true}$.*
3. *G contains a minor isomorphic to H if and only if some middle set $\mathbf{mid}(e)$ satisfies $\mathbf{mod}_e(\emptyset, V(H), \emptyset, \varnothing, \varnothing) = \mathtt{true}$.*

3.2   The Algorithm

Let us now see how the values of $\mathbf{mod}_e(\mathcal{A}, S, R, \psi, \chi)$ can be explicitly computed using dynamic programming over a branch decomposition of G.

First, let e, e_1, e_2 be three edges of T that are incident to the same vertex and such that e is closer to the root of T than the other two. The value of $\mathbf{mod}_e(\mathcal{A}, S, R, \psi, \chi)$ is then given by: $\mathbf{mod}_e(\mathcal{A}, S, R, \psi, \chi) =$

$$
= \begin{cases}
\mathtt{true}, & \text{if there exist two compatible rooted packings} \\
& \mathbf{rp}_1 = (\mathcal{A}_1, S_1, R_1, \psi_1, \chi_1) \text{ and } \mathbf{rp}_2 = (\mathcal{A}_2, S_2, R_2, \psi_2, \chi_2) \\
& \text{of } \mathbf{mid}(e_1) \text{ and } \mathbf{mid}(e_2), \text{ such that} \\
& \text{(i) } \mathbf{mod}_{e_1}\mathbf{rp}_1 = \mathbf{mod}_{e_2}\mathbf{rp}_2 = \mathtt{true}; \\
& \text{(ii) } \bigcup \mathcal{A}_1 \cap (\mathbf{mid}(e_1) \cap \mathbf{mid}(e_2)) = \bigcup \mathcal{A}_2 \cap (\mathbf{mid}(e_1) \cap \mathbf{mid}(e_2)); \\
& \text{(iii) } (\mathcal{A}, S, R, \psi, \chi) = \mathbf{rp}_1 \oplus \mathbf{rp}_2|_{\mathbf{mid}(e)}; \\
& \text{(iv) Let } (\mathcal{A}', S, R', \psi', \chi') = \mathbf{rp}_1 \oplus \mathbf{rp}_2. \\
& \quad \text{Then, for each } u \in (R_1 \cup R_2) \setminus R, \ |\psi'^{-1}(u)| = 1. \\
\mathtt{false}, & \text{otherwise.}
\end{cases}
$$

We have shown above how to compute $\mathbf{mod}_e(\mathcal{A}, S, R, \psi, \chi)$ for e being an internal edge of T. Finally, suppose that $e_{\mathrm{leaf}} = \{x, y\} \in E(T)$ is an edge such that x is a leaf of T. Let $\mu(x) = \{v_1, v_2\} \in E(G)$, and let u and v be two arbitrary distinct vertices of H. Then $\mathbf{mod}_{e_{\mathrm{leaf}}}(\mathcal{A}, S, R, \psi, \chi) =$

$$
= \begin{cases}
\mathtt{true}, & \text{if} \quad \mathcal{A} = \{\{v_1, v_2\}\}, \ S = R = \{u\}, \\
& \qquad \psi(\{v_1, v_2\}) = u, \text{ and } \chi(u, u) = 0, \\
& \text{or} \quad \mathcal{A} = \{\{v_i\}\}, \ S = \{u, v\}, \ R = \{u\}, \\
& \qquad \psi(\{v_i\}) = u, \text{ and } \chi(u, u) = 0, \text{ for } i \in \{1, 2\}, \\
& \text{or} \quad \mathcal{A} = \{\{v_i\}\}, \ S = R = \{u\}, \\
& \qquad \psi(\{v_i\}) = u, \text{ and } \chi(u, u) = 0, \text{ for } i \in \{1, 2\}, \\
& \text{or} \quad \mathcal{A} = \{\{v_1\}, \{v_2\}\}, \ S = R = \{u, v\}, \\
& \qquad \psi(\{v_1\}) = u, \ \psi(\{v_2\}) = v, \ \chi(u, u) = \chi(v, v) = 0, \\
& \qquad \text{and } \chi(u, v) = \chi(v, u) = 1 \text{ only if } \{u, v\} \in E(H), \\
& \text{or} \quad \mathcal{A} = \emptyset, \ S = \{u, v\}, \ R = \emptyset \text{ and } \psi = \chi = \varnothing, \\
& \text{or} \quad \mathcal{A} = \emptyset, \ S = \{u\}, \ R = \emptyset \text{ and } \psi = \chi = \varnothing, \\
& \text{or} \quad \mathcal{A} = S = R = \emptyset \text{ and } \psi = \chi = \varnothing. \\
\mathtt{false}, & \text{otherwise.}
\end{cases}
$$

Correctness of the algorithm. By Lemma 1, G contains a minor isomorphic to H if and only if for some middle set $\mathbf{mid}(e)$, $\mathbf{mod}_e(\emptyset, V(H), \emptyset, \varnothing, \varnothing) = \mathtt{true}$. Observe that if $e_r = \{v_{\mathrm{new}}, r\}$, we can assume that $\mathcal{A} = R = \emptyset$ and that $\psi = \chi = \varnothing$.

Given three edges e, e_1, e_2 as described above, we shall now see that the formula to compute $\mathbf{mod}_e(\mathcal{A}, S, R, \psi, \chi)$ is correct. Indeed, condition (i) guarantees that the required compatible models in G_{e_1} and G_{e_2} exist, while condition (ii) assures that the packings $\mathcal{A}_1$ and $\mathcal{A}_2$ contain the same vertices in the intersection of both middle sets. Condition (iii) says that the rooted packing of $\mathbf{mid}(e)$ can be obtained by first merging the two rooted packings of $\mathbf{mid}(e_1)$ and $\mathbf{mid}(e_2)$, and then projecting the obtained rooted packing to $\mathbf{mid}(e)$. Finally, condition (iv) imposes that each of the vertices in $R_1 \cup R_2$ that has been forgotten in $\mathbf{mid}(e)$ induces a single connected component in the desired potential model. This is indeed necessary, as the vertex-models of these forgotten vertices will not be updated anymore, so it is necessary that they are already connected. For each such vertex $u \in (R_1 \cup R_2) \setminus R$, the connectivity of the vertex-model of u is captured by the number of subsets colored u in the packing obtained by merging the packings $\mathcal{A}_1$ and $\mathcal{A}_2$. Indeed, the vertex-model of u is connected in M if and only if there is a single connected component colored u in the merged packing.

Suppose that $e_{\mathrm{leaf}} = \{x, y\} \in E(T)$ is a leaf-edge. Then $\mathbf{mid}(e_{\mathrm{leaf}}) \subseteq \{v_1, v_2\}$ and $|S| \leq 2$. Let us discuss the formula to compute $\mathbf{mod}_{e_{\mathrm{leaf}}}(\mathcal{A}, S, R, \psi, \chi)$. In the first case, $\mathcal{A} = \{\{v_1, v_2\}\}$, so both v_1 and v_2 must be mapped to the same vertex in S. The second and third case are similar, except that one of the two vertices v_1, v_2 is either not present in $\mathbf{mid}(e)$ or we omit it. In the forth case we have $\mathcal{A} = \{\{v_1\}, \{v_2\}\}$, so each vertex in $\mathbf{mid}(e)$ corresponds to a distinct vertex of H, say, to u and v, respectively. We must distinguish two cases. Namely, if $\{u, v\} \in E(H)$, then the edge $\{v_1, v_2\} \in E(G)$ is a realization of $\{u, v\} \in E(H)$, so in this case we can set $\chi(u, v) = \chi(v, u) = 1$. Otherwise, we set $\chi(u, v) = \chi(v, u) = 0$. Finally, in the cases $\mathcal{A} = \emptyset$, we omit the whole middle set, and we set $R = \emptyset$.

Running time. The size of the tables of the dynamic programming over a branch decomposition of the input graph G determines the running time of our algorithms. For $e \in E(T)$, let $|\mathbf{mid}(e)| \leq k$, and let $h = |V(H)|$. To bound the size of the tables in e, namely $|\mathscr{P}_e|$, we discuss each element appearing in a rooted packing $(\mathcal{A}, S, R, \psi, \chi)$ of $\mathbf{mid}(e)$ independently:

- Bound on the number of $\mathcal{A}$'s: The number of ways a set of k elements can be partitioned into non-empty subsets is well-known as the k-th *Bell number* [9], and it is denoted by B_k. The number of packings of a set of k elements can be expressed in terms of the Bell numbers as

$$\sum_{i=0}^{k} \binom{k}{i} B_{k-i} = B_{k+1} \leq 2^{k \cdot \log k},$$

where the equality is a well-known recursive formula of the Bell numbers, and the inequality follows from $B_k \leq \frac{e^k - 1}{(\log k)^k} \cdot k!$ [9].

- Bound on the number of S's: the number of subsets of $V(H)$ is $2^{|V(H)|} = 2^h$.
- Bound on the number of R's: for a fixed $S \subseteq V(H)$, the number of subsets of S is at most 2^h.
- Bound on the number of ψ's: ψ is a mapping from subsets of $\mathbf{mid}(e)$ to vertices in R, so the number of such mappings for a fixed packing of $\mathbf{mid}(e)$ is at most h^k.
- Bound on the number of χ's: χ is a symmetric function from $R \times R$ to $\{0, 1\}$, so for a fixed R with $|R| \leq h$, the number of choices for χ is at most $2^{h^2/2}$.

Summarizing, for each edge $e \in E(T)$, we have that

$$|\mathscr{P}_e| \leq 2^{k \cdot \log k} \cdot h^k \cdot 2^{h^2/2} \cdot 2^{2h} \leq 2^{k \cdot \log k} \cdot h^k \cdot 2^{h^2} \ .$$

At each edge e of the branch decomposition, in order to compute all the values $\mathbf{mod}_e(\mathcal{A}, S, R, \psi, \chi)$, we test all the possibilities of combining compatible rooted packings of the two middle sets $\mathbf{mid}(e_1)$ and $\mathbf{mid}(e_2)$. The operations $(\mathcal{A}_1, S_1, R_1, \psi_1, \chi_1) \oplus (\mathcal{A}_2, S_2, R_2, \psi_2, \chi_2)$ and $(\mathcal{A}, S, R, \psi, \chi)|_B$ take $\mathcal{O}(|\mathbf{mid}(e)|)$ time, as well as testing whether two rooted packings are compatible. That is, these operations just incur a multiplicative term $\mathcal{O}(k) = \mathcal{O}(2^{\log k})$ in the running time. Hence, from the above discussion we conclude the following theorem.

Theorem 1. *Given a general host graph G with $|E(G)| = m$, a pattern H with $|V(H)| = h$, and a branch decomposition of G of width at most k, we can decide whether G contains a minor isomorphic to H in $\mathcal{O}(2^{(2k+1) \cdot \log k} \cdot h^{2k} \cdot 2^{2h^2} \cdot m)$ time.*

4   Conclusions and Further Research

In this article we presented an algorithm to test whether an input host graph G contains a fixed graph H as a minor. Parameterizing the problem by the branchwidth of G (**bw**), we improved the best existing algorithm for general host graphs, and we provided the first algorithm with running time single-exponential on **bw** when the host graph can be embedded in a surface. Finally, we showed how to modify our algorithm to solve some related problems, like induced or contraction minor containment.

There are a number of interesting lines for further research concerning minor containment problems. First of all, it may be possible to improve the dependence on $h = |V(H)|$ of our algorithms. At least when H belongs to some restricted class, in the spirit of [7] for subgraph isomorphism. On the other hand, we believe that the dependence on **bw** of our algorithm for general host graphs (that is, $2^{\mathcal{O}(\mathbf{bw} \cdot \log \mathbf{bw})})$) is best possible.

We also believe that the approach we have used to obtain single-exponential algorithms when G is embedded in a surface (see [16, 1]) can be extended to more general classes of graphs, like apex-minor-free graphs or general minor-free graphs. In order to do so, a first step could be to generalize the framework developed in [16] to minor-free graphs, which looks like a promising (but highly non-trivial) direction.

Finally, a challenging problem concerning minor containment is to provide explicit and hopefully not too big constants depending on h in the polynomial-time algorithm of Robertson and Seymour [14]. Of course these constants must be superpolynomial on h unless $P = NP$, as when H is not fixed the problem of deciding whether G contains H as a minor is NP-complete [10].

References

1. Adler, I., Dorn, F., Fomin, F.V., Sau, I., Thilikos, D.M.: Faster Parameterized Algorithms for Minor Containment (2010),
   http://users.uoa.gr/~sedthilk/papers/minorch.pdf
2. Adler, I., Grohe, M., Kreutzer, S.: Computing excluded minors. In: Proc. of the 19th Annual ACM-SIAM Symposium on Discrete Algorithms (SODA), pp. 641–650 (2008)
3. Courcelle, B.: Graph rewriting: An algebraic and logic approach. In: Handbook of Theoretical Computer Science. Formal Models and Semantics (B), vol. B, pp. 193–242 (1990)
4. Dawar, A., Grohe, M., Kreutzer, S.: Locally Excluding a Minor. In: Proc. of the 22nd IEEE Symposium on Logic in Computer Science (LICS), pp. 270–279 (2007)
5. Demaine, E.D., Hajiaghayi, M.T., Kawarabayashi, K.-i.: Algorithmic Graph Minor Theory: Decomposition, Approximation, and Coloring. In: Proc. of the 46th Annual IEEE Symposium on Foundations of Computer Science (FOCS), pp. 637–646 (2005)
6. Diestel, R.: Graph Theory, vol. 173. Springer, Heidelberg (2005)
7. Dorn, F.: Planar Subgraph Isomorphism Revisited. In: Proc. of the 27th International Symposium on Theoretical Aspects of Computer Science (STACS), pp. 263–274 (2010)
8. Fellows, M.R., Langston, M.A.: Nonconstructive tools for proving polynomial-time decidability. Journal of the ACM 35(3), 727–739 (1988)
9. Flajolet, P., Sedgewick, R.: Analytic Combinatorics. Cambridge University Press, Cambridge (2008)
10. Garey, M.R., Johnson, D.S.: Computers and Intractability: A Guide to the Theory of NP-Completeness. W. H. Freeman, New York (1979)
11. Hicks, I.V.: Branch decompositions and minor containment. Networks 43(1), 1–9 (2004)
12. Kawarabayashi, K.-i., Reed, B.A.: Hadwiger's conjecture is decidable. In: Proc. of the 41st Annual ACM Symposium on Theory of Computing (STOC), pp. 445–454 (2009)
13. Matoušek, J., Thomas, R.: On the complexity of finding iso- and other morphisms for partial k-trees. Discrete Mathematics 108, 143–364 (1992)
14. Robertson, N., Seymour, P.: Graph Minors. XIII. The Disjoint Paths Problem. Journal of Combinatorial Theory, Series B 63(1), 65–110 (1995)
15. Robertson, N., Seymour, P.D.: Graph Minors. XX. Wagner's conjecture. J. Comb. Theory, Ser. B 92(2), 325–357 (2004)
16. Rué, J., Sau, I., Thilikos, D.M.: Dynamic Programming for Graphs on Surfaces. To appear in Proc. of the 37th International Colloquium on Automata, Languages and Programming, ICALP (2010)
17. Seymour, P.D., Thomas, R.: Call routing and the ratcatcher. Combinatorica 14(2), 217–241 (1994)

Fixed-Parameter Algorithms for Cochromatic Number and Disjoint Rectangle Stabbing*

Pinar Heggernes[1], Dieter Kratsch[2,**], Daniel Lokshtanov[1],
Venkatesh Raman[3], and Saket Saurabh[3]

[1] Department of Informatics, University of Bergen, N-5020 Bergen, Norway
{pinar,daniello}@ii.uib.no
[2] Université de Metz, France
kratsch@univ-metz.fr
[3] The Institute of Mathematical Sciences, Chennai, India
{vraman,saket}@imsc.res.in

Abstract. Given a permutation π of $\{1, \ldots, n\}$ and a positive integer k, we give an algorithm with running time $2^{O(k^2 \log k)} n^{O(1)}$ that decides whether π can be partitioned into at most k increasing or decreasing subsequences. Thus we resolve affirmatively the open question of whether the problem is fixed parameter tractable. This NP-complete problem is equivalent to deciding whether the cochromatic number (the minimum number of cliques and independent sets the vertices of the graph can be partitioned into) of a given permutation graph on n vertices is at most k. In fact, we give a more general result: within the mentioned running time, one can decide whether the cochromatic number of a given perfect graph on n vertices is at most k.

To obtain our result we use a combination of two well-known techniques within parameterized algorithms, namely greedy localization and iterative compression. We further demonstrate the power of this combination by giving a $2^{O(k^2 \log k)} n \log n$ time algorithm for deciding whether a given set of n non-overlapping axis-parallel rectangles can be stabbed by at most k of the given set of horizontal and vertical lines. Whether such an algorithm exists was mentioned as an open question in several papers.

1   Introduction

Given a permutation π on $[n] = \{1, \ldots, n\}$ and a positive integer k, a well known partitioning problem asks whether we can partition π into at most k monotone (increasing or decreasing) subsequences. This partition problem is NP-complete [31] and can be solved in time $n^{O(k)}$ [3]. Using the famous Erdős-Szekeres theorem [15] which states that every sequence of $p \cdot q + 1$ real numbers has a monotone subsequence of length either $p+1$ or $q+1$, an algorithm with running time $n^{O(k)}$

---

* This work is supported by the Research Council of Norway.
** Supported by ANR Blanc AGAPE.

H. Kaplan (Ed.): SWAT 2010, LNCS 6139, pp. 334–345, 2010.
© Springer-Verlag Berlin Heidelberg 2010

implies a subexponential-time algorithm with running time $n^{O(\sqrt{n})}$ for partitioning π into the minimum number of monotone subsequences [3]. A natural question which has been left open, and most recently stated at a 2008 Dagstuhl Seminar [18], is whether the problem is fixed parameter tractable (FPT) when parameterized by the number of monotone subsequences. I.e. is there an algorithm with running time $f(k) \cdot n^{O(1)}$ for partitioning a permutation π into at most k monotone subsequences? We answer this question affirmatively by giving an algorithm with running time $2^{O(k^2 \log k)} n^{O(1)}$.

Every permutation π on $[n]$ corresponds to a *permutation graph* $G(\pi)$ on n vertices. This graph has a vertex for each number $1, 2, \ldots, n$ and there is an edge between any two numbers that are in reversed order in the permutation, that is, we have an edge between i and j, $i < j$ if $\pi(i) > \pi(j)$. Hence, the above mentioned partitioning problem is equivalent to deciding whether the vertices of $G(\pi)$ can be partitioned into at most k independent sets or cliques. This brings us to the notion of *cochromatic number* of a graph. The *cochromatic number* of a graph $G = (V, E)$ is the minimum number of sets the vertex set V can be partitioned into, such that each set is either an independent set or a clique. Thus, the above mentioned partitioning problem is equivalent to finding the COCHROMATIC NUMBER of permutation graphs. Formally, the COCHROMATIC NUMBER problem is defined as follows.

COCHROMATIC NUMBER
Input: A graph G on n vertices, and an integer $k \geq 1$.
Parameter: k.
Question: Is the cochromatic number of G at most k?

COCHROMATIC NUMBER, being a natural extension of chromatic number and graph colorings has been well studied [13,14]. The COCHROMATIC NUMBER problem is NP-complete even on permutation graphs [31]. Brandstädt [2] showed that we can recognize in polynomial time whether the vertex set of a given undirected graph can be partitioned into one or two independent sets and one or two cliques. However, it remains NP-complete to check whether we can partition the given graph into at most κ independent sets and at most ℓ cliques if either $\kappa \geq 3$ or $\ell \geq 3$. It is easy to show that testing whether the cochromatic number of a given graph is at most 3 is NP-complete. Thus, we can not hope to solve COCHROMATIC NUMBER on general graphs even in time $n^{f(k)}$ for any arbitrary function f of k unless P=NP. We show that COCHROMATIC NUMBER is fixed parameter tractable on *perfect graphs*; a graph class that subsumes bipartite graphs, chordal graphs and permutation graphs, to name a few.

A graph is *perfect* if the chromatic number is equal to the clique number for each of its induced subgraphs. Perfect graphs were introduced by Berge in the early 60's, and is one of the well studied classes of graphs[4,16,22,28]. Perfect graphs have many nice algorithmic properties. They can be recognized in polynomial time [7] and one can find a maximum independent set, minimum coloring, maximum clique all in polynomial time [23]. Our algorithm solves COCHROMATIC NUMBER in $2^{O(k^2 \log k)} n^{O(1)}$ time on perfect graphs and crucially uses several algorithmic properties of perfect graphs. To the best of our knowledge even an $n^{O(k)}$ algorithm solving this problem on perfect graphs was not known before. The only known algorithmic results for the COCHROMATIC NUMBER problem are by Fomin et al. [19] who gave a factor 1.71 approximation algorithm for COCHROMATIC NUMBER on comparability (or cocomparability) graphs, and a factor $\log n$ approximation algorithm on perfect graphs.

To show our result, we use a combination of two well-known techniques within parameterized algorithms, namely, greedy localization and iterative compression. This combination follows the well known iterative compression paradigm in parameterized complexity, but finds a small witness to branch and move towards an optimal solution "for the compressed instance" at the compression step.

Using this new combination, we are also able to resolve another open question [10,11], namely whether DISJOINT RECTANGLE STABBING is fixed parameter tractable.

> DISJOINT RECTANGLE STABBING
> **Input:** A set R of n axis-parallel non-overlapping rectangles embedded in a plane, a set L of vertical and horizontal lines embedded in the plane, and an integer $k \geq 1$.
> **Parameter:** k.
> **Question:** Is there a set $L' \subseteq L$ with $|L'| \leq k$ such that every rectangle from R is stabbed by at least one line from L'?

Here we say that a rectangle is stabbed by a line if their intersection is nonempty. Also two rectangles are said to be overlapping if there exists a vertical line v and a horizontal line h such that both rectangles are stabbed by the lines v and h. For example, non intersecting rectangles are always non overlapping.

The RECTANGLE STABBING problem, the more general version of the DISJOINT RECTANGLE STABBING problem, where the rectangles can overlap is a generic geometric covering problem having wide applicability [24]. A number of polynomial-time approximation results for RECTANGLE STABBING and its variants are known [8,32,26,24]. In [11], the authors prove a W[1]-hardness result for a higher dimensional version of the RECTANGLE STABBING problem and show several restrictions of this two dimensional version fixed-parameter tractable. Recently, Dom et al. [10] and Giannopoulos et al. [21] independently considered the general two dimensional version and showed it to be complete for the parameterized complexity class W[1]. They also showed a restricted version of DISJOINT RECTANGLE STABBING, b-DISJOINT SQUARE STABBING fixed parameter tractable for a fixed b. All these papers leave the parameterized complexity of the general DISJOINT RECTANGLE STABBING problem open.

Our paper is organized as follows. In Section 2 we give necessary definitions and set up our notations. Section 3 gives an overview of the method we use to solve the two problems we address. The fixed parameter tractable algorithm for COCHROMATIC NUMBER on perfect graphs is given in Section 4 and for DISJOINT RECTANGLE STABBING in Section 5.

2   Definitions and Notation

All graphs in this paper are simple, undirected, and unweighted. For a graph $G = (V, E)$, we denote the size of the vertex set V by n and the edge set E by m. For a subset S of V, the *subgraph of G induced by S* is denoted by $G[S]$. The *complement* of $G = (V, E)$ is denoted by $\overline{G}$, it has the same vertex set V, and edge set $\{uv \mid u, v \in V \text{ and } u \neq v \text{ and } uv \notin E\}$.

A *(proper) coloring* of a graph is an assignment of colors to its vertices so that adjacent vertices receive different colors. A coloring is *minimum* if it uses the minimum number of colors. The *chromatic number* and the *clique number* of G are, respectively, the number of colors in a minimum coloring of G, and the size of a largest clique in G. A *clique cover* in a graph is a collection of cliques such that every vertex is in one of the cliques of the collection. A graph is *perfect* if the chromatic number is equal to the clique number for each of its induced subgraphs. The *cochromatic number* of a graph G is the minimum number of sets V can be partitioned into, such that each set is either an independent set or a clique.

A parameterized problem L takes two values as input – an input x and an integer parameter k, and is said to be fixed parameter tractable (FPT) if there is an algorithm that decides whether the input (x, k) is in L in $f(k)n^{O(1)}$ time where f is some function of k. We refer to [12,29,17] for more information on fixed-parameter algorithms and parameterized complexity.

3   Methodology

Our method can be viewed as a combination of two well known methods in obtaining fixed parameter tractable algorithms, that is, *greedy localization* and *iterative compression*. The method of greedy localization is primarily used for maximization problems. The idea is to start off by greedily computing a solution to the problem at hand and showing that the optimal solution must in some sense lie "close" to the current solution. For a concrete example consider the problem of finding k vertex disjoint P_3's, paths on three vertices, in a given graph G. We greedily compute a maximal collection of pairwise disjoint P_3's. If the number of P_3's in our collection is at least k, we are done. Else, observe that any collection of k pairwise disjoint P_3's must intersect with vertices of P_3's in our greedy solution. We refer to [9,25] for applications of greedy localization.

Iterative Compression is a technique primarily used for minimization problems. Algorithms based on iterative compression apply polynomially many "compression steps". In a compression step, we are given an instance I of the problem, a solution S' to the problem, and the objective is to check whether there exists a solution S for I such that $|S| < |S'|$ in $f(|S'|)|I|^{O(1)}$ time. The idea is to process the instance incrementally, maintaining a solution S to the intermediate instances. In each incremental step both the instance and the maintained solution increase in size. Then the compression algorithm is run to decrease the size of S again. Iterative Compression has proved very useful in the design of parameterized algorithms and is instrumental in the currently fastest parameterized algorithms for ODD CYCLE TRANSVERSAL [30], FEEDBACK VERTEX SET [6] and DIRECTED FEEDBACK VERTEX SET [5]. We refer to [29] for a more thorough introduction to Iterative Compression.

We combine the two methods, applying iterative compression to incrementally build the solution. In each compression step we search the solution space around the given solution similar to how it is done in greedy localization.

4 COCHROMATIC NUMBER OF PERFECT GRAPHS

In this section we give an algorithm for COCHROMATIC NUMBER on perfect graphs. We start by guessing the number α of cliques and the number β of independent sets such that $\alpha + \beta = k$, and for each of the $k+1$ guesses (of (α, β)) we will decide whether G can be partitioned into at most α cliques and at most β independent sets. We order V into $v_1 v_2 \ldots v_n$ and define $V_i = \{v_1, \ldots, v_i\}$ for every i. Notice that if V can be partitioned into at most α cliques and β independent sets, then for every i, so can $G[V_i]$. Also, given such a partitioning for $G[V_i]$ we can easily make a partitioning of $G[V_{i+1}]$ into $\alpha + 1$ cliques and β independent sets by letting v_{i+1} be a clique by itself. At this point we want to use the current partitioning to decide whether there is a partitioning of $G[V_{i+1}]$ into α cliques and β independent sets. This naturally leads to the definition of the compression version of the problem.

COMPRESSION COCHROMATIC NUMBER (CCN)
Input: A perfect graph $G = (V, E)$ on n vertices, a partition $\mathcal{P} = (C_1, \ldots, C_{\alpha+1}, I_1, \ldots, I_\beta)$ of V, where C_i, $1 \leq i \leq \alpha + 1$, are cliques, and I_j, $1 \leq j \leq \beta$, are independent sets.
Task: Find a partitioning of G into α cliques and β independent sets, or conclude that no such partitioning exists.

We will give a $2^{O(\alpha\beta \log(\alpha\beta))} n^{O(1)}$ time algorithm for CCN, which together with the discussion above yields a $2^{O(\alpha\beta \log(\alpha\beta))} n^{O(1)}$ time algorithm for COCHROMATIC NUMBER. To do that we use the following classical results.

Lemma 1 ([23]). *Let G be a perfect graph. Then there exist algorithms that can compute in polynomial time: (a) a minimum coloring of G, (b) a maximum independent set of G, and (c) a maximum clique of G.*

Lemma 2 ([27]). *G is a perfect graph if and only if $\overline{G}$ is a perfect graph.*

Using Lemmata 1 and 2, we can prove the following preliminary lemma which is integral to our algorithm.

Lemma 3. *Given a perfect graph $G = (V, E)$ and an integer ℓ, there is a polynomial time algorithm to output*

(a) either a partition of V into at most ℓ independent sets or a clique of size $\ell + 1$, and

(b) either a partition of V into at most ℓ cliques or an independent set of size $\ell + 1$.

Proof. We first give a proof for (a). We start by finding a minimum coloring of G in polynomial time using Lemma 1. If the number of colors required is at most ℓ, then we have our required partitioning of V into at most ℓ independent sets. Otherwise since G is a perfect graph, the chromatic number of G is the same as the clique number of G. Hence a maximum clique is of size at least $\ell + 1$. Now we find a maximum clique using Lemma 1. Let C be such a clique. Now choose an arbitrary subset $C' \subseteq C$ of size $\ell + 1$ and return C' as the desired clique of size $\ell + 1$. To prove part (b) we just need to observe that a clique in G is an independent set in $\overline{G}$. Now the proof follows by the proof of (a) and using the fact that $\overline{G}$ is also a perfect graph (by Lemma 2). □

We will now explain how the given partitioning $\mathcal{P}$ is useful in solving the compression step. The main idea is that while the partitioning $\mathcal{P}'$ we search for may differ significantly from $\mathcal{P}$, only a few vertices that were covered by cliques in $\mathcal{P}$ can be covered by independent sets in $\mathcal{P}'$ and vice versa. To formalize this idea we introduce the notion of a bitstring $B_{\mathcal{P}}$ of the partition $\mathcal{P}$.

Given $G = (V, E)$ with $V = v_1 \ldots v_n$, and a partition $\mathcal{P}$ of V into cliques and independent sets, let $B_{\mathcal{P}}$ be the binary vector of length n in which position i is 0 if v_i belongs to a clique in $\mathcal{P}$, and 1 if v_i belongs to an independent set in $\mathcal{P}$. Given a n-vertex graph G and a bitstring B of length n, define X_B to be the set of vertices of G whose corresponding entry in B is 0, and $Y_B = V \setminus X_B$. For two integers α and β we say that B is *valid* in G with respect to α and β if there exists a partition $\mathcal{P}$ of V into at most α cliques and at most β independent sets such that $B = B_{\mathcal{P}}$. Given two bitstrings B_1 and B_2 of equal length, the *hamming distance* between B_1 and B_2 is the number of positions on which the corresponding bits differ, and it is denoted by $\mathcal{H}(B_1, B_2)$.

First, in Lemma 4 we will show that for a perfect graph G a valid bitstring B is sufficient to reconstruct a partition of G into α cliques and β independent sets. Then, in Lemma 5 we will show that two valid bitstrings must be "similar".

Lemma 4. *There is a polynomial time algorithm that given a perfect graph $G = (V, E)$ on n vertices, a bitstring B of length n, and positive integers α and β tests whether B is valid in G with respect to α and β. If B is valid the algorithm outputs a partition $\mathcal{P}$ of V into α cliques and β independent sets. If not, the algorithm either outputs an independent set of size $\alpha + 1$ in $G[X_B]$ or a clique of size $\beta + 1$ in $G[Y_B]$.*

Proof. As G is perfect, $G[X_B]$ and $G[Y_B]$ the induced subgraphs on X_B and Y_B respectively are perfect. The algorithm first uses Lemma 3 (b) to either find a partitioning of $G[X_B]$ into α cliques or an independent set of size $\alpha + 1$ in $G[X_B]$. Then it uses Lemma 3 (a) to either find a partitioning of $G[Y_B]$ into β independent sets or a clique set of size $\beta + 1$ in $G[Y_B]$. □

Lemma 5. *Let $G = (V, E)$ be a graph, $\mathcal{P} = (C_1, C_2, \ldots, C_{\alpha}, I_1, \ldots, I_{\beta})$ be a partition of G into α cliques and β independent sets and $\mathcal{Q} = (C_1', C_2', \ldots, C_{\alpha'}', I_1', \ldots,$*

ALGO-CCN(G, B, α, β, μ))
(Here G is a perfect graph, B is a bit vector, α, β and μ are integers. ALGO-CCN outputs a partition $\mathcal{P}'$ of G into α cliques and β independent sets such that $\mathcal{H}(B_{\mathcal{P}'}, B) \leq \mu$ if such a partition exists, or answers "NO" otherwise.)

1. If $\mu < 0$ return "NO".
2. Use Lemma 4 to either find a partition of $\mathcal{P}'$ of G into α cliques and β independent sets with $B_{\mathcal{P}'} = B$ or find an independent set $I \subseteq X_B$ of size $\alpha + 1$ or find a clique $C \subseteq Y_B$ of size $\beta + 1$. If a partition was found answer "YES".
3. If an independent set I was found in step 2: For each vertex $v \in I$, let the bitvector $B(v)$ be obtained from B by flipping v's bit 0 to 1. For each $v \in I$, recursively solve the subproblem ALGO-CCN($G, B(v), \alpha, \beta, \mu - 1$). Return "YES" if any of the recursive calls returns "YES", otherwise return "NO".
4. If a clique C was found in step 2: For each vertex $v \in C$, let the bitvector $B(v)$ be obtained from B by flipping v's bit 1 to 0. For each $v \in C$, recursively solve the subproblem ALGO-CCN($G, B(v), \alpha, \beta, \mu - 1$). Return "YES" if any of the recursive calls returns "YES", otherwise return "NO".

Fig. 1. Description of Algorithm ALGO-CCN

$I'_{\beta'}$) be a partition of G into α' cliques and β' independent sets. Let $B_{\mathcal{P}}$ and $B_{\mathcal{Q}}$ be the bitstrings associated $\mathcal{P}$ and $\mathcal{Q}$ respectively. Then $\mathcal{H}(B_{\mathcal{P}}, B_{\mathcal{Q}}) \leq \alpha\beta' + \alpha'\beta$.

Proof. Observe that an independent set and a clique can intersect in at most one vertex. Hence

$$\left| \left(\bigcup_{i=1}^{\alpha} C_i \right) \cap \left(\bigcup_{j=1}^{\beta'} I'_j \right) \right| \leq \alpha\beta' \quad \text{and} \quad \left| \left(\bigcup_{i=1}^{\beta} I_i \right) \cap \left(\bigcup_{j=1}^{\alpha'} C'_j \right) \right| \leq \beta\alpha'.$$

This concludes the proof of the lemma.   □

We are now ready to present our algorithm for the compression step. Recall that we are given a perfect graph G as input, together with integers α and β and a partition $\mathcal{P}$ of G into $\alpha + 1$ cliques and β independent sets. The task is to find a partition $\mathcal{P}'$ of G into α cliques and β independent sets, if such a partition exists. Lemma 5 yields that it is sufficient to look for solutions with bitstrings 'close to' $B_{\mathcal{P}}$. Lemma 3 is used to pinpoint the "wrong" bits of $B_{\mathcal{P}}$. Formally, the algorithm ALGO-CCN takes as input a perfect graph G, a bitstring B and integers α, β and μ. It outputs a partition $\mathcal{P}'$ of G into α cliques and β independent sets such that $\mathcal{H}(B_{\mathcal{P}'}, B) \leq \mu$ if such a partition exists, and answers "NO" otherwise. To solve the problem CCN we call ALGO-CCN($G, B_{\mathcal{P}}, \alpha, \beta, 2\alpha\beta + \beta$). A formal description of the algorithm Algo-CCN is given in Figure 1.

Lemma 6. *The call to* ALGO-CCN *($G, B_{\mathcal{P}}, \alpha, \beta, 2\alpha\beta + \beta$) correctly solves the CCN instance $G, \mathcal{P}, \alpha, \beta$ in time $(\alpha + \beta)^{2\alpha\beta + \beta} n^{O(1)}$.*

Proof. We first argue about the correctness. By Lemma 5 it is sufficient to search for partitions $\mathcal{P}'$ such that $\mathcal{H}(B_{\mathcal{P}'}, B_{\mathcal{P}}) \leq 2\alpha\beta + \beta$. If the algorithm answers "YES" it also outputs a partition $\mathcal{P}'$ of G into α cliques and β independent sets such that $B_{\mathcal{P}'}$ was obtained from $B_{\mathcal{P}}$ by flipping at most $2\alpha\beta + \beta$ bits. Thus it remains to argue that if such a partition $\mathcal{P}'$ exists, the algorithm will find it. Let $B' = B_{\mathcal{P}'}$.

In any recursive call, if there exists an independent set I of size $\alpha+1$ in $G[X_B]$ then there is a vertex $v \in I$ whose corresponding bit is 1 in B'. The algorithm tries flipping the bit of each v in I, decreasing the hamming distance between B and B' by one in at least one recursive call. Similarly, if there exists a clique C of size $\beta + 1$ in $G[Y_B]$ then there is a vertex $v \in C$ whose corresponding bit is 0 in B'. Again, the algorithm tries flipping the bit of each v in C, decreasing the hamming distance between B and B' by one in at least one recursive call, and the correctness of the algorithm follows.

To argue the time bound, we consider a slight modification of the algorithm that if either $\alpha = 0$ or $\beta = 0$, applies Lemma 4 to solve the CCN instance in polynomial time. Hence without loss of generality $\alpha+1 \leq \alpha+\beta$ and $\beta+1 \leq \alpha+\beta$. Then every node in the branch tree has at most $\alpha + \beta$ children and the depth of the tree is at most $2\alpha\beta + \beta$ and hence the number of nodes in the branch tree is $O((\alpha + \beta)^{2\alpha\beta+\beta})$. Since the amount of work in each node is polynomial, the time bound follows. $\square$

Now we are ready to prove the main theorem of this section.

Theorem 1. *The* COCHROMATIC NUMBER *problem can be solved in* $2^{O(k^2 \log k)}$ $n^{O(1)}$ *time on perfect graphs.*

Proof. We apply iterative compression as described in the beginning of this section. In particular, the algorithm guesses the value of α and β and decides whether G can be partitioned into α cliques and β independent sets using $n - k - 1$ compression steps. The i'th compression step is resolved by calling ALGO-CCN$(G[V_i], B_{\mathcal{P}}, \alpha, \beta, 2\alpha\beta + \beta)$ where $\mathcal{P}$ is the partition into $\alpha + 1$ cliques and β independent sets. The correctness and time bound follow from Lemma 6. $\square$

5   DISJOINT RECTANGLE STABBING

In this section we give a fixed parameter tractable algorithm for DISJOINT RECT-ANGLE STABBING. Recall that a rectangle is stabbed by a line if their intersection is nonempty. In $O(n \log n)$ time we can sort the coordinates of the rectangles in non-decreasing order, and make two lists containing all the rectangles, one with the rectangles sorted in non-decreasing order by their x-coordinates and the other where the rectangles are sorted in non-decreasing order by ther y-coordinates of their top right corner. We also sort the set L_H of horizontal lines in L by their y-coordinates and the set L_V of vertical lines in L by their x-coordinates. *When-ever we speak of a subset of R (the set of all rectangles) or L we will assume that the corresponding sorted lists are given.*

For each of the $k + 1$ possible choices of non-negative integers α, β such that $\alpha + \beta = k$ we will run an algorithm that decides whether the rectangles can be stabbed by at most α horizontal and β vertical lines. In order to get an $O(n \log n)$ time bound for fixed k, we apply a *recursive* compression scheme rather than an iterative compression scheme. In particular, our algorithm partitions the n rectangles into two groups R_1 and R_2 with at most $\lceil n/2 \rceil$ rectangles in each group and runs recursively on the two groups. Now, if R can be stabbed by α horizontal and β vertical lines then so can R_1 and R_2, so if either of the recursive calls returns "NO" we return "NO" as well. Otherwise we combine the solutions to R_1 and R_2 to a solution with at most 2α horizontal and at most 2β vertical lines that stab R. We want to use this solution in order to decide whether R can be stabbed by at most α horizontal and at most β vertical lines. This leads to the definition of the compression version of DISJOINT RECTANGLE STABBING.

COMPRESSION DISJOINT RECTANGLE STABBING (CDRS)
Input: A set R of n axis-parallel non-overlapping rectangles embedded in the plane, a set $L = L_H \cup L_V$ of horizontal and vertical lines embedded in the plane, integers α and β and a set $Z \subseteq L$ with at most 2α horizontal and at most 2β vertical lines such that every rectangle from R is stabbed by at least one line from Z.
Task: Find a set $L' \subseteq L$ with at most α horizontal and at most β vertical lines such that every rectangle from R is stabbed by at least one line from L'. If no such set exists, return "NO".

The algorithm for CDRS has exactly the same structure as ALGO-CCN, but with the individual pieces tailored to fit the DISJOINT RECTANGLE STABBING problem. We start by proving a lemma analogous to Lemma 3.

Lemma 7. *Given a set R of axis-parallel rectangles in the plane, two sets L_V, L_H of vertical and horizontal lines respectively, and an integer k, there is an $O(n)$ time algorithm that outputs*

(a) *either a set $L'_H \subseteq L_H$ of lines with $|L_H| \leq k$ that stabs all rectangles in R or a collection H of $k + 1$ rectangles such that each horizontal line in L stabs at most one rectangle in H, and*

(b) *either a set $L'_V \subseteq L_V$ of lines with $|L_V| \leq k$ that stabs all rectangles in R or a collection V of $k + 1$ rectangles such that each vertical line in L stabs at most one rectangle in V.*

Proof. We only show (b) as the proof of (a) is identical. Initially $V = \emptyset$ and $L'_V = \emptyset$. We process R sorted in non-decreasing order of the x-coordinate of the top-right corner. At any step, if the considered rectangle r is not stabbed by any line in L'_V, add r to V and add the rightmost vertical line in L stabbing r to L'_V (to check whether r is not stabbed by any line in L'_V, we just keep track of the right most line in L'_V and check whether it stabs r; to find the right most vertical line in L that stabs r, we compute and keep this for every r in a preprocessing step). If at any step $|V| \geq k + 1$, output V otherwise all rectangles have been considered, $|L_V'| \le k$ and every rectangle is stabbed by some line in L_V. Clearly no vertical line in L can stab any two rectangles in V. The described algorithm can easily be implemented to run in $O(n)$ time. $\square$

Now we define the notion of a bitstring of a solution and prove lemmata analogous to Lemmata 4 and 5 for CDRS. Let $L' \subseteq L$ be a set of lines that stab all the rectangles of R. We define the bitstring $B_{L'}$ of L' as follows. Let R be ordered into $r_1, r_2, \ldots, r_n$. The i'th bit of $B_{L'}$ is set to 0 if r_i is stabbed by a horizontal line in L' and 1 otherwise. Observe that if r_i is stabbed both by a horizontal and by a vertical line of L', the i'th bit of $B_{L'}$ is 0. Given a collection R of rectangles, a set L of horizontal and vertical lines, integers α and β and a bitstring B of length n, we say that B is *valid in R with respect to α and β* if there exists a set $L' \subseteq L$ of at most α horizontal lines and at most β vertical lines stabbing all rectangles in R such that $B = B_{L'}$. Given R and B we define X_B to be the set of rectangles in R whose corresponding entry in B is 0, and $Y_B = R \setminus X_B$.

Lemma 8. *There is a $O(n)$ time algorithm that given a collection R of n axis-parallel rectangles, a collection L of horizontal and vertical lines, a bitstring B of length n, and positive integers α and β, tests whether B is valid in R with respect to α and β. If B is valid the algorithm outputs a set $L' \subseteq L$ of at most α horizontal and at most β vertical lines such that every rectangle in R is stabbed by a line in L' and $B_{L'} = B$. If B is not valid, the algorithm either outputs a set $H \subseteq R$ of size $\alpha + 1$ such that every horizontal line in L stabs at most 1 rectangle in H or a set $V \subseteq R$ of size $\beta + 1$ such that every vertical line in L stabs at most 1 rectangle in V.*

Proof. Let X_B be the set of rectangles in R whose bit in B is 0, and let $Y_B = R \setminus X_B$. Apply the first part of Lemma 7 to X_B and the second part of Lemma 7 to Y_B. $\square$

Now we are ready to show an analogue of Lemma 5 to the CDRS problem. To that end, for a line $l \in L$ let S_l be the set of rectangles in R stabbed by L. The proof of Lemma 9 relies on the fact that the rectangles in R are non-overlapping.

Lemma 9. *Let R be a collection of n non-overlapping axis-parallel rectangles, L be a collection of horizontal and vertical lines, $P \subseteq L$ and $Q \subseteq L$ be sets of lines so that every $r \in R$ is stabbed by a line in P and a line in Q. Suppose $|P \cap L_H| \le \alpha$, $|P \cap L_V| \le \beta$, $|Q \cap L_H| \le \alpha'$ and $|Q \cap L_V| \le \beta'$. Let B_P be the bitstring of P and B_Q be the bitstring of Q. Then $\mathcal{H}(B_P, B_Q) \le \alpha\beta' + \alpha'\beta$.*

Proof. Notice that since the rectangles are non-overlapping, if $l_H \in L_H$ and $l_V \in L_V$ then $|S_{l_H} \cap S_{l_V})| \le 1$. Hence

$$\left| \left(\bigcup_{p \in P \cap L_H} S_p \right) \cap \left(\bigcup_{q \in Q \cap L_V} S_q \right) \right| \le \alpha\beta' \text{ and } \left| \left(\bigcup_{p \in P \cap L_V} S_p \right) \cap \left(\bigcup_{q \in Q \cap L_H} S_q \right) \right| \le \alpha'\beta.$$

This concludes the proof of the lemma. $\square$

Now we have given all the ingredients required to solve the compression step of the disjoint rectangle stabbing problem. The proof of the following Lemma is identical to the proof of Lemma 6 and is therefore omitted.

Lemma 10. *There is an algorithm running in time $O((\alpha + \beta)^{4\alpha\beta+O(1)}n)$ which solves the* COMPRESSION DISJOINT RECTANGLE STABBING *problem.*

Theorem 2. *There is an algorithm solving the* DISJOINT RECTANGLE STABBING *problem that runs in time* $2^{O(k^2 \log k)}n \log n$.

Proof. The algorithm follows the recursive scheme described in the beginning of this section and uses the algorithm of Lemma 10 to solve the compression step. Correctness follows directly from Lemma 10, Let $T(n, k)$ be the time required to solve an instance with n rectangles and $\alpha + \beta = k$. Then

$$T(n, k) \leq 2T(n/2, k) + 2^{O(k^2 \log k)}n$$

which solves to $T(n, k) \leq 2^{O(k^2 \log k)}n \log n$ by the Master's Theorem. The $O(k)$ overhead of trying all possible values of α and β with $\alpha + \beta = k$ is subsumed in the asymptotic notation. □

References

1. Berge, C.: Färbung von Graphen, deren sämtliche bzw. deren ungeraden Kreise starr sind (Zusammenfassung), Wiss. Z. Martin-Luther-Univ. Halle-Wittenberg Math.-Natur., Reihe 10, 114 (1961)
2. Brandstädt, A.: Partitions of graphs into one or two independent sets and cliques. Discrete Mathematics 152, 47–54 (1996)
3. Brandstädt, A., Kratsch, D.: On the partition of permutations into increasing or decreasing subsequences. Elektron. Inform. Kybernet. 22, 263–273 (1986)
4. Brandstädt, A., Le, V.B., Spinrad, J.P.: Graph Classes: A Survey. SIAM, Philadelphia (1999)
5. Chen, J., Liu, Y., Lu, S., O'Sullivan, B., Razgon, I.: A fixed-parameter algorithm for the directed feedback vertex set problem. J. ACM 55 (2008)
6. Chen, J., Fomin, F.V., Liu, Y., Lu, S., Villanger, Y.: Improved algorithms for feedback vertex set problems. J. Comput. Syst. Sci. 74, 1188–1198 (2008)
7. Chudnovsky, M., Cornuejols, G., Liu, X., Seymour, P., Vuskovic, K.: Recognizing berge graphs. Combinatorica 25, 143–186 (2005)
8. Gaur, T.I.D.R., Krishnamurti, R.: Constant ratio approximation algorithms for the rectangle stabbing problem and the rectilinear partitioning problem. Journal of Algorithms 43, 138–152 (2002)
9. Dehne, F.K.H.A., Fellows, M.R., Rosamond, F.A., Shaw, P.: Greedy localization, iterative compression, modeled crown reductions: New FPT techniques, an improved algorithm for set splitting, and a novel 2k kernelization for vertex cover. In: Downey, R.G., Fellows, M.R., Dehne, F. (eds.) IWPEC 2004. LNCS, vol. 3162, pp. 271–280. Springer, Heidelberg (2004)
10. Dom, M., Fellows, M.R., Rosamond, F.A.: Parameterized complexity of stabbing rectangles and squares in the plane. In: Das, S., Uehara, R. (eds.) WALCOM 2009. LNCS, vol. 5431, pp. 298–309. Springer, Heidelberg (2009)

11. Dom, M., Sikdar, S.: The parameterized complexity of the rectangle stabbing problem and its variants. In: Preparata, F.P., Wu, X., Yin, J. (eds.) FAW 2008. LNCS, vol. 5059, pp. 288–299. Springer, Heidelberg (2008)
12. Downey, R.D., Fellows, M.R.: Parameterized Complexity. Springer, Heidelberg (1999)
13. Erdős, P., Gimbel, J.: Some problems and results in cochromatic theory. In: Quo Vadis, Graph Theory?, pp. 261–264. North-Holland, Amsterdam (1993)
14. Erdős, P., Gimbel, J., Kratsch, D.: Extremal results in cochromatic and dichromatic theory. Journal of Graph Theory 15, 579–585 (1991)
15. Erdős, P., Szekeres, G.: A combinatorial problem in geometry. Compositio Mathematica 2, 463–470 (1935)
16. Fishburn, P.C.: Interval Orders and Interval Graphs: A Study of Partially Ordered Sets. Wiley, Chichester (1985)
17. Flum, J., Grohe, M.: Parameterized Complexity Theory. Springer, Heidelberg (2006)
18. Fomin, F.V., Iwama, K., Kratsch, D., Kaski, P., Koivisto, M., Kowalik, L., Okamoto, Y., van Rooij, J., Williams, R.: 08431 Open problems – moderately exponential time algorithms. In: Fomin, F.V., Iwama, K., Kratsch, D. (eds.) Moderately Exponential Time Algorithms, Schloss Dagstuhl - Leibniz-Zentrum fuer Informatik, Germany. Dagstuhl Seminar Proceedings, vol. 08431 (2008)
19. Fomin, F.V., Kratsch, D., Novelli, J.-C.: Approximating minimum cocolorings. Inf. Process. Lett. 84, 285–290 (2002)
20. Frank, A.: On chain and antichain families of a partially ordered set. J. Comb. Theory, Ser. B 29, 176–184 (1980)
21. Giannopoulos, P., Knauer, C., Rote, G., Werner, D.: Fixed-parameter tractability and lower bounds for stabbing problems. In: Proceedings of the 25th European Workshop on Computational Geometry (EuroCG), pp. 281–284 (2009)
22. Golumbic, M.C.: Algorithmic Graph Theory and Perfect Graphs, 2nd edn., vol. 57. Elsevier, Amsterdam (2004)
23. Grötschel, M., Lovász, L., Schrijver, A.: Polynomial algorithms for perfect graph. Annals of Discrete Mathematics 21, 325–356 (1984)
24. Hassin, R., Megiddo, N.: Approximation algorithms for hitting objects with straight lines. Discrete Applied Mathematics 30, 29–42 (1991)
25. Jia, W., Zhang, C., Chen, J.: An efficient parameterized algorithm for -set packing. J. Algorithms 50, 106–117 (2004)
26. Kovaleva, S., Spieksma, F.C.R.: Approximation algorithms for rectangles tabbing and interval stabbing problems. SIAM J. Discrete Mathematics 20, 748–768 (2006)
27. Lovász, L.: A characterization of perfect graphs. J. Comb. Theory, Ser. B 13, 95–98 (1972)
28. Mahadev, N., Peled, U.: Threshold graphs and related topics, vol. 56. North-Holland, Amsterdam (1995)
29. Niedermeier, R.: Invitation to Fixed Parameter Algorithms. Oxford Lecture Series in Mathematics and Its Applications. Oxford University Press, USA (2006)
30. Reed, B.A., Smith, K., Vetta, A.: Finding odd cycle transversals. Oper. Res. Lett. 32, 299–301 (2004)
31. Wagner, K.: Monotonic coverings of finite sets. Elektron. Inform. Kybernet. 20, 633–639 (1984)
32. Xu, G., Xu, J.: Constant approximation algorithms for rectangle stabbing and related problems. Theory of Computing Systems 40, 187–204 (2007)

Dispatching Equal-Length Jobs to Parallel Machines to Maximize Throughput

David P. Bunde[1] and Michael H. Goldwasser[2]

[1] Dept. of Computer Science, Knox College
dbunde@knox.edu
[2] Dept. of Math. and Computer Science, Saint Louis University
goldwamh@slu.edu

Abstract. We consider online, nonpreemptive scheduling of equal-length jobs on parallel machines. Jobs have arbitrary release times and deadlines and a scheduler's goal is to maximize the number of completed jobs $(Pm \mid r_j, p_j = p \mid \sum 1 - U_j)$. This problem has been previously studied under two distinct models. In the first, a scheduler must provide *immediate notification* to a released job as to whether it is accepted into the system. In a stricter model, a scheduler must provide an *immediate decision* for an accepted job, selecting both the time interval and machine on which it will run. We examine an intermediate model in which a scheduler *immediately dispatches* an accepted job to a machine, but without committing it to a specific time interval. We present a natural algorithm that is optimally competitive for $m = 2$. For the special case of unit-length jobs, it achieves competitive ratios for $m \geq 2$ that are strictly better than lower bounds for the immediate decision model.

1 Introduction

We consider a model in which a scheduler manages a pool of parallel machines. Job requests arrive in an online fashion, and the scheduler receives credit for each job that is completed by its deadline. We assume that jobs have equal length and that the system is nonpreemptive. We examine a series of increasingly restrictive conditions on the timing of a scheduler's decisions.

unrestricted: In this most flexible model, all requests are pooled by a scheduler. Decisions are made in real-time, with jobs dropped only when it is clear they will not be completed on time.

immediate notification: In this model, the scheduler must decide whether a job will be admitted to the system when it arrives. Once admitted, a job must be completed on time. However, the scheduler retains flexibility by centrally pooling admitted jobs until they are executed.

immediate dispatch: In this model, a central scheduler must immediately assign an admitted job to a particular machine, but each machine retains autonomy in determining the order in which to execute the jobs assigned to it, provided they are completed on time.

H. Kaplan (Ed.): SWAT 2010, LNCS 6139, pp. 346–358, 2010.

immediate decision: In this model, a central scheduler must fully commit an admitted job to a particular machine and to a particular time interval for execution on that machine.

The problem has been previously studied in the unrestricted, immediate notification, and immediate decision models. Immediate dispatching is a natural model, for example when distributing incoming requests to a server farm or computer cluster to avoid a centralized queue [1,14]. Our work is the first to examine the effect of immediate dispatching on throughput maximization.

We introduce a natural FIRSTFIT algorithm for the immediate dispatch model. In short, it fixes an ordering of the m machines $M_1, \ldots, M_m$, and assigns a newly-arrived job to the lowest-indexed machine that can feasibly accept it (the job is rejected if it is infeasible on all machines). We present the following two results regarding the analysis of FIRSTFIT. For $m = 2$, we prove that FIRST-FIT is $\frac{5}{3}$-competitive and that this is the best possible ratio for a deterministic algorithm with immediate dispatch. This places the model strictly between the immediate notification model (deterministic competitiveness $\frac{3}{2}$) and the immediate decision model (deterministic competitiveness $\frac{9}{5}$). For the case of unit-length jobs, we show that FIRSTFIT has competitiveness $1/\left(1 - \left(\frac{m-1}{m}\right)^m\right)$ for $m \geq 1$. Again, the model lies strictly between the others; an EDF strategy gives an optimal solution in the immediate notification model, and our upper bound is less than a comparable lower bound with immediate decision for any m (both tend toward $\frac{e}{e-1} \approx 1.582$ as $m \to \infty$). In addition, we present a variety of deterministic and randomized lower bounds for both the immediate dispatch and unrestricted models. Most notably, we strengthen the deterministic lower bound for the unrestricted and immediate notification models from $\frac{6}{5}$ to $\frac{5}{4}$ for the asymptotic case as $m \to \infty$. A summary of results regarding the deterministic and

Table 1. A summary of deterministic lower and upper bounds on the achievable competitiveness for various models. Entries in bold are new results presented in this paper.

m:	1	2	3	4	5	6	7	8	∞
unit-length UB				1 (using EDF)					
equal-length LB	2	1.5	1.4	1.333	**1.333**	1.3	**1.294**	**1.308**	**1.25**
equal-length UB	2	1.5							

Unrestricted or Immediate Notification.

unit-length LB		**1.143**							
unit-length UB		**1.333**	**1.421**	**1.463**	**1.487**	**1.504**	**1.515**	**1.523**	**1.582**
equal-length LB		**1.667**	**1.5**	**1.5**	**1.429**	**1.444**	**1.4**	**1.417**	**1.333**
equal-length UB		**1.667**							

Immediate Dispatch.

unit-length LB		1.678	1.626	1.607	1.599	1.594	1.591	1.589	1.582
equal-length LB		1.8							
equal-length UB	2	1.8	1.730	1.694	1.672	1.657	1.647	1.639	1.582

Immediate Decision.

Table 2. A summary of randomized lower bounds for the problem with equal-length jobs. Entries in bold are new results presented in this paper. The only non-trivial upper bound using randomization is a $\frac{5}{3}$-competitive algorithm for the unrestricted model on a single machine [5].

m:	1	2	3	4	5	6	7	8	∞
Notification	1.333	**1.263**	**1.256**	**1.255**	**1.25**	**1.252**	**1.251**	**1.251**	**1.25**
Dispatch						**1.333**			
Decision						1.333			

randomized competitiveness of the models is given in Tables 1 and 2. Due to space limitations, some proofs are omitted from this version of the paper.

Previous Work. Baruah et al. consider an unrestricted model for scheduling jobs of varying length on a single machine to maximize the number of completed jobs, or the time spent on successful jobs [2]. Among their results, they prove that any reasonable nonpreemptive algorithm is 2-competitive with equal-length jobs, and that this is the best deterministic competitiveness. Two-competitive algorithms are known for the unrestricted model [9], the immediate notification model [10], and the immediate decision model [6]. We note that for $m = 1$, the immediate notification and immediate dispatch models are the same, as any accepted job is trivially dispatched to the sole machine. Goldman et al. [9] show that any *randomized* algorithm can be at best $\frac{4}{3}$-competitive, but no algorithm with this ratio has (yet) been found. Chrobak et al. present a $\frac{5}{3}$-competitive randomized algorithm that is *barely random*, as it uses a single bit to choose between two deterministic strategies [5]. They also prove a lower bound of $\frac{3}{2}$ for such barely random algorithms.

For the two-machine version of the problem, Goldwasser and Pedigo [12], and independently Ding and Zhang [7], present a $\frac{3}{2}$-competitive deterministic algorithm in the immediate notification model, and a matching lower bound that applies even for the unrestricted model. Ding and Zhang also present a deterministic lower bound for $m \geq 3$ that approaches $\frac{6}{5}$ as $m \to \infty$.

The immediate decision model was first suggested by Ding and Zhang, and formally studied by Ding et al. [6]. They provide an algorithm named BESTFIT, defined briefly as follows. Jobs assigned to a given machine are committed to being executed in FIFO order. A newly-released job is placed on the most *heavily-loaded* machine that can feasibly complete it (or rejected, if none suffice). They prove that BESTFIT is $1/\left(1 - (\frac{m}{m+1})^m\right)$-competitive for any m. This expression equals 1.8 for $m = 2$ and approaches $\frac{e}{e-1} \approx 1.582$ as $m \to \infty$. They show that their analysis is tight for this algorithm, and they present a general lower bound for $m = 2$ and $p \geq 4$, showing that 1.8 is the best deterministic competitiveness for the immediate decision model. For $m \geq 3$, it is currently the best-known algorithm, even for the unrestricted model. Finally, they adapt the $\frac{4}{3}$ randomized lower bound for the unrestricted, single-processor case to the immediate decision model for $m \geq 1$. In subsequent work, Ebenlendr and Sgall prove that as $m \to \infty$, the 1.582 ratio of BESTFIT is the strongest possible for deterministic algorithms in the immediate decision model, even with unit-length jobs [8]. Specifically, they provide a lower bound of $\left(e^{\frac{m-1}{m}}\right) / \left(e^{\frac{m-1}{m}} - \frac{m}{m-1}\right)$.

Motivated by buffer management, Chin et al. consider scheduling *weighted* unit-length jobs to maximize the weighted throughput [4]. They give a randomized algorithm for a single processor that is 1.582-competitive. For multiprocessors, they give a $1 / \left(1 - \left(\frac{m-1}{m}\right)^m\right)$-competitive deterministic algorithm for the unrestricted model. This is precisely our bound for FIRSTFIT in the unweighted case with immediate dispatch, though the algorithms are not similar.

Although there is no previous work on maximizing throughput with immediate dispatch, Avrahami and Azar compare immediate dispatch to the unrestricted model for multiprocessor scheduling to minimize flow time or completion time [1]. For those objectives, once jobs are assigned to processors, each machine can schedule its jobs in FIFO order (and thus immediately assign time intervals).

Model and Notations. A scheduler manages $m \geq 1$ machines $M_1, \ldots, M_m$. Job requests arrive, with job j specified by three nonnegative integer parameters: its release time r_j, its processing time p_j, and its deadline d_j. We assume all processing times are equal, thus $p_j = p$ for a fixed constant p. We consider a nonpreemptive model. To complete a job j, the scheduler must commit a machine to it for p consecutive time units during the interval $[r_j, d_j)$. The scheduler's goal is to maximize the number of jobs completed on time. We use competitive analysis, considering the worst-case over all instances of the ratio between the optimal throughput and that produced by an online policy [3,13,15]. We presume that an online scheduler has no knowledge of a job request until the job is released. Once released, all of a job's parameters become known to the scheduler[1]. We note the important distinction between having equal-length jobs and *unit-length* jobs. With $p > 1$, the algorithm may start (nonpreemptively) executing one job, and learn of another job that is released while the first is executing. In the unit-length model (i.e., $p = 1$), such a scenario is impossible.

2   The FIRSTFIT Algorithm

We define the FIRSTFIT algorithm as follows. Each machine maintains a queue of jobs that have been assigned to it but not yet completed. Let $Q_k(t)$ denote FIRSTFIT's queue for M_k at the onset of time-step t (including any job that is currently executing). We define FIRSTFIT so that it considers each arrival independently (i.e., the *online-list* model). To differentiate the changing state of the queues, we let $Q_k^j(t)$ denote the queue as it exists when job j with $r_j = t$ is considered. Note that $Q_k^j(t) \supseteq Q_k(t)$ may contain newly-accepted jobs that

---

[1] When jobs share a release time, there are two distinct models. FIRSTFIT operates in an *online-list* model in which those jobs arrive in arbitrary order and the scheduler dispatches or rejects each job before learning of the next. All except the last of our lower bounds apply in the more general *online-time* model, where a scheduler learns about all jobs released at a given time before making decisions about any of them.

were considered prior to j. For a job j arriving at time t, we dispatch it to the minimal M_k for which $Q_k^j(t) \cup \{j\}$ remains feasible, rejecting it if infeasible on all machines. Unlike the BESTFIT algorithm for the immediate decision model [6], FIRSTFIT allows each machine to reorder its queue using the Earliest-Deadline-First (EDF) rule each time it starts running a job from its queue (as an aside, EDF is also used to perform the feasibility test of $Q_k^j(t) \cup \{j\}$ when j is released).

In the remainder of this section, we prove two theorems about FIRSTFIT. In Section 2.1, we show that FIRSTFIT is $\frac{5}{3}$-competitive for equal-length jobs on two machines; this is the best-possible deterministic competitiveness, as later shown in Theorem 5. In Section 2.2 we show, for the special case of unit-length jobs, that FIRSTFIT is $1/\left(1 - \left(\frac{m-1}{m}\right)^m\right)$-competitive for any m.

2.1   Optimal Competitiveness for Two Machines

We use an analysis style akin to that of [11,12]. We fix a finite instance $\mathcal{I}$ and an optimal schedule OPT for that instance. Our analysis of the relative performance of FIRSTFIT versus OPT is based upon two potential functions Φ^{FF} and Φ^{OPT} that measure the respective progress of the developing schedules over time. We analyze the instance by partitioning time into consecutive regions of the form $[u, v)$ such that the increase in Φ^{FF} during a region is guaranteed to be at least that of Φ^{OPT}. Starting with $u = 0$, we end each region with the first time $v > u$ at which the set $Q_1(v)$ can be feasibly scheduled on M_1 starting at time $v + p$ (as opposed to simply v). Such a time is well defined, as the queue eventually becomes empty and thus trivially feasible.

We introduce the following notations. We let $S^{FF}(t)$ and $S^{OPT}(t)$ denote the sets of jobs started *strictly* before time t by FIRSTFIT and OPT respectively. We define $D^{FF}(t) = S^{OPT}(t) \cap S^{FF}(\infty) \setminus S^{FF}(t)$ as the set of "delayed" jobs. These are started prior to time t by OPT, yet on or after time t by FIRSTFIT. We define $D^{OPT}(t) = S^{FF}(t) \cap S^{OPT}(\infty) \setminus S^{OPT}(t)$ analogously. Lastly, we define a special set of "blocked" jobs for technical reasons that we will explain shortly. Formally, we let $B^{OPT}(t)$ denote those jobs that were not started by either algorithm prior to t, but are started by OPT while FIRSTFIT is still executing a job of $S^{FF}(t)$. Based on these sets, we define our potential functions as follows:

$$\Phi^{FF}(t) = 5 \cdot |S^{FF}(t)| \; + 2 \cdot |D^{FF}(t)|$$
$$\Phi^{OPT}(t) = 3 \cdot |S^{OPT}(t)| + 3 \cdot |D^{OPT}(t)| + 2 \cdot |B^{OPT}(t)|$$

Intuitively, these functions represent payments for work done in the respective schedules. In the end, we award 5 points to FIRSTFIT for each job completed and 3 points to OPT, thus giving a $\frac{5}{3}$ competitive ratio. However, at intermediate times we award some advance payment for accepted jobs that are not yet started. For example, we award FIRSTFIT an advanced credit of 2 points for a job in its queue that OPT has already started. The algorithm gets the 3 other points when it starts the delayed job. In contrast, we immediately award OPT its full 3 credits for delayed jobs. We will show that OPT has limited opportunities to carry jobs from one region to the next as delayed; we pay for those discrepancies in advance.

The payment of 2 for jobs in $B^{\text{OPT}}(t)$ is a technical requirement related to our division of time into regions. By definition, each job that FIRSTFIT starts on M_1 completes by the region's end. However, a job started on M_2 may execute past the region's end, possibly hurting it in the next region. We account for this by prepaying OPT during the earlier region for progress made during the overhang.

Lemma 1. *If FIRSTFIT rejects job j, all machines are busy during $[r_j, d_j - p)$.*

Proof. If M_k for $k \in \{1, 2\}$ were idle at a time t, its queue is empty. For $t \in [r_j, d_j - p)$, this contradicts j's rejection, as $Q_k^j(r_j) \cup \{j\}$ is feasible by scheduling $Q_k^j(r_j)$ during $[r_j, t)$ as done by the algorithm, and j from $[t, t + p)$. $\square$

Lemma 2. *Any job j started by FIRSTFIT during a region $[u, v)$ has $d_j < v + p$, with the possible exception of the job started by M_1 at time u.*

Proof. Consider j with $d_j \geq v + p$ started during $[u, v)$. The set $Q_1^j(r_j) \cup \{j\}$ must be feasible on M_1 at time r_j; this is demonstrated by using the algorithm's actual schedule for $[r_j, v)$, followed by j during $[v, v + p)$, and, based on our definition of v, set $Q_1(v)$ starting at $v + p$ Therefore, such j must have been assigned to M_1 and started at some time $u \leq t \leq v - p$. We note that $Q_1(t)$ could be feasibly scheduled starting at time $t + p$ by using the algorithm's schedule from $[t + p, v)$, running j from $[v, v + p)$, and the remaining $Q_1(v)$ starting at time $v + p$. If $t > u$, this feasibility of $Q_1(t)$ relative to time $t + p$ contradicts our choice of v (rather than t) as the region's end. Therefore, j must be started on M_1 at time u. $\square$

Lemma 3. *For a region $[u, v)$ in which M_1 idles at time u for FIRSTFIT, $\Phi^{\text{FF}}(u) \geq \Phi^{\text{OPT}}(u)$ implies $\Phi^{\text{FF}}(v) \geq \Phi^{\text{OPT}}(v)$.*

Proof. M_1's idleness implies that $Q_1(u) = Q_1(u + 1) = \emptyset$. Therefore, $v = u + 1$ by definition. Any job started by OPT at time u must have been earlier accepted and completed on M_1 by FIRSTFIT, given its feasibility at a time when M_1 idles. We conclude that $\Phi^{\text{FF}}(v) = \Phi^{\text{FF}}(u)$ and $\Phi^{\text{OPT}}(v) = \Phi^{\text{OPT}}(u)$ $\square$

Lemma 4. *For a region $[u, v)$ in which M_1 starts a job at time u for FIRSTFIT, $\Phi^{\text{FF}}(u) \geq \Phi^{\text{OPT}}(u)$ implies $\Phi^{\text{FF}}(v) \geq \Phi^{\text{OPT}}(v)$.*

Proof (sketch). Let $n_1 \geq 1$ denote the number of jobs started by FIRSTFIT on M_1 during the region, and $n_2 \geq 0$ denote the number of jobs started on M_2. Note that M_1 never idles during the region, for such a time would contradict our definition of v. Therefore, $v - u = p \cdot n_1$. We begin by considering possible contributions to $\Phi^{\text{OPT}}(v) - \Phi^{\text{OPT}}(u)$, partitioned as follows.

3 · d due to $d \geq 0$ jobs that are newly added to $D^{\text{OPT}}(v)$. Such delayed jobs must be started by FIRSTFIT during the region, yet held by OPT for a later region. By Lemma 2, there is at most one job started by FIRSTFIT with deadline of $v+p$ or later, thus $d \leq 1$.

3 · a due to $a \geq 0$ jobs that are newly added to $S^{\text{OPT}}(v)$, not previously credited as part of $D^{\text{OPT}}(u)$ or $B^{\text{OPT}}(u)$, and that were *accepted* by FIRSTFIT upon their release. Given that these jobs were accepted by FIRSTFIT and had not previously been started by OPT, they must either lie in $S^{\text{FF}}(v)$ or $D^{\text{FF}}(v)$.

3 · r due to $r \geq 0$ jobs that are newly added to $S^{\mathrm{Opt}}(v)$, not previously credited
as part of $B^{\mathrm{Opt}}(u)$, and that were *rejected* by FIRSTFIT upon their release.

1 · b_old due to $b_{\mathrm{old}} \geq 0$ jobs that are newly added to $S^{\mathrm{Opt}}(v)$ yet were previously
credited as part of $B^{\mathrm{Opt}}(u)$.

2 · b_new due to $b_{\mathrm{new}} \geq 0$ jobs that newly qualify as blocked in $B^{\mathrm{Opt}}(v)$. For such
jobs to exist, there must be a newly-started job by FIRSTFIT on M_2 whose
execution extends beyond v. Since jobs have equal length, OPT can run at
most one such blocked job per machine, thus $b_{\mathrm{new}} \leq 2$.

Based on these notations, we have that $\Phi^{\mathrm{Opt}}(v) - \Phi^{\mathrm{Opt}}(u) = 3(d + a + r) +$
$b_{\mathrm{old}} + 2 \cdot b_{\mathrm{new}}$. The remainder of our analysis depends upon the following two
inequalities that relate OPT's progress to that of FIRSTFIT.

2 · n_1 ≥ (a + r + b_old)
By definition, OPT must start the jobs denoted by a, r, and b_{old} strictly
within the range $[u, v)$. There can be at most $2 \cdot n_1$ such jobs, given that the
size of the region is known to be $v - u = p \cdot n_1$ and there are two machines.

2 · n_2 ≥ (r + b_new)
We claim that jobs denoted by r and b_{new} must be started by OPT at times
when FIRSTFIT is running one of the jobs denoted by n_2 on M_2, and thus
that $r + b_{\mathrm{new}} \leq 2 \cdot n_2$ since OPT may use two machines. Intuitively, this is
due to Lemma 1 for jobs of r, and by the definition of $B^{\mathrm{Opt}}(t)$ for jobs of
b_{new}. The only technical issue is that if OPT starts a job when M_2 is running
a job that started strictly before time u (but overhangs), the job of OPT
belongs to $B^{\mathrm{Opt}}(u)$, and thus does not contribute to r or b_{new}.

To complete the proof, we consider $\Phi^{\mathrm{FF}}(v) - \Phi^{\mathrm{FF}}(u)$. By our definitions, this is at
least $3(n_1 + n_2) + 2(a + d)$, as jobs for a and d were not credited within $D^{\mathrm{FF}}(u)$.
If $n_1 - n_2 \geq d$, these bounds suffice for proving the claim. If $n_1 - n_2 < d$, it must
be that $n_1 = n_2$ and $d = 1$. Extra contributions toward Φ^{FF} can be claimed by
a further case analysis depending on whether $n_1 = 1$. Details are omitted. $\square$

Theorem 1. FIRSTFIT *is* $\frac{5}{3}$*-competitive for* $m = 2$ *and equal-length jobs.*

Proof. Initially, $\Phi^{\mathrm{Opt}}(0) = \Phi^{\mathrm{FF}}(0) = 0$. Repeated applications of Lemma 3 or 4
for regions $[u, v)$ imply $\Phi^{\mathrm{Opt}}(\infty) \leq \Phi^{\mathrm{FF}}(\infty)$, thus $3 \cdot |S^{\mathrm{Opt}}(\infty)| \leq 5 \cdot |S^{\mathrm{FF}}(\infty)|$.
We conclude that $\frac{\mathrm{Opt}}{\mathrm{FF}} \leq \frac{5}{3}$. $\square$

2.2   Unit-Length Jobs

We consider a job j to be *regular* with respect to FIRSTFIT if the machine to
which it is dispatched (if any) never idles during the interval $[r_j, d_j)$. We consider
an instance $\mathcal{I}$ to be *regular* with respect to FIRSTFIT if all jobs are regular.

Lemma 5. *For* $p = 1$, *the worst case competitive ratio for* FIRSTFIT *occurs on
a regular instance.*

Proof. Consider an irregular instance $\mathcal{I}$, and let j on M_k be the last irregular job started by FIRSTFIT. Let s_j denote the time at which j starts executing. The idleness of M_k leading to j's irregularity cannot occur while j is in the queue, so it must occur within the interval $[s_j + 1, d_j)$. We claim that $j \in Q_k(t)$ has the largest deadline of jobs in the queue for any $r_j \leq t \leq s_j$. For the sake of contradiction, assume jobs j and j' are in the queue at such time, for a j' coming after j in EDF ordering. Job j' must also be irregular, since we know there is idleness within interval $[s_j + 1, d_j) \subseteq [r_{j'}, d_{j'})$. Since j' starts after j by EDF, this contradicts our choice of j as the last irregular job to be started.

Next, we claim that FIRSTFIT produces the exact schedule for $\mathcal{I}' = \mathcal{I} - \{j\}$ as it does for $\mathcal{I}$, except replacing j by an idle slot. In essence, we argue that j's existence never affects the treatment of other jobs. Since j always has a deadline that is at least one greater than the cardinality of Q_k while in the queue, it cannot adversely affect a feasibility test when considering the dispatch of another job to M_k. Also, since j has the largest deadline while in Q_k, its omission does not affect the choice of jobs that are started, other than by the time s_j when it is the EDF job, and therefore $Q_k(s_j) = \{j\}$. There are no other jobs to place in the time slot previously used for j.

To conclude, since FIRSTFIT completes one less job on $\mathcal{I}'$ than $\mathcal{I}$, and OPT loses at most one job, the competitive ratio on $\mathcal{I}'$ is at least as great as on $\mathcal{I}$. □

Theorem 2. *For $p = 1$, algorithm FIRSTFIT is $\dfrac{1}{1 - \left(\frac{m-1}{m}\right)^m}$-competitive.*

Proof. By Lemma 5, we can prove the competitiveness of FIRSTFIT by analyzing an arbitrary *regular* instance. We rely on a charging scheme inspired by the analysis of BESTFIT in the immediate decision model [6], but with a different sequence of charges. We define $Y_k = (m-1)^{m-k} \cdot m^{k-1}$ for $1 \leq k \leq m$. Note that $\sum_{k=1}^{m} Y_k = m^m - (m-1)^m$ is a geometric sum with ratio $\frac{m}{m-1}$. A job i started at time t by OPT will distribute $m^m - (m-1)^m$ units of charge by assigning $Y_1, Y_2, \ldots Y_k$ respectively to the jobs $j_1, j_2, \ldots, j_k$ run by FIRSTFIT at time t on machines $M_1, M_2, \ldots, M_k$ for some k. When $k < m$, the remaining charge of $\sum_{z=k+1}^{m} Y_z$ is assigned to i itself; this is well-defined, as i must have been accepted by FIRSTFIT since there is an idle machine at time t when i is feasible.

We complete our proof by showing that each job j run by FIRSTFIT collects at most m^m units of charge, thereby proving the competitiveness of $\frac{m^m}{m^m - (m-1)^m} = \frac{1}{1 - \left(\frac{m-1}{m}\right)^m}$. Consider a job j that is run by FIRSTFIT on M_k. By our definition of regularity, machine M_k (and hence machines M_1 through M_{k-1} by definition of FIRSTFIT) must be busy at a time when OPT starts j. Therefore, j receives at most $\sum_{z=k+1}^{m} Y_z$ units of supplemental charge from itself. In addition, j may collect up to $m \cdot Y_k$ from the jobs that OPT runs at the time FIRSTFIT runs j. So j collects at most $m \cdot Y_k + \sum_{z=k+1}^{m} Y_z = (m-1) \cdot Y_k + \sum_{z=k}^{m} Y_z$. We prove by induction on k that $(m-1) \cdot Y_k + \sum_{z=k}^{m} Y_z = (m-1) \cdot Y_1 + \sum_{z=1}^{m} Y_z$. This is trivially so for $k = 1$. For $k > 1$, $(m-1) \cdot Y_k = (m-1)^{m-(k-1)} \cdot m^{k-1} = m \cdot Y_{k-1}$. Therefore $(m-1) \cdot Y_k + \sum_{z=k}^{m} Y_z = m \cdot Y_{k-1} + \sum_{z=k}^{m} Y_z = (m-1) \cdot Y_{k-1} + \sum_{z=k-1}^{m} Y_k$, which by induction equals $(m-1) \cdot Y_1 + \sum_{z=1}^{m} Y_z$. Finally, we note that $(m-1) \cdot Y_1 =$

$(m-1)^m$ and $\sum_{z=1}^{m} Y_z = m^m - (m-1)^m$, thus each job j run by FIRSTFIT collects at most m^m units of charge.   □

Our analysis of FIRSTFIT is tight. Consider $m+1$ "waves" of jobs. For $1 \le w \le m$, wave w has $m \cdot Y_{m+1-w}$ jobs released at $\sum_{z=1}^{w-1} Y_{m+1-z}$ with deadline m^m. The last wave has $m \cdot (m-1)^m$ jobs released at time $m^m - (m-1)^m$ with deadline m^m. FIRSTFIT dispatches wave i to machine M_i, using it until time m^m. FIRSTFIT must reject the last $m(m-1)^m$ jobs and runs only $m \cdot \sum_{k=1}^{m} Y_k = m(m^m - (m-1)^m)$ jobs. OPT runs all $m \cdot m^m$ jobs by distributing each wave across all m machines, giving a competitive ratio of $\frac{m^m}{m^m-(m-1)^m}$.

3   Lower Bounds

In this section, we provide lower bounds on the competitiveness of randomized and deterministic algorithms for the immediate dispatch model, the unrestricted model, and the special case of $m = 2$ and $p = 1$. In our constructions, we use $\langle r_j, d_j \rangle$ to denote a job with release time r_j and deadline d_j. Goldman et al. provide a $\frac{4}{3}$-competitive lower bound for randomized algorithms on one machine in the unrestricted model [9]. Their construction does not apply to multiple machines in the unrestricted model, but Ding et al. use such a construction in the *immediate decision* model to provide a randomized lower bound of $\frac{4}{3}$ for any m [6]. We first show that this bound applies to the *immediate dispatch* model.

Theorem 3. *For the* immediate dispatch *model with $p \ge 2$, every randomized algorithm has a competitive ratio at least $\frac{4}{3}$ against an oblivious adversary.*

Proof. We apply Yao's principle [3], bounding the expected performance of a *deterministic* algorithm against a *random* distribution. In particular, we consider two possible instances, both beginning with m jobs denoted by $\langle 0, 2p+1 \rangle$. For a given deterministic algorithm, let α be the number of machines, at time 0, that start a job or have two jobs already assigned. Our first instance continues with m jobs having parameters $\langle p, 2p \rangle$. The $m - \alpha$ machines that were assigned less than two jobs and that idle at time 0 can run at most one job each. The other α machines run at most 2 jobs each. Overall, the algorithm runs at most $2 \cdot \alpha + (m - \alpha) = m + \alpha$ jobs, for a competitive ratio of at least $\frac{2m}{m+\alpha}$. Our second instance continues with m jobs having parameters $\langle 1, p+1 \rangle$. At least α of these are rejected, since none can run on the α machines that are otherwise committed, making the competitive ratio at least $\frac{2m}{2m-\alpha}$. On a uniform distribution over these two instances, the deterministic algorithm has an expected competitive ratio of at least $\frac{1}{2}\left(\frac{2m}{m+\alpha} + \frac{2m}{2m-\alpha}\right)$, which is minimized at $\frac{4}{3}$ when $\alpha = \frac{m}{2}$.   □

We can prove slightly stronger bounds for *deterministic* algorithms, since an adversary can apply the worse of two instances (rather than their average).

Theorem 4. *For the* immediate dispatch *model with $p \ge 2$ and m odd, no deterministic algorithm has a competitive ratio strictly better than $\frac{4m}{3m-1}$.*

Proof (sketch). For the same two instances as in Theorem 3, $\max(\frac{2m}{m+\alpha}, \frac{2m}{m-\alpha})$ is minimized at $\frac{4m}{3m-1}$ with $\alpha = \lfloor \frac{m}{2} \rfloor = \frac{m-1}{2}$ or $\alpha = \lceil \frac{m}{2} \rceil = \frac{m+1}{2}$.   □

Theorem 5. *For the* immediate dispatch *model with $p \geq 3$ and m even, no deterministic algorithm has a competitive ratio strictly better than $\frac{4m+2}{3m}$.*

Proof (sketch). We adapt our construction, starting with one job $\langle 0, 4p+1 \rangle$, which we assume is started at time t by the algorithm. We next release m jobs $\langle t+1, t+2p+2 \rangle$, and let α denote the number of machines at time $t+1$ that are either running a job or have two jobs already assigned. We release a final set of m jobs, either all $\langle t+p+1, t+2p+1 \rangle$ or all $\langle t+2, t+p+2 \rangle$. In the first case, an algorithm gets at most $m+\alpha$, and in the second at most $1+2m-\alpha$. The lower bound of $\max(\frac{1+2m}{m+\alpha}, \frac{1+2m}{1+2m-\alpha})$ is minimized at $\frac{4m+2}{3m}$ when $\alpha = \lfloor \frac{1+m}{2} \rfloor = \frac{m}{2}$.   □

Although the $\frac{4}{3}$-competitive lower bound construction for the single-machine case has been adapted to the multiple machine case in the immediate decision and immediate dispatch models, it does not directly apply to the less restrictive model of immediate notification or the original unrestricted model. If facing the construction used in Theorem 3, an optimal deterministic algorithm could accept the initial m jobs with parameters $\langle 0, 2p+1 \rangle$, starting $\frac{m}{3}$ of them at time 0 and centrally queuing the other $\frac{2m}{3}$. If at time 1 it faces the arrival of m additional jobs with parameters $\langle 1, p+1 \rangle$, it can accept $\frac{2m}{3}$ of them on idle machines, while still completing the remaining initial jobs at time $p+1$ on those machines. The competitive ratio in this setting is $2m/(m + \frac{2m}{3}) = \frac{6}{5}$. If no jobs arrive by time 1 for the given adversarial construction, it can commit another $\frac{m}{3}$ machines to run initial jobs from $[1, p+1)$, with the final third of the initial jobs slated on those same machines from $[p+1, 2p+1)$. In that way, it retains room for $\frac{2m}{3}$ jobs in a second wave during the interval $[p, 2p]$, by using the idle machines and the first third of the machines that will have completed their initial job, again leading to a competitive ratio of $\frac{6}{5}$. Ding and Zhang [7] provide a slightly stronger deterministic bound for fixed values of m, by releasing a single initial job with larger deadline, followed by the classic construction (akin to our construction from Theorem 5).

In our next series of results, we give a new construction that strengthens the randomized and deterministic lower bounds for these models, showing that competitiveness better than $\frac{5}{4}$ is impossible in general. We do this by doubling the size of the second wave in one of the two instances, thereby changing the balancing point of the optimal behavior for the construction.

Theorem 6. *For the* unrestricted *model with $p \geq 2$, no randomized algorithm has a competitive ratio strictly better than the following, with m given mod 5:*

$m \equiv 0$	$m \equiv 1$	$m \equiv 2$	$m \equiv 3$	$m \equiv 4$
$\frac{5}{4}$	$\frac{20m^2}{16m^2-1}$	$\frac{30m^2}{24m^2-1}$	$\frac{30m^2}{24m^2-1}$	$\frac{20m^2}{16m^2-1}$

Proof (sketch). We use Yao's principle with a distribution of two instances. Both instances begin with m jobs $\langle 0, 2p+1 \rangle$. For a fixed deterministic algorithm, let α

be the number of machines that start a job at time 0. Our first instance continues with $2m$ jobs $\langle p, 3p \rangle$. A machine that is not starting a job at time 0 can run at most 2 jobs. Therefore, an online algorithm completes at most $3\alpha + 2 \cdot (m - \alpha) = 2m + \alpha$ jobs, for a competitive ratio of at least $\frac{3m}{2m+\alpha}$ on this instance. Our second instance continues with m jobs $\langle 1, p + 1 \rangle$. An online algorithm runs at most $2m - \alpha$ jobs, as it must reject α of the jobs arriving at time 1. Thus, its competitive ratio is at least $\frac{2m}{2m-\alpha}$ on this instance. For $m \equiv 0 \pmod 5$, we select the first instance with probability $\frac{1}{2}$. The expected competitive ratio of a deterministic algorithm for this distribution is at least $\frac{1}{2} \left(\frac{3m}{2m+\alpha} + \frac{2m}{2m-\alpha} \right)$, minimized at $\frac{5}{4}$ when $\alpha = \frac{2m}{5}$. This completes the theorem for $m \equiv 0 \pmod 5$. For other modularities of m, an even stronger bound holds because the algorithm cannot choose $\alpha = \frac{2m}{5}$. $\qquad \square$

Our next theorem strengthens the bound for deterministic algorithms by first releasing a single job with large deadline (similar to Theorem 5).

Theorem 7. *For the* unrestricted *model with $p \geq 3$, no deterministic algorithm has a competitive ratio strictly better than the following, with m given mod 5:*

$m \equiv 0$	$m \equiv 1$	$m \equiv 2$	$m \equiv 3$	$m \equiv 4$
$\frac{5}{4}\left(1 + \frac{1}{3m}\right)$	$\frac{5}{4}\left(1 + \frac{1}{(4m+1)}\right)$	$\frac{5}{4}\left(1 + \frac{3}{(12m+1)}\right)$	$\frac{5}{4}\left(1 + \frac{3}{(8m+1)}\right)$	$\frac{5}{4}\left(1 + \frac{1}{(4m-1)}\right)$

Proof (sketch). We release a job $\langle 0, 5p - 1 \rangle$, which we assume is started at time t by the algorithm. Next, we release m' jobs $\langle t + 1, t + 2p + 2 \rangle$, where $m' = m - 1$ if $m = 4 \pmod 5$ and $m' = m$ otherwise. Let α be the number of jobs (including the first) started on or before time $t + 1$. Our adversary continues in one of two ways, releasing either $2m$ jobs with parameters $\langle t + p + 1, t + 3p + 1 \rangle$ or m jobs with parameters $\langle t + 2, t + p + 2 \rangle$. These choices give competitive ratios of at least $\frac{1+m'+2m}{2m+\alpha}$ and $\frac{1+m'+m}{1+m'+m-\alpha}$ respectively. The precise lower bounds come from optimizing α for varying values of m. $\qquad \square$

The construction of Theorem 7 requires $p \geq 3$, to leverage the introduction of the job $\langle 0, 5p - 1 \rangle$. For $p = 2$, the following bound can be shown using the construction from Theorem 6, and deterministic choice $\alpha = \lfloor \frac{2m}{5} \rfloor$ or $\alpha = \lceil \frac{2m}{5} \rceil$.

Theorem 8. *For the* unrestricted *model with $p = 2$, no deterministic algorithm has a competitive ratio strictly better than the following:*

$m \equiv 0$	$m \equiv 1$	$m \equiv 2$	$m \equiv 3$	$m \equiv 4$
$\frac{5}{4}$	$\frac{15m}{12m-2}$	$\frac{10m}{8m-1}$	$\frac{15m}{12m-1}$	$\frac{5m}{4m-1}$

Finally, we focus on the special case of $p = 1$ and $m = 2$. Our analysis in Section 2.2 shows that FIRSTFIT is precisely $\frac{4}{3}$-competitive in this setting. However, the $\frac{4}{3}$ lower bounds from the previous theorems do not apply to $p = 1$; an adversary cannot force the rejection of new jobs due to machines that are committed to other tasks. With the following theorems, we provide (weaker) lower bounds for unit-length jobs, drawing a distinction between the *online-time* and *online-list* models, as defined in the introduction.

Theorem 9. *For the immediate dispatch model with $p = 1$ and $m = 2$, a deterministic online-time algorithm cannot be better than $9/8$-competitive. A deterministic online-list algorithm cannot be better than $8/7$-competitive.*

4   Conclusions

In this paper, we have introduced a study of the *immediate dispatch* model when maximizing throughput with equal-length jobs. We demonstrate that this model is strictly more difficult than the *immediate notification* model, and strictly easier than the *immediate decision* model. The primary open problem is to develop stronger algorithms for $m \geq 3$ in any of these models.

Acknowledgments. We thank the referees for helpful comments. D.P. Bunde was supported in part by Howard Hughes Medical Institute grant 52005130.

References

1. Avrahami, N., Azar, Y.: Minimizing total flow time and total completion time with immediate dispatching. Algorithmica 47(3), 253–268 (2007)
2. Baruah, S.K., Haritsa, J.R., Sharma, N.: On-line scheduling to maximize task completions. J. Combin. Math. and Combin. Computing 39, 65–78 (2001)
3. Borodin, A., El-Yaniv, R.: Online Computation and Competitive Analysis. Cambridge University Press, New York (1998)
4. Chin, F.Y.L., Chrobak, M., Fung, S.P.Y., Jawor, W., Sgall, J., Tichý, T.: Online competitive algorithms for maximizing weighted throughput of unit jobs. J. Discrete Algorithms 4(2), 255–276 (2006)
5. Chrobak, M., Jawor, W., Sgall, J., Tichý, T.: Online scheduling of equal-length jobs: Randomization and restarts help. SIAM Journal on Computing 36(6), 1709–1728 (2007)
6. Ding, J., Ebenlendr, T., Sgall, J., Zhang, G.: Online scheduling of equal-length jobs on parallel machines. In: Arge, L., Hoffmann, M., Welzl, E. (eds.) ESA 2007. LNCS, vol. 4698, pp. 427–438. Springer, Heidelberg (2007)
7. Ding, J., Zhang, G.: Online scheduling with hard deadlines on parallel machines. In: Cheng, S.-W., Poon, C.K. (eds.) AAIM 2006. LNCS, vol. 4041, pp. 32–42. Springer, Heidelberg (2006)
8. Ebenlendr, T., Sgall, J.: A lower bound for scheduling of unit jobs with immediate decision on parallel machines. In: Bampis, E., Skutella, M. (eds.) WAOA 2008. LNCS, vol. 5426, pp. 43–52. Springer, Heidelberg (2009)
9. Goldman, S., Parwatikar, J., Suri, S.: On-line scheduling with hard deadlines. J. Algorithms 34(2), 370–389 (2000)
10. Goldwasser, M.H., Kerbikov, B.: Admission control with immediate notification. J. Scheduling 6(3), 269–285 (2003)
11. Goldwasser, M.H., Misra, A.B.: A simpler competitive analysis for scheduling equal-length jobs on one machine with restarts. Information Processing Letters 107(6), 240–245 (2008)

12. Goldwasser, M.H., Pedigo, M.: Online nonpreemptive scheduling of equal-length jobs on two identical machines. ACM Trans. on Algorithms 5(1), 18, Article 2 (2008)
13. Karlin, A., Manasse, M., Rudolph, L., Sleator, D.: Competitive snoopy paging. Algorithmica 3(1), 70–119 (1988)
14. Pruhs, K.: Competitive online scheduling for server systems. SIGMETRICS Perform. Eval. Rev. 34(4), 52–58 (2007)
15. Sleator, D., Tarjan, R.: Amortized efficiency of list update and paging rules. Communications of the ACM 28, 202–208 (1985)

Online Function Tracking
with Generalized Penalties*

Marcin Bienkowski[1] and Stefan Schmid[2]

[1] Institute of Computer Science, University of Wrocław, Poland
[2] Deutsche Telekom Laboratories / TU Berlin, Germany

Abstract. We attend to the classic setting where an observer needs to inform a tracker about an arbitrary time varying function $f : \mathbb{N}_0 \to \mathbb{Z}$. This is an optimization problem, where both wrong values at the tracker and sending updates entail a certain cost. We consider an online variant of this problem, i.e., at time t, the observer only knows $f(t')$ for all $t' \leq t$. In this paper, we generalize existing cost models (with an emphasis on concave and convex penalties) and present two online algorithms. Our analysis shows that these algorithms perform well in a large class of models, and are even optimal in some settings.

1 Introduction

Online function tracking has a wide range of applications. For instance, consider a sensor network where a node measures physical properties (e.g., oxygen levels) at a certain location, and needs to report this data to a sink node collecting the measurements of multiple nodes in order to, e.g., raise an alarm if necessary. There is a natural tradeoff between communication and energy costs (how often is the sink informed?) and accuracy (how accurate is the information at the sink?). Function tracking also finds applications in publish/subscribe systems or organization theory where similar tradeoffs exist.

This paper attends to a two-party version of the problem where a node observing a certain function f informs a tracking node. Our main objective is to devise online algorithms for the observing node, which guarantee that the overall cost (sum of update costs and penalties for inaccuracies) is *competitive* — for any possible sequence of function changes — to the cost of an optimal offline algorithm knowing all values of f in advance. This simple two-party instantiation already requires non-trivial solutions [12]. In this paper, we consider an arbitrary function f and different classes of penalty functions.

1.1 Model

We consider a situation where an observer node wants to keep a tracker node informed about a certain function $f :$ Time $(\mathbb{N}_0) \to \mathbb{Z}$ evolving over time in synchronous time steps (*rounds*). Let $f(t)$ be the actual function value observed

---

* Supported by MNiSW grants number N N206 2573 35 and N N206 1723 33.

H. Kaplan (Ed.): SWAT 2010, LNCS 6139, pp. 359–370, 2010.

at round t. Let ALG_t denote the value at the tracker at time t, specified by an algorithm ALG; we say that the algorithm is in state ALG_t. Initially, at time $t = 0$, $\text{ALG}_0 = f(0)$. When the time is clear from the context, we drop the time index.

We study the design of algorithms that allow the observer to inform the tracker about the current values of $f(t)$. In each round t, the following happens:

1. The function f can assume a new arbitrary value $f(t) \in \mathbb{Z}$.
2. The algorithm may change its state ALG_t to any integer paying fixed *update cost* C; otherwise $\text{ALG}_t = \text{ALG}_{t-1}$.
3. The algorithm pays *penalty* $\Psi(|\text{ALG}_t - f(t)|)$, where $\Psi : \mathbb{N}_0 \to \mathbb{N}_0$ is a general function that specifies the cost of a given inaccuracy (e.g., $\Psi(x) = x$).

For succinctness, we abuse notation and will sometimes write $\Psi(x, y)$ meaning $\Psi(|x-y|)$. In this paper, we use the reasonable assumption that $\Psi(x)$ grows monotonically in x, i.e., the penalty cost never decreases for larger errors. Moreover, without loss of generality, we assume that $\Psi(0) = 0$; otherwise, the competitive ratio only improves.

Our main objective is to find an ideal trade-off between *update cost* (informing the tracker about new values) and *penalty cost* (difference between $f(t)$ and ALG_t):

$$\text{Cost} = \text{Cost}_{\text{update}} + \text{Cost}_{\text{penalty}}$$
$$= C \cdot \sum_{t=0}^{T} (\text{ALG}_t \neq \text{ALG}_{t+1}) + \sum_{t=0}^{T} \Psi(\text{ALG}_t, f(t)) \ , \tag{1}$$

where T is the total number of rounds (chosen by the adversary). In other words, our cost function counts the number of updates made by an algorithm and accumulates penalties at the tracker over time. For any input sequence (i.e., the sequence of function changes over time) σ and algorithm ALG, by $\text{ALG}(\sigma)$ we denote the cost of ALG on σ.

We assume that at time t, the algorithm only knows the function values $f(t')$ for $t' \leq t$, but has no information about upcoming values. We are in the realm of online algorithms and competitive analysis [3], i.e., we want to compare the performance of an online algorithm ALG with an optimal offline algorithm OPT. An algorithm is ρ-competitive if there exists a constant γ, such that for any input σ, it holds that

$$\text{ALG}(\sigma) \leq \rho \cdot \text{OPT}(\sigma) + \gamma \ . \tag{2}$$

For a randomized algorithm, we replace the cost of ALG by its expected value and we consider *oblivious* adversaries [3], which do not have access to random bits of the algorithm. For succinctness, we will sometimes use the terminology from the request-answer games [3], saying that in round t a *request* occurred at $f(t)$.

1.2   Related Work

The tradeoff between accuracy and update or transmission cost has challenged researchers from various fields for many years. A classic example of this tradeoff is known as the TCP acknowledgement problem [6]. In the design of Internet transfer protocols such as the TCP protocol, an important question concerns the times when acknowledgments (ACKs) are sent from the receivers to a sender (to inform about the successful reception of packets). In many protocols, a delay algorithm is employed to acknowledge multiple ACK packets with a single message. The main goal of these protocols is to save bandwidth (and other overhead) while still guaranteeing small delays. Aggregating ACKs has similarities with function tracking as in some sense, the number of to be acknowledged packets can be regarded as the to be tracked function. Karlin et al. [8] gave an optimal $e/(e-1)$-competitive randomized online algorithm for a single link. There are also many variations of the theme, e.g., where the goal is to minimize the maximum delays of the packets [1], to minimize the total time elapsed while packets are waiting at the leaf node [7], to meet fixed deadlines [2] or to find schedules on tree topologies [9,11].

While our model is reminiscent of the TCP acknowledgment problem, there are crucial differences. First of all, we track an arbitrary function f that can both increase and decrease over time, whereas the number of ACKs can only become larger if no message is sent, which means that the to be tracked function is essentially monotonic. A more general aggregation function has already been proposed in [11]; however as there the value at the tracker is updated with a delay, the offline algorithm is unrealistically strong as it can always anticipate function changes and update the values before observing them. We note however that their *offline* solution, running in time quadratic in number of function changes, works also in our model.

In the field of distributed tracking (e.g., [4,5]), a coordinator seeks to keep track of the online inputs distributed over several sites. This problem can be regarded as a generalization of the model studied here. However, these results are still not applicable in our setting. For instance, [4] only considers monotonic functions, and [5] only allows a site to send the current function values, which is trivial in our case.

The closest work to ours is the SODA 2009 paper by Yi and Zhang [12]. In our terminology, they consider a special case with update cost $C = 1$, and the penalty function $\Psi(x) = 0$ for $x \leq \Delta$ and ∞ otherwise (Δ is a fixed constant). They present a deterministic algorithm, which achieves an asymptotically optimal competitive ratio of $\Theta(\log \Delta)$. They also generalize their algorithms to the multidimensional case, i.e., they are able to track functions whose values are integer vectors.

1.3   Our Contributions and Paper Organization

We present a simple online algorithm MED that achieves good competitive ratios for a large class of penalty functions (see Section 2.1). For example, our analysis shows that MED performs particularly well for concave penalty functions, where it achieves a ratio of $\mathcal{O}(\log C / \log \log C)$. This bound is matched for linear penalty functions, for which we show a lower bound of $\Omega(\log C / \log \log C)$ (see

Section 3.1). The same lower bound also holds for randomized algorithms (even against oblivious adversaries).

In Section 2.2, we propose an alternative algorithm SET which is $\mathcal{O}(\log \Delta)$-competitive for convex penalty functions, where $\Delta = \min\{x : \Psi(x) \geq C\}$ (see Section 2.2). This is a generalization of the bound in [12]; in their paper, Ψ can only assume values from $\{0, \infty\}$. We prove that for certain classes of convex functions, this bound is optimal (again, even for randomized algorithms).

Further, we observe that MED behaves well for a class of functions with "bounded growth". In particular, for polynomial penalty functions $\Psi(x) = x^\alpha$, MED is $\mathcal{O}(4^\alpha \cdot \log C / \log \log C)$-competitive and SET is $\mathcal{O}(\max\{1, \frac{1}{\alpha} \cdot \log C\})$-competitive. Thus, by choosing the better of the two algorithms MED and SET, we get a competitive ratio of $\mathcal{O}(\log C / \log \log \log C)$ for all choices of α, i.e., for all polynomial penalty functions.

2   Algorithms

All our algorithms follow the *accumulate-and-update paradigm*: they wait until the total penalty (since the last update) exceeds the threshold $\Theta(C)$ and then they update the value. Henceforth, such a subsequence between two consecutive updates is called a *phase*. In the simplest case, when f is non-decreasing, the problem becomes a discrete variant of the TCP acknowledgement problem [6].

Observation 1. *If f changes monotonically, then the algorithm which updates the value at the end of the phase to the last observed value is 4-competitive.*

The proof is similar to the one presented in [6] and is omitted. However, in the general case, updating always to the last observed value is bad, as the adversary can exploit this strategy.

One may see the choice of the new value as a pursuit of the optimal algorithm: we imagine that both the online as well as the optimal offline algorithm OPT are processing the input in parallel; then the algorithm wants to have a state as close to OPT's state as possible.

Where can OPT be found? A straightforward answer is that its state should be close to the recent requests. Indeed, if the penalty function grows fast at the beginning (e.g., it is concave), OPT has to be relatively close to the requests (otherwise, it accrues a high cost). For such functions, we construct the algorithm MED, which, roughly speaking, changes it state to the median of the recent requests and in this way decreases the distance between its state and the state of OPT. However, if the penalty function is relatively flat at the beginning (e.g., it is convex), then there are many states which are similarly well-suited for the optimal algorithm. In this case, in the construction of our second algorithm, SET, we use an approach which bears some resemblance to the *work function technique* (see, e.g., [10]). Namely, we track a set of states with the property that an algorithm which remains at such states pays little, i.e., the states are potential candidates for OPT. By choosing our position in the middle of such a set, in each phase the cardinality of the set decreases by a constant factor.

The intuitions above are formalized in the upcoming sections.

2.1   Concave Penalties and the Median Strategy

In this section, we present an online algorithm MED pursuing a median strategy and derive an upper bound on its competitive ratio on concave functions and functions of bounded growth. Later, we prove that its competitive ratio is asymptotically optimal for linear penalty functions.

Definition 1 (Growth). *Let $f : \mathbb{N}_0 \to \mathbb{N}_0$ be a monotonic function with $f(0) = 0$. The* growth *of f is defined as $\max_{x \geq 1}\{f(2x)/f(x)\}$*

For example, the growth of any concave function is at most 2. To give another example, $f(x) = c \cdot x^\alpha$ has growth 2^α.

Observation 2 (Triangle Inequality). *Let Ψ be a penalty function of growth at most β. For any three integers a, b, and c, it holds that $\Psi(a,c) \leq \beta \cdot (\Psi(a,b) + \Psi(b,c))$, since $\max\{|a - b|, |b - c|\} \geq |a - c|/2$ and since Ψ is monotonic. Consequently, $\Psi(a,b) \geq \Psi(a,c)/\beta - \Psi(b,c)$.*

The online algorithm MED we introduce here is based on a median strategy. MED partitions the input sequence into *phases*, each phase consisting of several rounds; the first phase starts with the beginning of the input, i.e., $f(0) = \text{MED}_0$. A phase is defined as a period of time, in which MED does not update the tracker but monitors the total penalty paid so far in this phase. Let t_0 be the first round of the current phase. If in one round t, the function $f(t)$ changes abruptly and is far away from MED_{t-1}, i.e., $\Psi(\text{MED}_{t-1}, f(t)) > C$, then MED updates $\text{MED}_t := f(t)$. Otherwise, if the accumulated sum of differences up to the current round would exceed or be equal to C, i.e., $\sum_{i=t_0}^{t} \Psi(\text{MED}_i, f(i)) \geq C$, then $\text{MED}_t := \widetilde{x}$ where $\widetilde{x}$ is the the *median* of of the function values in this phase. (In case of two medians the tie is broken arbitrarily.) In either case, if MED changes its state, the current phase ends and the new begins in the next round.

First, we bound the cost of MED in any phase.

Lemma 1. *Assume that the growth of the penalty function is bounded by β. Consider a phase P and let σ_P be the input sequence of P. Then, $\text{MED}(\sigma_P) \leq 2 \cdot (\beta + 1) \cdot C$.*

Proof. By the definition of MED, its accumulated penalty in all rounds except for the last one is at most C, and MED pays update cost C for changing its state in the last round t. In the case that MED changes its state to the last value $f(t)$, the total cost is $2C$ as no additional penalty accrues. Otherwise, MED updates to the median value $\widetilde{x}$ and in the last round it pays $\Psi(\widetilde{x}, f(t)) \leq \beta \cdot (\Psi(\text{MED}_{t-1}, \widetilde{x}) + \Psi(\text{MED}_{t-1}, f(t)))$. As the median is chosen among all the requests in P and for any $i \in P$, $\Psi(\text{MED}_{t-1}, f(i)) \leq C$, the total cost in the last round is at most $2\beta \cdot C$. $\qquad \square$

Next, we turn our attention to OPT. In the following, a phase in which OPT pays less than α is called α-*constrained*. The main idea for proving the competitiveness of MED is as follows. Phases which are not $\mathcal{O}(C)$-constrained — e.g., phases in which OPT updates — are trivial, as OPT incurs a cost of $\Omega(C)$ in them. On the other hand, in α-constrained phases with small α, the possible distance between OPT and MED becomes smaller: the less OPT pays, the faster the MED's state converges to the state of OPT. We show next that if OPT tries to pay $o(C)$ in a single phase, then after a sequence of $\mathcal{O}(\log C / \log \log C)$ phases, MED's state becomes equal to OPT's state, which entails a OPT cost $\Omega(C)$ in the next phase. This idea is formalized in the two lemmas below.

Lemma 2. *Assume that the growth of the penalty function is at most β. Fix any α-constrained phase P, starting at round t_0 and ending at t_1, for a given $\alpha < C/(3\beta)$. Assume that OPT is in state ξ throughout P. Then, it holds that $\Psi(\mathrm{MED}_{t_1}, \xi) \leq 2C$ and*

$$\frac{\Psi(\mathrm{MED}_{t_1}, \xi)}{\Psi(\mathrm{MED}_{t_0}, \xi)} \leq \frac{2 \cdot \alpha}{C/\beta - \alpha} \ .$$

Proof. First, we consider the case that MED updates its state to the last request in P, i.e., $\mathrm{MED}_{t_1} = f(t_1)$. By the definition, $\Psi(\xi, f(t)) \leq \alpha$ for all $t \in \{t_0, t_0 + 1, ..., t_1\}$. Then, by Observation 2, $\Psi(\mathrm{MED}_{t_0}, \xi) \geq \Psi(\mathrm{MED}_{t_0}, f(t_1))/\beta - \Psi(\xi, f(t_1)) > C/\beta - \alpha$. Finally, $\Psi(\mathrm{MED}_{t_1}, \xi) = \Psi(f(t_1), \xi) \leq \alpha$, and the lemma holds.

Second, we consider the case that P ends with MED updating its state to the median. Since all the requests are at distance at most C from MED_{t_0}, initially $\Psi(\mathrm{MED}_{t_0}, \xi) \leq 2C$ (as otherwise OPT would pay C for each request). Thus, in the following, we show that $\Psi(\mathrm{MED}_{t_1}, \xi)/\Psi(\mathrm{MED}_{t_0}, \xi) \leq 2\alpha/(C/\beta - \alpha)$. As $2\alpha/(C/\beta - \alpha) \leq 1$, this implies both parts of the claim. Let n be the number of rounds in P and let $\tilde{x}$ be the median of the corresponding n requests, denoted by $x_1, x_2, \ldots, x_n$. By Observation 2, we obtain a lower bound for $\Psi(\mathrm{MED}_{t_0}, \xi)$:

$$n \cdot \Psi(\mathrm{MED}_{t_0}, \xi) \geq \sum_{i=1}^{n} \left(\frac{1}{\beta} \cdot \Psi(\mathrm{MED}_{t_0}, x_i) - \Psi(\xi - x_i) \right)$$

$$= \frac{1}{\beta} \cdot \sum_{i=1}^{n} \Psi(\mathrm{MED}_{t_0}, x_i) - \sum_{i=1}^{n} \Psi(\xi, x_i)$$

$$\geq C/\beta - \alpha \ .$$

Moreover, by the median definition, it follows that at least half of the requests in P are further from ξ than the median is, and thus $(n/2) \cdot \Psi(\xi, \tilde{x}) \leq \mathrm{OPT}(\sigma_P) \leq \alpha$. This implies that $\Psi(\mathrm{MED}_{t_1}, \xi) = \Psi(\tilde{x}, \xi) \leq 2\alpha/n$. Comparing $\Psi(\mathrm{MED}_{t_1}, \xi)$ to $\Psi(\mathrm{MED}_{t_0}, \xi)$ immediately yields the lemma:

$$\frac{\Psi(\mathrm{MED}_{t_1}, \xi)}{\Psi(\mathrm{MED}_{t_0}, \xi)} \leq \frac{2\alpha/n}{(C/\beta - \alpha)/n} = \frac{2\alpha}{C/\beta - \alpha} \ . \qquad \square$$

Lemma 3. *Assume that the growth of the penalty function is at most β. There exists $\ell = \Theta(\log C/\log\log C)$, such that in any subsequence τ consisting of consecutive $2\ell + 1$ phases, $\mathrm{OPT}(\tau) = \Omega(C/\beta)$.*

Proof. Fix any input sequence σ and any contiguous subsequence τ consisting of $2\ell + 1$ phases. (The exact value of ℓ is discussed later.)

If OPT changes its state within τ, then the lemma follows trivially. Thus, in the remainder of the proof, we assume that throughout τ, OPT is in state ξ. We look at the prefix τ' of 2ℓ phases of τ (i.e., ignoring the last phase). Let $B = \frac{1}{\beta} \cdot C/\sqrt{\log C}$. We assume that C is sufficiently large, i.e., $B \leq C/(3\beta)$. We consider three cases:

1. τ' contains a phase for which OPT pays at least $\frac{1}{3} \cdot C/\beta$. In this case, the claim follows trivially.
2. τ' contains ℓ phases for which OPT pays at least B. Then, $\mathrm{OPT}(\tau') \geq \ell \cdot B = \Omega(C/\beta)$.
3. All phases of τ' are $(\frac{1}{3} \cdot C/\beta)$-constrained and at least $\ell + 1$ of them are additionally B-constrained. We show that this implies the existence of a phase in τ', at the end of which $\Psi(\mathrm{MED}, \xi) = 0$.

   By Lemma 2, we can make three key observations for this case: (1) in all phases of τ', the distance between MED and ξ does not increase; (2) after the first phase of τ', the distance between MED and ξ becomes at most $2C$; (3) in each of the next ℓ B-constrained phases, $\Psi(\mathrm{MED}, \xi)$ decreases by a factor of $q := 2B/(C/\beta - B) = \Theta(1/\sqrt{\log C})$. Let $\ell = \log_{1/q}(4C) = \Theta(\log C/\log\log C)$. Thus, at the end of these ℓ phases, $\Psi(\mathrm{MED}, \xi)$ decreases to at most $1/2$, i.e., it becomes 0.

   We consider the next phase of τ, during which $\Psi(\mathrm{MED}, \xi) = 0$ and we denote the requests in this phase by $x_1, x_2, \ldots, x_n$. By Observation 2, the cost of OPT in this phase is $\sum_{i=1}^{n} \Psi(\xi, x_i) \geq \sum_{i=1}^{n} (\Psi(\mathrm{MED}, x_i)/\beta - \Psi(\mathrm{MED}, \xi)) \geq C/\beta$ (as by the construction of MED, $\sum_{i=1}^{n} \Psi(\mathrm{MED}, x_i) \geq C$). $\qquad\square$

Theorem 3. MED *is $\mathcal{O}(\beta^2 \cdot \log C/\log\log C)$-competitive for penalty functions Ψ of growth at most β.*

Proof. Fix any input sequence σ and partition it into subsequences of length $2\ell + 1$ phases, where ℓ is as in the proof of Lemma 3. Fix any such subsequence τ. By Lemma 3, $\mathrm{OPT}(\tau) = \Omega(C/\beta)$ and by Lemma 1, $\mathrm{MED}(\tau) \leq (2\ell+1)\cdot 2(\beta+1)\cdot C$. Summing over all the subsequences of σ, we obtain that the competitive ratio is

$$\rho = \mathcal{O}\left(\frac{\beta C}{C/\beta} \cdot \log C/\log\log C\right) = \mathcal{O}(\beta^2 \log C/\log\log C) \ .$$

Finally, we observe that after partitioning σ, we might get a subsequence shorter than $2\ell + 1$ at the end. However, by Lemma 1, this contributes only a constant term to the overall cost, and hence it does not influence the competitive ratio (cf. Equation 2). $\qquad\square$

2.2 Convex Penalties and the Set Strategy

In the previous section, we have observed that MED performs particularly well for concave penalty functions. We now turn our attention to a different algorithm SET which is inspired by [12]: there, it is shown that such a strategy performs well under "0-or-∞" penalties, which is in some sense an extreme case of convexity. Thus, we seek to generalize the approach of [12] to an entire class of penalty functions, and provide a performance analysis.

The algorithm SET works as follows. First, on the basis of the penalty function Ψ, it computes a parameter $\Delta = \min\{x : \Psi(x) \geq C\}$. We call this value the C-gap of Ψ. This means that if an algorithm's state is at distance Δ from the request, then the algorithm pays at least C, and Δ is the smallest distance with this property.

SET keeps track of a set S, centered at its current state, consisting of consecutive integers. At the beginning, $S = [\text{SET}_0 - \Delta, \text{SET}_0 + \Delta] \cap \mathbb{Z}$, where $\text{SET}_0 = f(0)$. Again, in one *phase*, SET remains in the same state. Similarly to the MED algorithm, SET computes the penalties accumulated since the beginning of a phase. If this cost exceeds C, then SET changes its state as described below and a new phase starts.

For any point $x \in S$, SET computes the accumulated penalty of an algorithm A_x which remains at x during the whole phase. Among all points $x \in S$, we choose the leftmost (ℓ) and the rightmost (r) point for which $A_x \leq C/2$. Let S' be the set of all integers in $[\ell, r]$. Now SET distinguishes two cases. If S' is nonempty, then we set $S := S'$, otherwise we choose set S to contain all the integers from range $[z - \Delta, z + \Delta]$, where z is the latest request. In the second case, we say that an *epoch* has ended and a new epoch starts with the next phase. In either case, SET moves to the median of the new set S.

Below, we analyze the performance of the algorithm SET for convex penalty functions. We start with a simple property of set S' chosen at the end of each phase.

Observation 4. *Assume that the penalty function Ψ is convex. Let $S' = \{\ell, \ell + 1, \ldots, r\}$ be the set computed by SET at the end of phase P. Then $A_x(P) \leq C/2$ for any $x \in S'$ (and not only for $x \in \{\ell, r\}$).*

Proof. The function of the cumulative penalty over a fixed period is also convex (as the sum of convex functions is convex). This function is bitonic (i.e., first monotonically decreasing and then monotonically increasing), which implies the observation. □

In the lemmas below, we use the above observation, i.e., we assume that Ψ is a convex function.

Lemma 4. *In any phase, the cost of SET is at most $5C/2$.*

Proof. As in the proof of Lemma 1, the total penalty for all the requests but the last one is at most C. The cost of changing state is also C. Thus, we have to show that the penalty associated with the last request y is at most $C/2$. If set S' is non-empty, then SET changes its state to a median of S'. Hence, by Observation 4, the penalty associated with y is at most $C/2$. If S' is empty, then SET changes its state to y, in which case the penalty is zero.   □

Lemma 5. *In any two consecutive epochs E_{i-1} and E_i, the cost of OPT is at least $C/2$.*

Proof. The lemma follows trivially if OPT changes its state in these epochs, so we assume it does not. Let S_i be the set S of the algorithm SET at the beginning of epoch E_i. If OPT's state is in S_i, then its cost is at least $C/2$ in at least one phase of E_i. Otherwise, the OPT state is outside S_i. Then, we consider the last request of E_{i-1}, which, by the definition of SET is given at the center of S_i, i.e., at a distance of at least $\Delta + 1$ from the state of OPT. Thus, the penalty for OPT associated with this request is at least C.   □

Theorem 5. *For any convex penalty Ψ with C-gap equal to Δ, SET is $\mathcal{O}(\log \Delta)$-competitive.*

Proof. By Lemmas 4 and 5, it suffices to show that the number of phases in a single epoch is at most $O(\log \Delta)$. At the beginning of any epoch, $span(S)$ (defined as the distance between the rightmost and the leftmost point of S) is 2Δ. Fix any phase that is not the last phase in an epoch. Then a set S' chosen at the end is non-empty. Let x be the state of SET at the beginning of this phase. Since $A_x \geq C$, S' cannot contain x, i.e., the median of S. Thus, $span(S') \leq span(S)/2$, which means that $span(S)$ decreases at least by a factor of 2 in each phase. This may happen only $O(\log \Delta)$ times.   □

It follows from [12] that this result is asymptotically tight in the following sense: for any Δ there exists a convex penalty function Ψ, so that the competitive ratio of any online algorithm is $\Omega(\log \Delta)$. In Section 3, we will show that this lower bound holds also for randomized algorithms. The function Ψ for which our lower bound holds is any (possibly convex) function satisfying $\Psi(x) = 0$ for $x < \Delta$ and $\Psi(x) \geq C$ otherwise.

Remark. Note that our results are also applicable to convex penalty functions Ψ with the additional hard constraint that the difference between reported and observed value must not exceed T. Convexity was used only to obtain the property guaranteed in Observation 4; however, this property also holds for a function $\Psi(x)$ that is derived from a convex function Ψ' and a threshold T, such that $\Psi(x) = \Psi'(x)$ for $x \leq T$ and $\Psi(x) = \infty$ for $x > T$.

3   Lower Bounds

Next, we show that our algorithms MED and SET are asymptotically optimal in the classes of convex and concave functions, respectively. Note that we are not claiming their optimality for every such function, but for a quite broad subset of them. We emphasize that our lower bound holds even for randomized algorithms against oblivious adversaries.

3.1   Linear Penalties

We prove that our deterministic algorithm MED is asymptotically optimal for linear penalty functions.

Theorem 6. *For a given penalty function $\Psi(x) = a \cdot x$ (for any $a > 0$), the competitive ratio of any randomized algorithm is at least $\Omega(\log C/\log\log C)$.*

To prove this theorem, we employ a standard min-max approach. We fix an arbitrary deterministic online algorithm DET and generate a probability distribution π over input sequences in such a way that if an input sequence σ is chosen according to π, the following conditions hold:

1. $\text{OPT}(\sigma) = \mathcal{O}(C)$
2. $\mathbf{E}_\pi[\text{DET}(\sigma)] = \Omega(C \cdot \log C/\log\log C)$, where the expectation is taken over the random choice of the input;

Our construction below can be repeated an arbitrary number of times. Along with the second condition above, this ensures that the cost of the algorithm cannot be hidden in the additive constant in the definition of the competitive ratio (see Eq. 2). Then the lower bound for any randomized online algorithm follows immediately by the Yao min-max principle [3].

We now describe how to randomly choose an input σ, that is, we will implicitly create a probability distribution π over input sequences. Let $[a,b]_\mathbb{N}$ be the set $[a,b] \cap \mathbb{N} = \{a, a+1, \ldots, b\}$. Let s be the largest integer i for which $(\log C)^i \leq C$; clearly $s = \Theta(\log_{\log C} C) = \Theta(\log C/\log\log C)$. First, we create a sequence of $s+1$ random sets: $R_0 \supseteq R_1 \supseteq R_2 \supseteq \ldots \supseteq R_s$. $R_0 = [0, (\log C)^s - 1]_\mathbb{N}$. The remaining sets are chosen iteratively in the following manner. We partition R_i into $\log C$ disjoint contiguous subsets of the same size; R_{i+1} is chosen uniformly at random among them. For example, R_1 is chosen amongst the following sets: $[0, (\log C)^{s-1} - 1]_\mathbb{N}, [(\log C)^{s-1}, 2 \cdot (\log C)^{s-1} - 1]_\mathbb{N}, \ldots, [(\log C)^s - (\log C)^{s-1}, (\log C)^s - 1]_\mathbb{N}$. Note that the construction implies that R_s contains a single integer. The sequence σ associated with sets $R_0, R_1, \ldots, R_s$ consists of s phases, numbered from 1. In phase i there are $\lceil C/(a \cdot |R_{i-1}|)\rceil$ requests given at the leftmost integer from the set R_i.

Below, we present two lemmas, which directly imply the two conditions above, and thus also Theorem 6.

Lemma 6. *For any initial state of* OPT *and input σ generated in the way described above,* $\text{OPT}(\sigma) = O(C)$.

Proof. Let $R_0 \supseteq R_1 \supseteq \ldots \supseteq R_s$ be the sequence of sets associated with σ. Let x be the only element of the set R_s. The strategy for an offline (possibly not optimal) algorithm OFF is to change its state to x at the very beginning of σ and remain there for the whole σ.

Clearly, the update cost is C. In phase i, the distance between x and the requests is at most $|R_i|$, and thus the total penalty paid by OFF is at most $\mathcal{O}(a \cdot |R_i| \cdot \lceil C/(a \cdot |R_{i-1}|)\rceil) = \mathcal{O}(C/\log C)$. Thus, the total cost in the entire sequence is $\text{OFF}(\sigma) \leq C + s \cdot O(C/\log C) = \mathcal{O}(C)$. As $\text{OPT}(\sigma) \leq \text{OFF}(\sigma)$, the lemma follows. $\square$

Lemma 7. *For any deterministic online algorithm* DET *and input* σ *generated randomly in the way described above,* $\mathbf{E}_\pi[\mathrm{DET}(\sigma)] = \Omega(C \cdot \log C / \log \log C)$.

Proof. Any sequence σ consists of $s = \Theta(\log C / \log \log C)$ phases. It is therefore sufficient to show that the expected cost of DET in any phase is at least $\Omega(C)$.

We consider the moment at the very beginning of phase i, even before its first request is presented to DET. At that moment, DET knows the set R_{i-1} and is at some fixed state x (not necessarily but possibly from R_{i-1}). When the first request from phase i is revealed to DET, it immediately learns R_i, but it is already too late. For any fixed state x, the expected distance between x and the leftmost point of R_i is at least $\Omega(|R_{i-1}|)$. Thus, DET has two choices. It may change its state, paying C or it may remain at x paying in expectation the total penalty of at least $\Omega(a \cdot \lceil C/(a \cdot |R_{i-1}|) \rceil \cdot |R_{i-1}|) = \Omega(C)$. □

3.2 Convex Penalties

The following theorem shows the asymptotic optimality of the algorithm SET in the class of convex functions. We note that the deterministic variant of this theorem is already known (cf. Theorem 2.1 of [12]) and our proof can be viewed as its adaptation.

Theorem 7. *For any* Δ, *there exists a convex penalty function* Ψ, *whose C-gap is* Δ *and the competitive ratio of any randomized algorithm is at least* $\Omega(\log \Delta)$.

Proof. We consider the convex penalty function used in [12], i.e., $\Psi(x) = 0$ for $x \leq \Delta$ and $\Psi(x) = \infty$ otherwise. In fact, for our proof to work, we require only that $\Psi(x) = 0$ for $x \leq \Delta$ and $\Psi(x) = \Omega(C)$ otherwise.

Our approach is similar in flavor to the proof of Theorem 6, so we just concentrate on the differences. Once again, we randomly construct a family of sets $R_0, R_1, \ldots R_s$. This time $s = \lfloor \log \Delta \rfloor$ and $R_0 = [0, 2^s - 1]_\mathbb{N}$. To construct R_i out of R_{i-1}, we divide R_{i-1} into two equal contiguous halves and R_i is chosen randomly among them. The sequence consists now of s rounds, numbered from 1. In round i, the requests are given at x_i, such that the distance between x_i and any point of R_i is at most Δ and the distance between x_i and any point of $R_{i-1} \setminus R_i$ is greater than Δ.

OPT can serve the whole sequence without penalties changing its state to the only integer of R_s at the beginning (paying C for the state change). On the other hand, any deterministic algorithm DET at the beginning of round i knows set R_{i-1}, but does not know which half will be chosen as R_i, and with probability $1/2$ it is in "the wrong half". Thus, with probability $1/2$, when the request is presented to DET, it either has to pay the penalty $\Omega(C)$ or change its state paying C. Hence, the expected cost of DET on such a sequence is $\Omega(C \cdot s) = \Omega(C \log \Delta)$, i.e., the ratio of $\mathbf{E}[\mathrm{DET}]$ divided by OPT is at least $\Omega(\log \Delta)$. Again, using the Yao min-max principle, the same bound even holds for any randomized algorithm (against an oblivious adversary). □

4   Conclusions

This paper studies generalized penalty functions for the problem of approximating a function f (that is revealed gradually) by a piecewise constant function g, where the cost depends on the number of value changes of g plus the error cost summed over the discrete sampling points.

We believe that our work opens several interesting directions for future research. First, our results raise the question whether MED and SET can be combined in order to have the advantages of both worlds in penalty functions beyond concave and convex models. Another research direction is the study of different penalty functions in multi-dimensional tracking $f : \mathbb{N}_0 \to \mathbb{Z}^d$ and the analysis of the gains that can be obtained with the line predictions of [12]. Finally, distributed settings remain to be explored where there are multiple observers at different sites.

References

1. Albers, S., Bals, H.: Dynamic TCP acknowledgement: Penalizing long delays. In: Proc. of the 14th ACM-SIAM Symp. on Discrete Algorithms (SODA), pp. 47–55 (2003)
2. Becchetti, L., Korteweg, P., Marchetti-Spaccamela, A., Skutella, M., Stougie, L., Vitaletti, A.: Latency constrained aggregation in sensor networks. In: Proc. of the 14th European Symp. on Algorithms (ESA), pp. 88–99 (2006)
3. Borodin, A., El-Yaniv, R.: Online Computation and Competitive Analysis. Cambridge University Press, Cambridge (1998)
4. Cormode, G., Muthukrishnan, S., Yi, K.: Algorithms for distributed functional monitoring. In: Proc. of the 19th ACM-SIAM Symp. on Discrete Algorithms (SODA), pp. 1076–1085 (2008)
5. Davis, S., Edmonds, J., Impagliazzo, R.: Online algorithms to minimize resource reallocations and network communication. In: Proc. of the 9th Int. Workshop on Approximation Algorithms for Combinatorial Optimization (APPROX), pp. 104–115 (2006)
6. Dooly, D.R., Goldman, S.A., Scott, S.D.: On-line analysis of the TCP acknowledgment delay problem. Journal of the ACM 48(2), 243–273 (2001)
7. Frederiksen, J.S., Larsen, K.S.: Packet bundling. In: Proc. of the 8th Scandinavian Workshop on Algorithm Theory (SWAT), pp. 328–337 (2002)
8. Karlin, A.R., Kenyon, C., Randall, D.: Dynamic TCP acknowledgement and other stories about e/(e - 1). Algorithmica 36(3), 209–224 (2003)
9. Khanna, S., Naor, J., Raz, D.: Control message aggregation in group communication protocols. In: Proc. of the 29th Int. Colloq. on Automata, Languages and Programming (ICALP), pp. 135–146 (2002)
10. Koutsoupias, E., Papadimitriou, C.H.: On the k-server conjecture. Journal of the ACM 42(5), 971–983 (1995)
11. Pignolet, Y.A., Schmid, S., Wattenhofer, R.: Tight bounds for delay-sensitive aggregation. Discrete Mathematics and Theoretical Computer Science (DMTCS) 12(1) (2010)
12. Yi, K., Zhang, Q.: Multi-dimensional online tracking. In: Proc. of the 19th ACM-SIAM Symp. on Discrete Algorithms (SODA), pp. 1098–1107 (2009)

Better Bounds on Online Unit Clustering[*]

Martin R. Ehmsen and Kim S. Larsen

Department of Mathematics and Computer Science
University of Southern Denmark, Odense, Denmark
{ehmsen,kslarsen}@imada.sdu.dk

Abstract. Unit Clustering is the problem of dividing a set of points from a metric space into a minimal number of subsets such that the points in each subset are enclosable by a unit ball. We continue work initiated by Chan and Zarrabi-Zadeh on determining the competitive ratio of the online version of this problem. For the one-dimensional case, we develop a deterministic algorithm, improving the best known upper bound of 7/4 by Epstein and van Stee to 5/3. This narrows the gap to the best known lower bound of 8/5 to only 1/15. Our algorithm automatically leads to improvements in all higher dimensions as well. Finally, we strengthen the deterministic lower bound in two dimensions and higher from 2 to 13/6.

1 Introduction

Unit Clustering is the problem of dividing a set of points from a metric space into a minimal number of subsets such that the points in each subset are enclosable by a unit ball. The subsets are also referred to as *clusters*. Clustering algorithms have applications in for instance information retrieval, data mining, and facility location.

In the online version, we must treat a sequence of points one at a time, i.e., a point must be treated without knowing the remaining sequence of points to come, or even the length of this future sequence. When treating a point, we always have the option of opening a new cluster for the point, i.e., increasing the number of subsets by one new subset containing only the most recent point. We may also have the option of including the new point in an existing cluster, provided that the new point together with all other points already assigned to that cluster are still enclosable in a unit ball.

Each point must be assigned to exactly one cluster and this decision is irrevocable, i.e., we cannot at a later stage move a point from one cluster to another. Note that this problem is different from online covering [2]. In online covering, when a cluster is opened, a unit diameter is placed in a fixed location. In online clustering, points are assigned to a cluster, but the exact location of the cluster is not fixed. Another way to view this is that clusters open with size zero and then gradually have their sizes increased as points are assigned to the clusters.

To measure the quality of an online algorithm for this problem, we follow previous work on the same topic and use competitive analysis [6,8,7]. In unit

---

[*] This work was supported in part by the Danish Natural Science Research Council.

H. Kaplan (Ed.): SWAT 2010, LNCS 6139, pp. 371–382, 2010.

clustering, the cost of an algorithm is the number of clusters that have been opened. For the randomized results that we reference, the ratio is defined using the expected cost $E[A(I)]$ instead of the deterministic cost $A(I)$, and the results are with respect to an oblivious adversary.

We develop a deterministic algorithm for the one-dimensional Online Unit Clustering problem. The work on this problem was initiated in [1] by Chan and Zarrabi-Zadeh and the currently best known bounds are by Epstein and van Stee [5]. An overview of previous results, our results, and lower bounds for 1-dimensional unit cluster is given in Table 1. Thus, by this new upper bound of $\frac{5}{3}$, we narrow the gap to the best known lower bound of $\frac{8}{5}$ to only $\frac{1}{15}$.

Table 1. New and previous results for 1-dimensional unit clustering

	Paper	Result	Type
Upper Bounds	[1]	2	deterministic
	[1]	$\frac{15}{8} = 1.875$	randomized
	[9]	$\frac{11}{6} \approx 1.833$	randomized
	[5]	$\frac{7}{4} = 1.75$	deterministic
	This	$\frac{5}{3} \approx 1.667$	deterministic
Lower Bounds	[5]	$\frac{8}{5} = 1.6$	deterministic
	[5]	$\frac{3}{2} = 1.5$	randomized
	[1]	$\frac{3}{2} = 1.5$	deterministic
	[1]	$\frac{4}{3} \approx 1.333$	randomized

In higher dimensions, using an idea from [1], our 1-dimensional result improves the upper bound of $\frac{7}{8}2^d$ from [5] to $\frac{5}{6}2^d$ in d dimensions. As in the previous work, this is with respect to the L_∞ norm. Thus, the unit balls are really squares, cubes, etc. In one dimension, a 2-competitive algorithm is almost immediate. There are more than one easy formulation of such an algorithm. The greedy algorithm which only opens a new cluster when forced to do so is one of them. Thus, given a c-competitive algorithm for the 1-dimensional case, one can use that algorithm in the first dimension and combine that with an additional factor 2 for each additional dimension. So, in dimension $d \geq 2$, a $c2^{d-1}$-competitive algorithm can be designed from the 1-dimensional case, and we obtain an algorithm with competitive ratio $\frac{5}{3}2^{d-1} = \frac{5}{6}2^d$.

Finally, we strengthen the deterministic lower bound in dimension 2 and higher from two, obtained in [5], to $\frac{13}{6}$. The lower bound proof is carried out in dimension 2. The lower bound holds in higher dimensions by simply giving the corresponding points in a 2-dimensional subspace.

Many variants of online unit clustering have been studied by Epstein et al. [4].

This paper is an extended abstract of a 32 page full version [3]. It is not possible here to account for the full proof of the 1-dimensional upper bound; or even the full algorithm. We have tried to include material that gives a feel for the techniques involved. On the other hand, we have included the full proof of the lower bound for dimension 2 and higher.

2   The Algorithm

In this section we describe a new algorithm for online unit clustering. We structure our algorithm around components consisting of 2–4 clusters that we refer to as a *groups* with various characteristics. We then define behavior inside and outside these groups. We start with terminology and a few definitions of groups and other components.

We are working on the line, and define a *cluster* to be an interval with a maximal length of one. For a cluster C, let r_C denote the right endpoint of C and let l_C denote the left endpoint of C. For two points p_1 and p_2, we let $d(p_1, p_2)$ denote the distance between the two points.

We say that a cluster C can *cover* a point p, if $\max(d(p, r_C), d(p, l_C)) \leq 1$, i.e., if assigning p to C does not violate the restriction on the length of a cluster being at most one unit (see Fig. 1). A cluster C is said to be able to *reach* another cluster D, if C can cover all points on the line between C and D (see Fig. 2).

If, for some cluster D, there exist a cluster C to the left and a cluster E to the right of D, then D is said to be *underutilized* if $d(r_C, l_E) \leq 1$ (see Fig. 3). Two clusters C and D are said to be a *close pair* if $d(l_C, l_D) \leq 1$ and $d(r_C, r_D) \leq 1$ (see Fig. 4). Two clusters C and D are said to be a *far pair* if $d(l_C, l_D) > 1$, $d(r_C, r_D) > 1$, and $d(r_C, l_D) \leq 1$ (see Fig. 5).

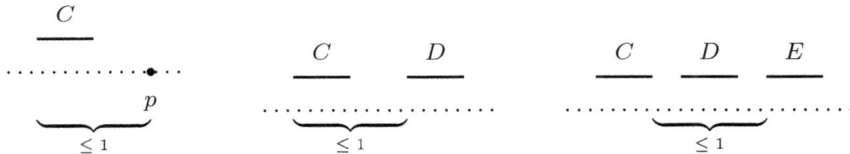

Fig. 1. C can cover p **Fig. 2.** C can reach D **Fig. 3.** D underutilized

We now start discussing groups. In the algorithm to be presented, we will ensure that groups satisfy certain specific criteria. However, satisfying these criteria does not define a group. For instance, we do not want overlapping groups, i.e., groups that share clusters. Thus, the definition of a group is a labelling issue, in the sense that when the situation is right, we may decide that a collection of clusters form a group.

We say that three consecutive clusters C, D, and E, none of which belong to another group, form a *potential regular group* if the clusters are related in the following way (see Fig. 6),

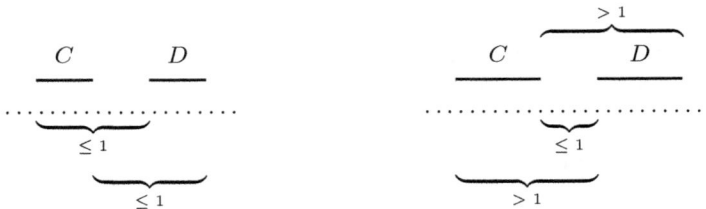

Fig. 4. A close pair   **Fig. 5.** A far pair

1. $d(r_C, l_D) \leq 1$ and $d(r_D, l_E) \leq 1$, and
2. $d(r_C, l_E) > 1$, and
3. $d(l_C, l_D) > 1$ or $d(r_D, r_E) > 1$, and
4. $d(r_C, r_D) \leq 1$ or $d(l_D, l_E) \leq 1$.

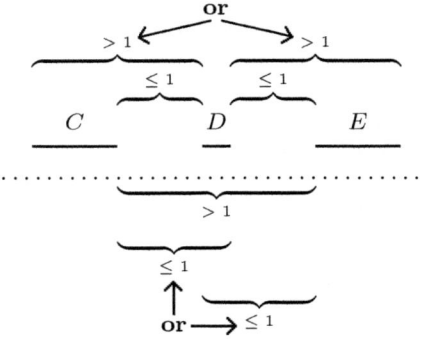

Fig. 6. A potential regular group

A regular group initially consists of three clusters. However, as the algorithm progresses, a fourth cluster might be needed at some point. On the other hand, the algorithm we develop ensures that a fifth cluster is never needed. We denote C and E *outermost* clusters in a regular group or a potential regular group. D and the possible fourth cluster are denoted *middle* clusters. We denote a regular group with three clusters a *regular 3-group* and a regular group with four clusters a *regular 4-group*. An outermost cluster in a regular group or a potential regular group that satisfies the third requirement, is said to be a *long cluster*, e.g., if $d(l_C, l_D) > 1$ in the above, then C is denoted a long cluster. Some regular groups have undesirable properties and we refer to a regular group with the following additional property as a *bad regular group*:

$$(d(r_C, r_D) > 1 \text{ and } d(r_D, r_E) \leq 1) \text{ or } (d(l_D, l_E) > 1 \text{ and } d(l_C, l_D) \leq 1)$$

Regular groups that are not bad are referred to as *good*.

A tight group (see Fig. 7) initially consists of two clusters. However, as the algorithm progresses, a third cluster might be needed at some point. We denote

Fig. 7. A tight group   **Fig. 8.** A close regular group

a tight group with two clusters a *tight 2-group* and a tight group with three clusters a *tight 3-group*. The two initial clusters in a tight group are denoted outermost clusters and the possible third cluster is denoted a middle cluster. A tight group is really just a far pair with a different label.

Finally, a regular group G_1 is said to be *close* to another group G_2, if a middle cluster from G_1 can reach a cluster in G_2; see Fig. 8.

We are now ready to define the algorithm. It is clear that if a new point falls in an already opened cluster (between its left and right endpoints), then that cluster covers the point. If the new point p falls inside a regular group, then Algorithm 2 handles the point, if p falls inside a tight group, then Algorithm 3 handles the point, and otherwise we let Algorithm 1 handle the point.

Algorithm 1. Main

Require: p falls outside any group
1. **if** p can be covered by a group cluster **then**
2.   Cover p by that cluster, avoid creating close regular groups if possible
3. **else if** covering p with some cluster creates a new good potential regular group **then**
4.   Cover p by the cluster and create a new regular group
5. **else if** opening a new cluster for p creates a new good potential regular group **then**
6.   **if** the new good potential regular group would be close to another group **then**
7.   Cover p and create a tight group
8.   **else**
9.   Open a new cluster for p and create the new regular group
10. **else if** p can be covered by a cluster **then**
11.   Cover p by a cluster, avoid creating a close pair if possible
12. **else**
13.   Open a new cluster for p

Algorithm 2. Regular Group

Require: p falls inside a regular group
1. **if** an outermost cluster can cover p without underutilizing a middle cluster **then**
2.   Cover p by that outermost cluster
3. **else if** p can be covered by a middle cluster and still reach an outermost cluster **then**
4.   Cover p by that middle cluster
5. **else**
6.   Open a new middle cluster for p

Algorithm 3. Tight Group
___
Require: p falls inside a tight group
 1. **if** p can be covered by an outermost cluster **then**
 2.   Cover p by that outermost cluster
 3. **else if** there is a middle cluster **then**
 4.   Cover p by the middle cluster
 5. **else**
 6.   Open a new middle cluster for p
___

3 Analysis

We first establish some properties of the algorithm, starting with correctness.

Lemma 1. *If two clusters are contained in a unit interval, then they are both part of the same regular group.*

Proof. Assume for the sake of contradiction that C_1 and C_2 are two clusters contained in a unit interval, and that they are not part of the same regular group. Assume without loss of generality that C_1 was opened before C_2. Consider the point p on which C_2 was opened. It is clear that C_1 can cover p.

By Algorithm 1, p can either be opened in Line 9 or in Line 13. We consider both choices and show that both lead to a contradiction, proving the lemma.

First, assume C_2 was opened in Line 9. Since, by assumption, C_1 and C_2 are not part of the same regular group, there must exist two other clusters, say C_3 and C_4, such that a new cluster for p can create a good regular group together with C_3 and C_4. It follows that if opening C_2 for p can create a good regular group of C_2, C_3, and C_4, then if C_1 covers p, then C_1, C_3, and C_4 would also be a good regular group. Hence, by Line 4 in Algorithm 1, C_1 would cover p and we reach a contradiction.

Next, assume C_2 was opened in Line 13 in Algorithm 1. Since, C_1 can cover p, we reach a contradiction, since p would have been covered in Line 11.

Lemma 2. *Algorithm 1 is well-defined.*

Proof. The only part which is not obviously well-defined is Line 7 in Algorithm 1. However, it is clear that a close regular group has two clusters contained in a unit interval. Hence, it follows from Lemma 1 that the new good potential regular group was about to be created because one of these two clusters was about to be opened. Thus, the other of these two clusters can cover the new point and create a tight group.

The following is an example of the many properties that must be established:

Lemma 3. *The middle cluster in a regular group cannot reach a non-group cluster.*

3.1   Overall Proof Structure

The basic idea of the overall proof is to divide the clustering produced by our algorithm into chunks, and show that if we process this division from left to right, then the algorithm is $\frac{5}{3}$-competitive after each chunk. In order to keep track of our status in this process, we introduce the notation $\left\{\begin{smallmatrix} a \\ b \end{smallmatrix}\right\}$ which means that after possibly having discarded a number of chunks where the ratio of clusters used by the online algorithm to clusters used by OPT is at least as good as $\frac{5}{3}$, we are currently in a situation where the online algorithm has used a clusters and OPT has used b clusters.

We define an ordering on pairs in the following way:

$$\left\{\begin{matrix} a \\ b \end{matrix}\right\} \succeq \left\{\begin{matrix} c \\ d \end{matrix}\right\} \iff \forall x, y : \frac{x+c}{y+d} \leq \frac{5}{3} \Rightarrow \frac{x+a}{y+b} \leq \frac{5}{3}.$$

where the latter is equivalent to $5(b-d) \geq 3(a-c)$.

From this we also define the relations $\preceq$, $\sim$, $\prec$, and $\succ$ in the straight forward way, and addition of pairs as

$$\left\{\begin{matrix} a \\ b \end{matrix}\right\} \oplus \left\{\begin{matrix} c \\ d \end{matrix}\right\} = \left\{\begin{matrix} a+c \\ b+d \end{matrix}\right\}.$$

Note that $\left\{\begin{smallmatrix} a+5 \\ b+3 \end{smallmatrix}\right\} \sim \left\{\begin{smallmatrix} a \\ b \end{smallmatrix}\right\}$ and that $\left\{\begin{smallmatrix} 1 \\ 1 \end{smallmatrix}\right\} \succ \left\{\begin{smallmatrix} 3 \\ 2 \end{smallmatrix}\right\}$, for example, expresses that in the $\left\{\begin{smallmatrix} 1 \\ 1 \end{smallmatrix}\right\}$ scenario we are in a better situation with respect to obtaining the $\frac{5}{3}$ bound than if we were in the $\left\{\begin{smallmatrix} 3 \\ 2 \end{smallmatrix}\right\}$ situation.

Observe that we can produce an OPT-clustering of a request sequence (offline, that is), by processing the points from left to right, and clustering points greedily: only open new clusters when the previously opened cluster cannot cover the next point. Consider the situation after we have produced an OPT-clustering in this manner (from left to right). Let a be the number of clusters used by our algorithm so far, and let b be the number of OPT-clusters used in the processing of the clusters from left to right. Consider the last opened OPT cluster, and the next cluster opened by our algorithm with respect to the processing of the clusters. We identify three states for the OPT cluster and associate a letter with each of these states. The OPT cluster might not be able to cover points from the next cluster opened by our algorithm. We use the letter N to denote this state. The remaining two states represent different situations where the OPT cluster can cover points from the next cluster opened by our algorithm. We use the letter A to denote the state where the OPT cluster covered all of the last cluster opened by our algorithm. Finally, we use the letter S to denote the state where the OPT cluster covered some (but not all) from the last cluster opened by our algorithm.

We show that if we are in state N, then $\left\{\begin{smallmatrix} a \\ b \end{smallmatrix}\right\} \succeq \left\{\begin{smallmatrix} 0 \\ 0 \end{smallmatrix}\right\}$, if we are in state A, then $\left\{\begin{smallmatrix} a \\ b \end{smallmatrix}\right\} \succeq \left\{\begin{smallmatrix} 3 \\ 2 \end{smallmatrix}\right\}$, and if we are in state S, then $\left\{\begin{smallmatrix} a \\ b \end{smallmatrix}\right\} \succeq \left\{\begin{smallmatrix} 2 \\ 2 \end{smallmatrix}\right\}$. With a slight abuse of notation we also use N, A, and S to denote $\left\{\begin{smallmatrix} 0 \\ 0 \end{smallmatrix}\right\}$, $\left\{\begin{smallmatrix} 3 \\ 2 \end{smallmatrix}\right\}$, and $\left\{\begin{smallmatrix} 2 \\ 2 \end{smallmatrix}\right\}$, respectively. Observe that $S \succ A \succ N$.

If we can show that the above is an invariant after each decision our algorithm makes, then we have shown that our algorithm is $\frac{5}{3}$-competitive.

In order to make the analysis easier to carry through, we mostly allow OPT to not cover points that are not start- or endpoints of a cluster opened by the algorithm. When we deviate from this, we explicitly discuss such a point and argue why we still have the necessary properties.

We describe how we divide the clustering into chunks. First, observe that we only need to consider sets of consecutive points where the distance between two consecutive points is at most one, since no cluster opened by the algorithm or by OPT can be shared between two such sets of points. In each set, each group is a chunk and anything between two consecutive groups or at the end of a set is a chunk. For each chunk and start state we analyze the chunk (in a worst case manner) and identify the end state and the number of new clusters used by our algorithm and OPT. If, for example, we are in state S before we process a chunk, the online algorithm uses four clusters to process the chunk, the OPT-clustering only needs to open two new clusters for the chunk, and the end state is A. We are then faced with the inequality $S \oplus \left\{ {4 \atop 2} \right\} \succeq A$, which is true since $S \oplus \left\{ {4 \atop 2} \right\} \sim \left\{ {2 \atop 2} \right\} \oplus \left\{ {4 \atop 2} \right\} \sim \left\{ {6 \atop 4} \right\} \sim \left\{ {1 \atop 1} \right\} \succ \left\{ {3 \atop 2} \right\} \sim A$.

Based on the properties of the algorithm established in the above, we now proceed to analyze groups and sequences of non-group clusters, based on the state classification N, A, and S.

3.2   Groups

For groups, the analysis can be completely captured in a collection of figures. We now describe how such figures should be read. Confer with Fig. 9 below as an example. Above the dotted line, we show the figure that we are analyzing. We do not know the placement of OPT's clusters, but know that we can assume a division of starting states according to the classification described above. Thus, each possible OPT starting state is analyzed separately, using one line on each. When this is not obvious from the distances given in the online configuration, a reference to a lemma on top of an OPT cluster explaining why that cluster cannot reach the next point, or why a given end state can be argued.

All the way to the right on an OPT line, we list a pair denoting how many new online and OPT clusters, respectively, were used on that line. As an example, in Fig. 9, in the second OPT line (start state A), the first OPT cluster in the line has already been counted because we analyze from left to right and the cluster shares a point with whatever online configuration is placed to the left of the current. Thus, three new online and two new OPT clusters are used. In all cases, we must verify (and have done so) that the start state plus the pair of clusters used sum up to at least the end state, e.g., $A \oplus \left\{ {3 \atop 2} \right\} \succeq A$.

We analyze the groups at the time of creation (for regular 3-groups and tight 2-groups), at the time when the fourth cluster is opened (for regular 4-groups), and at the time when the third cluster is opened (for tight 3-groups). In addition, we analyze the situation where a regular group is close to another group separately. Hence, in the analysis of regular groups and tight groups, we can assume they are not close to another group.

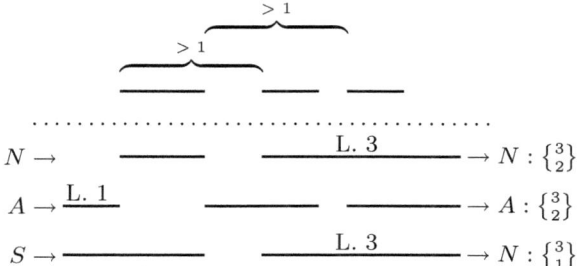

Fig. 9. Regular 3-Group Case 1: Long cluster to the left

In this extended abstract, we have included only this one example from the exhaustive case analysis of the different ¡group constructions. Analysis of non-group sections has been omitted completely. In the full paper [3], we give all the algorithmic details and the full proof of the main theorem:

Theorem 1. *There exists a 5/3-competitive deterministic algorithm for one-dimensional online unit clustering.*

4   Two-Dimensional Lower Bound

In this section, we establish the 2-dimensional lower bound which all higher dimensions inherit.

Theorem 2. *No deterministic on-line algorithm can have a competitive ratio less than $\frac{13}{6}$ in two dimensions.*

Proof. Observe that it is enough to consider on-line algorithms that never produce clusterings where a cluster is contained in another cluster. Consider any such deterministic on-line algorithm A, and assume by contradiction that it has a competitive ratio less than $\frac{13}{6}$.

First, four points arrive on the corners of a unit square. The algorithm A can either assign them all to one cluster or assign them to two clusters. Otherwise, since there exists a feasible solution using a single cluster, it has ratio of at least three and the input stops.

If A assigns all points to a single cluster, then four additional points arrive (see the first eight points of Fig. 10). The algorithm A must group at least one of the new pairs in one cluster, since it is possible to group all existing points with only two clusters. Otherwise, A would not stay below the $\frac{13}{6}$-bound.

If A groups the four points (points 5–8) in exactly two clusters, then eight additional points arrive (see the first 16 points of Fig. 10). Since it is possible to serve all these points with only four clusters, A must use at most five clusters to group the new points.

If A groups all eight points using four clusters, then six additional points arrive (see Fig. 10). The algorithm A must open six clusters for them and has

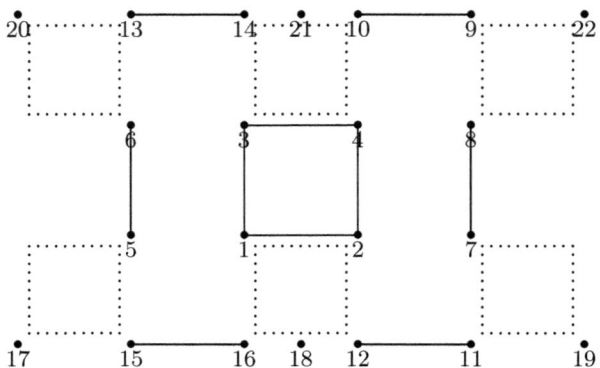

Fig. 10. Lower bound

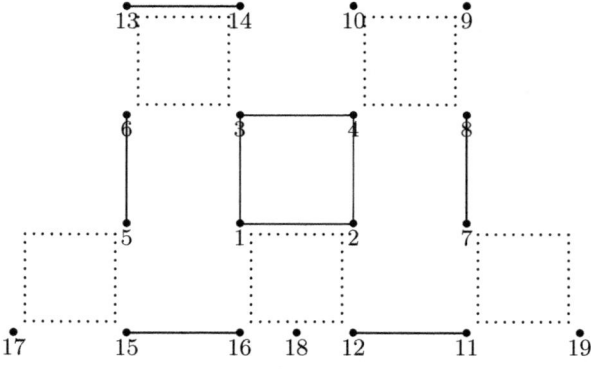

Fig. 11. Lower bound

now used 13 clusters while an optimal solution requires only six clusters, which is a contradiction.

If A groups the eight points using five clusters, then three additional points arrive (see Fig. 11). The algorithm A must open three clusters for them and has now used 11 clusters while an optimal solution requires only five clusters, which is a contradiction.

If A groups the four points (points 5–8) using three clusters, then the following sequences of points arrive (see Fig. 12).

Now give one additional point (point 9). Since the nine points can be grouped using two clusters A must assign the new point to an already open cluster.

Next, four points arrive (points 10–13). Since the 13 points can be grouped using three clusters A must group the new points into two clusters.

Now, six additional points arrive (points 14–19). Since the 18 points can be grouped using five clusters A must group the new points using four clusters.

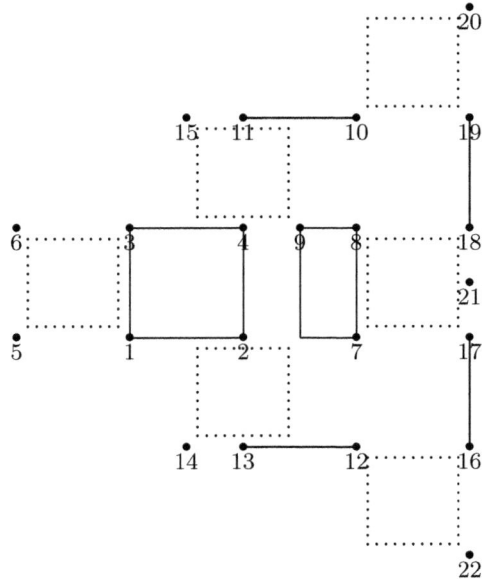

Fig. 12. Lower bound

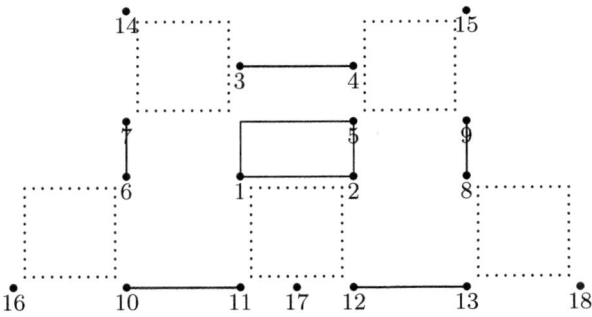

Fig. 13. Lower bound

Finally, three points arrive (points 20–22). The algorithm A must open three new clusters for them and has now used 13 clusters while an optimal solution only requires six clusters, which is a contradiction.

If A groups the first four points given using two clusters, then the following sequence of points arrive (see Fig. 13).

Now give one additional point (point 5). Since the five points can be grouped using one cluster, A must assign the new point to an already open cluster.

Next, four points arrive (points 6–9). Since, the nine points can be grouped using two clusters A must group the new points using two clusters.

Now, six additional points arrive (points 10–15). Since the 15 points can be grouped using four clusters, A must group the new points using four clusters.

Finally, three points arrive (points 16–18). The algorithm A must open three new clusters for them and has now used 11 clusters while an optimal solution only requires five clusters, which is a contradiction.

In all cases, we reach a contradiction, so 's competitive ratio is at least $\frac{13}{6}$.

References

1. Chan, T.M., Zarrabi-Zadeh, H.: A randomized algorithm for online unit clustering. Theory of Computing Systems 45(3), 486–496 (2009)
2. Charikar, M., Chekuri, C., Feder, T., Motwani, R.: Incremental clustering and dynamic information retrieval. SIAM Journal on Computing 33(6), 1417–1440 (2004)
3. Ehmsen, M.R., Larsen, K.S.: Better Bounds on Online Unit Clustering. Preprint 8, Department of Mathematics and Computer Science, University of Southern Denmark (2009)
4. Epstein, L., Levin, A., van Stee, R.: Online unit clustering: Variations on a theme. Theoretical Computer Science 407, 85–96 (2008)
5. Epstein, L., van Stee, R.: On the online unit clustering problem. In: Proceedings of the 5th International Workshop on Approximation and Online Algorithms, pp. 193–206 (2007)
6. Graham, R.L.: Bounds for certain multiprocessing anomalies. Bell Systems Technical Journal 45, 1563–1581 (1966)
7. Karlin, A.R., Manasse, M.S., Rudolph, L., Sleator, D.D.: Competitive snoopy caching. Algorithmica 3, 79–119 (1988)
8. Sleator, D.D., Tarjan, R.E.: Amortized efficiency of list update and paging rules. Communications of the ACM 28(2), 202–208 (1985)
9. Zarrabi-Zadeh, H., Chan, T.M.: An improved algorithm for online unit clustering. Algorithmica 54(4), 490–500 (2009)

Online Selection of Intervals and t-Intervals⋆

Unnar Th. Bachmann[1], Magnús M. Halldórsson[1], and Hadas Shachnai[2]

[1] School of Computer Science, Reykjavik University, 101 Reykjavik, Iceland
{mmh,unnar07}@ru.is
[2] Department of Computer Science, The Technion, Haifa 32000, Israel
hadas@cs.technion.ac.il

Abstract. A t-interval is a union of at most t half-open intervals on the real line. An interval is the special case where $t = 1$. Requests for contiguous allocation of a linear resource can be modeled as a sequence of t-intervals. We consider the problems of online selection of intervals and t-intervals, which show up in Video-on-Demand services, high speed networks and molecular biology, among others. We derive lower bounds and (almost) matching upper bounds on the competitive ratios of randomized algorithms for selecting intervals, 2-intervals and t-intervals, for any $t > 2$. While offline t-interval selection has been studied before, the online version is considered here for the first time.

1 Introduction

Interval scheduling is a form of a resource allocation problem, in which the machines are the resource. As argued by Kolen et al. [11], operations management has undergone a "transition in the last decennia from resource oriented logistics (where the availability of resources has dictated the planning and completion of jobs) to demand oriented logistics (where the jobs and their completion are more or less fixed and the appropriate resources must be found)." They suggest that this implies a move from traditional scheduling to *interval scheduling*.

Suppose you are running a resource online. Customers call and request to use it from time to time, for up to t time periods, not necessarily of same length. These requests must either be accepted or declined. If a request is accepted then it *occupies* the resource for these periods of time. A request cannot be accepted if one or more of its periods intersect a period of a previously accepted request. The goal is to accept as many requests as possible.

This can be modeled as the following *online t-interval selection (t-Isp)* problem. Let t be the maximum number of periods involved in any request. Each request is represented by a t-interval, namely, a union of at most t half-open intervals (referred to as *segments*) on the real line. The t-intervals arrive one by one and need to be scheduled non-preemptively on a single machine. Two t-intervals, I and J, are disjoint if none of their segments intersect, and intersect if a segment of one intersects a segment of the other. Upon arrival of a t-interval,

---

⋆ Supported by Icelandic Research Fund (grant 060034022).

H. Kaplan (Ed.): SWAT 2010, LNCS 6139, pp. 383–394, 2010.

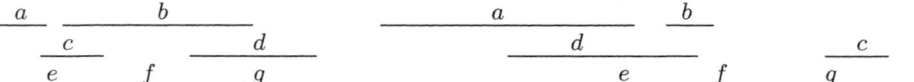

Fig. 1. A linear resource is requested by customers a, b, c, d, e, f and g in that order, for two periods each. If b is accepted then each of the following requests must be declined. An optimal selection consists of a, f and g.

the scheduler needs to decide whether it is accepted; if not, it is lost forever. The goal is to select a subset (or "form a schedule") of non-intersecting t-intervals of maximum cardinality. The special case where $t = 1$ is known as the *online interval selection problem (*Isp*)*. An example of an instance of online t-Isp is given in Figure 1.

The performance of an online algorithm is measured in terms of its competitive ratio. Formally, let OPT be an optimal offline algorithm for the problem. The competitive ratio of A is defined as $\sup_\sigma \frac{OPT(\sigma)}{A(\sigma)}$, where σ is an input sequence, and $OPT(\sigma), A(\sigma)$ are the number of t-intervals selected by OPT and A, respectively. For randomized algorithms, we replace $A(\sigma)$ with the expectation $\mathbb{E}[A(\sigma)]$ and define the competitive ratio as $\rho_A = \sup_\sigma \frac{OPT(\sigma)}{\mathbb{E}[A(\sigma)]}$. An algorithm with competitive ratio of at most ρ is called ρ-competitive. Let n be the number of intervals in the instance; also, denote by Δ the ratio between the longest and shortest segment lengths.

1.1   Related Work

Selecting intervals and t-intervals: We can view t-Isp as the problem of finding a *maximum independent set (IS)* in a t-interval graph. While for the special case of interval graphs the problem is known to be polynomially solvable (see, e.g., [10]), already for $t = 2$ the IS problem becomes APX-hard [4]. The paper [4] presents a $2t$-approximation algorithm for the offline weighted tISP. Later works extended the study to the selection of t-intervals with demands, where each interval is associated with a set of segments and a demand for machine capacity [5], as well as the study of other optimization problems on t-interval graphs (see, e.g., [6]).

There is a wide literature on the maximum independent set problem in various classes of graphs. The online version of the IS problem was studied in [7], where a $\Omega(n)$-lower bound on the competitive ratios of randomized algorithms was given, even for interval graphs (but not when the interval representation is given). A survey of other works is given in [2]. Numerous natural applications of our problems are described in [2] and [3].

Online interval selection: Lipton and Tomkins [12] considered an online interval selection problem where the intervals have weights proportional to their length and the intervals arrive by time (i.e., in order of their left endpoints). They showed that $\theta(\log \Delta)$-competitive factor was optimal, when Δ is known, and introduced a technique that gives a $O(\log^{1+\epsilon} \Delta)$-competitive factor when Δ is unknown. Woeginger [14] considered a preemptive version of weighted IsP and gave an optimal 4-competitive algorithm. Numerous results are known about interval scheduling under the objective of minimizing the number of machines, or alternatively, online coloring interval graphs. In particular, a 3-competitive algorithm was given by Kierstead and Trotter [9]. The t-IsP problem bears a resemblance to the JISP problem [13], where each job consists of several intervals and the task is to complete as many jobs as possible. The difference is that in JISP, it suffices to select only one of the possible segments of the job.

Call admission: Similar problems have been studied also in the area of call admission. We note that IsP can be viewed as call admission on a line, where the objective is to maximize the number of accepted calls. The paper [1] presents a strongly $\lceil \log N \rceil$-competitive algorithm for the problem, where N is the number of nodes on the line. This yields an $O(\log \Delta)$-competitive algorithm for general IsP instances when Δ is known a-priori. We give an algorithm that achieves (almost) the same ratio for the case where Δ is unknown.

1.2   Our Results

We derive the first lower and upper bounds on the competitive ratios of online algorithms for t-IsP and new or improved bounds for IsP. Table 1 summarizes the results for various classes of instances of IsP, 2-IsP and t-IsP. All of the results apply to randomized algorithms against oblivious adversary. In comparison, proving strong lower bounds for deterministic algorithms (including a lower bound of $\Delta + 1$ for IsP) is straightforward. The upper bounds for general inputs are for the case where Δ is unknown in advance.

Table 1. Results for randomized online interval and t-interval selection. Entries marked with · follow by inference. Entries marked with † were known; the lower bound for IsP follows from [1], while the upper bounds for unit lengths are trivial.

	IsP		2-IsP		t-IsP	
	$u.b.$	$l.b.$	$u.b.$	$l.b.$	$u.b.$	$l.b.$
General inputs	$O(\log^{1+\varepsilon} \Delta)$	$\Omega(\log \Delta)$†	$O(\log^{2+\varepsilon} \Delta)$	$\Omega(\log \Delta)$†	–	·
Two lengths	4	4	16	6	–	·
Unit length	2†	2	4†	3	$2t$†	$\Omega(t)$
Bounded depth s	$3/2$ $(s=2)$	$2-1/s$	–	–	–	–

Due to space constraints, some of the results (or proofs) are omitted. The detailed results appear in [3] (see also [2]).

2   Technique: Stacking Construction

When deriving lower bounds for randomized algorithms for (t-)interval selection, we use the following technique. The adversary can take advantage of the fact that he can foresee the probability with which the algorithm selects any given action, even if he doesn't know the outcome. He presents intervals on top of each others, or "stacks" them, until some interval is chosen with sufficiently low probability. The adversary uses that to force a desirably poor outcome for the algorithm. The general idea is similar to a lower bounding technique of Awerbuch et al. [1] for call control.

Let R be an IsP-algorithm and let parameters q and x be given. A (q,x)-*stacking construction* for R is a collection of intervals formed as follows, where q is an upper bound on the number of intervals stacked and x is the extent to which intervals can be shifted. Form q unit intervals I_1, ..., I_q that mutually overlap with left endpoints spaced x/q apart towards the left. Namely, $I_i = [x(1 - i/q), 1 + x(1 - i/q))$, for $i = 1, \ldots, q$. Let p_i be the (unconditional) probability that R selects I_i. The adversary knows the values p_i and forms its construction accordingly. Namely, let m be the smallest value such that $p_m \leq 1/q$. Since $\sum_i p_i \leq 1$, there must be at least one such value. The input sequence construction consists of $\langle I_1, I_2, \ldots, I_m, J_m \rangle$, where $J_m = [1 + x(1 - m/q), 2 + x(1 - m/q))$. This is illustrated in Fig. 2.

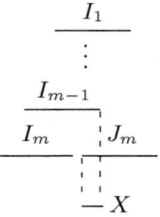

Fig. 2. (q, x)-stacking construction

Lemma 1. *A (q, x)-stacking construction $\mathcal{I}$ has the following properties.*

1. *All intervals in $\mathcal{I} \setminus \{I_m\}$ overlap the segment $[1, 1 + x)$.*
2. *All intervals in $\mathcal{I}$ are contained within the interval $[0, 2 + x)$.*
3. *The intervals in $\mathcal{I} \setminus \{I_m\}$ have a common intersection of length x/q, given by the segment $X = I_{m-1} \cap J_m = [1 + x(1 - m/q), 1 + x(1 - (m-1)/q))$.*
4. *$\mathbb{E}_R[I_m] = p_m \leq 1/q$. Thus, $\mathbb{E}_R[\mathcal{I}] \leq 1 + 1/q$,*
5. *$OPT(\mathcal{I}) = 2$. Thus, the performance ratio of R is at least $2/(1 + 1/q)$.*

By taking q arbitrarily large, we obtain the following performance bound.

Theorem 1. *Any randomized online algorithm for IsP with unit intervals has competitive ratio at least 2.*

We can imitate the stacking construction with 2-intervals by repeating the construction for both segments. We refer to this as a *2-interval (q, x)-stacking construction*.

We shall also use the stacking construction *shifted* by a displacement f, by adding f to the starting point of each interval. We may also use intervals of non-unit length.

3 Online Interval Selection

3.1 Unit Intervals and Depth

We give upper and lower bounds on the competitive ratios for ISP with unit intervals. We parameterize the problem in terms of the *depth* of the interval system, which is the maximum number of intervals that overlap a common point. This corresponds to the clique number of the corresponding interval graph.

Theorem 2. *The competitive ratio of any randomized algorithm for ISP of unit intervals is at least $2 - 1/s$, where s is the depth of the instance.*

Proof. We modify the $(s, 1)$-stacking construction slightly. Let p_i be the probability that the given algorithm R selects interval I_i. If $p_1 \leq 1/(2 - 1/s) = s/(2s - 1)$, then we conclude with the unit sequence $\langle I_1 \rangle$. The performance ratio is then at least $1/p_1 \geq 2 - 1/s$. Otherwise we stop the sequence at I_m, where m is the smallest number such that $p_m \leq 1/(2s - 1)$. This is well defined since $s/(2s - 1) + \sum_{i=2}^{s} 1/(2s - 1) = 1$. As before, this is followed by the interval J_m intersecting only the first $m - 1$ intervals. The algorithm obtains expected value at most $1 + p_m \leq 1 + 1/(2s - 1) = 2s/(2s - 1)$, versus 2 for the optimal solution. The above procedure can be repeated arbitrarily often, ensuring that the lower bound holds also in the asymptotic case.

We now describe a randomized algorithm that achieves the above ratio for $s = 2$. Consider the algorithm RoG (Random_or_Greedy), which handles an arriving interval as follows. If the interval does not overlap any previously presented interval, select it with probability $2/3$, else select the interval greedily.

Theorem 3. *Algorithm RoG is $3/2$-competitive for unit intervals with depth 2.*

Proof. Assume that the instance is connected; otherwise, we can argue the bound for each component separately.

The depth restriction means that each interval can intersect at most two other intervals: one from the left and one from the right. The instance is therefore a chain of unit intervals. We divide the intervals into three types, based on the number of previous intervals the given interval intersects. A type-i interval, for $i = 0, 1, 2$, intersects i previously presented intervals. Two type-2 intervals cannot intersect, as otherwise the one that appears earlier will have degree 3, leading to depth at least 3. Therefore, the instance consists of chains of type-0 and type-1 intervals attached together by type-2 intervals. Each chain is started by a type-0 interval, followed by type-1 intervals. Let n_i denote the number of intervals of type i, then we have that

$$n_0 \geq n_2 + 1. \tag{1}$$

Consider now the unconditional probability that intervals of each type are selected, i.e. the probability independent of other selections. The probability of type-0 intervals being selected is $2/3$. The probability of the selection of type-1 intervals alternates between $1/3$ and $2/3$. The expected number of intervals selected by the algorithm is then, using (1), bounded below by

$$\frac{2}{3}n_0 + \frac{1}{3}n_1 \geq \frac{1}{3}(n_0 + n_1 + n_2 + 1) = \frac{n+1}{3}.$$

On the other hand, the number of intervals in an optimal solution is the independence number of the path on n vertices, or $\lceil \frac{n}{2} \rceil \leq \frac{n+1}{2}$. Hence, the competitive ratio is at most $3/2$.

3.2 ISP with Intervals of Two Lengths

Consider now ISP instances where the intervals can be of two different lengths, 1 and d. It is easy to argue a 4-competitive algorithm by the classic Classify-and-Select approach: Flip a coin, choosing either the unit intervals or the length-d intervals, and then greedily adding intervals of that length only.

We find that it is not possible to significantly improve on that very simple approach (we omit the proof).

Theorem 4. *Any randomized online algorithm for* ISP *with intervals of two lengths 1 and d has performance ratio at least 4, asymptotically with d.*

3.3 ISP with Parameter n

ISP is easily seen to be difficult for deterministic algorithm on instances without constraints on the size of the intervals. The adversary keeps introducing disjoint intervals until the algorithm selects one of them, I; the remaining intervals presented will then be contained in I. This leaves the algorithm with a single interval, while the optimal solution contains the rest, for a ratio of $n-1$. It is less obvious that a linear lower bound holds also for randomized algorithms against oblivious adversary.

Theorem 5. *Any randomized online algorithm for* ISP *has competitive ratio $\Omega(n)$.*

Proof. Let $n > 1$ be an integer. Let $r_1, r_2, \ldots, r_{n-1}$ be a sequence of uniformly random bits. Let the sequence $x_1, x_2, \ldots, x_n$ of points be defined inductively by $x_1 = 0$ and $x_{i+1} = x_i + r_i \cdot 2^{n-i}$. We construct a sequence $\mathcal{I}_n$ of n intervals $I_1, \ldots, I_n$, where $I_i = [x_i, x_i + 2^{n-i})$, for $i = 1, \ldots n$.

The collection $A = \{I_i : r_i = 1\} \cup \{I_n\}$ forms an independent set, informally referred to as the "good" intervals. The set $B = \mathcal{I}_n \setminus A = \{I_i : r_i = 0\}$ forms a clique; informally, these are the "bad" intervals.

Consider a randomized algorithm R and the sequence of intervals chosen by R. The event that a chosen interval is good is a Bernoulli trial, and these events are independent. Thus, the number of intervals chosen until a bad one is chosen is a geometric random variable with a mean of 2. Even accounting for the last interval, which is known to be good, the expected number of accepted intervals $\mathbb{E}[(\sigma)]$ is at most 3.

On the other hand, the expected number of good intervals is $(n-1)/2 + 1$, and so the expected size of the optimal solution is $n/2$. By standard arguments, this holds also with high probability, up to lower order terms. The competitive ratio of R on $\mathcal{I}_n$ is therefore at least $n/6$.

Notice that in Theorem 5, the intervals are presented in order of increasing left endpoints. Thus, the bound holds also for the scheduling-by-time model. The adversary in Theorem 5 has also the property of being *transparent* [8] in the sense that as soon as the algorithm has made its decision on an interval, the adversary reveals his own choice.

4   Online 2-Interval Selection

4.1   Unit Segments

Theorem 6. *Any randomized online algorithm for* 2-ISP *of unit intervals has competitive ratio at least* 3.

Proof. Consider any randomized online 2-ISP algorithm R. Let q be an even number and let $q' = 3q/2$.

We start with 2-interval $(q', 1)$-stacking construction $\mathcal{I}$ for R. See the top half of Fig. 3. Recall that the expected gain of R on interval I_m is $\mathbb{E}_R[I_m] \leq 1/q'$. Let p be the probability that R selects some interval in $\mathcal{I}' = \mathcal{I} \setminus \{I_m\} = \{I_1, \ldots, I_{m-1}, I_{m+1}\}$. If $p < 2/3$, then we stop the construction. The expected solution size found by R is then $\mathbb{E}_R[\mathcal{I}] \leq p + 1/q'$, while the optimal solution is of size 2, for a ratio of $2/(p + 1/q') \geq 2/(2/3 + 2/(3q)) = 3/(1 + 1/q)$.

Assume therefore that $p \geq 2/3$. Let X_1 be the common intersection of the first segments of the 2-intervals in $\mathcal{I}'$, and X_2 be the common intersection of the second segments. Let f_i denote the starting point of X_i, $i = 1, 2$. By Lemma 1, the length of each X_i is $1/q'$.

We now form a (q, x)-stacking construction $\mathcal{I}_1$ of 2-intervals for R shifted by f_1, where $x = |X_1| = 1/q'$. The first segments are positioned to overlap X_1, where $x = |X_1| = 1/q'$; the second segments are immaterial as long as they do not intersect any previous intervals. This is shown in the bottom left of Fig. 3. We then do an identical construction $\mathcal{I}_2$ shifted by f_2; again, the second segments do not factor in. This completes the construction.

We can make the following observations about the combined construction $\mathcal{J} = \mathcal{I} \cup \mathcal{I}_1 \cup \mathcal{I}_2$.

Observation 1

1. *All intervals in $\mathcal{I}_1$ overlap X_1.*
2. *All intervals in $\mathcal{I}_2$ overlap X_2.*
3. *$OPT(\mathcal{J}) = 4$, given by $I_m^2, J_m^2, I_m^3, I_m^3$.*
4. *$\mathbb{E}_R[\mathcal{I}_1] \leq (1-p)(1+1/q)$, by Lemma 1 (3) and part 1. of this observation.*

It follows that

$$\mathbb{E}_R[\mathcal{J}] = \mathbb{E}_R[I_m] + \mathbb{E}_R[\mathcal{I}'] + \mathbb{E}_R[\mathcal{I}_1] + \mathbb{E}_R[\mathcal{I}_2] \leq 1/q' + p + 2(1-p)(1+1/q) = 2 - p + (4/3 - 2p)q \ .$$

Since $p \geq 2/3$, $\mathbb{E}_R[\mathcal{J}] \leq 2-p$, and the performance ratio of R on $\mathcal{J}$ is $OPT(\mathcal{J})/\mathbb{E}_R$ $[\mathcal{J}] \geq 4/(4/3) = 3$.

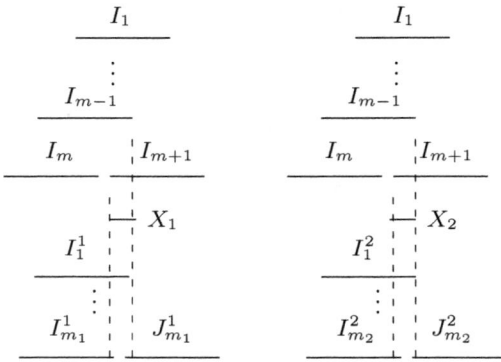

Fig. 3. Construction of a lower bound of 3 for unit 2-Isp

4.2   Segments of Two Lengths

In this section, we give a 16-competitive algorithm for 2-Isp where the 2-interval segments have lengths either 1 or d. A lower bound of 6 for $d \gg 1$ is omitted in this version.

Consider the following algorithm A_v, which either schedules (i.e., selects) a given 2-interval, rejects it, or schedules it *virtually*.[1] A virtually scheduled interval does not occupy the resource but blocks other 2-intervals from being scheduled. The length of each segment is either *short* (1) or *long* (d). A 2-interval is *short-short* (*long-long*) if both segments are short (long), respectively, and *short-long* if one is short and the other long. In processing a 2-interval I, A_v applies the following rules, which depend on the availability of the resource.

1. *I **is short-short.*** Schedule I greedily (with probability 1).
2. **A long segment of I intersects a virtually selected 2-interval.** Do nothing.
3. **Otherwise,** schedule I with probability 1/2 and schedule it virtually with probability 1/2.

---

[1] The term was used before, e.g., in [12].

Our analysis of A_v uses the following charging scheme. Let S_{OPT} be an optimal solution and S_{A_v} the set of 2-intervals selected by A_v. For any $I \in S_{A_v}$ and $J \in S_{\mathsf{OPT}}$, we assign $w(I,J) = 1/4 \cdot t(I,J)$, where $t(I,J)$ is the number of endpoints of I that intersect with J. In particular, $w(I,J) = 0$ when I and J do not overlap or if segments of I properly contain segments of J. Also, $w(J,J) = 1$. Since each 2-interval has 4 endpoints, and the 2-intervals in S_{OPT} are disjoint, it follows that $\sum_{J \in S_{\mathsf{OPT}}} w(I,J) \leq 1$. Intuitively, we distribute the value that A_v receives for selecting I among the 2-intervals in S_{OPT} that intersect it. Let $w(bucket(J)) = \sum_{I \in S_{A_v}} w(I,J)$, for $J \in S_{\mathsf{OPT}}$. To show that A_v is c-competitive it suffices to prove that, for any $J \in S_{\mathsf{OPT}}$, $\mathbb{E}[w(bucket(J))] \geq 1/c$.

Theorem 7. *Algorithm A_v is 16-competitive for online 2-Isp with segments of length 1 and d.*

Proof. Consider an interval $J \in S_{\mathsf{OPT}}$. We shall show that $\mathbb{E}[w(bucket(J)) \geq 1/16$, which yields the theorem. The argument is based on considering the various possible configurations of intervals overlapping J that were presented before J. In what follows, we shall say that an interval was *addressed* if it precedes J, overlaps J, and was either scheduled or virtually scheduled, i.e. was not blocked when presented. We say that I *dominates* J if a short segment of J is properly contained in a long segment of I.

We consider cases depending on the lengths of J's segments.

J **is long-long.** Consider the first interval addressed, I (which is possibly J itself). Then $w(I,J) \geq 1/4$ and I is scheduled with probability at least $1/2$. Hence, $\mathbb{E}[w(bucket(J))] \geq 1/2 \cdot 1/4 = 1/8$.

J **is short-long.** Then, there is at most one interval in S_{A_v} that dominates J. With probability at least $1/2$, this interval is not selected (so, either virtually selected or blocked). Some other interval I overlapping J (possibly J itself) is then selected with probability at least $1/2$, assigning a weight $w(I,J) \geq 1/4$. Thus, $\mathbb{E}[w(bucket(J))] \geq 1/2 \cdot 1/2 \cdot 1/4 = 1/16$.

J **is short-short.** At most two intervals are addressed that dominate J (if they dominate the same segment of J, then the latter interval is blocked by the former). With probability at least $1/4$, neither of them are selected. With probability 1, some other interval I intersecting J (possibly J itself) is selected, since J is short-short. A weight of at least $w(I,J) \geq 1/4$ is transferred. Hence, $\mathbb{E}[w(bucket(J))] \geq 1/4 \cdot 1 \cdot 1/4 = 1/16$.

4.3   Segments of Arbitrary Lengths

Consider now more general instances of 2-Isp, in which the ratio between the longest and shortest segment is Δ, for some $\Delta > 1$. W.l.o.g. we may assume that the short segment is of length 1. We partition the set of first segments to $K = \lceil \log \Delta \rceil$ groups, such that the segments in group i have lengths in $[2^{i-1}, 2^i)$, $1 \leq i \leq K$. Partition the second segments similarly into K groups. A 2-interval whose first segment is of length in $[2^{i-1}, 2^i)$, and whose second segment is of length $[2^{j-1}, 2^j)$, $1 \leq i,j \leq K$, is in group (i,j).

We now apply algorithm A_v to 2-Isp instances where the length of the short segment is in $[1, 2)$ and the long segment in $[d, 2d)$. A_v makes scheduling decisions as before, using the new definitions of 'short' and 'long' segments.

Theorem 8. *Algorithm A_v is 24-competitive for 2-Isp instances with segments of two types: short with lengths in $[1, 2)$, and long with lengths in $[d, 2d)$.*

Proof. Each interval I intersects now at most 6 intervals in S_{OPT} that it does not dominate. For instance, a long segment can now contain one long segment from S_{OPT} and properly overlap two other segments. Thus, we change the charging scheme to $w(I, J) = 1/6 \cdot t(I, J)$. The rest of the proof of Theorem 7 is unchanged.

Now, given a general instance of 2-Isp, suppose that Δ is known a-priori. Consider algorithm A_{vg} which applies A_v on groups of 2-intervals. The instance is partitioned to $K^2 = \lceil \log \Delta \rceil^2$ groups, depending on the lengths of the first and second segments of each 2-interval. A_{vg} selects uniformly at random a group (i, j), $1 \leq i, j, \leq K$ and considers scheduling only 2-intervals in this group. All other 2-intervals are declined. The next result follows from Theorem 8.

Theorem 9. *A_{vg} is $O(\log^2 \Delta)$-competitive for 2-Isp with intervals of various lengths, where Δ is known in advance.*

For the case where Δ is *a priori unknown*, consider algorithm $\tilde{A}_{vg}$, which proceeds as follows.[2] A presented 2-interval, I, is in the same group as a previously presented 2-interval, I', if the ratio between the length of the first/second segment of I and I' is between 1 and 2. If not, I belongs to a new group. Thus, each group has an index $i \in \{1, \ldots, \lceil \log \Delta \rceil^2\}$. The algorithm chooses randomly at most one group and selects only 2-intervals from that group, using algorithm A_v. Define

$$c_i = \frac{1}{\zeta(1 + \epsilon/2)i^{1+\epsilon/2}}, \quad \text{and} \quad p_i = \frac{c_i}{\Pi_{j=1}^{i-1}(1 - p_j)}, \tag{2}$$

where $\zeta(x) = \sum_{i=1}^{\infty} i^{-x}$ is the Riemann zeta function. Recall that $\zeta(x) < \infty$, for $x > 1$.

If a given 2-interval belongs to a new group i, and none of the groups $1, 2, \ldots, i-1$ has been selected, then group i is chosen with probability p_i and rejected with probability $1 - p_i$. If a given 2-interval belongs to an already selected group i, it is scheduled using algorithm A_v; if the given 2-interval belongs to an already rejected group then it is rejected. Note that by the definition of p_i, as given in (2), it follows that c_i is the unconditional probability that $\tilde{A}_{vg}$ chooses the i-th group.

In analyzing $\tilde{A}_{vg}$ we first show that the values p_i form valid probabilities, and that the c_i values give a probability distribution.

Lemma 2. $\sum_{i=1}^{\infty} c_i = 1$. *Also, $p_i \leq 1$, for all $i \geq 1$.*

---

[2] W.l.o.g., we assume that Δ is an integral power of 2.

Proof. Observe that $\sum_{i=1}^{\infty} c_i = \frac{1}{\zeta(1+\epsilon/2)} \sum_{i=1}^{\infty} \frac{1}{i^{1+\epsilon/2}} = 1$, proving the first half of the lemma. It follows that $c_i \leq 1 - \sum_{j=1}^{i-1} c_j$. To prove the second half of the lemma, it suffices then to prove the following claim that

$$p_i = \frac{c_i}{1 - \sum_{j=1}^{i-1} c_j}, \tag{3}$$

for each $i \geq 1$. We prove the claim by induction on i. The base case then holds since $p_1 = c_1$. Suppose now that

$$p_{k-1} = \frac{c_{k-1}}{1 - c_1 - c_2 - \dots - c_{k-2}} \tag{4}$$

then using (2) we have that

$$p_k = \frac{c_k}{c_{k-1}} \cdot \frac{p_{k-1}}{1 - p_{k-1}} .$$

Plugging in the value of p_{k-1} in (4) we get the claim.

Theorem 10. $\tilde{A}_{vg}$ *is* $O(\log^{2+\epsilon} \Delta)$*-competitive for* 2-ISP *with intervals of various lengths, where* Δ *is unknown in advance.*

Proof. Let S_i denote the set of 2-intervals in group i, $1 \leq i \leq \log^2 \Delta$.

The probability that $\tilde{A}_{vg}$ chooses any given group S_i is at least $c_{\log^2 \Delta}$. After selecting the group, $\tilde{A}_{vg}$ uses A_v to schedule the 2-intervals in the group. For a given group, S_i, we have:

$$\mathbb{E}[\tilde{A}_{vg}(S_i)] \geq c_{\log^2 \Delta} \cdot \mathbb{E}[A_v(S_i)] \geq \frac{1}{\zeta(1 + \epsilon/2)(\log \Delta)^{2+\epsilon}} \cdot \frac{1}{12} \cdot \mathbb{E}[OPT(S_i)].$$

Thus, by linearity of expectation $\tilde{A}_{vg}$ is $O(\log^{2+\epsilon} \Delta)$-competitive.

5   Online t-Interval Selection

We show here that any online algorithm for t-ISP has competitive ratio $\Omega(t)$. This is done by a *reduction* to a known problem; this is standard for offline problems but rather unusual approach in the online case. We reduce the problem to the online version of the independent set (IS) problem in graphs: given vertices one by one, along with edges to previous vertices, determine for each vertex whether to add it to a set of independent vertices.

Theorem 11. *Any randomized online algorithm for* t-ISP *with unit segments has competitive ratio* $\Omega(t)$.

Proof. Let n be a positive integer. We show that any graph on n vertices, presented vertex by vertex, can be converted on-the-fly to an n-interval representation. Then, an $f(t)$-competitive online algorithm for t-ISP applied to the

n-interval representation yields an $f(n)$-competitive algorithm for the independent set problem. As shown in [7] (and follows also from Theorem 5), there is no cn-competitive algorithm for the online IS problem, for some fixed $c > 0$. The theorem then follows.

Let $G = (V, E)$ be a graph on n vertices with vertex sequence $\langle v_1, v_2, \ldots, v_n \rangle$. Given vertex v_k and the induced subgraph $G[\langle v_1, v_2, \ldots, v_i \rangle]$, form the n-interval I_i by

$$I_i = \bigcup_{j=1} X_{ij}, \quad \text{where} \quad X_{ij} = \begin{cases} [nj + i, nj + i + 1) \text{ if } j < i \text{ and } (i, j) \in E \\ [ni + j, ni + j + 1) \text{ otherwise.} \end{cases}$$

Observe that $I_i \cap I_j \neq \emptyset$ iff $(i, j) \in E$. Hence, solutions to the t-Isp instance are in one-one correspondence with independent sets in G.

A greedy selection of t-intervals yields a $2t$-competitive algorithm for unit t-Isp, implying that the bound above is tight.

References

1. Awerbuch, B., Bartal, Y., Fiat, A., Rosén, A.: Competitive non-preemptive call control. In: SODA, pp. 312–320 (1994)
2. Bachmann, U.T.: Online t-Interval Scheduling. MSc thesis, School of CS, Reykjavik Univ. (December 2009)
3. Bachmann, U.T., Halldórsson, M.M., Shachnai, H.: Online selection of intervals and t-intervals (2010), http://www.cs.technion.ac.il/~hadas/PUB/onint_full.pdf
4. Bar-Yehuda, R., Halldórsson, M.M., Naor, J.S., Shachnai, H., Shapira, I.: Scheduling split intervals. In: SODA, pp. 732–741 (2002)
5. Bar-Yehuda, R., Rawitz, D.: Using fractional primal-dual to schedule split intervals with demands. Discrete Optimization 3(4), 275–287 (2006)
6. Butman, A., Hermelin, D., Lewenstein, M., Rawitz, D.: Optimization problems in multiple-interval graphs. In: SODA (2007)
7. Halldórsson, M.M., Iwama, K., Miyazaki, S., Taketomi, S.: Online independent sets. Theoretical Computer Science 289(2), 953–962 (2002)
8. Halldórsson, M.M., Szegedy, M.: Lower bounds for on-line graph coloring. Theoretical Comput. Sci. 130, 163–174 (1994)
9. Kierstead, H.A., Trotter, W.T.: An extremal problem in recursive combinatorics. In: Congr. Numer., vol. 33, pp. 143–153 (1981)
10. Kleinberg, J., Tardos, E.: Algorithm Design. Addison-Wesley, Reading (2005)
11. Kolen, A.W., Lenstra, J.K., Papadimitriou, C.H., Spieksma, F.C.: Interval scheduling: A survey. Naval Research Logistics 54, 530–543 (2007)
12. Lipton, R.J., Tomkins, A.: Online interval scheduling. In: SODA, pp. 302–311 (1994)
13. Spieksma, F.: On the approximability of an interval scheduling problem. J. Sched. 2, 215–227 (1999)
14. Woeginger, G.J.: On-line scheduling of jobs with fixed start and end times. Theor. Comput. Sci. 130(1), 5–16 (1994)

Approximating the Maximum 3- and 4-Edge-Colorable Subgraph

(Extended Abstract)

Marcin Kamiński[1] and Łukasz Kowalik[2,*]

[1] Département d'Informatique, Université Libre de Bruxelles
[2] Institute of Informatics, University of Warsaw

Abstract. We study large k-edge-colorable subgraphs of simple graphs and multigraphs. We show that:

- every simple subcubic graph G has a 3-edge-colorable subgraph (3-ECS) with at least $\frac{13}{15}|E(G)|$ edges, unless G is isomorphic to K_4 with one edge subdivided,
- every subcubic multigraph G has a 3-ECS with at least $\frac{7}{9}|E(G)|$ edges, unless G is isomorphic to K_3 with one edge doubled,
- every simple graph G of maximum degree 4 has a 4-ECS with at least $\frac{5}{6}|E(G)|$ edges, unless G is isomorphic to K_5.

We use these combinatorial results to design new approximation algorithms for the Maximum k-Edge-Colorable Subgraph problem. In particular, for $k = 3$ we obtain a $\frac{13}{15}$-approximation for simple graphs and a $\frac{7}{9}$-approximation for multigraphs; and for $k = 4$, we obtain a $\frac{9}{11}$-approximation. We achieve this by presenting a general framework of approximation algorithms that can be used for any value of k.

1 Introduction

A graph is said to be k-edge-colorable if there exists an assignment of colors from the set $\{1, \ldots, k\}$ to the edges of the graph, such that every two edges sharing a vertex receive different colors. For a graph G, let $\Delta(G)$ denote the maximum degree of G. Clearly, we need at least $\Delta(G)$ colors to color all edges of graph G. On the other hand, the celebrated Vizing's Theorem [14] states that for simple graphs $\Delta + 1$ colors always suffice. However, if $k < \Delta + 1$ it is an interesting question how many edges of G can be colored in k colors. The maximum k-edge-colorable subgraph of G (maximum k-ECS in short) is a k-edge-colorable subgraph H of G with maximum number of edges. By $\gamma_k(G)$ we denote the ratio $|E(H)|/|E(G)|$. The MAXIMUM k-EDGE-COLORABLE SUBGRAPH problem (aka Maximum Edge k-coloring [6]) is to compute a maximum k-ECS of a given graph. It is known to be APX-hard when $k \geq 2$ [4,7,9,6].

The research on approximation algorithms for max k-ECS problem was initiated by Feige, Ofek and Wieder [6]. Among other results, they suggested the

* Supported by Polish Ministry of Science and Higher Education grant N206 355636.

H. Kaplan (Ed.): SWAT 2010, LNCS 6139, pp. 395–407, 2010.

following simple strategy. Begin with a maximum k-matching F of the input graph, i.e. a subgraph of maximum degree k which has maximum number of edges. This can be done in polynomial time (see e.g. [12]). Since a k-ECS is a k-matching itself, F has at least as many edges as the maximum k-ECS. Hence, if we color $\rho|E(F)|$ edges of F we get a ρ-approximation. It follows that studying large k-edge-colorable subgraphs of graphs of maximum degree k is particularly interesting.

1.1   Large k-Edge-Colorable Subgraphs of Maximum Degree k Graphs

As observed in [6], if we have a simple G, $\Delta(G) = k$, and we find its $(k+1)$-edge-coloring by the algorithm which follows from the proof of Vizing's Theorem, we can simply choose the k largest color classes to k-color at least $\frac{k}{k+1}$ edges of G. Can we do better? In general: no, and the tight examples are the graphs K_{k+1}, for even values of k. Indeed, since each color class is a matching, it can cover at most $\frac{k}{2}$ edges, so at most $\frac{k^2}{2}$ out of $\binom{k+1}{2}$ edges can be colored, which gives $\gamma_k(G) \leq \frac{k^2}{2}/\binom{k+1}{2} = \frac{k}{k+1}$. For odd values of k the best upper bound we know is for graphs K_{k+1} with one edge subdivided and it is $\gamma_k(G) \leq \binom{k+1}{2}/\left(\binom{k+1}{2} + 1\right)$, by a similar argument as above. This raises two natural questions.

Question 1. When k is odd, can we obtain a better lower bound than $\frac{k}{k+1}$ for simple graphs?

Question 2. When k is even and $G \neq K_{k+1}$, can we obtain a better lower bound than $\frac{k}{k+1}$ for simple graphs?

Previous Work.   Question 1 has been answered in affirmative for $k = 3$ by Albertson and Haas [1], namely they showed that $\gamma_3(G) \geq \frac{26}{31}$ for simple graphs. They also showed that $\gamma_3(G) \geq \frac{13}{15}$ when G is cubic (and not subcubic) simple graph. Recently, Rizzi [10] showed that $\gamma_3(G) \geq \frac{6}{7}$ when G is a simple subcubic graph. The bound is tight by a K_4 with an arbitrary edge subdivided (we denote it by G_5). Rizzi also showed that when G is a multigraph with no cycles of length 3, then $\gamma_3(G) \geq \frac{13}{15}$, which is tight by the Petersen graph. We are not aware of any results for k bigger than 3.

Our Contribution.   In the view of the result of Rizzi it is natural to ask whether G_5 is the only subcubic simple graph G with $\gamma_3(G) = \frac{6}{7}$. We answer this question in affirmative, namely we show that $\gamma_3(G) \geq \frac{13}{15}$ when G is a simple subcubic graph different from G_5. This generalizes both the bound of Rizzi for triangle-free graphs and the bound of Albertson and Haas [1] for cubic graphs. For a subcubic multigraph, the bound $\gamma_3(G) \geq \frac{3}{4}$ (Vizing's Theorem holds for subcubic multigraphs) is tight by the K_3 with an arbitrary edge doubled (we denote it by G_3). Again, we show that G_3 is the only tight example: $\gamma_3(G) \geq \frac{7}{9}$ when G is a subcubic multigraph different from G_3. In fact, for the above two cases we show stronger statements: graph G can be 4-colored so that the largest

Table 1. Best approximation ratios for the Maximum k-Edge-Colorable Subgraph problem (μ is the maximum multiplicity of an edge)

k	2	3	4	≥ 5
simple graphs	$\frac{5}{6}$	$\frac{13}{15}$	$\frac{9}{11}$	$\frac{k}{k+1}$
source	Kosowski [8]	**this work**	**this work**	Feige et al. [6]
multigraphs	$\frac{10}{13}$	$\frac{7}{9}$	$\frac{2}{3}$	$\max\{\frac{2}{3}, \frac{k}{k+\mu(G)}, \xi(k)\}$
source	Feige et al. [6]	**this work**	Feige et al. [6]	Feige et al. [6]

three color classes contain the relevant fraction of all edges. This can be viewed as a strengthening of Vizing's Theorem. Our main technical contribution is the positive answer to Question 2 for $k = 4$. Namely, we show that $\gamma_4(G) \geq \frac{5}{6}$ when G is a simple graph of maximum degree 4 different from K_5.

1.2 Approximation Algorithms for the Max k-ECS Problem

Previous work. As observed in [6], the k-matching technique mentioned in the beginning of this section together with the bound $\gamma_k(G) \geq \frac{k}{k+1}$ of Vizing's Theorem gives a $\frac{k}{k+1}$-approximation algorithm for simple graphs and every $k \geq 2$. For multigraphs we use the same method and we get three different approximation ratios, depending on which algorithm is used for edge-coloring the k-matching. Namely, we get a $\frac{k}{k+\mu(G)}$-approximation by Vizing's Theorem, a $\frac{2}{3}$-approximation by Shannon's Theorem [13] and an algorithm with approximation ratio of $\xi(k) = k / \left\lceil k + 2 + \sqrt{k+1} + \sqrt{\frac{9}{2}(k + 2 + \sqrt{k+1})} \right\rceil$ by the recent algorithm of Sanders and Steurer [11], which is the best known result for large values of k (note that $\lim_{k\to\infty} \xi(k) = 1$). Nothing better is known for large values of k, even for $k \geq 4$. The most intensively studied case is $k = 2$. The research of this basic variant was initiated by Feige et al. [6], who proposed an algorithm for multigraphs based on an LP relaxation with an approximation ratio of $\frac{10}{13} \approx 0.7692$. They also pointed out a simple $\frac{4}{5}$-approximation for simple graphs. This was later improved several times [3,2]. The current best approximation algorithm for 2-ECS problem in simple graphs, with the ratio of $\frac{5}{6} \approx 0.833$, is due to Kosowski [8] and is achieved by a very interesting extension of the k-matching technique (see Section 3).

Kosowski [8] studied also the case of $k = 3$ and obtained a $\frac{4}{5}$-approximation for simple graphs, which was later improved by a $\frac{6}{7}$-approximation resulting from the mentioned result of Rizzi [10].

Our contribution. We generalize the technique that Kosowski used in his algorithm for the max 2-ECS problem so that it may be applied for arbitrary number of colors. Roughly, we deal with the situation when for a graph G of maximum degree k one can find in polynomial time a k-edge colorable subgraph H with at least $\alpha|E(G)|$ edges, unless G belongs to a family $\mathcal{F}$ of "exception graphs", i.e. $\gamma(G) < \alpha$. As we have seen in the case of $k = 3, 4$ the set of exception graphs is small and in the case of $k = 2$ the exceptions form a very simple family of graphs (odd cycles). The exception graphs are the only obstacles which prevent us from obtaining an α-approximation algorithm (for general graphs) by using the k-matching approach. In such situation we provide a general framework, which allows for obtaining approximation algorithms with approximation ratio better than $\min_{A \in \mathcal{F}} \gamma_k(A)$. See Theorem 3 for the precise description of our general framework.

By combining the framework and our combinatorial results described in Section 1.1 we get the following new results (see Table 1): a $\frac{7}{9}$-approximation of the max-3-ECS problem for multigraphs, a $\frac{13}{15}$-approximation of the max-3-ECS problem for simple graphs, and a $\frac{9}{11}$-approximation of the max-4-ECS problem for simple graphs. Note that the last algorithm is the first result which breaks the barrier of Vizing Theorem for $k \geq 4$.

1.3   Notation

We use standard terminology; for notions not defined here, we refer the reader to [5]. By $N(x)$ we denote the set of neighbors of x. A graph with maximum degree 3 is called *subcubic*. Following [1], let $c_k(G)$ be the maximum number of edges of a k-edge-colorable subgraph of G. We also denote $\bar{c}_k(G) = |E(G)| - c_k(G)$, $c(G) = c_{\Delta(G)}(G)$ and $\bar{c}(G) = \bar{c}_{\Delta(G)}(G)$.

2   Combinatorial Results

In this section we will work with multigraphs (though for simplicity we will call them graphs). We use a result on triangle-free graphs from Rizzi [10].

Lemma 1 (Implicit in Corollary 2 in [10]). *Every subcubic, triangle-free multigraph G has a 4-edge-coloring in which the union of the three largest color classes has size at least $\frac{13}{15}|E(G)|$.*

We need one more definition. Let G_5^* be the graph on 5 vertices obtained from the four-vertex cycle by doubling one edge and adding a vertex of degree two adjacent to the two vertices of the cycle not incident with the double edge.

Theorem 1. *Let G be a biconnected subcubic multigraph different from G_3, G_5 and G_5^*. There exists a 4-edge-coloring of G in which the union of the three largest color classes has size at least $\frac{13}{15}|E(G)|$.*

Proof. We will prove the theorem by induction on the number of vertices of the graph. We introduce the operation of *triangle contraction* which is to contract the three edges of a triangle (order of contracting is inessential) keeping multiple edges that appear. Note that since G is biconnected and $G \neq G_3$, no triangle in G has a double edge, so loops do not appear after the triangle contraction operation. If a graph is subcubic, then it will be subcubic after a triangle contraction. Notice that if a graph has at least five vertices, the operation of triangle contraction in subcubic graphs preserves biconnectivity. It is easy to verify the claim for all subcubic graphs on at most 4 vertices, so in what follows we assume $|V(G)| \geq 5$. W.l.o.g. G has at least one triangle T for otherwise we just apply Lemma 1. Let G' be the graph obtained from G by contracting T. Then G' is subcubic and biconnected.

First, let us assume that G' is not isomorphic to G_3, G_5, or G_5^*. G' has less vertices than G so by the induction hypothesis it has a 4-edge-coloring in which the union of the three largest color classes has size at least $\frac{13}{15}|E(G')|$. Notice that it can always be extended to contain all three edges of T. Hence, G has a 4-edge-coloring in which the union of the three largest color classes has size at least $\frac{13}{15}|E(G')| + 3 \geq \frac{13}{15}|E(G)|$.

Now we consider the case when G' is isomorphic to G_3, G_5 or G_5^*. In fact, G' cannot be isomorphic to G_3, because then G would be G_5 or G_5^*. There are only three graphs from which G_5 can be obtained after triangle contraction; they all have 10 edges and a 3-edge-colorable subgraph with $9 > \frac{13}{15} \cdot 10$ edges. Similarly, there are only three graphs from which G_5^* can be obtained after triangle contraction; they all have 10 edges and a 3-edge-colorable subgraph with $9 > \frac{13}{15} \cdot 10$ edges.   □

Corollary 1. *Let G be a subcubic multigraph not containing G_3 as a subgraph and different from G_5 and G_5^*. There exists a 4-edge-coloring of G in which the union of the three largest color classes has size at least $\frac{13}{15}|E(G)|$.*

Proof. Suppose that the theorem is not true. Let G be a counter-example with the smallest number of vertices. Clearly, G is connected.

It is easy to check that if every biconnected component of G has a 4-edge-coloring in which the union of the three largest color classes contains at least $\frac{13}{15}$ of its edges, then so does G (note that in a subcubic graph if two biconnected components share a vertex then one of the components is a bridge). Thus, by Theorem 1 we can assume that there exists a biconnected component C of G which is isomorphic to G_5 or G_5^*. Since $C \neq G$, there is exactly one edge vw with $v \in V(C)$ and $w \notin V(C)$. If $G - C$ is isomorphic to G_5 or G_5^* then G is a cubic graph with 15 edges and has a a 4-edge-coloring in which the union of the three largest color classes has size 13, a contradiction. Hence, by minimality of G, $G - C$ has a 4-edge-coloring in which the union of the three largest color classes has size at least $\frac{13}{15}|E(G - C)| = \frac{13}{15}(|E(G)| - 8)$. But then it is easy to extend this coloring to G so that the union of the three largest color classes grows by 7 edges. Since $\frac{13}{15}(|E(G)| - 8) + 7 > \frac{13}{15}|E(G)|$, we get a contradiction.   □

Corollary 2. *Every subcubic simple graph G different from G_5 has a 4-edge-coloring in which the union of the three largest color classes has size $\geq \frac{13}{15}|E(G)|$.*

Corollary 3. *Every subcubic multigraph G different from G_3 has a 4-edge-coloring in which the union of the three largest color classes has size $\geq \frac{7}{9}|E(G)|$.*

Proof. Assume G is biconnected. If G is isomorphic to G_5 or G_5^*, then it has a 4-edge-coloring in which the union of the three largest color classes contains at least $\frac{6}{7}|E(G)| \geq \frac{7}{9}|E(G)|$ edges. Otherwise, we apply Theorem 1. Hence the claim holds for biconnected graphs.

Let G be a counter-example with the smallest number of vertices. If every biconnected component of G has a 4-edge-coloring in which the union of the three largest color classes has size at least $\frac{7}{9}$ of its edges, then so does G. Therefore, one of the biconnected components C is isomorphic to G_3. Then, there is exactly one edge vw with $v \in V(C)$ and $w \notin V(C)$. If $G - C$ is isomorphic to G_3, then G has 9 edges and a 4-edge-coloring in which the union of the three largest color classes has 7 edges, a contradiction. Hence, by the minimality of G, $G - C$ has a 4-edge-coloring in which the union of the three largest color classes has size at least $\frac{7}{9}(|E(G)| - 5)$. But then it is easy to extend this coloring to G so that the union of the three largest color classes grows by 4 edges. Since $\frac{7}{9}(|E(G)| - 5) + 4 > \frac{7}{9}|E(G)|$, we get a contradiction.   □

Finally we state our result on large 4-edge-colorable subgraphs.

Theorem 2. *A simple graph G with maximum degree 4 different from K_5 has a 4-colorable subgraph with at least $\frac{5}{6}|E(G)|$ edges.*

Our approach for proving Theorem 2 is as follows. We begin with an empty partial coloring and extend it to a "better" partial coloring step by step. Of course when a coloring c' has more colored edges than a coloring c, we consider c' as better than c. It turns out that if the subgraph induced by uncolored edges has a connected component of at least 3 edges, one can find a better coloring in polynomial time. Hence we can focus on uncolored single edges and pairs of incident edges. The goal is to find a highly structured partial coloring so that in the neighborhood of each uncolored edge or pair of incident edges there are many colored edges, which are not too close to other uncolored edges. It turns out that if a pair of incident uncolored edges does not satisfy this property then G is isomorphic to K_5 or one can get a better coloring. Motivated by this, we say that a coloring c' is better than a coloring c also when they have the same number of colored edges, but c' has more pairs of incident uncolored edges. Then, we infer that a single uncolored edge also has a few uncolored edges nearby for otherwise one can perform some local recoloring and get a new pair of uncolored incident edges, obtaining a better coloring. Since the number of the partial coloring improvements is $O(|E| \cdot |V|)$, after polynomial time we get the desired highly structured partial coloring. The space limitations force us to skip the full proof in this extended abstract. It will appear in the journal version.

3   Approximation Algorithms

In this section we describe a meta-algorithm for the maximum k-edge-colorable subgraph problem. It is inspired by a method of Kosowski [8] developed originally for $k = 2$. In the end of the section we show that the meta-algorithm gives new approximation algorithms for $k = 3$ in the case of multigraphs and for $k = 3, 4$ in the case of simple graphs.

Throughout this section $G = (V, E)$ is the input multigraph from a family of graphs $\mathcal{G}$ (later on, we will use $\mathcal{G}$ as the family of all simple graphs or of all multigraphs). We fix a maximum k-edge-colorable subgraph OPT of G.

As many previous algorithms, our method begins with finding a maximum k-matching F of G in polynomial time. Clearly, $|E(\text{OPT})| \leq |E(F)|$. Now, if we manage to color $\rho|E(F)|$ edges of F, we get a ρ-approximation. Unfortunately this way we can get a low upper bound on the approximation ratio. Consider for instance the case of $k = 3$ and $\mathcal{G}$ being the family of multigraphs. Then, if all connected components Q of F are isomorphic to G_3, we get $\rho = \frac{3}{4}$. In the view of Corollary 3 this is very annoying, since G_3 is the only graph which prevents us for obtaining the $\frac{7}{9}$ ratio here. However, we can take a closer look at the relation of a component Q isomorphic to G_3 and OPT. Observe that if OPT does not leave Q, i.e. OPT contains no edge with exactly one endpoint in Q then $|E(\text{OPT})| = |E(\text{OPT}[V \setminus V(Q)])| + |E(\text{OPT}[V(Q)])|$. Note also that $|E(\text{OPT}[V(Q)])| = 3$, so if we take only three of the four edges of Q to our solution we do not loose anything — locally our approximation ratio is 1. It follows that if there are many components of this kind, the approximation ratio is better than $\frac{3}{4}$. What can we do if there are many components isomorphic to G_3 with an incident edge of OPT? The problem is that we do not know OPT. However, then there are many components isomorphic to G_3 with an incident edge of the input graph G. The idea is to add some of these edges in order to form bigger components (possibly with maximum degree bigger than k) which have larger k-colorable subgraphs than the original components.

In the general setting, we consider a family of graphs $\mathcal{F} \subset \mathcal{G}$ such that for every graph $A \in \mathcal{F}$,

(F1) $\Delta(A) = k$ and A has at most one vertex of degree smaller than k,
(F2) $c_k(A) = c_k(K_{|V(A)|})$,
(F3) for every edge $uv \in E(A)$, a maximum k-edge colorable subgraph of A or $A - uv$ can be found in polynomial time;
(F4) for a given graph B one can check whether A is isomorphic to B in polynomial time,
(F5) A is 2-edge-connected,
(F6) for every edge $uv \in A$, we have $c(A - uv) = c(A)$.

A family that satisfies the above properties will be called a *k-normal family.* We assume there is a number $\alpha \in (0, 1]$ and a polynomial-time algorithm $\mathcal{A}$ which for every k-matching H of a graph in $\mathcal{G}$, such that $H \notin \mathcal{F}$ finds its k-edge colorable subgraph with at least $\alpha|E(H)|$ edges. Intuitively, $\mathcal{F}$ is a family of "bad exceptions" meaning that for every graph A in $\mathcal{F}$, there is $c(A) < \alpha|E(H)|$, e.g. in the above example of subcubic multigraphs $\mathcal{F} = \{G_3\}$. We note that the family $\mathcal{F}$ needs not be finite, e.g. in the work [8] of Kosowski $\mathcal{F}$ contains all odd cycles. Now we can state the main result of this section.

Theorem 3. *Let $\mathcal{G}$ be a family of graphs and let $\mathcal{F}$ be a k-normal family of graphs. Assume there is a polynomial-time algorithm which for every k-matching*

H of a graph in $\mathcal{G}$, such that $H \notin \mathcal{F}$ finds its k-edge colorable subgraph with at least $\alpha|E(H)|$ edges. Moreover, let

$$\beta = \inf_{\substack{A,B \in \mathcal{F} \\ A \text{ is not } k\text{-regular}}} \frac{c_k(A) + c_k(B) + 1}{|E(A)| + |E(B)| + 1}, \quad \gamma = \inf_{A \in \mathcal{F}} \frac{c_k(A) + 1}{|E(A)| + 1}, \quad and$$

$$\delta = \inf_{A,B \in \mathcal{F}} \frac{c_k(A) + c_k(B) + 2}{|E(A)| + |E(B)| + 1}.$$

Then, there is an approximation algorithm for the maximum k-ECS problem with approximation ratio $\min\{\alpha, \beta, \gamma, \delta\}$.

In what follows, we prove Theorem 3. Let Γ be the set of all connected components of F that are isomorphic to a graph in $\mathcal{F}$.

Observation 1. *Without loss of generality, there is no edge $xy \in E(G)$ such that for some $Q \in \Gamma$, $x \in V(Q)$, $y \notin V(Q)$ and $\deg(y) < k$.*

Proof. If such an edge exists, we replace in F any edge of Q incident with x with the edge xy. The new F is still a maximum k-matching in G. By (F5) the number of connected components of F increases, so the procedure eventually stops with a k-matching satisfying the desired property. □

When H is a subgraph of G we denote $\Gamma(H)$ as the set of components Q in Γ such that H contains an edge xy with $x \in V(Q)$ and $y \notin V(Q)$. We denote $\overline{\Gamma}(H) = \Gamma \setminus \Gamma(H)$. The following lemma, a generalization of Lemma 2.1 from [8], motivates the whole approach.

Lemma 2. $|E(\text{OPT})| \leq |E(F)| - \displaystyle\sum_{Q \in \overline{\Gamma}(\text{OPT})} \overline{c}_k(Q).$

Proof. Since for every component $Q \in \overline{\Gamma}$ the graph OPT has no edges with exactly one endpoint in Q,

$$|E(\text{OPT})| = |E(\text{OPT}[V'])| + \sum_{Q \in \overline{\Gamma}(\text{OPT})} |E(\text{OPT}[V(Q)])|, \qquad (1)$$

where $V' = V \setminus \bigcup_{Q \in \overline{\Gamma}(\text{OPT})} V(Q)$. Since obviously for every Q in Γ we have $c_k(Q) \leq |E(\text{OPT}[V(Q)])| \leq c_k(K_{|V(Q)|})$ and by (F2), $c_k(Q) = c_k(K_{|V(Q)|})$, we get

$$|E(\text{OPT}[V(Q)])| = c_k(Q). \qquad (2)$$

Since OPT is k-edge-colorable, $E(\text{OPT}[V'])$ is a k-matching. Since F is maximal, $|E(\text{OPT}[V'])| \leq |E(F[V'])|$. This, together with (1) and (2) gives the desired inequality as follows.

$$|E(\text{OPT})| \leq |E(F[V'])| + \sum_{Q \in \overline{\Gamma}(\text{OPT})} c_k(Q) = |E(F)| - \sum_{Q \in \overline{\Gamma}(\text{OPT})} \overline{c}_k(Q). \qquad (3)$$

□

The above lemma allows us to leave up to $\sum_{Q \in \overline{\Gamma}(\text{OPT})} \overline{c}_k(Q)$ edges of components in Γ uncolored for free, i.e. without obtaining approximation factor worse than α. In what follows we "cure" some components in Γ by joining them with other components by edges of G. We want to do it in such a way that the remaining, "ill", components have a partial k-edge-coloring with no more than $\sum_{Q \in \overline{\Gamma}(\text{OPT})} \overline{c}_k(Q)$ uncolored edges. To this end, we find a k-matching $R \subseteq G$ which fulfills the following conditions:

(M1) for each edge $xy \in R$ there is a component $Q \in \Gamma$ such that $x \in V(Q)$ and $y \notin V(Q)$,

(M2) R maximizes $\sum_{Q \in \Gamma(R)} \overline{c}(Q)$,

(M3) R is inclusion-wise minimal k-matching subject to (M1) and (M2).

Lemma 3. *R can be found in polynomial time.*

Proof. We use a slightly modified algorithm from the proof of Proposition 2.2 in [8]. We define graph $G' = (V', E')$ as follows. Let $V' = V \cup \{u_Q, w_Q \, : \, Q \in \Gamma\}$. Then, for each $Q \in \Gamma$, the set E' contains three types of edges:

- all edges $xy \in E(G)$ such that $x \in V(Q)$ and $y \notin V(Q)$,
- an edge vu_Q for every vertex $v \in V(Q)$, and
- an edge $u_Q w_Q$.

Next we define functions $f, g : V' \to \mathbb{N}$ as follows: for every $v \in \bigcup_{Q \in \Gamma} V(Q)$ we set $f(v) = 1$, $g(v) = k$; for every $v \in V \setminus \bigcup_{Q \in \Gamma} V(Q)$ we set $f(v) = 0$, $g(v) = k$; for every $Q \in \Gamma$ we set $f(u_Q) = 0$, $g(u_Q) = |V(Q)|$ and $f(w_Q) = 0$, $g(w_Q) = 1$. Additionally, all edges $u_Q w_Q$ have weight $\overline{c}(Q)$ while all the other edges have weight 0. Then we find a maximum weight $[f, g]$-factor R' in G', which can be done in polynomial time (see e.g. [12]). It is easy to see that $R = E(R') \cap E(G)$ fulfills (M1) and (M2). Next, as long as R contains an edge xy such that $R - xy$ still fulfills (M1) and (M2), we replace R by $R - xy$. □

The following lemma shows why the k-matching R is useful.

Lemma 4. $\displaystyle \sum_{Q \in \overline{\Gamma}(R)} \overline{c}_k(Q) \le \sum_{Q \in \overline{\Gamma}(\text{OPT})} \overline{c}_k(Q).$

Proof. Let $R_{\text{OPT}} = \{xy \in E(\text{OPT}) : \text{for some } Q \in \Gamma, \ x \in Q \text{ and } y \notin Q\}$. Since OPT is k-edge-colorable, R_{OPT} is a k-matching. By (M2) it follows that

$$\sum_{Q \in \Gamma(R)} \overline{c}_k(Q) \ge^{(\text{M2})} \sum_{Q \in \Gamma(R_{\text{OPT}})} \overline{c}_k(Q) = \sum_{Q \in \Gamma(\text{OPT})} \overline{c}_k(Q), \qquad (4)$$

and next

$$\sum_{Q \in \overline{\Gamma}(R)} \overline{c}_k(Q) = \sum_{Q \in \Gamma} \overline{c}_k(Q) - \sum_{Q \in \Gamma(R)} \overline{c}_k(Q) \le^{(4)}$$

$$\sum_{Q \in \Gamma} \overline{c}_k(Q) - \sum_{Q \in \Gamma(\text{OPT})} \overline{c}_k(Q) = \sum_{Q \in \overline{\Gamma}(\text{OPT})} \overline{c}_k(Q).$$

□

The following observation is immediate from the minimality of R, i.e. from condition (M3).

Observation 2. *Let H_F be a graph with vertex set $\{Q : Q$ is a connected component of $F\}$ and the edge set $\{PQ$: there is an edge $xy \in R$ incident with both P and $Q\}$. Then H_F is a forest, and every connected component of H_F is a star.*

In what follows, the components of F corresponding to leafs in H_F are called *leaf components*. Now we proceed with finding a k-edge-colorable subgraph S of G together with its coloring, using the algorithm described below. In the course of the algorithm, we maintain the following invariants:

Invariant 1. *For every $v \in V$, $\deg_R(v) \leq \deg_F(v)$.*

Invariant 2. *If F contains a connected component Q isomorphic to a graph in $\mathcal{F}$, then $Q \in \Gamma$, in other words a new component isomorphic to a graph in $\mathcal{F}$ cannot appear.*

By Observation 2, each edge of R connects a vertex x of a leaf component and a vertex y of another component. Hence $\deg_R(x) = 1 \leq \deg_F(x)$. By Observation 1, initially $\deg_F(y) = k$, so also $\deg_R(y) \leq \deg_F(y)$. It follows that Invariant 1 holds at the beginning, as well as Invariant 2, the latter being trivial. Now we describe the coloring algorithm.

Step 1: Begin with the empty subgraph $S = (V, \emptyset)$.

Step 2: As long as F contains a leaf component $Q \in \Gamma$ and a component P, such that
   - there is an edge $xy \in R$ with $x \in Q$ and $y \in P$,
   - there is an edge $yz \in E(P)$ such that no connected component of $P - e$ is isomorphic to a graph in $\mathcal{F}$,

   then we remove xy from R and both Q and yz from F. Notice that if z was incident with an edge $zw \in R$ then by Observation 2, w belongs to another leaf component Q'. Then we also remove zw from R and Q' from F. It follows that Invariants 1 and 2 hold.

Step 3: As long as there is a leaf component $Q \in \Gamma(R)$ we do the following. Let P be the component of F such that there is an edge $xy \in R$ with $x \in Q$ and $y \in P$. Then, by Step 2, for each edge $yz \in E(P)$ in graph $P - yz$ there is a connected component isomorphic to a graph in $\mathcal{F}$. In particular, by (F1) every edge $yz \in E(P)$ is a bridge in P. Let yz be any any edge incident with z in P, which exists by Invariant 1. Note that if $P - yz$ has a connected component C isomorphic to a graph in $\mathcal{F}$ and containing y then every edge of C incident with y is a bridge in C, a contradiction with (F5). Hence $P - yz$ has exactly one connected component, call it P_{yz}, isomorphic to a graph in $\mathcal{F}$ and $V(P_{yz})$ contains z. By the same argument, P_{yz} is not incident with an edge of R. Then we remove Q, yz and P_{yz} from F and xy from R. The above discussion shows that Invariants 1 and 2 hold.

Step 4: Process each of the remaining components Q of F, depending on its kind.

(a) If $Q \in \Gamma$, it means that $Q \in \overline{\Gamma}(R)$, because otherwise there are leaf components in $\Gamma(R)$, which contradicts Step 3. Then we find a maximum k-edge-colorable-subgraph $S_Q \subseteq Q$, which is possible in polynomial time by (F3), and add it to S with the relevant k-edge-coloring.

(b) If $Q \notin \Gamma$ we use the algorithm $\mathcal{A}$ to color at least $\alpha|E(Q)|$ edges of Q and we add the colored edges to S.

(c) For every Q, yz and P_{yz} deleted in Step 3, we find the maximum edge colorable subgraph Q^* of Q and P^* of P_{yz}. Note that the coloring of P^* can be extended to $P^* + yz$ since $\deg_{P^*}(z) < k$. Next we add Q^*, P^* and yz to S (clearly we can rename the colors of $P^* + yz$ so that we avoid conflicts with the already colored edges incident to y). To sum up, we added $c_k(Q) + c_k(P_{yz}) + 1$ edges to S, which is $\frac{c_k(Q)+c_k(P_{yz})+1}{|E(Q)|+|E(P_{yz})|+1} \geq \beta$ of the edges of F deleted in Step 3.

(d) For every xy and Q deleted in Step 2, let zw be any edge of Q incident with x and then we find the maximum k-edge-colorable subgraph Q^* of $Q - zw$ using the algorithm guaranteed by (F3). Next we add Q^* and xy to S (similarly as before, we can rename the colors of $Q^* + xy$ so that we avoid conflicts with the already colored edges incident to y). By (F6), $c_k(Q - zw) = c_k(Q)$. Recall that in Step 2 two cases might happen: either we deleted only Q and yz from F, or we deleted Q, yz and Q'. In the former case we add $c_k(Q) + 1$ edges to S, which is $\frac{c_k(Q)+1}{|E(Q)|+1} \geq \gamma$ of the edges removed from F. In the latter case we add $c_k(Q) + c_k(Q') + 2$ edges to S, which is $\frac{c_k(Q)+c_k(Q')+2}{|E(Q)|+|E(Q')|+1} \geq \delta$ of the edges removed from F.

Proposition 1. *Our algorithm has approximation ratio of* $\min\{\alpha, \beta, \gamma, \delta\}$.

Proof. Let $\rho = \min\{\alpha, \beta, \gamma, \delta\}$.

$$|S| \geq \rho(|E(F)| - \sum_{Q \in \overline{\Gamma}(R)} |E(Q)|) + \sum_{Q \in \overline{\Gamma}(R)} c_k(Q) \geq$$

$$\rho(|E(F)| - \sum_{Q \in \overline{\Gamma}(R)} |E(Q)| + \sum_{Q \in \overline{\Gamma}(R)} c_k(Q)) =$$

$$\rho(|E(F) - \sum_{Q \in \overline{\Gamma}(R)} \overline{c}_k(Q)) \geq^{\text{(Lemma 4)}}$$

$$\rho(|E(F) - \sum_{Q \in \overline{\Gamma}(\text{OPT})} \overline{c}_k(Q)) \geq^{\text{(Lemma 2)}} \rho|E(\text{OPT})|.$$

$\square$

This finishes the proof of Theorem 3. Now we apply it to particular cases.

Theorem 4. *The maximum 3-ECS problem has a $\frac{7}{9}$-approximation algorithm for multigraphs.*

Proof. Let $\mathcal{F} = \{G_3\}$. It is easy to check that $\mathcal{F}$ is 3-normal. Now we give the values of parameters α, β, γ and δ from Theorem 3. By Corollary 3, $\alpha = \frac{7}{9}$. Notice that $c_3(G_3) = 3$ and $|E(G_3)| = 4$. Hence, $\beta = \frac{7}{9}$, $\gamma = \frac{4}{5}$ and $\delta = \frac{8}{9}$. By Theorem 3 the claim follows. $\square$

Theorem 5. *The maximum 3-ECS problem has a $\frac{13}{15}$-approximation algorithm for simple graphs.*

Proof. Let $\mathcal{F} = \{G_5\}$. It is easy to check that $\mathcal{F}$ is 3-normal. Now we give the values of parameters α, β, γ and δ from Theorem 3. By Corollary 1, $\alpha = \frac{13}{15}$. Notice that $c_3(G_5) = 6$ and $|E(G_5)| = 7$. Hence, $\beta = \frac{13}{15}$, $\gamma = \frac{7}{8}$ and $\delta = \frac{14}{15}$. By Theorem 3 the claim follows. $\square$

Theorem 6. *The maximum 4-ECS problem has a $\frac{9}{11}$-approximation algorithm for simple graphs.*

Proof. Let $\mathcal{F} = \{K_5\}$. It is easy to check that $\mathcal{F}$ is 4-normal. Now we give the values of parameters α, β, γ and δ from Theorem 3. By Theorem 2, $\alpha = \frac{5}{6}$. Observe that $\beta = \infty$, since $\mathcal{F}$ contains only K_5 which is 4-regular. Notice that $c_4(K_5) = 8$ and $|E(K_5)| = 10$. Hence, $\gamma = \frac{9}{11}$ and $\delta = \frac{18}{21}$. By Theorem 3 the claim follows. $\square$

References

1. Albertson, M.O., Haas, R.: Parsimonious edge coloring. Discrete Mathematics 148(1-3), 1–7 (1996)
2. Chen, Z.Z., Tanahashi, R.: Approximating maximum edge 2-coloring in simple graphs via local improvement. Theor. Comput. Sci. 410(45), 4543–4553 (2009)
3. Chen, Z.Z., Tanahashi, R., Wang, L.: An improved approximation algorithm for maximum edge 2-coloring in simple graphs. J. Discrete Algorithms 6(2), 205–215 (2008)
4. Cornuéjols, G., Pulleyblank, W.: A matching problem with side conditions. Discrete Mathematics 29, 135–159 (1980)
5. Diestel, R.: Graph Theory, Electronic edn. Springer, Heidelberg (2005)
6. Feige, U., Ofek, E., Wieder, U.: Approximating maximum edge coloring in multigraphs. In: Jansen, K., Leonardi, S., Vazirani, V.V. (eds.) APPROX 2002. LNCS, vol. 2462, pp. 108–121. Springer, Heidelberg (2002)
7. Holyer, I.: The NP-completeness of edge coloring. SIAM Journal on Computing 10, 718–720 (1981)
8. Kosowski, A.: Approximating the maximum 2− and 3−edge-colorable subgraph problems. Discrete Applied Mathematics 157, 3593–3600 (2009)
9. Leven, D., Galil, Z.: NP-completeness of finding the chromatic index of regular graphs. Journal of Algorithms 4, 35–44 (1983)
10. Rizzi, R.: Approximating the maximum 3-edge-colorable subgraph problem. Discrete Mathematics 309, 4166–4170 (2009)

11. Sanders, P., Steurer, D.: An asymptotic approximation scheme for multigraph edge coloring. ACM Trans. Algorithms 4(2), 1–24 (2008)
12. Schrijver, A.: Combinatorial Optimization: Polyhedra and Efficiency. Springer, Heidelberg (2003)
13. Shannon, C.E.: A theorem on coloring the lines of a network. J. Math. Phys. 28, 148–151 (1949)
14. Vizing, V.G.: On the estimate of the chromatic class of a p-graph. Diskret. Analiz 3, 25–30 (1964)

Improved Algorithm for Degree Bounded Survivable Network Design Problem

Anand Louis[1] and Nisheeth K. Vishnoi[2]

[1] College of Computing, Georgia Tech., Atlanta
anand.louis@cc.gatech.edu
[2] Microsoft Research India, Bangalore
nisheeth.vishnoi@gmail.com

Abstract. We consider the *Degree-Bounded Survivable Network Design Problem*: the objective is to find a minimum cost subgraph satisfying the given connectivity requirements as well as the degree bounds on the vertices. If we denote the upper bound on the degree of a vertex v by $b(v)$, then we present an algorithm that finds a solution whose cost is at most twice the cost of the optimal solution while the degree of a degree constrained vertex v is at most $2b(v) + 2$. This improves upon the results of Lau and Singh [13] and Lau, Naor, Salavatipour and Singh [12].

1 Introduction

The degree-bounded survivable network design problem. In the Survivable Network Design Problem (SNDP), the input is an undirected graph $G = (V, E)$, connectivity requirements $\rho(u, v)$ for all pairs of vertices u, v, and a function $u : E \rightarrow \mathbb{Z}_{\geqslant 0} \cup \{\infty\}$ stating an upper bound on the number of copies of e we are allowed to use (if $u(e) = \infty$, then there is no upper bound for edge e). The goal is to select a multiset of edges from E such that there are $\rho(u, v)$ edge disjoint paths between every $u, v \in V$. In addition, if each edge has an associated cost given by $c : E \rightarrow R_{\geqslant 0}$, then the goal is to find the minimum cost solution satisfying the connectivity requirements (each copy of edge e used for this construction will cost $c(e)$). Interest in this problem derives from its applications in algorithm design, networking, graph theory and operations research. This problem is NP-hard and the best approximation algorithm is, a 2-approximation, due to Jain [8]. In this paper we will consider the degree bounded variant of the SNDP. Here, in addition we are given degree constraints $b : W_0 \rightarrow \mathbb{Z}_{\geqslant 0}$ on a subset of vertices W_0. The goal is to find a network of minimum cost satisfying the connectivity requirements and, in addition, ensuring that every vertex in W_0 has degree at most $b(v)$. This problem is referred to as Minimum Degree-Bounded Survivable Network Design Problem (Deg-SNDP).

The iterative rounding approach. Jain's approach for SNDP starts off by writing a natural linear programming relaxation for it, with a variable for each edge, and considering an optimal solution to it. In fact, the key to his approach is to consider a *vertex optimal solution* to his LP and derive enough structural information from this solution to prove

H. Kaplan (Ed.): SWAT 2010, LNCS 6139, pp. 408–419, 2010.
© Springer-Verlag Berlin Heidelberg 2010

that it has a variable of value at least $1/2$.[1] Then, picking this edge and noticing that the residual problem is also an instance of a (slight generalization) of **SNDP**, one could iterate and obtain a solution with cost at most 2 times the cost of the optimal solution. Fleischer *et. al* [4] have generalized this to the element connectivity **SNDP** problem.

For **Deg-SNDP**, one can easily augment Jain's **LP** with degree constraints on the vertices. It is no longer clear that a vertex optimal solution to this **LP** should have a variable of value at least $1/2$. It was proved by Lau, Naor, Salavatipour and Singh [12] that in any vertex optimal solution to this degree-constrained **LP**, either there is an edge whose variable has value at least $1/2$ or there is a degree constrained vertex such that at most 4 edges are incident to it in this solution. Using this iteratively[2], they obtain an algorithm which outputs a solution of cost at most 2 times the optimal and such that the degree of every vertex in this solution is at most $2b(v) + 3$.

Necessity of degree bound violations? Improving the 2 in the approximation factor would result in improving the approximation for the **SNDP** problem itself and it seems out of bound. We study to what extent can the $2b(v) + 3$ be improved. As shown by Lau *et al.* [12], the issue with degree constraints here is that there are instances such that this **LP** may be feasible even when there may be no feasible integral solution to the problem: consider a 3-regular, 3-edge-connected graph without a Hamiltonian Path. Let $W_0 = V$, $b(v) = 1$ for all $v \in V$ and $\rho(u,v) = 1$ for all $u \neq v \in V$. It can be seen that $x_e = 1/3$ is a feasible solution to this **LP**. On the other hand, it is easy to observe that there is no integral solution even when one relaxes the degree bound function from $b(v) = 1$ to $2b(v) = 2$ for every $v \in V$. This is because such a feasible integral solution would correspond to a Hamiltonian Path. Hence, in any feasible integral solution, there must be a vertex v of degree at least $2b(v) + 1$ (or $b(v) + 2$). Further, Lau and Singh [13] show a family of instances which have a feasible fractional solution but every integral solution to an instance from their family has a degree constrained vertex v with degree $b(v) + \Omega(\max_{u,v \in V} \rho(u,v))$. Hence, the best we can hope for, in general, is an approximation algorithm based on this **LP** which outputs a solution of cost at most 2 times that of the optimal of this **LP** and though it satisfies all the connectivity requirements, it satisfies the relaxed degree constraints: for every vertex, the degree in solution obtained is at most $2b(v) + 1$.

Our result. We improve the result of [12,13] for **Deg-SNDP** to $2b(v)+2$. More precisely, we present an algorithm that finds a solution whose cost is at most twice the cost of the optimal solution while the degree of a degree constrained vertex v is at most $2b(v) + 2$.

An important special case of the **Deg-SNDP** problem is the Minimum Bounded Degree Spanning Tree Problem. The breakthrough work of Goemans [6] was followed up by Singh and Lau [17] to provide an optimal, in terms of the degree violations, **LP**-based result for this problem. For **Deg-SNDP**, proving the exact limits of the **LP**-based approach has been significantly more challenging as is evident from the work of [12,13]. This paper leaves us just one step away from this goal.

---

[1] This is not guaranteed at a non-vertex optimal solution.

[2] Since intermediate instances arising in their algorithms could have semi-integral degree bounds, they need to consider a more general input where the degree bounds could be half-integral.

Overview and technique. We follow the iterative rounding approach of Jain. We start with the LP augmented with degree constraints on a set of vertices W_0 given as input. Initially $b(v)$ is an integer for all $v \in W_0$. Since in an intermediate iteration we may be forced to pick an edge of value $1/2$, we will have to work with the degree constraint function b which is half-integral. The instance at any intermediate iteration consists of the edges which have neither been *picked* in or *dropped* from any previous iteration along with the suitably modified connectivity requirement function and, the set of vertices with degree constraints is a subset of W_0. The latter are vertices from which the degree constraint has not been removed in any previous iteration.

As is usual, during any iteration, we first find a vertex optimal solution $x^\star$ to the current instance. One way to prove our result would have been to tighten the Lau *et al.* result to show that in any intermediate iteration we can either find an edge e such that $x_e^\star \geq 1/2$ or that there is some degree constrained vertex with at most 3 edges in $x^\star$ with value strictly bigger than 0. Unfortunately, we do not know how to prove that. Instead, we are able to prove the following which suffices to prove our result. Let $E_{>0}$ denote those edges with $x_e^\star > 0$ and let W denote the current set of degree constrained vertices and b be the degree constraint function on them. Then, one of the following holds:

- there is an edge e such that $x_e^\star = 0$,
- or there is an edge e such that $x_e^\star = 1$,
- or there is an edge $e = \{u, v\}$ such that $1/2 \leq x_e^\star < 1$, and if $u \in W$, then $b(u) > 1$, and if $v \in W$, then $b(v) > 1$,
- or there is a vertex $v \in W$ such that $\deg_{E_{>0}}(v) \leq 2b(v) + 2$.

When we find an edge of value at least $1/2$ but strictly less than 1, and we decide to pick it, we need to reduce the degree of its endpoint(s) if they are constrained. To maintain feasibility, one can only reduce these degree constraints by at most the value of the edge picked. Since, we know that this is at least $1/2$, we instead choose to reduce the degree constraint by exactly $1/2$. Our invariant allows us to make sure that for any such edge we find, its endpoints, if degree constrained, never drop below 1. Hence, $b(v)$ never becomes less than 1. This is not guaranteed by the invariant of Lau *et al.* Moreover, we remove the degree constraint from a degree constrained vertex v when $\deg_{E_{>0}}(v)$ falls below $2b(v) + 2$. The invariant along with this *iterative relaxation* of constraints are crucial to the analysis of the algorithm. The main task then becomes to prove this invariant. This is done by a new *token distribution* argument the details of which, due to their highly technical nature, are left to a later section.

Other related work. Network design problems with degree constraints have been extensively studied by researchers in the computer science and operations research community. A well known problem is the Minimum Bounded Degree Spanning Tree (MBDST) problem where the objective is to minimize the cost of the spanning tree while satisfying the degree constraints on the vertices. This problem is NP-hard as solving an instance of this problem having degree bound of 2 for each vertex would be equivalent to solving the Hamilton Path problem on that graph. This problem had been studied extensively in [5,6,17,3]. The best known algorithm is due to Singh and Lau [17] which returns a solution of cost atmost the cost of OPT and degree of a degree bounded vertex is at most 1 more than its degree bound (see [17]).

The technique of iterative rounding was first introduced by Jain [8] and has been used subsequently in deriving several other important results such as [9,12,13,1,15]. The special case of SNDP with metric costs has been studied in [7,2]. Other related network design problems have been studied in [10,11,5]. For a detailed review of iterative rounding, the reader is referred to [14,16].

Organization. We start with formal definitions, the LP relaxation for Deg-SNDP in Section 2. We present our main theorem and our algorithm in Section 3. In this section we prove how the main theorem is implied by our main lemma about the vertex optimal solution. Section 4 is devoted to the proof of the main lemma.

2   Preliminaries

Given an undirected graph $G = (V, E)$ with $|V| = n$, a subset of vertices W_0, a cost function on the edges $c \colon E \to \mathbb{R}_{\geqslant 0}$, a *edge-multiplicity* function $u \colon E \to \mathbb{Z}_{\geqslant 0}$, a *connectivity requirement* function $\rho \colon V \times V \to \mathbb{Z}_{\geqslant 0}$, and a *degree bound* function $b \colon W_0 \to \mathbb{Z}_{\geqslant 0}$, the goal in the Minimum Bounded Degree Survival Network Design Problem (Deg-SNDP) is to find a (multiset) subset of edges H of G of minimum cost such that, at most $u(e)$ copies of edge $e \in E$ are present in H, for every $u, v \in V$, there are $\rho(u, v)$ edge-disjoint paths connecting u and v in the graph on edges in H and the degree of every $v \in W_0$ in H is at most $b(v)$. Here, $\mathsf{cost}(H) \overset{\text{def}}{=} \sum_{e \in H} c(e)$. For a set $S \subseteq V$, let $R(S) \overset{\text{def}}{=} \max_{u \in S, v \in \bar{S}} \rho(u, v)$. The function R is *weakly super-modular*[3], and in general we will assume the weaker property that the function R is weakly super-modular, rather than derived from some connectivity function ρ. Hence, we will often denote an instance of Deg-SNDP by a tuple by $(G(V, E), W_0, c, u, R, b)$ where R is any weakly super-modular function from $2^{[V]}$ to $\mathbb{Z}_{\geqslant 0}$.

For an instance $\mathcal{I} = (G(V, E), W_0, c, u, R, b)$ of Deg-SNDP, let $\mathsf{opt}(\mathcal{I})$ denote the cost of the optimal solution. Let $\mathsf{lp}(\mathcal{I})$ denote the value of the LP of Figure 1 for $\mathcal{I}$.

$$\mathsf{lp}(\mathcal{I}) \overset{\text{def}}{=} \text{minimize } \sum_{e \in E} c(e) x_e$$
$$\text{subject to}$$
$$\forall S \subseteq V, \quad x(\delta_E(S)) \; \geqslant R(S)$$
$$\forall v \in W_0, \quad x(\delta_E(v)) \; \leqslant b(v)$$
$$\forall e \in E, \quad u(e) \geqslant x_e \geqslant 0$$

Fig. 1. The LP for Deg-SNDP Problem

---

[3] A function $f \colon 2^{[V]} \mapsto \mathbb{Z}_{\geqslant 0}$ is said to be *weakly super-modular* if $f(V) = 0$, and for every two sets $S, T \subseteq V$, at least one of the following conditions holds:

- $f(S) + f(T) \leqslant f(S \setminus T) + f(T \setminus S)$
- $f(S) + f(T) \leqslant f(S \cap T) + f(S \cup T)$.

Here $x(\delta(S)) \stackrel{\text{def}}{=} \sum_{e \in \delta_E(S)} x_e$ where for a set $S \subset V$, and a collection of edges F on V, $\delta_F(S)$ denotes the subset of edges of F with one endpoint in S and the other in its complement. Also, for a set $S \subseteq V$, let $\chi_S \in \{0, 1\}^E$ be the vector such that $\chi_S(e = \{i, j\}) = 1$ if $|\{i, j\} \cap S| = 1$ and $\chi_S(e) = 0$ otherwise. It is easily seen that this LP is a relaxation to Deg-SNDP. Also, it is known (see [8]) that this LP has a polynomial time separation oracle.

We call an algorithm a (α, β, γ)-approximation for Deg-SNDP if on every instance $\mathcal{I}$ of it, the algorithm outputs a collection of edges which have cost at most $\alpha \cdot \text{opt}(\mathcal{I})$, satisfy the R constraints, and the degree of every vertex in $v \in W$ is at most $\beta \cdot b(v) + \gamma$. As mentioned before, the best result known for this problem is due to [12,13] who gave a $(2, 2, 3)$-approximation. In this paper we give a new iterative rounding algorithm which results in a $(2, 2, 2)$-approximation algorithm for Deg-SNDP. We leave open the possibility of a $(2, 2, 1)$-approximation algorithm.

3   Main Theorem and the Algorithm

In this section we prove our main theorem.

Theorem 1 (Main Theorem). *There is a polynomial time $(2, 2, 2)$-approximation algorithm for* Deg-SNDP.

The algorithm used appears in Figure 2. We assume that R is weakly super-modular. A *vertex optimal solution* of a LP is an optimal solution which cannot be written as a non-trivial convex combination of two or more feasible solutions to the LP.

It follows from a result of Jain that each iteration of this algorithm can be implemented in polynomial time. It remains to be proved that the algorithm is correct. We first state the main lemma of the paper and then show how it implies the main theorem. The proof of the main lemma is the technical core of the paper and appears in Section 4.

Lemma 1 (Main Lemma). *Given an instance $\mathcal{I} = (G(V, E), W, c, u, R, b)$, where $R \in \mathbb{Z}_{\geqslant 0}$ is weakly-super-modular, $b \in \mathbb{Z}_{\geqslant 0} \cup \{\mathbb{Z} + 1/2\}_{\geqslant 0}$ (i.e. b is semi-integral), let $(x_e)_{e \in E}$ be a vertex optimal solution to the LP of Figure 1 for this instance. Let $E_{>0}$ denote those edges with $x_e > 0$. Then one of the following holds:*

- *there is an edge e^* such that $x_{e^*} = 0$,*
- *or there is an edge e^* such that $x_{e^*} \geqslant 1$,*
- *or there is an edge $e^* = \{u, v\} \in E$ such that $1 > x_{e^*} \geqslant 1/2$, and if $u \in W$, then $b(u) > 1$, and if $v \in W$, then $b(v) > 1$,*
- *or there is a vertex $v \in W$ such that $\deg_{E_{>0}}(v) \leqslant 2b(v) + 2$.*

Now we see how this lemma implies Theorem 1.

Proof. (of Main Theorem) Lemma 1 implies that each iteration of the algorithm is successful. Further, note that the set of edges F satisfies the connectivity requirement function R of instance $\mathcal{I}$. Hence, when the algorithm ends, since it only picked edges with value at least $1/2$, by a standard argument, the cost of the solution produced by the algorithm is at most $2 \cdot \text{lp}(\mathcal{I}) \leqslant 2 \cdot \text{opt}(\mathcal{I})$. Hence, the only thing left to prove is that for

1. Given an instance $\mathcal{I} = (G(V,E), W_0, c, u, R, b)$, initialize
   (a) $F := \emptyset$, $W' := W_0$, $E' := E$, $R'(S) := R(S)$ for all $S \subseteq V$, $b'(v) := b(v)$ for all $v \in W'$, $u'(e) = u(e)$ for all $e \in E$
2. While F is not a feasible solution for G satisfying the connectivity function R do
   (a) Compute a vertex optimal solution $(x_e)_{e \in E}$ to the LP of Figure 1 for the instance $(G(V,E'), W', c, u', R', b')$. Let $H := \{e \in E' : x_e > 0\}$.
   (b) For every edge e with $x_e = 0$, let $E' := E' \backslash \{e\}$.
   (c) For every edge $e = \{u,v\}$ with $x_e \geqslant 1$,
       i. add $\lfloor x_e \rfloor$ copies of e to F
       ii. $u'(e) = u'(e) - \lfloor x_e \rfloor$
       iii. If $u \in W'$,
           A. if $b'(u) \neq {}^3/_2$ then $b'(u) := b'(u) - \lfloor x_e \rfloor$,
           B. else if $b'(u) = {}^3/_2$ then $b'(u) := 1$,
       iv. If $v \in W'$,
           A. if $b'(v) \neq {}^3/_2$ then $b'(v) := b'(v) - 1$,
           B. else if $b'(v) = {}^3/_2$ then $b'(v) := 1$,
   (d) For every edge $e = \{u,v\}$ such that $^1/_2 \leqslant x_e < 1$ and if $u \in W'$, then $b'(u) > 1$, and if $v \in W'$, then $b'(v) > 1$, let
       i. $F := F \cup \{e\}$,
       ii. $u'(e) := u'(e) - 1$
       iii. If $u \in W'$, $b'(u) := b'(u) - ^1/_2$,
       iv. If $v \in W'$, $b'(v) := b'(v) - ^1/_2$.
   (e) For every degree constrained vertex $v \in W'$ such that $\deg_H(v) \leqslant 2b'(v) + 2$, let $W' := W' \backslash \{v\}$.
   (f) For every $S \subseteq V$, let $R'(S) := R(S) - |\delta_F(S)|$.

Fig. 2. An Iterative Rounding based algorithm for Deg-SNDP

Note that for a vertex $u \in W'$, if $b'(u) = {}^3/_2$ then we set $b'(u) = 1$ on picking an edge adjacent to u. We do this because our analysis crucially depends on the fact that $b'(u) \geqslant 1 \; \forall u \in W'$.

every $v \in W_0$, its degree in the final integral solution produced by the algorithm is at most $2b(v) + 2$.

Consider $v \in W_0$ with degree bound $b(v)$. Suppose that in all the iterations of the algorithm, we picked n_1 edges adjacent to v with value 1, and $n_{1/2}$ edges with value in $[^1/_2, 1)$ adjacent to v. In case we picked a 1-edge adjacent to v when $b'(v)$ was $^3/_2$, then we had decreased $b'(v)$ by $^1/_2$ and not by 1. Hence, we will count this edge in $n_{1/2}$ and not in n_1.

There are two cases:

- If at the point when the algorithm terminated v was still a degree constrained vertex, then its degree in the solution produced by the algorithm would be $n_1 + n_{1/2} < 2b(v)$ as we decreased the degree bound by at least $^1/_2$ every time we picked an edge adjacent to v and $b'(v) > 0$ at the point of termination of the algorithm.

- If $v \in W_0$ was not a degree constrained vertex when the algorithm terminated, then at some iteration the degree constraint on it had been removed. This happened when in the set of edges H in that iteration $\deg_H(v) \leq 2b'(v) + 2$. Since v had a degree constraint in that iteration, by a similar argument as above we also get that $n_1 + n_{1/2} \leq 2(b(v) - b'(v))$. Now, $\deg_F(v) \leq n_1 + n_{1/2} + 2b'(v) + 2 \leq 2b(v) + 2$.

This completes the proof of Theorem 1.

4   Proof of Main Lemma

In this section we prove Lemma 1. Consider a vertex optimal solution $(x_e)_{e \in E}$ to the LP for an instance $\mathcal{I} = (G(V, E), W, c, u, R, b)$. Let $E_{>0} \stackrel{\text{def}}{=} \{e : x_e > 0\}$. To prove the lemma we will assume on the contrary all of the following:

1. $0 < x_e < 1$ for each $e \in E_{>0}$.
2. If there is an edge $e = \{u, v\}$ such that $1 > x_e \geq 1/2$, and if $u \in W$ then $b(u) \leq 1$, or if $v \in W$ then $b(v) \leq 1$.
3. For every $v \in W$, $\deg_{E_{>0}}(v) \geq (2b(v) + 2) + 1 \geq 5$.

The proof of this lemma starts with a well known characterization of the vertex optimal solution. The following is standard notation in this setting. A family of sets $\mathcal{L} \subseteq 2^{[V]}$ is said to be *laminar* if for any two sets $S, T \in \mathcal{L}$, either one of them is contained in the other or they are disjoint. In a laminar family, a set S is said to be *child* of T if T is the smallest set containing S. (T is called the *parent* of S.) Thus, a laminar family can be represented by a forest where the nodes are sets and there is an edge between two sets S and T if one is the child of the other. Let $C(S)$ denote the set of children of S. The maximal elements of the laminar family are referred to as the *roots*. The following lemma shows how to derive a laminar family of sets from a vertex optimal solution to the LP of Figure 1.

Lemma 2. *Given an instance $\mathcal{I} = (G(V, E), W, c, u, R, b)$, where R is weakly-super-modular, $b \in \mathbb{Z}_{\geq 0} \cup \{\mathbb{Z} + 1/2\}_{\geq 0}$, let $(x_e)_{e \in E}$ be a vertex optimal solution to the LP of Figure 1 for this instance. Then, there exists a laminar family of sets $\mathcal{L}$ which partitions into $\mathcal{S}$ and $\mathcal{V}$ such that*

1. *For every $v \in \mathcal{V} \subseteq W$, $x(\delta_E(v)) = b(v) \geq 1$ and every $S \in \mathcal{S}$, $x(\delta_E(S)) = R(S) \geq 1$.*
2. *$|\mathcal{L}| = |E_{>0}|$.*
3. *The vectors χ_S, for $S \in \mathcal{L}$, are linearly independent over the reals.*

Proof. The proof follows from the uncrossing method, see [12] and [13]. $\qquad\blacksquare$

Notation. Before we continue with the proof of Lemma 1, we need some notation. Let $\mathcal{L}$ be the laminar family associated to the vertex solution $(x_e)_{e \in E}$ as promised by Lemma 2. We will refer to a member of $\mathcal{S}$ as a *set* and to a member of $\mathcal{V}$ as a *tight vertex*. An edge e is said to be *heavy* is $1 > x_e \geq 1/2$ and *light* if $0 < x_e < 1/2$. We define the *corequirement* (**coreq**) of an edge e as $1/2 - x_e$ if e is light and $1 - x_e$ if it is heavy. For a set S, $\text{coreq}(S) = \sum_{e \in \delta(S)} \text{coreq}(e)$. We will say that a set S is *odd* if $\text{coreq}(S) = a + 1/2$, where $a \in \mathbb{Z}_{\geqslant 0}$, and that it is *even* if $\text{coreq}(S) \in \mathbb{Z}_{\geqslant 0}$.

We say that a set *owns* an endpoint u of an edge $e = \{u, v\}$ if S is the smallest set in $\mathcal{L}$ containing u. For a set S let $c(S)$ denote the number of children of S, $l(S)$ the number of endpoints of light edges owned by it, $h(S)$ the number of endpoints of heavy edges owned by it, $l'(S)$ the number of light edges in $\delta(S)$ and $h'(S)$ denote the number of heavy edges in $\delta(S)$. Note that $l'(S) + h'(S) = |\delta(S)|$.

We say that an edge e is incident on a set S if $e \in \delta(S)$. The degree of a set S is defined as the number of edges incident on it, i.e., $\text{degree}(S) \stackrel{\text{def}}{=} |\delta(S)|$. The following fact is easy to see now.

Fact 2. *A set S has semi-integral corequirement only if $l'(S)$ is odd.*

Proof. $\text{coreq}(S) = \sum_{e \in \delta(S)} \text{coreq}(e) = \sum_{e \in \delta(S) \text{ and } e \text{ is light}}(1/2 - x_e) + \sum_{e \in \delta(S) \text{ and } e \text{ is heavy}}(1 - x_e) = l'(S)/2 + h'(S) - f(S)$. All the three terms $l'(S), h'(S), f(S) \in \mathbb{Z}_{\geqslant 0}$. Therefore, S is semi-integral only if $l'(S)$ is odd.

Proof. (of Main Lemma) We will prove Lemma 1 using a counting argument. Initially, we will assign two tokens to each edge. We will redistribute the tokens in such a manner that each member of $\mathcal{L}$ gets at least 2 tokens while the roots get at least 3 tokens. This will give us a contradiction to the fact that $|\mathcal{L}| = |E_{>0}|$ of Lemma 2.

Token distribution scheme. Initially, we will assign two tokens to each edge. If $e = \{u, v\}$ is a light edge then one of the two tokens assigned to e goes to the smallest set containing u and the other to the smallest set containing v. If e is a heavy edge, then w.l.o.g. assume that $u \in W, b(u) = 1$. In this case assign both tokens of e to the smallest set containing v.

Claim 3. *[Token Redistribution] Consider a tree $\mathcal{T}$ in $\mathcal{L}$ rooted at S. The tokens owned by $\mathcal{T}$ can be redistributed in such a way that S gets at least 3 tokens, and each of its descendants gets at least 2 tokens. Moreover, if $\text{coreq}(S) \neq 1/2$, then S gets at least 4 tokens.*

Proof. We will prove this claim by induction on the height of $\mathcal{T}$. We start with the base case.

1. If S is a leaf set and it has no heavy edges incident on it then, since $f(S) = \sum_{e \in \delta(S)} x_e \geqslant 1$ and $\forall e \in \delta(S) : x_e < 1/2$, therefore S has at least 3 edges incident on it and, hence, will collect at least 3 tokens. It will collect exactly 3 tokens when its degree is 3 in which case $\text{coreq}(S) = \sum_{e \in \delta(S)}(1/2 - x_e) = 3/2 - \sum_{e \in \delta(S)} x_e = 3/2 - f(S)$; since $f(S) \in \mathbb{Z}^+$ and $\text{coreq}(S) > 0$ therefore $\text{coreq}(S) = 1/2$. Hence, by the inductive hypothesis it suffices for S to collect only 3 tokens.

2. In the case when S has a heavy edge, say e_1, incident on it, it will still have at least 1 other edge incident on it as $f(S) = \sum_{e \in \delta(S)} x_e$ is a positive integer and $\forall e \in \delta(S), x_e < 1$. Recall that, by our assumption a heavy edge must have a tight vertex with degree bound equal to 1 as one of its end points. Since S is a leaf set, the tight vertex cannot be contained in S. Therefore, the tight vertex must be that end point of the heavy edge which is not in S. By our token distribution scheme, S would

get both the tokens from the heavy edge and at least one token from the other edge. Hence, it will get at least 3 tokens. S will get exactly 3 only when it has only one other light edge, say e_2, incident on it. In such a case $\mathsf{coreq}(S) = \sum_{e \in \delta(S)} \mathsf{coreq}(e) = (1 - x_{e_1}) + (1/2 - x_{e_2}) = 3/2 - f(S)$; since $f(S) \in \mathbb{Z}_{>0}$ and $\mathsf{coreq}(S) > 0$, therefore $\mathsf{coreq}(S) = 1/2$. Hence, by the inductive hypothesis it suffices for S to collect only 3 tokens.

3. In the case of a tight vertex v, by our assumption it has degree 5 and, hence, will collect at least 5 tokens unless it has a heavy edge incident on it, in which case it might have to give both tokens to the smallest set containing the other endpoint but will still be able to collect at least 4 tokens.

This proves the base case. Now, we move on to the general case. Let us consider the case when S is not a leaf in $\mathcal{L}$. If a set has collected t tokens, we will say that it has a *surplus* of $t - 2$. There are four cases:

1. If S has 4 children (either sets or vertices), then S can collect 1 token from each one of its children, as from the inductive hypothesis each one of its children has a surplus of at least 1. Thus, S can collect 4 tokens for itself.

2. If S has 3 children (either sets or vertices) and if at least one of them has surplus 2, then S can collect 4 tokens for itself. If S owns any end points then again it can collect at least 4 tokens: 1 from each its children and at least 1 from the end point(s) it owns. If all children have a surplus of exactly 1 then S is still able to collect at least 3 tokens and moreover, by the induction hypothesis, all the children of S must have a corequirement of $1/2$. Furthermore, if S owns no endpoints then, using Claim 4, we get that S also has a corequirement of $1/2$. Hence, by the induction hypothesis it suffices for S to collect 3 tokens only.

3. If S has 2 children (either sets or vertices) and both of the children have surplus at least 2 then S can collect 4 tokens from them. If one of them, say S_1 has surplus exactly 1, then by the induction hypothesis $\mathsf{coreq}(S_1) = 1/2$. In such a case, by using Claim 6, it must own at least 1 end point and, hence, can collect at least 3 tokens. It will collect exactly 3 tokens when both the children have a surplus of exactly one (and, hence, both have a corequirement of $1/2$) and S owns exactly one end point (in which case $l(S) + 2h(S) = 1$). In such a case, by Claim 4, it suffices for S to collect 3 tokens only.

4. If S has exactly one child, two cases arise:
   - If the child is a set then, by Claim 5, S must own at least 2 end points and, hence, it can collect at least 3 tokens: at least one from the surplus of the child and at least 2 from end points it owns. S will collect exactly 3 tokens if its child has a surplus of exactly 1 (which can happen only when the child has a coreqirement of $1/2$) and S owns exactly 2 end points (in which case $l(S) + 2h(S) = 2$). In such a case, by Claim 4, it suffices for S to collect 3 tokens only.
   - If the child is a tight vertex v then again 2 cases arise:
     • v has an integral degree constraint: this case can be handled akin to case when the child is a set.
     • v has semi-integral degree constraint : In this case $b(v) \geqslant 3/2$ as our algorithm maintains the invariant that $b'(v) \in \mathbb{Z}^+ \cup \{\mathbb{Z}^+ + 1/2\}$. By our assumption

degree(v) $\geqslant 2b(v) + 3 \geqslant 6$. Hence, $\{v\}$ will be able to collect 6 tokens. By the induction hypothesis it requires only 2 tokens for itself and, hence, can give 4 tokens to S.

Hence, we have proved Lemma 1.

Now we present the proofs of the claims used in this proof.

Claim 4. *Let $S \in S$ and suppose $c(S) + l(S) + 2h(S) = 3$. Furthermore, assume that each child of S, if any, has a corequirement of $1/2$. Then* $\mathsf{coreq}(S) = 1/2$.

Proof. For a light edge e, $\mathsf{coreq}(e) = 1/2 - x_e < 1/2$. For a heavy edge e, $\mathsf{coreq}(e) = 1 - x_e \leqslant 1/2$ as $x_e \geqslant 1/2$.

An argument similar to one used in Exercise 23.3 of [18] can be used to show that S is also odd. Using Fact 2, $\mathsf{coreq}(S)$ is semi-integral. Now, $\mathsf{coreq}(S) \leqslant \sum_{C \in C(S)} \mathsf{coreq}(C) + \sum_e \mathsf{coreq}(e)$ where the second summation is over all those edges whose 1 endpoint is owned by S. Every term in the first summation is $1/2$ and every term in the second summation is at most $1/2$ (by definition of corequirement, $\mathsf{coreq}(e) \leqslant 1/2 \ \forall e \in E$). Note that $\mathsf{coreq}(S)$ is a sum of at most 3 terms each of which is at most $1/2$. Therefore, $\mathsf{coreq}(S) \leqslant 3/2$. Hence, proving that $\mathsf{coreq}(S) < 3/2$ suffices. We do this next.

A term in the second summation will be exactly equal to $1/2$ if the edge corresponding to it is a heavy edge, i.e. $h(S) = 1$. Hence, from the premises of this claim we get that $c(S) + l(S) = 1$. This means that there can be either 1 light edge or 1 child or S contributing to the summation. Either of them will contribute at most half to the summation and, therefore, $\mathsf{coreq}(S) \leqslant 1$. Considering the case when every term in the second summation is strictly less than $1/2$, $\mathsf{coreq}(S) < 3/2$ and, hence, $\mathsf{coreq}(S) = 1/2$. Now the second summation cannot be empty as then the set of edges incident on S would be exactly all the edges incident on its children. This would contradict the linear independence of the vectors $\chi(\delta(S)) \cup \{\chi(\delta(C)\}_{C \in C(S)}$.

Claim 5. *If a set S has only 1 child which is a set, then S owns at least two end points.*

Proof. Let S_1 be a set which is the only child of S. If S owned no end point, that would contradict the linear independence of $\chi(\delta(S))$ and $\chi(\delta(S_1))$. Therefore, S owns at least one endpoint. If S owned exactly one end point (associated with an edge, say e) then $x(\delta(S))$ and $x(\delta(S_1))$ would differ by a x_e, a fraction, which would contradict the fact that S and S_1 are tight sets having integral connectivity requirements. Hence, S owns at least two endpoints, which proves this claim.

Claim 6. *If S has two children (either sets or vertices), one of which has a corequirement of $1/2$, then S must own at least one end point.*

Proof. Let C_1 and C_2 be the children of S with $\mathsf{coreq}(C_1) = 1/2$. If C_1 were a tight vertex, say v, then $\mathsf{coreq}(v) = 1/2$ is equal to $(|\delta(v)|/2 - b(v))$ if v has no tight edge incident on it and is equal to $(|\delta(v)|/2 + 1/2 - b(v))$ if it has a tight edge incident on it. In either case, $|\delta(v)| \leqslant 2b(v) + 1$ and, hence, we would have removed the degree constraint from v. Therefore, C_1 cannot be a tight vertex and, hence, has to be a tight set.

Suppose S does not own any end point. Let $f_1 \stackrel{\text{def}}{=} \sum_{e \in \delta(S) \cap \delta(C_1)} \text{coreq}(e)$, $f_2 \stackrel{\text{def}}{=} \sum_{e \in \delta(C_1) \cap \delta(C_2)}$ $\text{coreq}(e)$ and $f_3 \stackrel{\text{def}}{=} \sum_{e \in \delta(S) \cap \delta(C_2)} \text{coreq}(e)$. By definition, $f_1, f_2, f_3 \geqslant 0$. Since $\chi_{\delta(S)}, \chi_{\delta(C_1)}$ and $\chi_{\delta(C_2)}$ are linearly independent, there has to be at least one edge incident on C_1 and C_2, i.e., $|\delta(C_1) \cap \delta(C_2)| \geqslant 1$. Therefore, $f_2 > 0$. $f_1 + f_2 = \text{coreq}(C_1) = 1/2$. Now, since $\text{coreq}(C_1)$ is semi-integral and that C_1 is a set, $l(C_1)$ must be odd by Fact 2.

Further, $l'(S) = l'(C_1) + l'(C_2) - 2l'(\delta(C_1) \cap \delta(C_2))$. Therefore, if $l'(C_2)$ is odd then $l'(S)$ will be even and if $l'(C_2)$ is even then $l'(S)$ will be odd. $l'(C_2)$ and $l'(S)$ having different parities implies that S and C_2 have different corequirements: one of them being integral and one being semi-integral. Now, $\text{coreq}(S) - \text{coreq}(C_2) = (f_1 + f_3) - (f_2 + f_3) = f_1 - f_2$. But $f_1, f_2 \geqslant 0$, $f_1 + f_2 = 1/2$ and $f_2 > 0$ and, hence, $-1/2 < f_1 - f_2 < 1/2$, which implies that $\text{coreq}(S)$ and $\text{coreq}(C_2)$ (both being half-integral and differing by less than $1/2$) must be equal, which contradicts the fact S and C_2 have different corequirements. Hence, it cannot be the case that S owns no end point.

This completes the proofs.

Extensions. Our techniques trivially extend to the case when there is a lower bound $l(v)$ on the degree of each vertex $v \in V$. Any degree lower bound constraint can be considered as a connectivity constraint $R(v) = l(v)$ (for the cut $(\{v\}, V \setminus \{v\})$). It can be easily verified that the augmented connectivity function R still remains weakly supermodular. Therefore, any feasible solution to LP of Figure 1 with the augmented R will satisfy all degree lower bounds implicitly.

References

1. Bansal, N., Khandekar, R., Nagarajan, V.: Additive guarantees for degree bounded directed network design. In: STOC, pp. 769–778 (2008)
2. Chan, Y.H., Fung, W.S., Lau, L.C., Yung, C.K.: Degree bounded network design with metric costs. In: FOCS, pp. 125–134 (2008)
3. Chaudhuri, K., Rao, S., Riesenfeld, S., Talwar, K.: What would edmonds do? augmenting paths and witnesses for degree-bounded msts. Algorithmica 55(1), 157–189 (2009)
4. Fleischer, L., Jain, K., Williamson, D.P.: Iterative rounding 2-approximation algorithms for minimum-cost vertex connectivity problems. J. Comput. Syst. Sci. 72(5), 838–867 (2006)
5. Fürer, M., Raghavachari, B.: Approximating the minimum-degree steiner tree to within one of optimal. J. Algorithms 17(3), 409–423 (1994)
6. Goemans, M.X.: Minimum bounded degree spanning trees. In: FOCS, pp. 273–282 (2006)
7. Goemans, M.X., Bertsimas, D.: Survivable networks, linear programming relaxations and the parsimonious property. Math. Program. 60, 145–166 (1993)
8. Jain, K.: Factor 2 approximation algorithm for the generalized steiner network problem. In: FOCS, pp. 448–457 (1998)
9. Király, T., Lau, L.C., Singh, M.: Degree bounded matroids and submodular flows. In: IPCO, pp. 259–272 (2008)
10. Klein, P.N., Krishnan, R., Raghavachari, B., Ravi, R.: Approximation algorithms for finding low-degree subgraphs. Networks 44(3), 203–215 (2004)
11. Könemann, J., Ravi, R.: A matter of degree: improved approximation algorithms for degree-bounded minimum spanning trees. In: STOC, pp. 537–546 (2000)

12. Lau, L.C., Naor, J., Salavatipour, M.R., Singh, M.: Survivable network design with degree or order constraints. In: Johnson, D.S., Feige, U. (eds.) STOC, pp. 651–660. ACM, New York (2007)
13. Lau, L.C., Singh, M.: Additive approximation for bounded degree survivable network design. In: Ladner, R.E., Dwork, C. (eds.) STOC, pp. 759–768. ACM, New York (2008)
14. Lau, L.C., Singh, M.: Iterative rounding and relaxation. In: RIMS Kokyuroku Bessatsu, Kyoto (2009)
15. Melkonian, V., Tardos, É.: Algorithms for a network design problem with crossing supermodular demands. Networks 43(4), 256–265 (2004)
16. Singh, M.: Iterative methods in combinatorial optimization. PhD Thesis, Tepper School of Business, Carnegie Mellon University (2008)
17. Singh, M., Lau, L.C.: Approximating minimum bounded degree spanning trees to within one of optimal. In: STOC, pp. 661–670 (2007)
18. Vazirani, V.V.: Approximation Algorithms. Springer, Heidelberg (March 2001)

Minimizing the Diameter of a Network Using Shortcut Edges

Erik D. Demaine and Morteza Zadimoghaddam

MIT Computer Science and Artificial Intelligence Laboratory
32 Vassar St., Cambridge, MA 02139, USA
{edemaine,morteza}@mit.edu

Abstract. We study the problem of minimizing the diameter of a graph by adding k shortcut edges, for speeding up communication in an existing network design. We develop constant-factor approximation algorithms for different variations of this problem. We also show how to improve the approximation ratios using resource augmentation to allow more than k shortcut edges. We observe a close relation between the single-source version of the problem, where we want to minimize the largest distance from a given source vertex, and the well-known k-*median* problem. First we show that our constant-factor approximation algorithms for the general case solve the single-source problem within a constant factor. Then, using a linear-programming formulation for the single-source version, we find a $(1 + \varepsilon)$-approximation using $O(k \log n)$ shortcut edges. To show the tightness of our result, we prove that any $(\frac{3}{2} - \varepsilon)$-approximation for the single-source version must use $\Omega(k \log n)$ shortcut edges assuming P $\neq$ NP.

Keywords: Approximation algorithms, network design, network repair.

1 Introduction

Diameter is an important metric of network performance, measuring the worst-case cost of routing a point-to-point message or a broadcast. Such communication operations are ubiquitous in a variety of networks, such as information networks, data networks, telephone networks, multicore networks, and transportations networks. In information networks, search engines need to access all nodes (or sometimes just "important" nodes) in the shortest possible time; nodes might represent webpages and edges links. We can also see this problem as the information diffusion time in information networks [8,9]. In transportation networks, passengers want short commutes. In telephone networks, we want to reduce the length of the paths between the nodes to reduce connection lag. In multicore processors, we want to build an underlying network to have short paths between different cores [2]; in many cases, the bottleneck in running time is the time spent on communication between cores.

Each of these applications has several constraints on the network design, from existing infrastructure to connectivity or fault tolerance. Minimizing diameter

H. Kaplan (Ed.): SWAT 2010, LNCS 6139, pp. 420–431, 2010.
© Springer-Verlag Berlin Heidelberg 2010

may be at odds with some of these constraints, yet small diameter remains important. Thus we consider the problem of augmenting an existing network design with a limited number of additional edges to reduce the diameter as much as possible.

Many variations are also of interest. We might want to reduce the shortest-path distances among the nodes in a just special subset of the nodes. In a telephone or transportation network, a company might only care about distances among their own nodes (e.g., phone numbers or airports), though doing so might require adding edges (e.g., cables or flights) anywhere in the network. In an information network, we might ignore all spam pages.

In the single-source version of the problem, a node wants to construct edges in order to minimize its distances from the other nodes. This problem has been considered in selfish agent networks [1,4,5,6], where every node simultaneously tries to solve the single-source problem. Agents have high incentive to join social networks with low diameter because messages spread in short time with small delays. Because the budget of these selfish agents is limited in many applications [10], they can not add more than a few edges, and they want to minimize their distances to the other nodes.

Model. In all these applications, we can assume that we are given a weighted undirected graph $G = (V, E, \ell)$, a positive integer k, and a nonnegative real number δ. The length of edge e is represented by $\ell(e)$. Our goal is to add k shortcut edges of length δ in order to minimize the diameter of the resulting graph. Recall that the diameter of a graph is the maximum distance between two nodes, and the distance between two nodes is the length of the shortest path between them. In most applications, including [11], δ is a small constant compared to the diameter of the graph.

Related Work. Meyerson and Tagiku [11] considered the problem of minimizing the average distance between the nodes instead of the maximum distance. This is the only work that considers the problem with a hard limit on the budget (the number of edges we can add). They obtained several constant-factor approximations using the *k-median with penalties* problem. They also improved the best known approximation ratio for metric k-median with penalties, to get better approximation factors for the other problems they consider. If α denotes the best approximation known for metric k-median with penalties, they presented an α-approximation for the single-source average-shortest-path problem, and a 2α-approximation for the general average shortest-path problem.

Our Results. We start with a simple clustering algorithm, and find a lower bound on the diameter of optimum solution. In Section 2, we find a $(4 + \varepsilon)$-approximation algorithm (using at most k shortcut edges).

Next we study approximation algorithms with *resource augmentation*: by allowing the algorithm to add more than k edges, but still comparing to the optimal solution with just k edges, we can decrease the approximation ratio. To do so, we study the structure of the optimum solutions in more detail to get a better lower bound. In Section 3, we obtain a $(2 + \varepsilon)$-approximation using at most $2k$ shortcut edges for small values of δ. Previous work assumes that δ is zero for simplicity [11], and in most of the applications, it is negligible comparing to the diameter of the graph.

In Section 4, we study the single-source version of the problem in which we want to minimize the maximum distance of all nodes from a specified node, called the *source*, by adding k shortcut edges. We prove that any α-approximation algorithm for the original problem (minimizing diameter of the whole graph) can be seen as a 2α-approximation for the single-source version of the problem. We present linear-programming approaches to get better approximations for the single-source version with resource augmentation. We obtain a $(1 + \varepsilon)$-approximation for the single-source problem using $O(k \log n)$ shortcut edges. Our linear program is similar to that for the k-median problem studied in [3]. To show the optimality of our algorithm, we prove that any $(\frac{3}{2} - \varepsilon)$-approximation for the single-source problem uses at least $\Omega(k \log n)$ shortcut edges assuming P $\neq$ NP.

In Section 5, we consider the multicast version of our problems in which only a given subset of nodes is important, and we want to reduce the maximum distance between the nodes in the given subset. This problem also has a single-source variant in which we want to minimize the maximum distance of the nodes in the subset from a given source node. We show that all of our results apply just as well to these multicast variations.

2 $(4 + \varepsilon)$-Approximation Using k Shortcut Edges

Let D be the diameter of the current graph G (without any shortcut edges). The diameter of the optimum solution is a value $D' \leq D$. In fact, there exists a set of at most k edges S such that the diameter of graph $G = (V, E \cup S)$ is D'. We want to find and add k shortcut edges, such that the new graph after adding our edges has diameter at most a constant times D'.

At first, we estimate the value of D' with an iterative process as follows. We need to find a lower bound and an upper bound for D'. We know that D' is not more than D. So D can be seen as an upper bound. Let a be the minimum length of the edges in G. We know that D' can not be less than $\min\{a, \delta\}$ because every path between any pair of vertices should use at least either one of the current edges in G or one of the shortcut edges. We can also assume that a is not zero otherwise we can contract the zero edges, and solve the problem for the graph after these contractions. If δ is also nonzero, we can use $\min\{a, \delta\}$ as a nonzero lower bound for D'. Otherwise if k is at least $n - 1$, we can add $n - 1$ shortcut zero edges to build a spanning tree with only zero edges. This way, the diameter of graph would be zero. So the problem is solved. Otherwise we can not reduce the diameter to $\delta = 0$, and the distance between some pairs of vertices would be at least a after adding $k < n - 1$ shortcut edges. So in all cases, we can find a positive lower bound for D' which is either $\min\{a, \delta\}$ or a. Define L to be this lower bound. So we can assume that $0 < L \leq D' \leq D$.

Choose an arbitrary small $\varepsilon > 0$. There exists an $0 \leq i \leq \log_{1+\varepsilon}(D/L)$ such that $D/(1 + \varepsilon)^{(i+1)} < D' \leq D/(1 + \varepsilon)^i$.

Clustering algorithm. This algorithm receives an input parameter $x \geq 0$. We partition the vertices of our graph into clusters of diameter at most $2x$ as follows. At first we pick a subset of vertices S as the centers of our clusters. This set should satisfy the following properties. The distance between any pair of vertices of set S in graph G should be greater than $2x - \delta$. For every vertex $u \notin S$, there should be a vertex v in S whose distance to v is at most $2x - \delta$. We find a set S with above properties as follows. Choose an arbitrary vertex from G like v and put it in S. While there exists a vertex like u outside S whose distance to every vertex in S is greater than $2x - \delta$, we add u to S. Clearly this iterative process finishes in at most n iterations because there are n vertices in G. For every vertex u outside S, there exists a vertex v in S such that $\text{dist}(u, v)$ is at most $2x - \delta$ where $\text{dist}(u, v)$ is the distance between u and v. Otherwise we would add u to S. Let k' be $|S|$, and $v_1, v_2, \ldots, v_{k'}$ be the vertices in S.

If we add $k' - 1$ shortcut edges from v_1 to all other center vertices in set S $(v_2, v_3, \ldots, v_{k'})$, the diameter of the new graph is at most $2[2x - \delta] + 2\delta = 4x$. Consider two vertices u and w in the new graph. There are two vertices in v_i and v_j in S such that $\text{dist}(v_i, u) \leq 2x - \delta$ and $\text{dist}(v_j, w) \leq 2x - \delta$. We also know that the distance between v_i and v_j is at most 2δ in the new graph because they are both connected to v_1 using two shortcut edges. We conclude that we can reduce the diameter of G to a value at most $4x$ using $k' - 1$ edges. Following we show how to use this clustering algorithm to solve our problem. Note that D' is the diameter of the optimum solution. We show that without using at least $k' - 1$ edges, the diameter of graph can not be reduced to x or less. So the number of edges used in the optimum solution, which is k, should be at least $k' - 1$ if D' is at most x.

Lemma 1. *If D' is at most x, the number of edges used in the optimum solution is at least $k' - 1$, and therefore k is at least $k' - 1$.*

Proof. Assume that there are less than $k' - 1$ edges used in the optimum solution. Let G' be the new graph after addition of shortcut edges of the optimum solution. Consider a shortest path tree T from vertex v_1 in G'. The shortest path tree T is a tree that contains a shortest path from v_1 to every other vertex of graph G', i.e., the result of the Dijkstra's algorithm in graph G' from source v_1. Note that the distance between v_1 and v_i in G is greater than $2x - \delta$ because they are both in S. Their distance in G' is at most D'. Note that $x \geq D'$, and δ can not be greater than D' otherwise addition of some edges with length δ does not reduce the diameter of the graph to some value less than D'. So $2x - \delta$ is at least D', and therefore the distance between v_i and v_1 is reduced during addition of the new edges. So there exists at least one shortcut edge in the new shortest path from v_i to v_1. Let e_i be the first shortcut edge in the shortest path from v_i to v_1 in T'. For each $2 \leq i \leq k'$, we have a shortcut edge e_i. But there are less than $k' - 1$ shortcut edges in the whole graph. So some of these edges must be the same.

Suppose e_i and e_j are the same edge (u', w'). Let P_i and P_j be the shortest paths from v_i and v_j to v_1 that both use the edge (u', w'). Without loss of generality, assume that the distance from u' to v_1 is less than the distance from

w' to v_1 in graph G'. So both paths P_i and P_j should use this edge in direction u' to w'. Let Q_i and Q_j be the first parts of the paths P_i and P_j that connects v_i and v_j to u' respectively. So Q_i and Q_j are two paths that do not use any shortcut edge (note that we picked the first shortcut edge in each path). Because the length of P_i and P_j are both at most D', and they both use at least one shortcut edge, the lengths of paths Q_i and Q_j are at most $D' - \delta$. Because Q_i and Q_j are two paths from v_i and v_j to the same destination u', and they do not use any shortcut edge, the length of the shortest path between v_i and v_j in graph G is at most the sum of the lengths of Q_i and Q_j which is $2(D' - \delta) \leq 2x - 2\delta$. This is a contradiction because $\text{dist}(v_i, v_j)$ should be greater than $2x - \delta$. This contradiction shows that the number of shortcut edges in the optimum solution is at least $k' - 1$.   □

Now we can use the clustering algorithm iteratively to find a $(4 + \varepsilon)$-approximation algorithm. Recall that an instance of the problem consists of a graph G and two parameters k and δ. We should use k shortcut edges of length δ to minimize the diameter of the graph.

Theorem 1. *For any $\varepsilon' > 0$, there exists a polynomial-time $(4 + \varepsilon')$-approximation algorithm that uses at most k shortcut edges.*

Proof. We choose an arbitrary small $\varepsilon > 0$. There exists an $0 \leq i \leq \log_{1+\varepsilon}(D/L)$ such that $D/(1 + \varepsilon)^{(i+1)} < D' \leq D/(1 + \varepsilon)^i$. We can run the clustering algorithm with $x = D/(1 + \varepsilon)^j$ for different values of $0 \leq j \leq \log_{1+\varepsilon}(D/L)$, so for one of these values, the above inequality holds, and we can estimate D' with multiplicative error ε. If the number of clusters k' is at most $k + 1$, we can find a solution with diameter $4x$ by adding $k' - 1 \leq k$ edges. If the number of clusters is more than $k + 1$, using Lemma 1, we know that $D' > x$.

Let $x = D/(1+\varepsilon)^{j'}$ be the smallest value of x for which the number of clusters is at most x. We can find a solution with diameter $4x = 4D/(1+\varepsilon)^{j'}$ in this case. On the other hand, we know that the number of clusters for $x = D/(1+\varepsilon)^{j'+1}$ is more than $k+1$, so $D/(1+\varepsilon)^{j'+1}$ is less than D'. We conclude that the diameter of our solution is at most $4D/(1 + \varepsilon)^{j'} \leq 4(1 + \varepsilon)D'$. By choosing $\varepsilon = \varepsilon'/4$, we find a $(4 + \varepsilon')$-approximation.   □

3   Improving the Approximation Ratio Using $2k$ Edges

In this section, we show how to use the clustering algorithm to find a solution with at most $2k$ additional edges having diameter at most $(2 + \varepsilon)D' + 2\delta$, where D' is the diameter of the optimum solution using k additional edges.

We change our clustering algorithm slightly as follows. We pick the vertices of S such that their distance is greater than $2x$ instead of $2x - \delta$. Like before, we iteratively add a vertex u to S if its distance to all vertices in S is more than $2x$. We stop when we can not insert anything to S. Again we run the clustering algorithm with $x = D/(1 + \varepsilon)^j$ for different values of $0 \leq j \leq \log_{1+\varepsilon}(D/L)$. We prove that this time the number of clusters is not more than $2k$ when x is at least $D'/2$ as follows.

Lemma 2. *If x is at least $D'/2$, the number of clusters in our algorithm is not more than $2k + 1$.*

Proof. Here we show why our algorithm acts like a clustering algorithm. Let $v_1, v_2, \ldots, v_{k'}$ be the k' centers we pick in our algorithm. Partition the vertices of graph G into k' clusters as follows. Put $v_1, v_2, \ldots, v_{k'}$ in clusters $C_1, C_2, \ldots, C_{k'}$ respectively. For every other vertex u, find the center v_i $(1 \leq i \leq k')$ with minimum distance $dist(v_i, u)$, and put u in cluster C_i (remember $dist(v_i, u)$ is the distance between v_i and u in graph G). This distance is at most $2x$ for every vertex u by definition of our algorithm (otherwise we could add u to S in the clustering algorithm). The distance between each pair of the k' centers is greater than $2x$. We conclude that all vertices in graph G whose distance to v_i is at most x are in cluster C_i for each $1 \leq i \leq k'$. We can prove this by contradiction. Assume vertex u whose distance to v_i is at most x in in another cluster C_j. It means that center v_j is the closest center to u so the distance between u and v_j is also at most x. Therefore vertex u has distance at most x from both centers v_i and v_j. So the distance between two centers v_i and v_j is at most $x + x = 2x$ which is a contradiction. We conclude that each cluster contains the vertices around its own center with radius x (and probably some other vertices as well).

Now consider the optimum solution that uses at most k edges. If the number of clusters k' is at most $2k + 1$, the claim is proved. Otherwise there exists at least two clusters like C_i and C_j such that the additional edges are not incident to the vertices of $C_i \cup C_j$. Because every additional edge has two endpoints, therefore it can be incident to at most two clusters. So the number of clusters that are incident to some additional edges is not more than twice the number of additional edges. Because we have at least $2k + 2$ clusters, there exists two clusters like C_i and C_j whose vertices are not incident to any additional edge.

Let G' be the new graph after adding additional edges in the optimum solution. The distance between v_i and v_j should be at most D' in G'. Their distance is greater than $2x \geq D'$ in graph G. This means that the shortest path between v_i and v_j in G' should use at least one of the additional edges. Let P be this shortest path between v_i and v_j in G'. Suppose $v_i = u_1, u_2, u_3, \ldots, u_l = v_j$ are the vertices of this path. Assume that edge (u_a, u_{a+1}) is the first additional edge in this path, and edge (u_b, u_{b+1}) is the last additional edge in this path where $a \leq b$. Let P_1 be the first part of the path P before the first additional edge, i.e., $P_1 = (v_i = u_1, u_2, \ldots, u_a)$. And let P_2 be the last part of the path P after the last additional edge, i.e., $P_2 = (u_{b+1}, u_{b+2}, \ldots, u_l = v_j)$.

The paths P_1 and P_2 are two shortest paths in graph G' and G between pairs of vertices (v_i, u_a) and (u_{b+1}, v_j). Vertices u_a and u_{b+1} are outside clusters C_i and C_j because these two clusters are not incident to any additional edge, but vertices u_a and u_{b+1} are both incident to some additional edges. So the distance between u_a and v_i is greater than x otherwise u_a would be in cluster C_i (as proved above). So the length of path P_1 is more than x. With the same proof, the length of path P_2 is also more than x. So the length of P is greater than $x + x + \delta \geq 2x \geq D'$. This is a contradiction because the distance between v_i

and v_j should be at most D' in the new graph G'. This shows that the number of clusters k' is at most $2k + 1$.   □

Now if we add $k' - 1$ from v_1 to all other centers, the diameter of the new graph would be at most $2x + 2\delta$. If we choose x such that $D'/2 \leq x \leq D'(1+\varepsilon)/2$, the number of additional edges in our solution would not be more than $2k$, and the diameter of our graph would be at most $(2 + \varepsilon)$ times the optimum solution.

Theorem 2. *There is a polynomial-time algorithm which finds a solution with diameter at most $(2 + \varepsilon)D' + 2\delta$ using at most $2k$ edges.*

Proof. Similar to our previous approach, we know that there exists an $0 \leq i \leq \log_{1+\varepsilon}(D/L)$ such that $D/(1 + \varepsilon)^{(i+1)} < D' \leq D/(1 + \varepsilon)^i$. We can run our clustering algorithm with $x = D/(1 + \varepsilon)^j$ for different values of $0 \leq j \leq \log_{1+\varepsilon}(D/L)$. This way we can find x such that $D'/2 \leq x \leq D'(1+\varepsilon)/2$ in one of our runs. In that run, we find the desired solution.   □

4   Single-Source Version

In this section, we study the problem of adding k shortcut edges in order to minimize the maximum distance of all vertices from a given source s. Let $e(s)$ denote the maximum distance of vertex s from all other vertices, known as the *eccentricity* of vertex s.

At first we show that constant-factor approximations for the diameter-minimization problem can be converted to constant-factor approximations for the single-source version.

Lemma 3. *If there exists an α-approximation for diameter minimization problem, then there also exists a 2α-approximation for the single-source version.*

Proof. Consider a graph G. Let D' be the minimum possible diameter of G after adding k shortcut edges. Let r_s be the minimum possible $e(s)$ after adding k shortcut edges. We show that $D'/2 \leq r_s \leq D'$. If we add the k edges to reduce the diameter of G to D', the eccentricity of s would be also at most D'. So r_s can not be greater than D'. On the other hand, if we add the k edges such that the distance of every vertex to s is at most r_s, the distance between any pair of vertices can not be greater than $r_s + r_s = 2r_s$ using the triangle inequality. So the diameter of this graph is at most $2r_s$. We conclude that $D' \leq 2r_s$.

So if we use the α-approximation to minimize diameter, we get a graph with diameter at most $\alpha D'$. The eccentricity of s is also at most $\alpha D'$ which is at most $2\alpha r_s$. So using the same algorithm we get a 2α-approximation for the single-source version of the problem.   □

In the remainder of this section, we show how to obtain a $(1 + \varepsilon)$-approximation algorithms using more additional edges. We use a linear-programming formulation similar to the linear-programming formulations of facility location and k-median problems. Then we show how the single-source version of the problem can solve the set-cover problem. We conclude that any $(\frac{3}{2} - \varepsilon)$-approximation for the single-source version should use at least $\Omega(k \log n)$ edges. This shows the optimality of our linear-programming algorithm.

We need the following lemma in our algorithm.

Lemma 4. *There exists an optimal solution for the single-source version in which all k shortcut edges are incident to the given source s.*

Proof. The proof is similar to the proof of Lemma 1 in [11].   □

First we solve a decision problem using linear programming. The decision problem is the following: can we add k shortcut edges to reduce the eccentricity of vertex s to some value at most x where x is a given value in the input graph? In the other words, is r_s at most x? If r_s is at most x, then the following linear program has a feasible integral solution.

Let $v_1, v_2, \ldots, v_n$ be the vertices of the input graph G. Without loss of generality assume that s is v_1. For every vertex v_i put a variable y_i in the linear program where $2 \leq i \leq n$. For every pair of vertices v_i and v_j put a variable $x_{i,j}$ where $2 \leq i, j \leq n$. All these variables are between 0 and 1. If we assume the integer programming version of this linear program. The variable y_i is equal to 1 if there is a shortcut edge from v_i to $s = v_1$, and it is zero when there is no such an edge. We add the constraint $\sum_{i=2}^{n} y_i \leq k$ because we know that the number of shortcut edges is at most k in the optimum solution. Variable $x_{i,j}$ is equal to 1 if the shortest path from vertex v_j to s in the optimum solution uses the shortcut edge v_i to v_1. If $x_{i,j}$ is 1, there should be an edge from v_i to s. So we add the constraint $x_{i,j} \geq y_i$ for any $2 \leq i, j \leq n$. On the other hand, vertex v_j can not use shortcut edge (v_i, s) if the distance between v_j and v_i is more than $x - \delta$. So we define the variable $x_{i,j}$ only for a pair of vertices (v_i, v_j) such that $\text{dist}(v_i, v_j)$ is at most $x - \delta$. If the distance between s and v_j is at most x, the vertex v_j does not need to use any shortcut edge to reach s. But if $\text{dist}(v_j, s)$ is greater than x, it has to use one of this edges, so we have

$$\text{dist}(v_j, s) > x : \sum_{i:\text{dist}(v_i,v_j)\leq x-\delta} x_{i,j} = 1 \qquad \text{for } 1 \leq j \leq n.$$

Here is the whole formulation of our linear program, or more precisely, our linear feasibility problem, as we do not want to minimize or maximize anything.

$$\sum_{i=2}^{n} y_i \leq k,$$

$$x_{i,j} \leq y_i \qquad \text{for } 2 \leq i, j \leq n,$$

$$\sum_{i:\text{dist}(v_i,v_j)\leq x-\delta} x_{i,j} = 1 \qquad \text{for } 1 \leq j \leq n \text{ with } \text{dist}(v_j, s) > x.$$

As described above, if x is at least the optimum solution r_s, then the above linear program has a feasible solution. Next we show how to solve our problem using this linear program, and then show how to find the best x.

Lemma 5. *If there exists a feasible solution to the above linear program, then there exists a polynomial-time algorithm that finds $O(k \log n)$ shortcut edges whose addition reduces the eccentricity of vertex s to some value at most x.*

Proof. The algorithm proceeds as follows. Solve the linear program and find a feasible solution. While there exists some vertex in graph whose distance to s is greater than x, do the following. Add a shortcut edge (v_i, s) to the graph with probability y_i for each $2 \leq i \leq n$. After adding these edges, if there still exist a set of vertices whose distance to s is greater than x, do the same. Iteratively add edges to the graph until the distances of all vertices to s are at most x.

When this algorithm stops, the eccentricity of s is at most x. Now we show that the algorithm does not add a lot of edges compared to the optimum solution. The expected number of edges we add in each phase is $\sum_{i=2}^{n} y_i$, which is at most k. So we do not add more than k edges in each iteration in expectation. We now prove that the number of iterations is at most $O(\log n)$ with high probability (probability $1 - 1/n^c$ for arbitrary constant c).

Consider a vertex v_j whose distance to s is greater than x. Let $\{v_{a_1}, v_{a_2}, \ldots, v_{a_l}\}$ be the set of vertices whose distance to v_j is at most $x - \delta$. We know that $\sum_{i=1}^{l} x_{a_i,j}$ is equal to 1. On the other hand, we have that y_{a_i} is at least $x_{a_i,j}$. So $\sum_{i=1}^{l} y_{a_i}$ is at least 1. If we add a shortcut edge from one of these l vertices to s, the distance of v_j to s reduces to at most x. So if its distance to s is still greater than x after one iteration, it means that we did not add any of these edges in this iteration.

The probability of the event that we do not add any edge from these k edges to s in one iteration is $\prod_{i=1}^{k}(1-y_{a_i})$ which is at most $[1-(\sum_{i=1}^{l} y_{a_i}/l)]^l$. Because $\sum_{i=1}^{l} y_{a_i}$ is at least 1. This probability is at most $(1-1/l)^l$ which is at most $1/e$. So the probability of not choosing any of these edges in p iterations is at most $1/e^p$. If we do this iterative process for $(c+1)\ln(n)$ times, the distance of v_j to s is greater than x with probability at most $1/e^{(c+1)\ln(n)} = 1/n^{(c+1)}$. Using the union bound, we can prove that there exists a vertex with distance greater than x from s with probability at most $n \cdot (1/n^{(c+1)}) = 1/n^c$. So with high probability (at least $1 - 1/n^c$) all distances from s are at most x, and the algorithm stops after $(c+1)\ln(n)$ iterations. □

Theorem 3. *For any $\varepsilon > 0$, there exists a polynomial-time algorithm that adds $O(k \log n)$ edges to reduce the eccentricity of s to at most $1+\varepsilon$ times the optimum eccentricity r_s for k shortcut edges.*

Proof. Let r be the eccentricity of the source in the input graph G. The optimum eccentricity r_s is at most r. We also know that r_s can not be less than δ. So r_s is in range $[\delta, r]$. Run the above algorithm for $x = r/(1+\varepsilon)^i$ for $0 \leq i \leq \log_{(1+\varepsilon)}(r/\delta)$. Consider the smallest x for which the linear program has a feasible solution, and return the result of the above algorithm in that case. There exists a j for which r_s is in range $[r/(1+\varepsilon)^{j+1}, r/(1+\varepsilon)^j]$. The linear program has a feasible solution for $x = r/(1+\varepsilon)^j$ because $r/(1+\varepsilon)^j$ is at least r_s. So we can find a solution with eccentricity x which is at most $(1+\varepsilon)r_s$ using $O(k \log n)$ shortcut edges. □

Now we reduce the set-cover problem to our single-source problem in order to show that using $o(k \log n)$ shortcut edges, we can not get an approximation ratio of better than $\frac{3}{2}$.

Theorem 4. *Any polynomial-time $(\frac{3}{2} - \varepsilon)$-approximation algorithm for the single-source version of our problem needs $\Omega(k \log n)$ shortcut edges assuming* $\mathrm{P} \neq \mathrm{NP}$.

Proof. Consider an instance of set cover. There are m sets $S_1, S_2, \ldots, S_m$, and we want to find the smallest collection of these sets whose union is equal to the union of all these m sets. Suppose there are n items in the union of these m sets. Construct a graph as follows. Let $v_1, v_2, \ldots, v_n$ be the items. Put a vertex for each of these n items. For each set S_j, put a vertex u_j. Connect u_j and v_i for each $1 \leq i \leq n$, and $1 \leq j \leq m$ if item v_i is in set S_j. Add two other vertices s and s'. Vertex s is the source of our instance, and is only connected to s'. The vertex s' has also m edges to all vertices $u_1, u_2, \ldots, u_m$. Set the length of all edges, and δ to be equal to 1. Now the problem is to add k shortcut edges in order to minimize the eccentricity of source s in this graph.

The eccentricity of s is now equal to 3. We prove that there exists a set of k shortcut edges whose addition reduces the eccentricity of s to at most 2 if and only if there is a solution for set-cover instance with size at most k. Assume that the set-cover instance has a solution with size at most k. We can add k edges from s to the vertices associated with these k sets (k subsets in the solution of the set-cover instance). This way the eccentricity of s would be at most 2. We now prove that if we can reduce the eccentricity of s using only k edges, the set-cover instance has a solution of size k. Note that the only vertices whose distance to s is more than 2 are $v_1, v_2, \ldots, v_n$. Using Lemma 4, we know that the k edges are all incident to s. We prove that there is an optimal solution in which all edges are between s and the vertices $u_1, u_2, \ldots, u_m$. Assume that there is an additional edge from s to v_i. Any path from s to v_j $(j \neq i)$ that uses this edge, has length at least 3. So the only usage of this edge is reducing the distance of v_i to s from 3 to 1. There exists a vertex u_l such that there is an edge from v_i to u_l (the item v_i is in some set S_l). We can add an edge from u_l to s instead of an edge from v_i to s. The eccentricity of s would be still at most 2. So we can assume that all edges are from s to vertices $u_1, u_2, \ldots, u_m$. The distances of all vertices $v_1, v_2, \ldots, v_n$ are also at most 2. So for each v_i there exists an additional edge (s, u_l) such that item v_i is in set S_l. So the k additional edges form a solution of size k for the set-cover instance.

Assume that there exists a $(\frac{3}{2} - \varepsilon)$-approximation algorithm A for the single-source problem. For any k, if there is a solution of size at most k for the set-cover instance, the eccentricity of s in the optimum solution of the single-source instance is at most 2. So the eccentricity of s in the solution of algorithm A can not be 3 or more. Because algorithm A is a $(\frac{3}{2} - \varepsilon)$-approximation, so it has to find a solution with eccentricity at most $2 \cdot (\frac{3}{2} - \varepsilon) < 3$. The fact that the eccentricity can only take integer values shows that the result of algorithm A also has eccentricity at most 2. The result of algorithm A can be converted to a solution for the set-cover instance. We also know that there is no $o(\log n)$-approximation for set-cover problem [7], so algorithm A uses at least $O(k \log n)$ in its solution. This completes the proof. $\square$

5   Multicast Version

In this section we show that our results and techniques are all applicable in the multicast version of the problem in which we just care about a subset of the nodes. Formally we are given an undirected weighted graph $G = (V, E, \ell)$, and a subset of vertices $V' \subseteq V$. We want to add k shortcut edges (of a fixed given length δ) in order to minimize the maximum distance between the nodes in set V'. In previous parts, we showed how to solve this problem when V' is equal to V. In the single-source version of the multicast problem, we are also given a source node s, and we want to minimize the maximum distance of nodes in V' from source s.

For our clustering algorithm, we just need to pick centers from vertex set V'. So we do not select any vertex outside V' as a center in our algorithm. We stop when we can not select any vertex of set V'. We get the same approximation ratio by this method, and all proofs and claims work similarly in this case as well.

For the linear-programming approach, we have to write the constraint $\sum_{i:\mathrm{dist}(v_i,v_j)\leq x-\delta} x_{i,j} = 1$ for vertex v_j if v_j is in set V', and its distance from s is greater than x, i.e., $\mathrm{dist}(v_j, s) > x$. Because the distances of vertices outside V' from s are not important for us. Again all claims can be proved in the same way in this case as well. To make it more clear, the new linear-programming formulation is the following:

$$\sum_{i=2}^{n} y_i \leq k,$$

$$x_{i,j} \leq y_i \qquad \text{for } 2 \leq i, j \leq n,$$

$$\sum_{i:\mathrm{dist}(v_i,v_j)\leq x-\delta} x_{i,j} = 1 \qquad \text{for } j \in V' \text{ with } \mathrm{dist}(v_j, s) > x.$$

References

1. Albers, S., Eilts, S., Even-Dar, E., Mansour, Y., Roditty, L.: On Nash equilibria for a network creation game. In: Proceedings of the 17th Annual ACM-SIAM Symposium on Discrete Algorithms, Miami, FL, pp. 89–98 (2006)
2. Benini, L., De Micheli, G.: Networks on chips: A new SoC paradigm. Computer 35(1), 70–78 (2002)
3. Charikar, M., Guha, S.: Improved combinatorial algorithms for the facility location and k-median problems. In: Proceedings of the 40th Annual IEEE Symposium on Foundations of Computer Science, pp. 378–388 (1999)
4. Demaine, E.D., Hajiaghayi, M.T., Mahini, H., Zadimoghaddam, M.: The price of anarchy in cooperative network creation games. SIGecom Exchanges 8(2) (December 2009)

 5. Demaine, E.D., Hajiaghayi, M.T., Mahini, H., Zadimoghaddam, M.: The price of anarchy in network creation games. In: Proceedings of the 26th Annual ACM SIGACT-SIGOPS Symposium on Principles of Distributed Computing, pp. 292–298 (2007); To appear in ACM Transactions on Algorithms
 6. Fabrikant, A., Luthra, A., Maneva, E., Papadimitriou, C.H., Shenker, S.: On a network creation game. In: Proceedings of the 22nd Annual Symposium on Principles of Distributed Computing, Boston, MA, pp. 347–351 (2003)
 7. Feige, U.: A threshold of $\ln n$ for approximating set cover. Journal of the ACM 45(4), 634–652 (1998)
 8. Kleinberg, J.: Small-world phenomena and the dynamics of information. Advances in Neural Information Processing Systems 14, 431–438 (2001)
 9. Kleinberg, J.: The small-world phenomenon: An algorithmic perspective. In: Proceedings of the 32nd ACM Symposium on Theory of Computing, pp. 163–170 (2000)
10. Laoutaris, N., Poplawski, L., Rajaraman, R., Sundaram, R., Teng, S.-H.: Bounded budget connection (BBC) games or how to make friends and influence people, on a budget. In: Proceedings of the 27th ACM Symposium on Principles of Distributed Computing, pp. 165–174 (2008)
11. Meyerson, A., Tagiku, B.: Minimizing average shortest path distances via shortcut edge addition. In: Dinur, I., Jansen, K., Naor, J., Rolim, J. (eds.) Approximation, Randomization, and Combinatorial Optimization. Algorithms and Techniques. LNCS, vol. 5687, pp. 272–285. Springer, Heidelberg (2009)

Author Index